Lecture Notes in Mathematics

2126

Editors-in-Chief:
J.-M. Morel, Cachan
B. Teissier, Paris

Advisory Board:
Camillo De Lellis (Zürich)
Mario di Bernardo (Bristol)
Alessio Figalli (Austin)
Davar Khoshnevisan (Salt Lake City)
Ioannis Kontoyiannis (Athens)
Gabor Lugosi (Barcelona)
Mark Podolskij (Heidelberg)
Sylvia Serfaty (Paris and NY)
Catharina Stroppel (Bonn)
Anna Wienhard (Heidelberg)

AF063080

More information about this series at
http://www.springer.com/series/304

Jacek Banasiak • Mustapha Mokhtar-Kharroubi
Editors

Evolutionary Equations with Applications in Natural Sciences

Springer

Editors
Jacek Banasiak
School of Mathematics, Statistics
and Computer Science
University of KwaZulu-Natal
Durban, South Africa

Mustapha Mokhtar-Kharroubi
Département de Mathématiques
UFR des Sciences et Techniques
Besançon, France

ISBN 978-3-319-11321-0 ISBN 978-3-319-11322-7 (eBook)
DOI 10.1007/978-3-319-11322-7
Springer Cham Heidelberg New York Dordrecht London

Lecture Notes in Mathematics ISSN print edition: 0075-8434
ISSN electronic edition: 1617-9692

Library of Congress Control Number: 2014955203

Mathematics Subject Classification (2010): 34G, 35Q, 47D, 92B, 65M, 35K, 60J

© Springer International Publishing Switzerland 2015
This work is subject to copyright. All rights are reserved by the Publisher, whether the whole or part of the material is concerned, specifically the rights of translation, reprinting, reuse of illustrations, recitation, broadcasting, reproduction on microfilms or in any other physical way, and transmission or information storage and retrieval, electronic adaptation, computer software, or by similar or dissimilar methodology now known or hereafter developed. Exempted from this legal reservation are brief excerpts in connection with reviews or scholarly analysis or material supplied specifically for the purpose of being entered and executed on a computer system, for exclusive use by the purchaser of the work. Duplication of this publication or parts thereof is permitted only under the provisions of the Copyright Law of the Publisher's location, in its current version, and permission for use must always be obtained from Springer. Permissions for use may be obtained through RightsLink at the Copyright Clearance Center. Violations are liable to prosecution under the respective Copyright Law.
The use of general descriptive names, registered names, trademarks, service marks, etc. in this publication does not imply, even in the absence of a specific statement, that such names are exempt from the relevant protective laws and regulations and therefore free for general use.
While the advice and information in this book are believed to be true and accurate at the date of publication, neither the authors nor the editors nor the publisher can accept any legal responsibility for any errors or omissions that may be made. The publisher makes no warranty, express or implied, with respect to the material contained herein.

Printed on acid-free paper

Springer is part of Springer Science+Business Media (www.springer.com)

Introduction

This book contains a collection of lectures presented at the 2013 CIMPA-UNESCO-South Africa School *Evolutionary Equations with Applications in Natural Sciences*. The School was part of the initiative *Mathematics of Planet Earth* and it was organized at the African Institute of Mathematical Sciences, http://www.aims.ac.za/, in Muizenberg (Cape Town), South Africa, from 22nd July to 2nd August 2013, under the auspices of the Centre International de Mathématiques Pures et Appliquées (CIMPA). CIMPA is a non-profit international organization whose aim is to promote international cooperation in higher education and research in mathematics for the benefit of developing countries. In accordance with CIMPA's mission, the School was aimed at postgraduate students and young researchers from such countries.

The School consisted of 9 courses delivered by invited lecturers coming from France, Germany, Poland, South Africa and Scotland, and attracted 29 participants from countries as diverse as Algeria, Benin, Cameroon, Chad, Congo, DRC, Ethiopia, Ivory Coast, Kenya, Lesotho, Mauritania, Morocco, Nigeria, Philippines, Pakistan, Sudan, South Africa, and Zimbabwe.

The School was a truly multidisciplinary event, spanning the fields of theoretical and applied functional analysis, partial differential equations, probability theory and numerical analysis applied to various models coming from theoretical physics, biology, engineering and complexity theory. The main emphasis was on the development of modelling, analytical and computational skills in a range of disciplines vital for the advancement of physical and natural sciences. The models discussed were time dependent and this led naturally to evolutionary equations which included ordinary, partial, integral or integro-differential equations. Particular examples of such equations were reaction-diffusion equations, transport equations coupled with Boltzmann type models and fragmentation-coagulation type equations, and the lectures gave detailed accounts of the functional-analytic, probabilistic and numerical frameworks for the analysis of such equations. This choice of programme was highly relevant to emerging researchers in many applied sciences as it exposed them to the diversity of applications of time dependent transport and kinetic type models and also provided a panorama of techniques for their analysis.

This volume contains the lectures given during the School, with each chapter devoted to the material presented by a specified lecturer. While each chapter is different and focuses on a specific topic, there is a common thread joining all of them—they are all concerned with evolution problems in a complex context and a significant part of each chapter deals with deep analytical methods for solving them. The lectures targeted postgraduate students and young researchers and thus the volume contains an appropriate blend of material, from an introductory and educational level at the beginning to a survey of cutting edge research at the end.

The foundation for all chapters is given by Wilson Lamb's *Applying functional analytic techniques to evolution equations*. This provides a gentle introduction to the basic tools needed for evolution equations and demonstrates their applicability. First the author discusses linear finite dimensional models, where the concept of the operator semigroup is introduced. Then he moves to nonlinear systems and discusses basic stability concepts. Finally, he discusses infinite dimensional models, both linear and nonlinear, illustrating abstract concepts with the discrete fragmentation-coagulation equation. We shall encounter the latter in a couple of other chapters, where the state-of-the-art theory of fragmentation and coagulation problems is developed.

The second chapter, *Boundary conditions in evolutionary equations in biology* by Adam Bobrowski, joins the narrative of the first at the concept of semigroups of operators, but takes this in a more general direction. First the author introduces basic models such as the transport and diffusion equations on the whole space and, writing them as Cauchy problems for abstract ordinary differential equations, discusses their solvability using the Hille–Yosida theory which is covered in more detail than in the first chapter. Next the author moves to the main subject of his lectures; that is, how to incorporate boundary conditions for the discussed equations so that they can be treated within the framework of semigroups of operators. Here he describes Greiner's approach and the method of images which are then applied to McKendrick models for population dynamics and to Feller–Wentzell boundary conditions. Finally, the author considers some singularly perturbed problems related to the previous models and discusses their small parameter limits using the powerful Sova–Kurtz approach.

The following chapter, *Introduction to complex networks: structure and dynamics* by Ernesto Estrada, takes us in the direction of discrete evolutionary problems. While, on the one hand, the author gives an introduction to the basic graph-theoretical concepts, including the 'small-world' and 'scale-free' networks, thus providing the mathematical foundations for the network transport problems considered in the next chapter, on the other hand he develops the theory of dynamical processes on networks, beginning with the consensus model, synchronization and Kuramoto models. The chapter is concluded by a discussion of a network version of the epidemiological SIR model introduced in the first chapter and the replicator-mutator model considered later in the lectures of R. Rudnicki.

In the next chapter, *Kinetic models in natural sciences* by Jacek Banasiak, we return to semigroup theory moving, however, to more complex models. In the study of so-called kinetic type equations (similar to Master Equations in Markov processes) one considers the loss and gain of agents at a particular state and the solvability of such problems depends on a delicate balance of these terms. In this chapter the author considers transport processes on networks, building on the introduction by E. Estrada, discusses the nonlinear versions of the McKendrick model presented in the lectures of A. Bobrowski and also extends the theory of fragmentation-coagulation processes introduced in the lectures of W. Lamb. In the process, the author further develops semigroup theory, discussed in the first two chapters, by introducing concepts of positivity and analyticity of semigroups and employing these ideas in various ways to arrive at solutions of the problems formulated in the introductory part of the chapter. In particular, the author provides the first proof of the existence of global classical solutions to fragmentation-coagulation equation with unbounded rates.

Following these results, Philippe Laurençot in *Weak compactness techniques and coagulation equations* presents an alternative way of approaching coagulation problems. Whereas in previous chapters in which coagulation-fragmentation processes are discussed, the coagulation part is treated as a perturbation of the linear fragmentation part, thus enabling semigroup theory to be applied, here the centre stage is taken by the continuous coagulation operator. The author presents a powerful weak compactness technique, first discussing intricacies of weak L_1 convergence and state-of-the-art methods of dealing with them. Next he moves to the continuous coagulation equation, where he considers in detail the cases in which the weak solutions are mass conserving and the cases where a phase transition, called gelation, occurs. These results provide a powerful counterpoint to the semigroup approach described in the chapter by J. Banasiak, and it is particularly noteworthy that this book is the first in which both approaches are presented together.

The next two chapters, although similar in that they present theories concerning the long-term behaviour of solutions to abstract evolution equations are nevertheless quite diverse in their approach. The chapter *Stochastic operators and semigroups and their applications in physics and biology* by Ryszard Rudnicki focusses on methods that have their origins in probability theory. The author begins with the concept of stochastic operators and semigroups and introduces some examples such as the Frobenius–Perron operators and stochastic integral operators. For the former, he discusses their relations with the ergodic properties of the described system. In the next step, he discusses the concept of substochastic and stochastic semigroups, linking it nicely with the theory developed in the lectures of J. Banasiak. The examples that follow include birth-and-death type problems, the continuity equation, usual and degenerate diffusions (where the representation theorem by Hörmander is presented), as well as piecewise deterministic Markov processes (a special case of which is given by coupled systems of McKendrick models discussed in the lectures of A. Bobrowski and J. Banasiak). The final part of the chapter is devoted to the analysis of long-term properties of models fitting into the developed theory. Here the main roles are played by the Lasota–Yorke lower function method, partially integral semigroups, the Foguel alternative and the Hasminskiĭ function, applied, among others, to cell cycle models, birth-and-death processes, and the description of the gene expression.

The chapter *Spectral theory for neutron transport* by Mustapha Mokhtar-Kharroubi focusses on a single model of neutron transport and is devoted to the classical approach to its analysis, spectral theory, which here is pushed to its limits. The author builds on the semigroup results developed in the lectures of W. Lamb, A. Bobrowski and J. Banasiak but leans more towards results on compactness of the semigroups and spectral mapping theorems which are further fine-tuned to cater for the advection and transport equations. The theory is first built for the advection equation where, when dealing with the natural L_1 space, the author makes a brief excursion into the so-called sun-dual theory. Then, to deal with the full equation, including the collision operator, he makes extensive use of perturbation theory and, in particular, Dyson–Phillips equations. Again, the L_1 theory is more difficult and is linked with the weak compactness results described in more detail in the lectures of P. Laurençot.

With the last two chapters, we move towards more concrete applications. The penultimate chapter *Reaction-diffusion-ODE models of pattern formation* by Anna Marciniak-Czochra begins with an introduction to pattern formation through Turing instabilities and discusses the example of an activator-inhibitor model, pointing out some limitations of the classical reaction-diffusion model. These limitations are addressed by introducing reaction-diffusion-ODE models for which an extensive theory of instabilities is developed. The theory presents a nontrivial extension of the semigroup methods for semilinear equations introduced in the lectures of W. Lamb and J. Banasiak, and of the spectral results discussed in detail in the preceding chapter by M. Mokhtar-Kharroubi. The author applies it to models such as early cancerogenesis and activator-inhibitor systems with non-diffusing activators. The final part of the chapter is devoted to pattern formation in systems with bistability and hysteresis and, in particular, to discontinuous patterns.

The final chapter, *Nonlinear Hyperbolic Systems of Conservation Laws and Related Applications* by Mapundi Banda, presents a vista into a numerical world. It begins with scalar conservation laws, the linear versions of which were discussed in the lectures of A. Bobrowski, J. Banasiak, R. Rudnicki and M. Mokhtar-Kharroubi, but quickly focuses on problems specific to quasi-linear cases such as weak, discontinuous solutions, formation of shock waves and entropy conditions. After introducing the basic toolbox for this field, the author moves to viscosity and entropy solutions for scalar conservation laws and then for systems. The main contents of the chapter are numerical methods for conservation laws. Here Godunov, Lax–Friedrichs and relaxation schemes are described in detail and applied to particular models. The chapter is concluded with numerical simulations of the flow in a network, thus providing an extension of some results discussed in the lectures of J. Banasiak to nonlinear models. The author also considers some aspects of the boundary stabilization method.

The Co-Directors of the School, Jacek Banasiak and Mustapha Mokhtar-Kharroubi, are deeply grateful to CIMPA for awarding the organization of the School to them, and to the Director of the African Institute of Mathematical Sciences

(AIMS), Professor Barry Green, for allowing the use of the excellent facilities of AIMS. We are also grateful to Rene January from AIMS whose expertise and enthusiasm greatly helped to make the School a success. The School would not have happened without the generous support received from many institutions. Apart from CIMPA, they are (in alphabetical order): AIMS, London Mathematical Society–African Mathematics Millennium Science Initiative, Commission for Developing Countries of the International Mathematical Union, French Embassy in Pretoria, Hanno Rund Fund of the School of Mathematics, Statistics and Computer Science of the University of KwaZulu-Natal, International Institute of Theoretical Physics, Office of Naval Research Global, National Research Foundation of South Africa and the National Institute of Theoretical Physics.

Finally our special thanks are due to our colleagues M.K. Banda, A. Bobrowski, E. Estrada, W. Lamb, P. Laurençot, A. Marciniak–Czochra and R. Rudnicki for the careful preparation and stimulating presentation of their lectures.

Durban, South Africa | Jacek Banasiak
Besançon, France | Mustapha Mokhtar-Kharroubi

Contents

Applying Functional Analytic Techniques to Evolution Equations

Wilson Lamb

1 Preliminaries

1.1 Introduction

Mathematical models arising in the natural sciences often involve equations that describe how the phenomena under investigation evolve in time. Such evolution equations can arise in a number of different forms; for example, the assumption that time is a discrete variable could lead to difference equations, whereas continuous-time models are often expressed in terms of differential equations.

The construction and application of a mathematical model usually proceeds in the following manner.

- We make assumptions on the various factors that influence the evolution of the time-dependent process that we are interested in.
- We obtain a 'model' by expressing these assumptions in terms of mathematics.
- We use mathematical techniques to analyse our model. If the model takes the form of an equation, then ideally we would like to obtain an explicit formula for its solution (unfortunately, this is impossible in the majority of cases).
- Finally, we examine the outcome of our mathematical analysis and translate this back into the real world situation to find out how closely the predictions from our model agree with actual observations.

In the case of kinetic models, where the interest is in describing, in mathematical terms, the evolution of some population of objects, the modelling process usually

W. Lamb (✉)
Department of Mathematics and Statistics, University of Strathclyde, Glasgow, UK
e-mail: w.lamb@strath.ac.uk

© Springer International Publishing Switzerland 2015

J. Banasiak, M. Mokhtar-Kharroubi (eds.), *Evolutionary Equations with Applications in Natural Sciences*, Lecture Notes in Mathematics 2126,
DOI 10.1007/978-3-319-11322-7_1

results in a so-called Kinetic (or Master) Equation. A nice account of the typical steps involved in deriving such an equation is given in Sect. 1.1 of the contribution to this volume by Jacek Banasiak [3].

Note that a mathematical model will usually be only an approximation to what is actually happening in reality. Highly detailed models, incorporating many different factors, inevitably mean very complicated mathematical equations which are difficult to analyse, whereas crude models, which are easy to analyse, are most likely to provide poor predictions of actual behaviour. In practice, a compromise has to be reached; a small number of key factors are identified and used to produce a model which is not excessively complicated.

When faced with a specific mathematical problem that has emerged from the modelling process, an important part of the mathematical analysis is to establish that the problem has been correctly formulated. The usual requirements for this to be the case are the following.

1. **Existence of Solutions.** We require at least one solution to exist.
2. **Uniqueness of Solutions.** There must be no more than one solution.
3. **Continuous Dependence on the Problem Data.** The solution should depend continuously on any input data, such as initial or boundary conditions.

Problems that meet these requirements are said to be well-posed. Note that implicit in the above statements is that we know exactly what is meant by a solution to the problem. Often there will be physical, as well as mathematical, constraints that have to be satisfied. For example we may be only interested in solutions which take the form of non-negative, differentiable functions. Also, in some cases, it may be possible to define a solution in different ways, and this could lead to a well-posed problem if we work with one type of solution but an ill-posed problem if we adopt a different definition of a solution. When multiple solutions exist, then we may be prepared to accept this provided a satisfactory explanation can be provided for the non-uniqueness condition being violated.

In these notes, we shall present some techniques that have proved to be effective in establishing the well-posedness of problems involving evolution equations. We shall illustrate how these techniques can be applied to standard problems that arise in population dynamics, beginning initially with the simple case of initial-value problems (IVPs) for scalar ordinary differential equations (ODEs) (the Malthus and Verhulst models of single-species population growth), and then going on to IVPs for finite systems of ODEs (e.g. models of interacting species and epidemics). We conclude by discussing and analysing models of coagulation–fragmentation processes that are expressed in terms of an infinite system of differential equations. To enable these problems to be treated in a unified manner, the techniques used will be developed from a dynamical systems point of view and concepts and results from the related theory of semigroups of operators will be introduced at appropriate stages.

1.2 Dynamical Systems

From a mathematical viewpoint, a dynamical system consists of the following two parts:

- a state vector that describes the state of the system at a given time,
- a function that maps the state at one instant of time to the state at a later time.

The following definition expresses this more precisely; see [15, p.160].

Definition 1 Let X represent the state space (i.e. the space of all state vectors) and let J be a subset of $\mathbb{R}$ (which we assume contains 0). A function $\phi : J \times X \to X$ that has the two properties

(i) $\phi(0, \overset{\circ}{u}) = \overset{\circ}{u}$
(ii) $\phi(s, \phi(t, \overset{\circ}{u})) = \phi(t + s, \overset{\circ}{u})$, for $t, s, t + s \in J$, (the semigroup property)

is called a dynamical system on X.

Remarks

1. Throughout, we assume that X is a Banach space (i.e. a complete normed vector space); see Sect. 1.3.2 for details.
2. We can regard $\phi(t, \overset{\circ}{u})$ as the state at time t of the system that initially was at state $\overset{\circ}{u}$. The semigroup property then has the following interpretation: let the system evolve from its initial state $\overset{\circ}{u}$ to state $\phi(t, \overset{\circ}{u})$ at time t, and then allow it to evolve from this state for a further time s. The system will then arrive at precisely the state $\phi(t + s, \overset{\circ}{u})$ that it would have reached through a single-stage evolution of $t + s$ from state $\overset{\circ}{u}$.
3. In these notes, we shall consider only the case when J is an interval in $\mathbb{R}$, usually $J = \mathbb{R}^+ = [0, \infty)$. The dynamical system is then called a continuous-time (semi- or forward) dynamical system. We shall abbreviate this to CDS.

   In operator form, we can write

$$\phi(t, \overset{\circ}{u}) = S(t)\overset{\circ}{u},$$

where $S(t)$ is an operator mapping the state space X into X. Note that $S(0) = I$ (the identity operator on X) and the semigroup property (in the case when $J = \mathbb{R}_+$) becomes

$$S(t)S(s) = S(t + s), \ \forall t, s \geq 0.$$

The family of operators $\mathbf{S} = \{S(t)\}_{t\geq 0}$ is said to be a semigroup of operators on X (algebraically, $\mathbf{S}$ is a semigroup under the associative binary operation of composition of operators).

Example 1 As a simple illustration of how a CDS arises from a differential equation, consider the initial value problem

$$u'(t) = lu(t), \quad u(0) = \mathring{u}, \tag{1}$$

where l is a real constant. Routine methods show that a solution to (1) is $u(t) = e^{tl}\mathring{u}$. This establishes that there exists at least one solution to (1). To prove that there is no other differentiable solution (i.e. to establish the uniqueness of the solution that we have produced), we argue as follows. Suppose that another solution v exists and let $t > 0$ be arbitrarily fixed. Then, for $0 < s \leq t$, we have

$$\begin{aligned}\frac{d}{ds}(e^{(t-s)l}v(s)) &= -le^{(t-s)l}v(s) + e^{(t-s)l}v'(s)\\ &= -le^{(t-s)l}v(s) + e^{(t-s)l}lv(s) = 0.\end{aligned}$$

It follows from this that $e^{(t-s)l}v(s)$ is a constant function of s on $[0, t]$. On choosing $s = 0$ and $s = t$, we obtain

$$e^{tl}v(0) = e^{(t-t)l}v(t) = v(t).$$

Since this argument works for any $t > 0$ and we already know that $v(0) = u(0) = \mathring{u}$, we deduce that $v(t) = u(t) = e^{tl}\mathring{u}$ for all $t \geq 0$. Now let $\phi : \mathbb{R} \times \mathbb{R} \to \mathbb{R}$ be defined by

$$\phi(t, \mathring{u}) = e^{tl}\mathring{u}, \quad t, \mathring{u} \in \mathbb{R},$$

that is, $\phi(t, \mathring{u})$ denotes the value at time t of the solution of the IVP (1). Clearly

(i) $\phi(0, \mathring{u}) = \mathring{u}$
(ii) $\phi(s, \phi(t, \mathring{u})) = \phi(s, e^{tl}\mathring{u}) = e^{sl}e^{tl}\mathring{u} = e^{(t+s)l}\mathring{u} = \phi(t + s, \mathring{u})$

and so $\phi : \mathbb{R} \times \mathbb{R} \to \mathbb{R}$ is a CDS (by Definition 1).

Example 2 To make things a bit more interesting, we shall add a time-dependent forcing term to the IVP (1) and consider the non-homogeneous problem

$$u'(t) = lu(t) + g(t), \ t > 0, \quad u(0) = \mathring{u}, \tag{2}$$

where g is some known, and suitably restricted, function of t. To find a solution of (2), we use the following trick to reduce the problem to one that is more straightforward. Suppose that the solution u can be written as $u(t) = e^{tl}v(t)$. On substituting into the non-homogeneous ODE, we obtain

$$le^{tl}v(t) + e^{tl}v'(t) = le^{tl}v(t) + g(t).$$

It follows that v satisfies the ODE $v'(t) = e^{-tl}g(t)$ and therefore, from basic calculus,

$$v(t) - v(0) = \int_0^t v'(s)\,ds = \int_0^t e^{-sl}g(s)\,ds.$$

Rearranging terms, and using the fact that $v(0) = \overset{\circ}{u}$, produces

$$v(t) = \overset{\circ}{u} + \int_0^t e^{-sl}g(s)\,ds,$$

and therefore a solution of the IVP (2) is given by

$$u(t) = e^{tl}\overset{\circ}{u} + \int_0^t e^{(t-s)l}g(s)\,ds. \tag{3}$$

Formula (3) is sometimes referred to as Duhamel's (or the variation of constants) formula. As we have actually found a solution, we have resolved the question of existence of solutions to (2). But what about uniqueness? Do solutions to (2) exist other than that given by the Duhamel formula? The following argument shows that (3) is the only solution. Suppose that another solution, say w, of (2) exists and consider $z = u - w$, where u is the solution given by (3). Then z must satisfy the IVP $z'(t) = lz(t),\ z(0) = 0$, and therefore, by the previous example, is given by $z(t) = e^{tl}0 = 0$ for all $t \geq 0$. Consequently, $w(t) = u(t)$ for all $t \geq 0$. In this case, if we define

$$\phi(t, \overset{\circ}{u}) := e^{tl}\overset{\circ}{u} + \int_0^t e^{(t-s)l}g(s)\,ds,$$

then we do not obtain a CDS as the semigroup property is not satisfied. The reason for this is that the right-hand side of (2) depends explicitly on t through the function g; i.e. the equation is non-autonomous. In the previous example, where the solution of the IVP led to a CDS, the equation is autonomous since the right-hand side depends on t only through the solution u.

In the sequel, we shall consider only autonomous differential equations. When existence and uniqueness of solutions can be established for IVPs associated with an equation of this type, then we end up with a CDS $\phi : J \times X \to X$ which we can go on to investigate further. Typical questions that we would like to answer are the following.

1. Given an initial value $\overset{\circ}{u}$, can we determine the asymptotic (long-term) behaviour of $\phi(t, \overset{\circ}{u})$ as $t \to \infty$?
2. Can we identify particular initial values which give rise to the same asymptotic behaviour?

3. Can we say anything about the stability of the system? For example, if $\overset{\circ}{u}$ is "close to" $\overset{\circ}{v}$ in X, what can be said about the distance between $\phi(t, \overset{\circ}{u})$ and $\phi(t, \overset{\circ}{v})$ for future values of t?

In many situations, a dynamical system may also depend on a parameter (or several parameters), that is, the system takes the form $\phi_\mu : J \times X \to X$ where $\mu \in \mathbb{R}$ represents the parameter. In such cases, the following questions would also be of interest.

4. Can we determine what happens to the behaviour of the dynamical system as the parameter varies?
5. Can we identify the values of the parameter at which changes in the behaviour of the system occur (bifurcation values)?

In some special cases, it is possible to find an explicit formula for the dynamical system. For example, $\phi(t, \overset{\circ}{u}) = e^{tl}\overset{\circ}{u}$ (where l can be regarded as a parameter). The formula can then be used to answer questions 1–5 above. Unfortunately, in most cases no such formula can be found and analysing the dynamical system becomes more complicated.

1.3 Some Basic Concepts from Functional Analysis

The definition we gave of a dynamical system in Sect. 1.2, involved a state space X. Recall that, from a mathematical point of view, a dynamical system is a function ϕ of time t and the state variable $\overset{\circ}{u} \in X$. In the context of evolution equations, $\overset{\circ}{u}$ represents the initial state of the system (physical, biological, economic, etc.) that is being investigated. We now examine the algebraic and analytical structure of the state spaces that will be used in these notes. For a more detailed account, see any standard book on Functional Analysis such as [16].

1.3.1 Vector Spaces

A complex vector space (or complex linear space) is a non-empty set X of elements $f, g \ldots$ (often called vectors) together with two algebraic operations, namely vector addition and multiplication of vectors by scalars (complex numbers). Vector addition associates with each ordered pair $(f, g) \in X \times X$ a uniquely defined vector $f + g \in X$ (the sum of f and g) such that

$$f + g = g + f \text{ and } f + (g + h) = (f + g) + h \qquad \forall f, g, h \in X.$$

Moreover there exists a zero element O_X and, for each $f \in X$, there exists $-f \in X$, such that

$$f + O_X = f \text{ and } f + (-f) = O_X.$$

Multiplication by scalars associates with each $f \in X$ and scalar $\alpha \in \mathbb{C}$ a uniquely defined vector $\alpha f \in X$ such that for all $f, g \in X$ and scalars α, β we have

$$\alpha(\beta f) = (\alpha\beta) f, \quad 1x = x, \quad \alpha(f + g) = \alpha f + \alpha g, \quad (\alpha + \beta) f = \alpha f + \beta f.$$

Note that a real vector space, in which the scalars are restricted to be real numbers, is defined analogously.

A linear combination of $\{f_1, f_2, \ldots, f_m\} \subset X$ is an expression of the form

$$\alpha_1 f_1 + \alpha_2 f_2 + \ldots + \alpha_m f_m = \sum_{j=1}^{m} \alpha_j f_j$$

where the coefficients $\alpha_1, \alpha_2, \ldots, \alpha_m$ are any scalars. For any (non-empty subset) $M \subset X$, the set of all linear combinations of elements in M is called the span of M, written span (M) (or sp (M)).

The vectors $f_1, f_2, \ldots, f_m$ are said to be linearly independent if

$$\alpha_1 f_1 + \alpha_2 f_2 + \ldots + \alpha_m f_m = O_X \quad \Leftrightarrow \alpha_1 = \alpha_2 = \ldots = \alpha_m = 0;$$

otherwise the vectors are linearly dependent. An arbitrary subset M of X is linearly independent if every non-empty finite subset of M is linearly independent; M is linearly dependent if it is not linearly independent.

A vector space X is said to be finite-dimensional if there is a positive integer n such that X contains a linearly independent set of n vectors whereas any set of $n+1$ or more vectors of X is linearly dependent—in this case X is said to have dimension n and we write $\dim X = n$. By definition, if $X = \{O_X\}$, then $\dim X = 0$. If $\dim X = n$, then any linearly independent set of n vectors from X forms a basis for X. If $e_1, e_2, \ldots, e_n$ is a basis for X then each $f \in X$ has a unique representation as a linear combination of the basis vectors; i.e.

$$f = \alpha_1 e_1 + \alpha_2 e_2 + \ldots + \alpha_n e_n,$$

with the scalars $\alpha_1, \alpha_2, \ldots \alpha_n$ uniquely determined by f.

1.3.2 Normed Vector Spaces and Banach Spaces

A norm on a vector space X is a mapping from X into $\mathbb{R}$ satisfying the conditions

- $\|f\| \geq 0$ for all $f \in X$ and $\|f\| = 0 \Leftrightarrow f = O_X$;

- $\|\alpha f\| = |\alpha| \, \|f\|$ for all scalars α and $f \in X$;
- $\|f + g\| \leq \|f\| + \|g\|$ for all $f, g \in X$ (the Triangle Inequality).

A vector space X, equipped with a norm $\| \cdot \|$, is called a normed vector space, denoted by $(X, \|\cdot\|)$ (or simply by X when it is clear which norm is being used). Note that a norm can be regarded as a generalisation to a vector space of the familiar idea of the modulus of a number. Moreover, just as $|\alpha - \beta|$ gives the distance between two numbers, we can use $\|f - g\|$ to measure the distance between two elements f, g in $(X, \| \cdot \|)$. This then enables us to discuss convergence of sequences of elements and continuity of functions in a normed vector space setting.

We say that a sequence $(f_n)_{n=1}^{\infty}$ in a normed vector space X (with norm $\| \cdot \|$) is convergent in X if there exists $f \in X$ (the limit of the sequence) such that

$$\lim_{n \to \infty} \|f_n - f\| = 0.$$

In this case we write $f_n \to f$ as $n \to \infty$. Note that a convergent sequence $(f_n)_{n=1}^{\infty}$ in X has a uniquely defined limit.

A sequence $(f_n)_{n=1}^{\infty}$ in a normed vector space X is a Cauchy sequence if for every $\epsilon > 0$, there exists $N \in \mathbb{N}$ such that

$$\|f_m - f_n\| < \epsilon \qquad \text{for all } m, n \geq N.$$

The normed vector space X is said to be complete if every Cauchy sequence in X is convergent, and we refer to a complete normed vector space as a Banach space. Note that every finite-dimensional normed vector space is complete and hence a Banach space.

Example 3 Let

$$\mathbb{C}^n := \{f = (f_1, \ldots, f_n) : \ f_i \in \mathbb{C} \text{ for } i = 1, \ldots n\}.$$

We say that two vectors $f = (f_1, \ldots, f_n)$ and $g = (g_1, \ldots, g_n)$ are equal in $\mathbb{C}^n$ if

$$f_1 = g_1, \ldots, f_n = g_n.$$

Also, if we define

$$f + g := (f_1 + g_1, \ldots, f_n + g_n), \qquad f, g \in \mathbb{C}^n,$$
$$\alpha f := (\alpha f_1, \ldots, \alpha f_n), \quad \alpha \in \mathbb{C}, \ f \in \mathbb{C}^n,$$

and

$$\|f\| := \sqrt{|f_1|^2 + \cdots + |f_n|^2}, \quad f = (f_1, \ldots, f_n) \in \mathbb{C}^n,$$

then $(\mathbb{C}^n, \|\cdot\|)$ is a normed vector space with dimension n. Consequently $(\mathbb{C}^n, \|\cdot\|)$ is a Banach space. The Banach space $(\mathbb{R}^n, \|\cdot\|)$ consisting of all ordered n-tuples of real numbers is defined in an analogous manner.

Example 4 For fixed $\mu \geq 0$, we define a vector space of scalar-valued sequences $(f_i)_{i=1}^{\infty}$ by

$$\ell_\mu^1 := \{f = (f_i)_{i=1}^{\infty} : \sum_{i=1}^{\infty} i^\mu |f_i| < \infty\}.$$

Equality, addition and multiplication by a scalar are defined pointwise in much the same way as in $\mathbb{C}^n$ (e.g. $(f_i)_{i=1}^{\infty} + (g_i)_{i=1}^{\infty} = (f_i + g_i)_{i=1}^{\infty}$) and if we define a norm on ℓ_μ^1 by

$$\|f\|_{1,\mu} = \sum_{i=1}^{\infty} i^\mu |f_i|,$$

then $(\ell_\mu^1, \|\cdot\|_{1,\mu})$ can be shown to be an infinite-dimensional Banach space.

1.3.3 Operators on Normed Vector Spaces

We now introduce some concepts related to functions that are defined on a normed vector space X. Functions of this type are often referred to as operators (or transformations) and we shall denote these by capital letters, such as L, S and T. We shall concentrate only on cases where the operator, say T, maps each vector $f \in D(T) \subseteq X$ onto another (uniquely defined) vector $T(f) \in X$. Note that $T(f)$ is often abbreviated to Tf and $D(T)$ is the domain of T.

The simplest type of operator on a normed space X is an operator L that satisfies the algebraic condition

$$L(\alpha_1 f_1 + \alpha_2 f_2) = \alpha_1 L(f_1) + \alpha_2 L(f_2), \quad \forall\, f_1, f_2 \in X \text{ and scalars } \alpha_1, \alpha_2. \tag{4}$$

Any operator L that satisfies (4) is said to be a linear operator on X. The set of all linear operators mapping X into X will be denoted by $L(X)$ and, defining $L_1 + L_2$ and αL in $L(X)$ by $(L_1 + L_2)(f) := L_1(f) + L_2(f)$ and $(\alpha L)(f) := \alpha L(f)$, where $L_1, L_2, L \in L(X)$, $f \in X$ and α is a scalar, $L(X)$ is a vector space.

An operator $T : X \to X$ (T not necessarily linear) is said to be continuous at a given $f \in X$ if and only if

$$f_n \to f \text{ in } X \Rightarrow T(f_n) \to T(f) \text{ in } X.$$

We say that T is continuous on X if it is continuous at each $f \in X$.

Another important concept is that of a bounded operator. We say that the operator $T : X \to X$ is bounded on the normed vector space X if

$$\|T(f)\| \leq M \|f\| \text{ for all } f \in X, \tag{5}$$

where M is a positive constant that is independent of f; i.e. the same constant M works for all $f \in X$. In the case of a linear operator $L : X \to X$, continuity and boundedness are equivalent as it can be proved that

$$\text{the linear operator } L : X \to X \text{ is continuous on } X \Leftrightarrow L \text{ is bounded on } X.$$

We shall denote the collection of bounded linear operators on X by $B(X)$. It is straightforward to verify that $B(X)$ is a subspace of $L(X)$. Moreover, if X is a finite-dimensional normed vector space, then all operators in $L(X)$ are bounded (so that, as sets, $L(X) = B(X)$).

It follows from (5) that, if L is bounded, then

$$\sup \{\|L(f)\| : f \in X \text{ and } \|f\| \leq 1\}$$

exists as a finite non-negative number. This supremum is used to define the norm of a bounded linear operator in the vector space $B(X)$; i.e.

$$\|L\| := \sup \{\|L(f)\| : f \in X \text{ and } \|f\| \leq 1\}.$$

Equipped with this norm, $B(X)$ is a normed vector space in its own right, and is a Banach space whenever X is a Banach space. Specific examples of bounded and unbounded linear operators can be found in Sect. 1.1 of the contribution to this volume by Adam Bobrowski [8].

1.3.4 Calculus of Vector-Valued Functions

The basic operations of differentiation and integration of scalar-valued functions can be extended to the case of functions which take values in a normed vector space $(X, \|\cdot\|)$. A function of this type is said to be vector-valued because each value taken by the function is an element in a vector space. In the sequel, we shall encounter functions of the form $u : J \to X$ where $J \subset \mathbb{R}$ is an interval. Thus, $u(t) \in X$ for all $t \in J$. Such a function u is said to be strongly continuous at $c \in J$ if, for each $\varepsilon > 0$, a positive δ can be found such that

$$\|u(t) - u(c)\| < \varepsilon \text{ whenever } t \in J \text{ and } |t - c| < \delta.$$

If u is strongly continuous at each point in J, then u is said to be strongly continuous on J. Similarly, u is said to be strongly differentiable at $c \in J$ if there exists an element $u'(c) \in X$ such that

$$\lim_{h \to 0} \frac{u(c+h) - u(c)}{h} = u'(c), \tag{6}$$

where the limit is with respect to the norm defined on X; i.e. given $\varepsilon > 0$, there exists $\delta > 0$ such that

$$\left\| \frac{u(c+h) - u(c)}{h} - u'(c) \right\| < \varepsilon \text{ whenever } c + h \in J \text{ and } 0 < |h| < \delta. \tag{7}$$

If u is strongly differentiable at each point in J then we say that u is strongly differentiable on J.

As regards integration of a vector-valued function $u : J \to X$, it is a straightforward task to extend the familiar definition of the Riemann integral of a scalar-valued function. For example, if $J = [a, b]$, then, for each partition P_n of J of the form

$$a = t_0 < t_1 < t_2 < \ldots < t_n = b,$$

there is a corresponding Riemann sum

$$S(u; P_n) := \sum_{k=1}^{n} u(\xi_k)(t_k - t_{k-1}),$$

in which ξ_k is arbitrarily chosen in the sub-interval $[t_{k-1}, t_k]$. We then define

$$\int_a^b u(t)\, dt := \lim_{\|P_n\| \to 0} S(u, P_n),$$

whenever this limit exists in X (and is independent of the sequence (P_n) of partitions and choice of ξ_k). Here

$$\|P_n\| := \max_{1 \le k \le n} (t_k - t_{k-1}).$$

We refer to this integral as the strong (Riemann) integral of u over the interval $[a, b]$. The strong Riemann integral has similar properties to its scalar version. For example, suppose that $u : [a, b] \to X$ is strongly continuous on $[a, b]$. Then it can be shown that, for each $t \in [a, b]$,

$$\int_a^t u(s)\, ds \text{ exists}, \quad \left\| \int_a^t u(s)\, ds \right\| \le \int_a^t \|u(s)\|\, ds, \quad \frac{d}{dt}\left(\int_a^t u(s)\, ds \right) = u(t);$$

see [7, Section 1.6] and also [4, Subsection 2.1.5].

1.3.5 The Contraction Mapping Principle

As discussed earlier, when carrying out a rigorous investigation into problems arising from mathematical models, the first step is usually to show that solutions actually exist. Moreover, such solutions should be uniquely determined by the problem data. Theoretical results which establish these properties are often referred to as Existence-Uniqueness Results. To end this section, we present one of the most important results of this type. We shall also supply a proof as this provides concrete motivation for working with Banach spaces.

Theorem 1 (Banach Contraction Mapping Principle) *Let $(X, \|\cdot\|)$ be a Banach space and let $T : X \to X$ be an operator with the property that*

$$\|Tf - Tg\| \leq \alpha \|f - g\| \quad \forall\, f, g \in X,$$

for some constant $\alpha < 1$ (such an operator T is said to be a (strict) contraction). Then the equation

$$Tf = f$$

has exactly one solution (called a fixed point of T) in X. Moreover, if we denote this unique solution by $\overline{f}$ and use T iteratively to generate a sequence of vectors $(f_1, Tf_1, T^2 f_1, T^3 f_1, \ldots)$, where f_1 is any given vector in X, then

$$T^n f_1 \to \overline{f} \text{ as } n \to \infty.$$

Proof Let the sequence $(f_n)_{n=1}^{\infty}$ be defined as in the statement of the theorem. Then, for $n \geq 2$,

$$\|f_{n+1} - f_n\| = \|Tf_n - Tf_{n-1}\| \leq \alpha\|f_n - f_{n-1}\| \leq \cdots \leq \alpha^{n-1}\|f_2 - f_1\|.$$

Note that the above inequality trivially holds for $n = 1$ as well. Hence, for any $m > n \geq 1$, we have

$$\begin{aligned}
\|f_m - f_n\| &= \|f_m - f_{m-1}\| + \|f_{m-1} - f_{m-2}\| + \cdots + \|f_{n+1} - f_n\| \\
&\leq (\alpha^{m-2} + \alpha^{m-3} + \cdots + \alpha^{n-1})\|f_2 - f_1\| \\
&< \alpha^{n-1}(1 + \alpha + \alpha^2 + \cdots)\|f_2 - f_1\| = \frac{\alpha^{n-1}}{1-\alpha}\|f_2 - f_1\|.
\end{aligned}$$

Since

$$\frac{\alpha^{n-1}}{1-\alpha}\,\|f_2 - f_1\| \to 0 \text{ as } n \to \infty,$$

it follows that $(f_n)_{n=1}^{\infty}$ is a Cauchy, and hence convergent, sequence in the Banach space $(X, \|\cdot\|)$. Let $f \in X$ be the limit of this convergent sequence. Then, by continuity of the operator T, we obtain

$$f_{n+1} = Tf_n \Rightarrow \lim_{n\to\infty} f_{n+1} = \lim_{n\to\infty} Tf_n = T\left(\lim_{n\to\infty} f_n\right) \Rightarrow f = Tf,$$

and so f is a fixed point of T. To show that no other fixed point exists, suppose that both f and g are fixed points, with $f \neq g$. Then

$$\|f - g\| = \|Tf - Tg\| \le \alpha\|f - g\|.$$

Dividing each side by $\|f - g\|$ ($\neq 0$) leads to $1 \le \alpha$, which is a contradiction. □

2 Finite-Dimensional State Space

In this section we give a brief account of some aspects of the theory associated with autonomous finite-dimensional systems of ODEs and will explain how continuous-time dynamical systems defined on the finite-dimensional state-space $\mathbb{R}^n$ arise naturally from such systems. This will pave the way for the discussion on infinite-dimensional dynamical systems that will follow in the next section. Note that the intention with these lectures is not to provide an exhaustive treatment of systems of ODEs. Instead, we concentrate only on those results which will be needed to analyse some selected problems arising in population dynamics. We begin by examining the most straightforward case where we have a linear system of constant-coefficient ODEs. We will then move on to systems involving nonlinear equations and describe how, through the process of linearisation, useful information on the long-time behaviour of solutions near an equilibrium solution can be obtained from a related linear, constant-coefficient system. Obviously, before we can talk about the long-time behaviour of solutions, we should make sure that solutions do, in fact, exist. Hence, we shall highlight some conditions which, thanks to the Contraction Mapping Principle, guarantee the existence and uniqueness of solutions to systems of ODEs.

2.1 Linear Constant-Coefficient Systems of ODEs

2.1.1 Matrix Exponentials

Consider the following IVP involving a linear system of n constant-coefficient ODEs:

$$u_1'(t) = l_{11}u_1(t) + l_{12}u_2(t) + \cdots + l_{1n}u_n(t),\ u_1(0) = \mathring{u}_1,$$

$$u_2'(t) = l_{21}u_1(t) + l_{22}u_2(t) + \cdots + l_{2n}u_n(t),\ u_2(0) = \mathring{u}_2,$$

$$\vdots$$

$$u_n'(t) = l_{n1}u_1(t) + l_{n2}u_2(t) + \cdots + l_{nn}u_n(t),\ u_n(0) = \mathring{u}_n,$$

where $l_{11}, l_{12}, \ldots, l_{n,n}$ and $\mathring{u}_1, \ldots, \mathring{u}_n$ are real constants. The problem is to find n differentiable functions $u_1, u_2, \ldots, u_n$ of the variable t that satisfy the n equations in the system. Obviously, before seeking solutions, we have to know that solutions actually exist, and it is here that considerable progress can be made if we adopt the strategy of working with matrix exponentials that was pioneered by the Italian mathematician Guiseppe Peano in 1887 (see [13, pp. 503–504]).

The first step is to express the IVP system in the matrix–vector form

$$u'(t) = Lu(t),\ u(0) = \mathring{u}, \tag{8}$$

where L is the $n \times n$ constant real matrix

$$L = \begin{bmatrix} l_{11} & l_{12} & \cdots & l_{1n} \\ l_{21} & l_{22} & \cdots & l_{2n} \\ \vdots & \vdots & \ddots & \vdots \\ l_{n1} & l_{n2} & \cdots & l_{nn} \end{bmatrix},$$

and $u(t) = (u_1(t), \ldots, u_n(t))$ is interpreted as a column vector. A solution of (8) will be a vector-valued function in the sense that $u(t)$ lies in the n-dimensional Banach space $\mathbb{R}^n$ for each t. This means that our state space X is $\mathbb{R}^n$.

Note that $u'(t) = (u_1'(t), \ldots, u_n'(t))$ with integrals of the vector-valued function u being interpreted similarly; e.g.

$$\int_0^t u(s)\,ds = (\int_0^t u_1(s)\,ds, \ldots, \int_0^t u_n(s)\,ds).$$

Expressed as (8), the linear system of ODEs bears a striking resemblance to the scalar equation $u'(t) = lu(t)$, and so it is tempting to write down a solution in the form

$$u(t) = e^{tL}\mathring{u}\,. \tag{9}$$

It turns out that the unique solution of (8) can indeed be written as (9), but this obviously leads to the following questions.

Q1. What does e^{tL} mean when L is an $n \times n$ constant matrix?
Q2. How do we verify that (9) is a solution of (8)?
Q3. How do we prove that (9) is the only differentiable solution of (8) that satisfies the initial condition $u(0) = \mathring{u} \in \mathbb{R}^n$?
Q4. For a given $n \times n$ constant matrix L, can we actually express e^{tL} in terms of standard scalar-valued functions of t?

To answer Q1, we consider the power series definition of the scalar exponential e^l, i.e.

$$e^l = 1 + l + \frac{l^2}{2!} + \frac{l^3}{3!} + \cdots\,. \tag{10}$$

This infinite series converges to the number $e^l = \exp(l)$ for each fixed $l \in \mathbb{R}$. Motivated by this, Peano defined the exponential of an $n \times n$ constant matrix L by a formula, which, in modern notation, takes the form

$$e^L = \exp(L) = I + L + \frac{L^2}{2!} + \frac{L^3}{3!} + \cdots\,. \tag{11}$$

Here I is the $n \times n$ identity matrix, L^2 represents the matrix product LL, L^3 is the product $LLL = L^2L = LL^2$ and so on. Note that the operation $f \mapsto Lf$, where f is a column vector in $\mathbb{R}^n$, defines a bounded linear transformation that maps $\mathbb{R}^n$ into $\mathbb{R}^n$. If we use L to represent both the matrix and the bounded linear operator that it defines, then it can be shown that the infinite series of $n \times n$ matrices (or, equivalently, bounded linear operators in $B(\mathbb{R}^n)$) will always converge (with respect to the norm on $B(\mathbb{R}^n)$) to a uniquely defined $n \times n$ matrix (which, as before, can be interpreted as an operator in $B(\mathbb{R}^n)$). Moreover

$$\|e^L\| \le e^{\|L\|},$$

where $\|L\| := \sup\{\|Lf\| : f \in \mathbb{R}^n \text{ and } \|f\| \le 1\}$ for any $L \in B(\mathbb{R}^n)$; see [15, pp. 82–84] and [13, p. 6]. It follows from (11) that, for any $n \times n$ constant matrix L and any scalar t,

$$e^{tL} = \exp(tL) = I + tL + \frac{t^2L^2}{2!} + \frac{t^3L^3}{3!} + \cdots\,, \tag{12}$$

and

$$\|e^{tL}\| \leq e^{|t|\,\|L\|}.$$

The time-dependent matrix exponential defined by (12) has similar properties to its one-dimensional "little brother". For example, if L is any $n \times n$ constant matrix, then

(P1) $e^{0L} = I$;
(P2) $e^{sL}e^{tL} = e^{(s+t)L}$ for all $s, t \in \mathbb{R}$;
(P3) $\frac{d}{dt}\left(e^{tL} f\right) = L\,e^{tL} f$ for any given vector $f \in \mathbb{R}^n$.

The derivative in (P3) is interpreted as a strong derivative with respect to the norm on $\mathbb{R}^n$, so that

$$\left\| \frac{e^{(t+h)L} f - e^{tL} f}{h} - L\,e^{tL} f \right\| \to 0 \text{ as } h \to 0.$$

Note that an $\mathbb{R}^n$-valued function u is strongly differentiable at $c \in \mathbb{R}$ if and only if each of its scalar-valued components u_k, $k = 1, 2, \ldots, n$, is differentiable at c; the strong and pointwise (or component-wise) derivatives are then identical. In other words, the notions of strong derivative and pointwise derivative coincide in this n-dimensional case. It should also be remarked that a stronger version of (P3) can be established. Since

$$\left\| \frac{1}{h}\left(e^{hL} - I\right) - L \right\| \leq \sum_{k=2}^{\infty} \frac{|h|^{k-1}\,\|L\|^k}{k!} = \frac{e^{|h|\,\|L\|} - 1}{|h|} - \|L\| \to 0 \text{ as } h \to 0,$$

and

$$e^{(t+h)L} - e^{tL} = \left(e^{hL} - I\right) e^{tL},$$

it follows that the operator-valued function $t \mapsto e^{tL}$ is strongly differentiable in $B(\mathbb{R}^n)$.

2.1.2 Existence and Uniqueness of Solutions

We can now answer Q2 and Q3. On setting $u(t) = e^{tL}\mathring{u}$, it follows immediately from properties (P1) and (P3) that

$$u(0) = I\,\mathring{u} = \mathring{u}$$

and

$$u'(t) = Lu(t).$$

Therefore $u(t) = e^{tL}\overset{\circ}{u}$ is a solution of the IVP

$$u'(t) = Lu(t), \quad u(0) = \overset{\circ}{u}\,.$$

To show that this IVP has no other differentiable solutions, we argue in exactly the same way as for the scalar case. Suppose that another solution v exists; i.e. $v'(t) = Lv(t)$ and $v(0) = \overset{\circ}{u}$, and let $t > 0$ be arbitrarily fixed. Then, for $0 < s \leq t$, we have

$$\begin{aligned}\frac{d}{ds}(e^{(t-s)L}v(s)) &= -Le^{(t-s)L}v(s) + e^{(t-s)L}v'(s)\\ &= -Le^{(t-s)L}v(s) + e^{(t-s)L}Lv(s) = \mathbf{0},\end{aligned}$$

where $\mathbf{0}$ is the zero vector in $\mathbb{R}^n$. It follows from this that $e^{(t-s)L}v(s)$ is a constant vector for all $s \in [0,t]$. On choosing $s = 0$ and $s = t$, we obtain

$$e^{tL}\overset{\circ}{u} = e^{tL}v(0) = e^{(t-t)L}v(t) = e^{0L}v(t) = v(t).$$

Since this argument works for any $t > 0$ and we already know that $v(0) = u(0) = \overset{\circ}{u}$, we deduce that $v(t) = u(t) = e^{tL}\overset{\circ}{u}$ for all $t \geq 0$.

Note that this solution can be used to define the n-dimensional CDS $\phi : [0,\infty) \times \mathbb{R}^n \to \mathbb{R}^n$, where $\phi(t, \overset{\circ}{u}) := e^{tL}\overset{\circ}{u}$. The associated semigroup of operators $\{S(t)\}_{t\geq 0}$, $S(t) := e^{tl}$, is referred to as the semigroup generated by the matrix L, and, for each $\overset{\circ}{u}$, the set $\{S(t)\overset{\circ}{u} \colon t \geq 0\} \subset \mathbb{R}^n$ is called the (positive semi-) orbit of $\overset{\circ}{u}$. Geometrically, we can regard the orbit as a continuous (with respect to t) "curve" (or path or trajectory), emanating from $\overset{\circ}{u}$, that lies in the state-space $\mathbb{R}^n$ for all $t \geq 0$. The continuity property follows from the fact that

$$\|e^{hL} - I\| \leq e^{|h|\,\|L\|} - 1 \to 0 \text{ as } h \to 0.$$

A constant solution, $u(t) \equiv \overline{u}$ for all t, where $\overline{u} = (\overline{u}_1, \ldots, \overline{u}_n) \in \mathbb{R}^n$ is called an equilibrium solution or steady state solution. The orbit of such a solution is the single element (or point) $\overline{u} \in \mathbb{R}^n$; $\overline{u}$ is called an equilibrium point (or rest point, stationary point or critical point). If $\overline{u}$ is an equilibrium point, then $L\overline{u} = \mathbf{0}$. We shall only consider the case when the matrix L is non-singular and therefore the only equilibrium point of the system $u'(t) = Lu(t)$ is $\overline{u} = \mathbf{0}$. When each eigenvalue of L has a negative real part, the equilibrium point $\mathbf{0}$ is globally attractive (or globally asymptotically stable) since $\|e^{tL}f\| \to 0$ as $t \to \infty$ for all $f \in \mathbb{R}^n$; see [13, p.12].

In principle, e^{tL} can be computed by using the fact that, if P is a non-singular matrix and $L = P\Lambda P^{-1}$, then $e^{tL} = Pe^{t\Lambda}P^{-1}$. For example, if L has n distinct real eigenvalues $\lambda_1, \ldots, \lambda_n$, then the corresponding eigenvectors can be used as the columns of a matrix P such that $L = P\Lambda P^{-1}$, where $\Lambda = \mathrm{diag}\{\lambda_1, \ldots, \lambda_n\}$, in which case

$$e^{tL} = P\, e^{t\Lambda} P^{-1} \text{ with } e^{t\Lambda} = \mathrm{diag}\{e^{\lambda_1 t}, \ldots, e^{\lambda_n t}\}.$$

More generally, it can be shown that the components $u_j(t),\ j = 1, \ldots, n$, of any given solution $u(t)$ can be written as a linear combination of the functions

$$t^k e^{\mu t} \cos(\nu t), \qquad t^\ell e^{\mu t} \sin(\nu t),$$

where $\mu + i\nu$ runs through all the eigenvalues of L, and k, ℓ are suitably restricted non-negative integers; see [15, p.135].

One final remark in this subsection is that it should be clear that the restriction $t \geq 0$ is unnecessary in all of the above, and that we could just as easily have defined a group $\{e^{tL}\}_{t \in \mathbb{R}}$. We have focussed only on the semigroup case since this is usually the best that we can hope to obtain when we look at the more complicated setting of semigroups generated by operators defined in an infinite-dimensional state space.

2.2 *Nonlinear Autonomous Systems of ODEs*

We have seen that IVPs involving constant-coefficient linear systems of ODEs have unique, globally defined solutions that can be expressed in terms of matrix exponentials. For more general systems of ODEs, life becomes a bit more complicated and it is usually difficult to obtain exact solutions. However, useful qualitative results can sometimes be obtained. We shall consider the IVP

$$u'(t) = F(u(t)), \quad u(0) = \mathring{u}, \tag{13}$$

where $u(t) = (u_1(t), \ldots, u_n(t))$, $\mathring{u} = (\mathring{u}_1, \ldots, \mathring{u}_n)$ and $F : \mathbb{R}^n \supseteq W \to \mathbb{R}^n$ is a vector-valued function $F = (F_1, \ldots, F_n)$ defined on an open subset W of $\mathbb{R}^n$. A solution of (13) is a differentiable function $u : J \to W$ defined on some interval $J \subset \mathbb{R}$, with $0 \in J$, such that

$$u'(t) = F(u(t))\ \forall\, t \in J, \text{ and } u(0) = \mathring{u}\,.$$

2.2.1 Existence and Uniqueness of Solutions

The following theorem provides sufficient conditions for the existence of a unique solution to (13) on some interval $J = (-a, a)$. We shall denote such a solution by $\phi(\cdot, \mathring{u})$, i.e. at time $t \in (-a, a)$, the solution is $u(t) = \phi(t, \mathring{u})$. We shall also express $\phi(t, \mathring{u})$ as $S(t)\mathring{u}$.

Theorem 2 *Let F be continuously differentiable on W.*

(i) ***(Local Existence and Uniqueness)*** *For each $\mathring{u} \in W$, there exists a unique solution $\phi(\cdot, \mathring{u})$ of the IVP (13) defined on some interval $(-a, a)$ where $a > 0$.*
(ii) ***(Continuous Dependence on Initial Conditions)*** *Let the unique solution $\phi(\cdot, \mathring{u})$ be defined on some closed interval $[0, b]$. Then there exists a neighbourhood U of $\mathring{u}$ and a positive constant K such that if $\mathring{v} \in U$, then the corresponding IVP $v' = F(v)$, $v(0) = \mathring{v}$, has a unique solution also defined on $[0, b]$ and*

$$\|\phi(t, \mathring{u}) - \phi(t, \mathring{v})\| = \|S(t)\mathring{u} - S(t)\mathring{v}\| \le e^{Kt} \| \mathring{u} - \mathring{v} \| \quad \forall t \in [0, b].$$

(iii) ***(Maximal Interval of Existence)*** *For each $\mathring{u} \in W$, there exists a maximal open interval $J_{max} = (\alpha, \beta)$ containing 0 (with α and β depending on $\mathring{u}$) on which the unique solution $\phi(t, \mathring{u})$ is defined. If $\beta < \infty$, then, given any compact subset K of W, there is some $t \in (\alpha, \beta)$ such that $u(t) \notin K$.*

Remarks

(a) Proofs of these results can be found in [15, Chapter 8].
(b) The vector function F is said to be differentiable at $g \in W$ if there exists a linear operator $F_g \in B(\mathbb{R}^n)$ such that

$$F(g + h) = F(g) + F_g(h) + E(g, h), \ h \in \mathbb{R}^n,$$

where

$$\lim_{\|h\| \to 0} \frac{\|E(g, h)\|}{\|h\|} = 0.$$

It can be shown that F_g can be represented by the $n \times n$ Jacobian matrix

$$DF = \begin{bmatrix} \partial_1 F_1 & \partial_2 F_1 & \cdots & \partial_n F_1 \\ \partial_1 F_2 & \partial_2 F_2 & \cdots & \partial_n F_2 \\ \vdots & \vdots & \ddots & \vdots \\ \partial_1 F_n & \partial_2 F_n & \cdots & \partial_n F_n \end{bmatrix}$$

evaluated at g. The function F is continuously differentiable on W if all the partial derivatives $\partial_j F_i$ exist and are continuous on W.
(c) The fact that F is continuously differentiable on W means that F satisfies a local Lipschitz condition on W; i.e. for each $\mathring{u} \in W$ there is a closed ball

$$\overline{B}_r(\mathring{u}) := \{f \in \mathbb{R}^n : \|f - \mathring{u}\| \le r\} \subset W$$

and a constant k, which may depend on $\mathring{u}$ and r, such that

$$\|F(f) - F(g)\| \leq k \|f - g\| \qquad \forall f, g \in \overline{B}_r(\mathring{u}).$$

(d) The proof of Theorem 2(i) involves the Banach Contraction Mapping Principle. The first step is to note that the IVP (13) is equivalent to the fixed point problem $u = Tu$, where T is the operator defined by

$$(Tu)(t) = \mathring{u} + \int_0^t F(u(s))ds,$$

i.e. u is a solution of (13) if and only if u satisfies the integral equation

$$u(t) = \mathring{u} + \int_0^t F(u(s))ds.$$

The local Lipschitz continuity of F can then be used to establish that T is a contraction on a suitably defined Banach space of functions; this yields existence and uniqueness. It is also possible to produce a sequence of iterates (u_n) convergent to the unique solution $\phi(\cdot, \mathring{u})$ by using the Picard successive approximation scheme. We simply take $u_1(t) \equiv \mathring{u}$ and then set

$$u_n(t) = \mathring{u} + \int_0^t F(u_{n-1}(s))ds, \qquad n = 2, 3, \ldots$$

(e) The proof of Theorem 2(ii) relies on Gronwall's inequality which states that if $\psi : [0, b] \to \mathbb{R}$ is continuous, non-negative and satisfies

$$\psi(t) \leq C + K \int_0^t \psi(s)\, ds \;\; \forall t \in [0, b],$$

for constants $C \geq 0,\ K \geq 0$, then

$$\psi(t) \leq Ce^{Kt} \;\; \forall t \in [0, b].$$

(f) It can be shown that the operators $S(t)$ have the following semigroup property:

$$S(t)S(s)\,\mathring{u} = S(t + s)\,\mathring{u},$$

where this identity is valid whenever one side exists (in which case, the other side will also exist).

2.2.2 Equilibrium Points

When analysing the nonlinear autonomous system of ODEs

$$u' = F(u), \tag{14}$$

the starting point is usually to look for equilibrium points (corresponding to constant, or steady-state solutions). In this case $\bar{u}$ is an equilibrium point if

$$F_1(\bar{u}) = 0, \ \ldots, F_n(\bar{u}) = 0,$$

and the local stability properties of the equilibrium $\bar{u}$ are usually determined by the eigenvalues of the Jacobian matrix $(DF)(\bar{u})$. The equilibrium $\bar{u}$ is hyperbolic if $(DF)(\bar{u})$ has no eigenvalues with zero real part.

An equilibrium $\bar{u}$ is said to be stable if nearby solutions remain nearby for all future time. More precisely, $\bar{u}$ is stable if, for any given neighbourhood U of $\bar{u}$, there is a neighbourhood U_1 of $\bar{u}$ in U such that

$$\mathring{u} \in U_1 \Rightarrow \phi(t, \mathring{u}) \text{ exists for all } t \geq 0 \text{ and } \phi(t, \mathring{u}) \in U \text{ for all } t \geq 0.$$

If, in addition,

$$\mathring{u} \in U_1 \Rightarrow \phi(t, \mathring{u}) \to \bar{u} \text{ as } t \to \infty,$$

then $\bar{u}$ is (locally) asymptotically stable. Any equilibrium which is not stable is said to be unstable. When $\bar{u}$ is hyperbolic then it is either asymptotically stable (when all eigenvalues of $(DF)(\bar{u})$ have negative real parts) or unstable (when $(DF)(\bar{u})$ has at least one eigenvalue with positive real part).

The basic idea behind the proof of these stability results is that of linearisation. Suppose that $\bar{u}$ is an equilibrium point and that $\mathring{u}$ is sufficiently close to $\bar{u}$. On setting $v(t) = \phi(t, \mathring{u}) - \bar{u}$, we obtain

$$\begin{aligned} & v'(t) = F(\bar{u} + v(t)) \approx F(\bar{u}) + (DF)(\bar{u})\, v(t) \\ \text{i.e.} \quad & v'(t) \approx (DF)(\bar{u})\, v(t). \end{aligned}$$

Thus, in the immediate vicinity of $\bar{u}$, the nonlinear ODE $u' = F(u)$ can be approximated by the linear equation

$$v' = Lv, \quad \text{where } L = (DF)(\bar{u}).$$

In effect, this means that in order to understand the stability of a hyperbolic equilibrium point $\bar{u}$ of $u' = F(u)$, we need only consider the linearised equation $v' = (DF)(\bar{u})v$.

2.2.3 Graphical Approach in One and Two Dimensions

In the scalar case,

$$u'(t) = F(u(t)),\ u(0) = \overset{\circ}{u},$$

we can represent the asymptotic behaviour of solutions using a phase portrait. Geometrically, the state space $\mathbb{R}^1$ can be identified with the real line which, in this context, is called the phase line, and so the value $u(t)$ of a solution u at time t defines a point on the phase line. As t varies, the solution $u(t)$ traces out a trajectory, emanating from the initial point $\overset{\circ}{u}$, that lies completely on the phase line. If we regard $u(t)$ as the position of a particle on the phase line at time t, then the direction of motion of the particle is governed by the sign of $F(u(t))$. If $F(u(t)) > 0$ the motion at time t is to the right; if $F(u(t)) < 0$, then motion is to the left.

In two-dimensions, we use the phase plane. Here, we interpret the components $u_1(t)$ and $u_2(t)$ of any solution $u(t)$ as the coordinates of a curve defined parametrically (in terms of t) in the u_1–u_2 phase plane. Each solution curve plotted on the phase plane is a trajectory. A trajectory can also be regarded as the projection of a solution curve which "lives" in the three-dimensional space $\mathbb{R}^3$ (with coordinates u_1, u_2 and t) onto the two-dimensional u_1–u_2 plane. Phase plane trajectories have the following important properties.

1. Each trajectory corresponds to infinitely many solutions.
2. Through each point of the u_1–u_2 phase plane there passes a unique trajectory and therefore trajectories cannot intersect.
3. On the phase plane, an equilibrium point $\bar{u} = (\bar{u}_1, \bar{u}_2)$ is the trajectory of the constant solution

$$u_1(t) = \bar{u}_1, \quad u_2(t) = \bar{u}_2, \quad t \in \mathbb{R}.$$

4. The trajectory of a non-constant periodic solution is a closed curve called a cycle.

The key to establishing these properties is to use the uniqueness of solutions to IVPs. For example, suppose that the point $\overset{\circ}{u}$ lies, not only on the trajectory $C(\overset{\circ}{u})$, but also on the trajectory $C(\overset{\circ}{v})$ corresponding to the solution $\phi(\cdot, \overset{\circ}{v})$. Then, $\overset{\circ}{u} = \phi(t_0, \overset{\circ}{v})$ for some t_0 and therefore the function $\psi(t) = \phi(t - t_0, \overset{\circ}{v})$ is a solution of the system that satisfies the initial condition $\psi(0) = \overset{\circ}{u}$. By uniqueness of solutions, $\psi(t) = \phi(t, \overset{\circ}{u})$. Therefore, the trajectories corresponding to ψ and $\phi(\cdot, \overset{\circ}{u})$ (and hence $\phi(\cdot, \overset{\circ}{v})$ and $\phi(\cdot, \overset{\circ}{u})$) are identical.

2.3 Dynamical Systems and Population Models

Suppose we are interested in the long-term behaviour of the population of a particular species (or the populations of several inter-related species). By a "population" we mean an assembly of individual organisms which can be regarded as being alike. What is required is a mathematical model that contains certain observed or experimentally determined parameters such as the number of predators, severity of climate, availability of food etc. This model may take the form of a differential equation or a difference equation, depending upon whether the population is assumed to change continuously or discretely. We shall restrict our attention to the case of continuous time. We can attempt to use the model to answer questions such as:

1. Does the population $\to 0$ as $t \to \infty$ (extinction)?
2. Does the population become arbitrarily large as $t \to \infty$ (eventual overcrowding)?
3. Does the population fluctuate periodically or even randomly?

Example 5 **Single Species Population Dynamics** (see [14, Section 2.1]). When all individuals in the population behave in the same manner, then the net effect of this behaviour on the total population is given by the product of the population size with the per capita effect (i.e. the effect due to the behaviour of a typical individual in the population). For example, if we consider the case of the production of new individuals, then the rate of change of the population size $N(t)$ at time t in a continuous-time model can be expressed as

$$\frac{dN}{dt} = N \times \text{per capita reproduction rate}. \tag{15}$$

This can be written as

$$\frac{1}{N}\frac{dN}{dt} = \text{per capita reproduction rate}$$

or, equivalently,

$$\frac{d}{dt}\ln(N) = \text{per capita reproduction rate}.$$

(i) **The Malthus Model.** In this extremely simple model, the per capita reproduction rate is assumed to be a constant, say β, in which case Eq. (15) becomes

$$\frac{dN}{dt} = \beta N,$$

and so $N(t) = e^{\beta t}\overset{\circ}{N}$, where $\overset{\circ}{N} = N(0)$. This type of population growth is often referred to as Malthusian growth. The Malthus model can easily be adapted to include the effect of deaths in the population. If we also assume that the mortality rate is proportional to the population size, then we obtain

$$\frac{dN}{dt} = \beta N - \delta N = rN,$$

where $-\delta N$ represents the decline in population size due to deaths, and the parameter $r = \beta - \delta$ is the net per capita "growth" rate. The solution now is given by

$$N(t) = e^{rt}\,\overset{\circ}{N}, \tag{16}$$

where $N(0) = \overset{\circ}{N}$ is the initial size of the population. It follows that:

$$\begin{array}{lll} r > 0 \Rightarrow N(t) \to \infty & \text{as } t \to \infty & \text{(overcrowding)} \\ r < 0 \Rightarrow N(t) \to 0 & \text{as } t \to \infty & \text{(extinction)} \\ r = 0 \Rightarrow N(t) = \overset{\circ}{N} & \forall t \geq 0. & \end{array}$$

Clearly, the solution (16) leads to an unrealistic prediction of what will happen to the size of the population in the long term and so we must include other (nonlinear) effects to improve the model.

(ii) The Verhulst Model. A slightly more realistic model is given by

$$\frac{dN}{dt} = G(N)N, \quad t > 0; \quad N(0) = \overset{\circ}{N},$$

with a variable net growth rate G depending on the population size N. In some cases we would expect G to reflect the fact that there is likely to be some intra-specific competition for a limited supply of resources. This would require a growth rate, $G(N)$, that would lead to a model predicting a small population growth when N is small, followed by more rapid population growth until N hits a saturation value, say K, beyond which N will level off. If N ever manages to exceed K, then $G(N)$ should be such that N rapidly decreases towards K.

For example, the equation of limited growth is

$$\frac{dN}{dt} = r\left(1 - \frac{N}{K}\right)N, \quad N(0) = \overset{\circ}{N}, \tag{17}$$

where K and r are positive constants. To obtain this equation, we have set $G(N) = r(1 - N/K)$. Note that K is the population size at which G is zero and therefore $dN/dt = 0$ when $N = K$. Equation (17) is called the (continuous time) logistic growth equation or Verhulst equation, the constant K is called the carrying capacity of the environment, and r is the unrestricted growth rate. The method of separation of variables can be used to show that the solution of (17) is

$$N(t) = \frac{K}{1 - (1 - K/\overset{\circ}{N})\exp(-rt)}, \tag{18}$$

and therefore $N(t) \to K$ as $t \to \infty$.

Example 6 **Models of Two Interacting Species** (see [14, Section 2.2]). We now consider how interactions between pairs of species affect the population dynamics of both species. The type of interactions that can occur can be classified as follows:

- **Competition**: each species has an inhibitory effect on the other;
- **Commensalism**: each species benefits from the presence of others (symbiosis);
- **Predation**: one species benefits and the other is inhibited by interactions between them.

In any given habitat, such as a lake, an island or a Petri dish, it is likely that a number of different species will live together. A common strategy is to identify two species as being the most important to each other, and then to ignore the effect on them of all the other species in the habitat.

In the case when the two species are in competition for the same resources, any increase in the numbers of one species will have an adverse effect on the growth rate of the other. The competitive Lotka–Volterra system of equations used to model this situation is given by

$$u_1' = u_1(r_1 - l_{11}u_1 - l_{12}u_2), \quad u_2' = u_2(r_2 - l_{21}u_1 - l_{22}u_2), \tag{19}$$

where

- $u_1(t), u_2(t)$ are the sizes of the two species at time t;
- r_1, r_2 are the intrinsic growth rates of the respective species;
- l_{11}, l_{22} represent the strength of the intraspecific competition within each species, with r_1/l_{11} and r_2/l_{22} the carrying capacities of the respective species;
- l_{12}, l_{21} represent the strength of the interspecific competition (i.e. competition between the species).

Each of the constants $r_1, r_2, l_{11}, l_{12}, l_{21}, l_{22}$ is positive.

It follows from the existence-uniqueness theorem that, for each initial state $\overset{\circ}{u}$, there exists a unique solution $u(t) = S(t)\overset{\circ}{u}$ defined on some interval $[0, t_{max})$, where $t_{max} < \infty$ only if $\|u(t)\|$ diverges to infinity in finite time. Moreover, since the non-negative u_1 and u_2 axes are composed of complete trajectories, any trajectory that starts off in the positive first quadrant must remain there; i.e. solutions that start off at positive values stay positive (recall from phase plane analysis that trajectories in the phase plane cannot intersect).

Let L be the matrix

$$L = \begin{bmatrix} l_{11} & l_{12} \\ l_{21} & l_{22} \end{bmatrix},$$

and assume that $|L| \neq 0$. The system of Eq. (19) has four equilibria, namely

$$U_1 = (0,0),\ U_2 = (r_1/l_{11}, 0),\ U_3 = (0, r_2/l_{22}) \text{ and } U_4 = (u_1^*, u_2^*),$$

where

$$\begin{bmatrix} u_1^* \\ u_2^* \end{bmatrix} = L^{-1} \begin{bmatrix} r_1 \\ r_2 \end{bmatrix} = \frac{1}{|L|} \begin{bmatrix} r_1 l_{22} - r_2 l_{12} \\ r_2 l_{11} - r_1 l_{21} \end{bmatrix}.$$

Note that

$$|L| > 0 \text{ when } \frac{l_{12}}{l_{22}} < \frac{l_{11}}{l_{21}}$$

$$|L| < 0 \text{ when } \frac{l_{12}}{l_{22}} > \frac{l_{11}}{l_{21}}.$$

From this, we can deduce that there are two scenarios that result in $u_1^* > 0$ and $u_2^* > 0$, namely

$$\text{Case I}: \frac{l_{12}}{l_{22}} < \frac{r_1}{r_2} < \frac{l_{11}}{l_{21}}$$

$$\text{Case II}: \frac{l_{11}}{l_{21}} < \frac{r_1}{r_2} < \frac{l_{12}}{l_{22}}.$$

The Jacobian matrix at (x, y) is given by

$$(DF)(x,y) = \begin{bmatrix} r_1 - 2l_{11}x - l_{12}y & -l_{12}x \\ -l_{21}y & r_2 - l_{21}x - 2l_{22}y \end{bmatrix}.$$

For the equilibrium U_1, we have

$$(DF)(0,0) = \begin{bmatrix} r_1 & 0 \\ 0 & r_2 \end{bmatrix},$$

and it follows immediately that U_1 is unstable in each of Case I and Case II.

Consider now the other three equilibria when Case I applies. To determine the stability properties of these, we note first that the characteristic equation of a real 2×2 matrix, say A, can be written in the form

$$\lambda^2 - trace(A)\,\lambda + |A| = 0.$$

It follows that a non-singular matrix A will have two eigenvalues with negative real parts when $|A| > 0$ and $trace(A) < 0$, and will have exactly one positive eigenvalue when $|A| < 0$. At U_2 we have

$$(DF)(r_1/l_{11}, 0) = \begin{bmatrix} -r_1 & -l_{12}\, r_1/\, l_{11} \\ 0 & r_2 - l_{21}\, r_1/\, l_{11} \end{bmatrix}.$$

As the determinant of this Jacobian matrix is

$$-r_1\left(r_2 - \frac{l_{21}\, r_1}{l_{11}}\right) < 0,$$

the equilibrium U_2 is unstable. Similarly, U_3 is unstable. Now consider U_4. In this case,

$$(DF)(u_1^*, u_2^*) = \begin{bmatrix} r_1 - 2l_{11}u_1^* - l_{12}u_2^* & -l_{12}u_1^* \\ -l_{21}u_2^* & r_2 - l_{21}u_1^* - 2l_{22}u_2^* \end{bmatrix} = \begin{bmatrix} -l_{11}u_1^* & -l_{12}u_1^* \\ -l_{21}u_2^* & -l_{22}u_2^* \end{bmatrix}.$$

Consequently, the characteristic equation takes the form

$$\lambda^2 + \lambda(l_{11}u_1^* + l_{22}u_2^*) + u_1^*u_2^*\,|L| = 0,$$

and therefore U_4 is locally asymptotically stable (since the trace of the Jacobian matrix is negative and the determinant is positive). In fact, it can be shown that all trajectories in the positive first quadrant converge to U_4 as $t \to \infty$; see [14, p. 32]. Thus, in Case I, the competing species may coexist in the long term. Note that the condition $l_{11}l_{22} > l_{12}l_{21}$, which holds here, can be interpreted as stating that the overall intraspecific competition is stronger than the overall interspecific competition.

In Case II, a similar analysis shows that U_2 and U_3 are both asymptotically stable, with U_4 unstable (in fact U_4 is a saddle point). It follows that, in the long term, one of the species will die out. The species that survives is determined by the initial conditions. Since U_4 is a saddle point, there exist stable and unstable orbits emanating from U_4; see [24, p. 21]. These orbits are referred to as separatrices. As discussed in [14, p. 31], if the initial point on a trajectory lies above the stable separatrix, then the trajectory converges to U_3 (i.e. species u_1 dies out). If $\mathring{u}$ lies below this separatrix, then the trajectory converges to U_2 (i.e. species u_2 dies out).

For an analysis of the case when U_4 does not lie in the first quadrant of the phase plane, see [14, Section 2.3]. Note also that the equations used to model two species which are interacting in a co-operative manner are also given by (19), but now we have $l_{12} < 0, l_{21} < 0, l_{11} > 0$ and $l_{22} > 0$.

Example 7 **The SIR Models of Infectious Diseases** (see [14, Chapter 3], [9, Chapter 3] and [12, Chapter 6]). In simple epidemic models, it is often assumed that the total population size remains constant. At any fixed time, each individual within this population will be in one (and only one) of the following classes.

- Class S : this consists of individuals who are susceptible to being infected (i.e. can catch the disease).
- Class I : this consists of infected individuals (i.e. individuals who have the disease and can transmit it to susceptibles).
- Class R : this consists of individuals who have recovered from the disease and are now immune.

The class R is sometimes regarded as the Removed Class as it can also include those individuals who have died of the disease or are isolated until recovery. The SIR model was pioneered in a paper "Contribution to the Mathematical Theory of Epidemics" published in 1927 by two scientists, William Kermack and Anderson McKendrick, working in Edinburgh. In searching for a mechanism that would explain when and why an epidemic terminates, they concluded that: "In general a threshold density of population is found to exist, which depends upon the infectivity, recovery and death rates peculiar to the epidemic. No epidemic can occur if the population density is below this threshold value."

If we let $S(t)$, $I(t)$ and $R(t)$ denote the sizes of each class, then the following system of differential equations can be used to describe how these sizes change with time:

$$\frac{ds}{dt} = -\beta SI \tag{20}$$

$$\frac{dI}{dt} = \beta SI - \gamma I \tag{21}$$

$$\frac{dR}{dt} = \gamma I. \tag{22}$$

Here we are making the following assumptions.

- The gain in the infective class is proportional to the number of infectives and the number of susceptibles; i.e. is given by βSI, where β is a positive constant. The susceptibles are lost at the same rate.
- The rate of removal of infectives to the recovered class is proportional to the number of infectives; i.e. is given by γI, where γ is a positive constant.

We refer to γ as the recovery rate and β as the transmission (or infection) rate.

Note that, when analysing this system of equations, we are only interested in non-negative solutions for $S(t), I(t)$ and $R(t)$. Moreover, the constant population size is built into the system (20)–(22) since adding the equations gives

$$\frac{dS}{dt} + \frac{dI}{dt} + \frac{dR}{dt} = 0,$$

showing that, for each t,

$$S(t) + I(t) + R(t) = N,$$

where N is the fixed total population size. The model is now completed by imposing initial conditions of the form

$$S(0) = \mathring{S} \approx N, \quad I(0) = \mathring{I} = N - \mathring{S} > 0, \quad R(0) = 0.$$

Given particular values of β, γ, $\mathring{S}$ and $\mathring{I}$, we can use the model to predict whether the infection will spread or not, and if it does spread, in what manner it will grow with time. One observation that can be made more or less immediately is that the infectious class will grow in size if $dI/dt > 0$. Since we are assuming that there are infectious individuals in the population at time $t = 0$, Eq. (21) shows that $I(t)$ will increase from its initial value provided $\mathring{S} > \gamma/\beta$. The parameter $R_0 = \beta/\gamma$ is called the Basic Reproductive Ratio and is defined as the average number of secondary cases produced by an average infectious individual in a totally susceptible population.

We shall determine the long term behaviour of solutions by arguing as follows.

- Since $S(t) + I(t) + R(t) = N$ for all t, the system is really only a 2-D system and so we shall concentrate on the equations governing the evolution of S and I. For this 2-D system, we have an infinite number of equilibria, namely $(\overline{S}, 0)$, where $\overline{S}$ can be any non-negative number in the interval $[0, N]$. Note that these equilibria are not isolated (i.e. for each of these equilibria, no open ball centred at the equilibrium can be found that contains no other equilibrium). This means that the customary local-linearisation at an isolated equilibrium cannot be used to determine the stability of the equilibria of this 2-D system.
- The non-negative S axis consists entirely of equilibrium points and the non-negative I axis is composed of two complete trajectories, namely the equilibrium $(0, 0)$ and the positive I axis. This means that solutions that start off with $\mathring{S} > 0$ and $\mathring{I} > 0$ remain positive.
- Since $S(t) > 0$ and $I(t) > 0$, it follows (from the equation for S) that $S(t)$ is strictly decreasing. Hence $S(t) < \mathring{S}$ for any $t > 0$ for which $S(t)$ exists. Note that it is impossible for $I(t)$ to blow up in finite time since

$$\begin{aligned}
I'(t) &\le (\beta\mathring{S} - \gamma)I(t) \\
&\Rightarrow \int_0^t \frac{I'(s)}{I(s)}\,ds \le \int_0^t (\beta\mathring{S} - \gamma)\,ds = (\beta\mathring{S} - \gamma)t \\
&\Rightarrow \ln(I(t)) \le \ln(\mathring{I}) + (\beta\mathring{S} - \gamma)t \\
&\Rightarrow 0 < I(t) \le \exp((\beta\mathring{S} - \gamma)t)\,\mathring{I}\,.
\end{aligned}$$

Therefore both $S(t)$ and $I(t)$ exist globally in time. Moreover, if $\mathring{S} < \gamma/\beta$, then $I(t) \to 0$ as $t \to \infty$.

- For the epidemic to spread initially, we require $\overset{\circ}{S} > \gamma/\beta$, since we will then have $I'(0) > 0$. However, in this case there will exist some finite time, say t^*, such that $S(t^*) < \gamma/\beta$. To see this, simply observe that if we assume that $S(t) \geq \gamma/\beta$ for all t then we obtain $I(t) \geq \overset{\circ}{I}$ and $S'(t) \leq -\gamma \overset{\circ}{I}$ for all t. From this it follows that

$$S(t) \leq -\gamma \overset{\circ}{I} t + \overset{\circ}{S} \rightarrow -\infty \text{ as } t \rightarrow \infty,$$

which clearly is a contradiction. Arguing as before (but now with $\overset{\circ}{S}$ replaced by $S(t^*)$) shows that once again $I(t) \rightarrow 0$ as $t \rightarrow \infty$, despite $I(t)$ initially increasing.
- Since $S(t)$ is a strictly decreasing function that is bounded below (by zero), $S(t)$ must converge to some limit $S_\infty \geq 0$ as $t \rightarrow \infty$. We now establish that $S_\infty > 0$, showing that although the epidemic ultimately dies out, this is not caused by the number of available susceptibles decreasing to zero. Here we make use of the equation for R. We have

$$\frac{dS}{dR} = \frac{dS/dt}{dR/dt} = -\frac{\beta}{\gamma} S \Rightarrow S = \exp(-\beta R/\gamma)\, \overset{\circ}{S}\, .$$

Since $R \leq N$, we deduce that S is always greater than the positive constant $\exp(-\beta N/\gamma)\, \overset{\circ}{S}$ and therefore $S_\infty > 0$.
- Finally the trajectories in the $S - I$ phase plane can be obtained from the ODE

$$\frac{dI}{ds} = -1 + \frac{\gamma}{\beta S}.$$

This has solution given by

$$I = N - S + (\gamma/\beta)\ln(S/\overset{\circ}{S});$$

here we have used the fact that $\overset{\circ}{S} + \overset{\circ}{I} = N$. Consequently, on taking limits $(t \rightarrow \infty)$ on each side, and rearranging, we obtain

$$S_\infty = N + (\gamma/\beta)\ln(S_\infty/\overset{\circ}{S}).$$

For each given $\overset{\circ}{S}$, this equation has only one positive solution S_∞.

To summarise, we have shown that each solution $(S(t), I(t))$ will converge to an equilibrium $(S_\infty, 0)$, with $S_\infty > 0$, which is determined by the initial value of S. From this, it follows that $(S(t), I(t), R(t)) \rightarrow (S_\infty, 0, N - S_\infty)$ as $t \rightarrow \infty$. The value of $N - S_\infty$ shows the extent to which the infection has affected the population.

3 Infinite-Dimensional State Space

We now move into the realm of infinite-dimensional dynamical systems. Therefore, in the following discussion, we shall assume that the state space X is an infinite-dimensional Banach space with norm $\|\cdot\|$. The aim now is to express evolution equations in operator form as ordinary differential equations which are posed in X. We shall consider only problems of the type

$$u'(t) = L(u(t)) + N(u(t)),\ t > 0, \quad u(0) = \overset{\circ}{u}, \tag{23}$$

where $L : X \supseteq D(L) \to X$ and $N : X \to X$ are, respectively, linear and nonlinear operators, with $D(L)$ a linear subspace of X. In (23), the derivative is interpreted as a strong derivative, defined via (6) and (7), and a solution $u : [0, \infty) \to X$ is sought. The operator $L + N$ that appears in (23) governs the time-evolution of the infinite-dimensional state vector $u(\cdot)$, and the initial-value problem (23) is usually called a (semi-linear) abstract Cauchy problem (ACP).

To provide some motivation for looking at infinite-dimensional dynamical systems, we shall investigate a particular mathematical model of a system of particles that can coagulate to form larger particles, or fragment into smaller particles. Coagulation and fragmentation (C–F) processes of this type can be found in many important areas of science and engineering. Examples range from astrophysics, blood clotting, colloidal chemistry and polymer science to molecular beam epitaxy and mathematical ecology. An efficient way of modelling the dynamical behaviour of these processes is to use a rate equation which describes the evolution of the distribution of the interacting particles with respect to their size or mass; see [10,23] and also Section 1 of the contribution to this volume by Philippe Laurençot [18].

Suppose that we regard the system under consideration as one consisting of a large number of clusters (often referred to as mers) that can coagulate to form larger clusters or fragment into a number of smaller clusters. Under the assumption that each cluster of size n (n-mer) is composed of n identical fundamental units (monomers), the mass of each cluster is simply an integer multiple of the mass of a monomer. By appropriate scaling, each monomer can be assumed to have unit mass. This leads to a so-called discrete model of coagulation–fragmentation, with discrete indicating that cluster mass is a discrete variable which, in view of the above, can be assumed to take positive integer values.

In many theoretical investigations into discrete coagulation–fragmentation models, both coagulation and fragmentation have been assumed to be binary processes. Thus a j-mer can bind with an n-mer to form a $(j+n)$-mer or can break up into only two mers of smaller sizes; see the review article [10] by Collet for further details. However, a model of multiple fragmentation processes in which the break-up of a n-mer can lead to more than two mers has also been developed by Ziff; for example, see [25]. Consequently, we shall consider the more general model of binary coagulation combined with multiple fragmentation in the work we present here. In this case, the kinetic equation describing the time-evolution of the clusters is given by

$$u_n'(t) = -a_n u_n(t) + \sum_{j=n+1}^{\infty} a_j b_{n,j} u_j(t)$$
$$+\frac{1}{2}\sum_{j=1}^{n-1} k_{n-j,j} u_{n-j}(t) u_j(t) - \sum_{j=1}^{\infty} k_{n,j} u_n(t) u_j(t) , \qquad (24)$$

$$u_n(0) = \overset{\circ}{u}_n , \qquad n = 1, 2, 3, \ldots , \qquad (25)$$

where $u_n(t)$ is the concentration of n-mers at time t (where t is assumed to be a continuous variable), a_n is the net rate of break-up of an n-mer, $b_{n,j}$ gives the average number of n-mers produced upon the break-up of a j-mer, and $k_{n,j} = k_{j,n}$ represents the coagulation rate of an n-mer with a j-mer. Note that the total mass in the system at time t is given by

$$M(t) = \sum_{n=1}^{\infty} n u_n(t) ,$$

and for mass to be conserved we require

$$\sum_{n=1}^{j-1} n b_{n,j} = j , \quad j = 2, 3, \ldots . \qquad (26)$$

On using this condition together with (24), a formal calculation establishes that $M'(t) = 0$.

When the fragmentation process is binary, the C-F equation is usually expressed in the form

$$u_n'(t) = -\frac{1}{2} u_n(t) \sum_{j=1}^{n-1} F_{j,n-j} + \sum_{j=n+1}^{\infty} F_{n,j-n} u_j(t)$$
$$+\frac{1}{2}\sum_{j=1}^{n-1} k_{n-j,j} u_{n-j}(t) u_j(t) - \sum_{j=1}^{\infty} k_{n,j} u_n(t) u_j(t) , \qquad (27)$$

where $F_{n,j} = F_{j,n}$ represents the rate at which an $(n + j)$-mer breaks up into an n-mer and a j-mer. In this case,

$$2a_n = \sum_{j=1}^{n-1} F_{j,n-j} , \quad b_{n,j}\, a_j = F_{n,j-n}$$

and so

$$b_{n,j} = \frac{F_{n,j-n}}{a_j} = \frac{2F_{n,j-n}}{\sum_{r=1}^{j-1} F_{r,j-r}} \Rightarrow \sum_{n=1}^{j-1} b_{n,j} = 2;$$

i.e. the number of clusters produced in any fragmentation event is always two.

Equation (27) is the binary model that has been studied in [1] and [11], where existence and uniqueness results are presented for various rate coefficients. The underlying strategy common to each of these is to consider finite-dimensional truncations of (27). Standard methods from the theory of ordinary differential equations then lead to the existence of a sequence of solutions to these truncated equations. It is then shown, via Helly's theorem, that a subsequence exists that converges to a function u that satisfies an integral version of (27). A solution obtained in this way is called an admissible solution. A similar approach has been used by Laurençot in [17] to prove the existence of appropriately defined global mass-conserving solutions of the more general Eq. (24), and also, in [18], of the continuous-size coagulation equation, which takes the form of an integro-differential equation.

In contrast to the truncation approach used in the aforementioned papers, here we shall show how results from the theory of semigroups of operators can be used to establish the existence and uniqueness of solutions to (24). For simplicity, we shall assume that $k_{n,j} = k$ for all $n,\ j$ where k is a non-negative constant. Note, however, that a semigroup approach can also deal with more general coagulation kernels. In particular, results related to the concept of an analytic semigroup play an important role. We shall not discuss analytic semigroups in these notes, but the interested reader should consult the contribution to this volume by Banasiak [3] where the continuous size C-F equation is investigated via analytic semigroups.

To see how an IVP for the discrete C–F equation can be expressed as an ACP, we define $u(t)$ to be the sequence $(u_1(t), u_2(t), \ldots, u_j(t), \ldots)$. Then $u(t)$ is a sequence-valued function of t for each $t \geq 0$, and it therefore makes sense to seek a function u, defined on $[0, \infty)$, that takes values in an infinite-dimensional state space consisting of sequences. The state space that is most often used due to its physical relevance is the Banach space ℓ_1^1 discussed in Example 4. The ℓ_1^1-norm of a non-negative element $f \in \ell_1^1$ (i.e. $f = (f_1, f_2, \ldots)$ with $f_j \geq 0$ for all j), given by $\sum_{j=1}^{\infty} jf_j$, represents the total mass of the system. Similarly, the ℓ_0^1-norm of such an f gives the total number of particles in the system. Note that ℓ_1^1 is continuously imbedded in ℓ_0^1 since

$$\|f\|_{0,1} \leq \|f\|_{1,1} \quad \forall\, f \in \ell_1^1.$$

The function u will be required to satisfy an ACP of the form

$$u'(t) = L\,(u(t)) + N\,(u(t)),\ u(0) = \overset{\circ}{u},$$

where L and N are appropriately defined operator versions of the respective mappings

$$f_n \to -a_n f_n + \sum_{j=n+1}^{\infty} a_j b_{n,j} f_j \text{ and}$$

$$f_n \to \frac{k}{2}\sum_{j=1}^{n-1} f_{n-j} f_j - k \sum_{j=1}^{\infty} f_n f_j, \quad (n = 1, 2, 3, \ldots).$$

We begin our investigation into (23) by considering the case when only the linear operator L appears on the right-hand side of the equation; i.e. (23) takes the form

$$u'(t) = L\,(u(t)),\ t > 0, \quad u(0) = \mathring{u}\ . \tag{28}$$

In the context of our C–F model, this will represent a situation when no coagulation is occurring; i.e. the coagulation rate constant k is zero.

A function $u : [0, \infty) \to X$ is said to be a strong solution to (28) if

(i) u is strongly continuous on $[0, \infty)$;
(ii) the strong derivative u' exists and is strongly continuous on $(0, \infty)$;
(iii) $u(t) \in D(L)$ for each $t > 0$;
(iv) the equations in (28) are satisfied.

3.1 *Linear Infinite-Dimensional Evolution Equations*

3.1.1 Bounded Infinitesimal Generators

Although an infinite-dimensional setting may seem a bit daunting, it turns out that, for a bounded linear operator L, the methods discussed earlier in finite dimensions continue to work. Indeed, when L is bounded and linear on X, then the unique strong solution of the linear infinite-dimensional ACP (28) is given by

$$u(t) = e^{tL}\mathring{u}, \tag{29}$$

where the operator exponential is defined by

$$e^{tL} = I + tL + \frac{t^2L^2}{2!} + \frac{t^3L^3}{3!} + \cdots, \tag{30}$$

with I denoting the identity operator on X. This infinite series of bounded, linear operators on X always converges in $B(X)$ to a bounded, linear operator on X. Moreover,

$$e^{0L} = I;\ e^{sL}e^{tL} = e^{(s+t)L} \text{ for all } s, t \in \mathbb{R};\ e^{tL}\mathring{u} \to \mathring{u} \text{ in } X \text{ as } t \to 0; \tag{31}$$

see [19, Theorem 2.10]. It can easily be verified that the function $\phi(t, \overset{\circ}{u}) = e^{tL}\overset{\circ}{u}$ defines a continuous, infinite-dimensional dynamical system on X.

The person who appears to have been the first to generalise the use of matrix exponentials for finite-dimensional systems of ODEs to operator exponentials in infinite-dimensional spaces is Maria Gramegna, a student of Peano, in 1910; see [13]. Peano had considered some special types of infinite systems of ODEs in 1894, but it was Gramegna who demonstrated that operator exponentials could be applied more generally, not only to infinite systems of ODEs, but also to integro-differential equations.

Example 8 We examine the simple case of an IVP for a fragmentation equation in which $a_j = a$ for all $j \geq 2$, where a is a positive constant. We shall show that the corresponding linear fragmentation operator L is bounded on ℓ_1^1. If we recall that $a_1 = 0$, and also that the mass-conservation condition (26) holds, then we obtain, for each $f \in \ell_1^1$,

$$\begin{aligned}
\|Lf\|_{1,1} &= \sum_{n=1}^{\infty} n \left| -a_n f_n + \sum_{j=n+1}^{\infty} a_j b_{n,j} f_j \right| \\
&\leq \sum_{n=1}^{\infty} n a_n |f_n| + \sum_{n=1}^{\infty} \sum_{j=n+1}^{\infty} n a_j b_{n,j} |f_j| \\
&= \sum_{n=1}^{\infty} n a_n |f_n| + \sum_{j=2}^{\infty} \left(\sum_{n=1}^{j-1} n b_{n,j} \right) a_j |f_j| \\
&= \sum_{n=1}^{\infty} n a_n |f_n| + \sum_{j=1}^{\infty} j a_j |f_j| \\
&= 2a \sum_{n=1}^{\infty} n |f_n| = 2a \|f\|_{1,1}.
\end{aligned}$$

It follows that $L \in B(\ell_1^1)$ and so the ACP

$$u'(t) = L\,(u(t)),\ u(0) = \overset{\circ}{u},$$

has a strong, globally-defined, solution given by

$$u(t) = e^{tL}\overset{\circ}{u}\,.$$

As we shall demonstrate later when we consider the fragmentation equation with less restrictive conditions imposed on the rate coefficients a_n, this strong solution is non-negative whenever $\overset{\circ}{u}$ is non-negative, and $\|u(t)\|_{1,1} = \|\overset{\circ}{u}\|_{1,1}$ for all $t > 0$, showing that mass is conserved.

3.1.2 Unbounded Infinitesimal Generators: The Hille–Yosida Theorem

In many applications that involve the analysis of a linear evolution equation, posed in an infinite-dimensional setting, when an approach involving semigroups of operators and exponentials of operators is tried, the restriction that L is bounded and defined on all of the state space X is frequently too severe. In most cases, L is unlikely to be bounded and is usually only defined on elements in X which have specific properties. Is it possible that a family of exponential operators $\{e^{tL}\}_{t\geq 0}$ can be generated from an unbounded linear operator L and yield a unique solution to the IVP (28) via (29)? The answer to this is yes. In 1948, Einar Hille and Kôsaku Yosida, simultaneously and independently, proved a theorem (the Hille–Yosida theorem) that forms the cornerstone of the Theory of Strongly Continuous Semigroups of Operators. Since then, there has been a great deal of research activity in the theory and application of semigroups of operators. Amongst many other important developments, the Hille–Yosida theorem was extended in 1952 to a result that completely characterises the operators L that generate strongly continuous semigroups on a Banach space X. What this means is that, when a natural interpretation of "solution" is adopted, a unique solution to (28) exists if and only if the operator satisfies the conditions of this more general version of the Hille–Yosida theorem. Moreover, the solution is still given by (29), although, for unbounded linear operators L, a different exponential formula has to be used to define e^{tL}. One such formula is

$$e^{tL}f := \lim_{n\to\infty}\left[\frac{n}{t}R(n/t, L)\right]^n f = \lim_{n\to\infty}\left(I - \frac{t}{n}L\right)^{-n} f, \tag{32}$$

where $R(\lambda, L)$ denotes the inverse of $\lambda I - L$. Compare this with the scalar sequential formula for e^{tl},

$$e^{tl} = \lim_{n\to\infty}(1 + tl/n)^n.$$

There are many excellent books devoted to the theory of strongly continuous semigroups; for example [6, 19, 21] and [13]. Important details can also be found in the lecture notes by Banasiak [3, Section 2.5] and a nice gentle introduction to the theory is given by Bobrowski [8, Section 1]. As in [8], the account of semigroups that is presented here is not intended to be comprehensive; instead we merely summarise several key results from this very elegant, and applicable, theory. We begin with the following fundamental definition.

Definition 2 Let $\{S(t)\}_{t\geq 0}$ be a family of bounded linear operators on a complex Banach space X. Then $\{S(t)\}_{t\geq 0}$ is said to be a strongly continuous semigroup (or C_0- semigroup) in $B(X)$ if the following conditions are satisfied.

S1. $S(0) = I$, where I is the identity operator on X.
S2. $S(t)S(s) = S(t+s)$ for all $t, s \geq 0$.
S3. $S(t)f \to f$ in X as $t \to 0^+$ for all $f \in X$.

Associated with each strongly continuous semigroup $\{S(t)\}_{t\geq 0}$ is a unique linear operator L defined by

$$Lf := \lim_{h\to 0+} \frac{S(h)f - f}{h}, \quad D(L) := \left\{ f \in X : \lim_{h\to 0+} \frac{S(h)f - f}{h} \text{ exists in } X \right\}. \quad (33)$$

The operator L is called the infinitesimal generator of the semigroup $\{S(t)\}_{t\geq 0}$. For example, the infinitesimal generator of the semigroup given by $S(t) = e^{tL}$, where $L \in B(X)$, is the operator L.

Before stating some important properties of strongly continuous semigroups and their generators, we require some terminology.

Definition 3 Let $L : X \supseteq D(L) \to X$ be a linear operator.

(i) The resolvent set, $\rho(L)$, of L is the set of complex numbers

$$\rho(L) := \{\lambda \in \mathbb{C} : R(\lambda, L) := (\lambda I - L)^{-1} \in B(X)\};$$

$R(\lambda, L)$ is called the resolvent operator of L (at λ).

(ii) L is a closed operator (or L is closed) if whenever $(f_n)_{n=1}^{\infty} \subset D(L)$ is such that $f_n \to f$ and $Lf_n \to g$ in X as $n \to \infty$, then $g \in D(L)$ and $Lf = g$.

(iii) An operator $L_1 : X \supset D(L_1) \to X$ is an extension of L, written $L \subset L_1$, if $D(L) \subset D(L_1)$ and $Lf = L_1 f$ for all $f \in D(L)$. The operator L is closable if it has a closed extension, in which case the closure $\overline{L}$ of L is defined to be the smallest closed extension of L.

(iv) L is said to be densely defined if $\overline{D(L)} = X$, i.e. if the closure of the set $D(L)$ (with respect to the norm in X) is X. This means that, for each $f \in X$, there exists a sequence $(f_n)_{n=1}^{\infty} \subset D(L)$ such that $\|f - f_n\| \to 0$ as $n \to \infty$.

Theorem 3 (Some Semigroup Results) *Let $\{S(t)\}_{t\geq 0} \subset B(X)$ be a strongly continuous semigroup with infinitesimal generator L. Then*

(i) $S(t)f \to S(t_0)f$ in X as $t \to t_0$ for any $t_0 > 0$ and $f \in X$;

(ii) $S(t)f \to f$ in X as $t \to 0^+$;

(iii) there are real constants $M \geq 1$ and ω such that

$$\|S(t)\| \leq Me^{\omega t} \text{ for all } t \geq 0; \quad (34)$$

(iv) $f \in D(L) \Rightarrow S(t)f \in D(L)$ for all $t > 0$ and

$$\frac{d}{dt}S(t)f = LS(t)f = S(t)Lf \text{ for all } t > 0 \text{ and } f \in D(L); \quad (35)$$

(iv) the infinitesimal generator L is closed and densely defined.

We shall write $L \in \mathscr{G}(M, \omega; X)$ when L is the infinitesimal generator of a strongly continuous semigroups of operators satisfying (34) on a Banach space X. When the operator $L \in \mathscr{G}(1, 0; X)$, L is said to generate a strongly continuous semigroup of contractions on X.

Theorem 4

> ***(Hille–Yosida)*** *The operator L is the infinitesimal generator of a strongly continuous semigroup of contractions on X if and only if*
>
> *(i) L is a closed, linear and densely-defined operator in X;*
> *(ii) $\lambda \in \rho(L)$ for all $\lambda > 0$;*
> *(iii) $\|R(\lambda, L)\| \leq 1/\lambda$ for all $\lambda > 0$.*
>
> ***(Hille–Yosida–Phillips–Miyadera–Feller)*** *$L \in \mathscr{G}(M, \omega; X)$ if and only if*
>
> *(i) L is a closed, linear and densely-defined operator in X;*
> *(ii) $\lambda \in \rho(L)$ for all $\lambda > \omega$;*
> *(iii) $\|(R(\lambda, L))^n\| \leq M/(\lambda - \omega)^n$ for all $\lambda > \omega$, $n = 1, 2, \ldots$.*

Proofs of these extremely important results can be found in [19, Chapter 3].

We can now state the following existence/uniqueness theorem for the linear ACP

$$u'(t) = L\,(u(t)),\ t > 0;\ \ u(0) = \overset{\circ}{u} \in D(L). \tag{36}$$

Theorem 5 *Let L be the infinitesimal generator of a strongly continuous semigroup $\{S(t)\}_{t \geq 0} \subset B(X)$. Then (36) has one and only one strong solution $u : [0, \infty) \to X$ and this is given by $u(t) = S(t)\overset{\circ}{u}$.*

The operator $S(t)$ can be interpreted as the exponential e^{tL} if we define the latter by (32); see [19, Chaper 6] for a proof.

3.1.3 The Kato–Voigt Perturbation Theorem

Although the Hille–Yosida theorem and the generalisation due to Phillips et al. are extremely elegant results, in practice it is often difficult to check that the resolvent conditions are satisfied for a given linear operator L. One way to get round this is to make use of perturbation theorems for infinitesimal generators; see the book by Banasiak and Arlotti [4]. The basic idea is to treat, if possible, the linear operator governing the dynamics of the system as the sum of two linear operators, say $A + B$, where A is an operator which can easily be shown to generate a strongly continuous semigroup $\{S_A(t)\}_{t \geq 0}$ on a Banach space X, and B is regarded as a perturbation of A. The question then is to identify sufficient conditions on B which will guarantee that $A + B$ (or some extension of $A + B$) also generates a strongly continuous semigroup on X. A number of perturbation results of this type have been established. We shall focus on just one of these, namely the Kato–Voigt Perturbation theorem, but only for the specific case when the state space is the Banach space ℓ^1_μ

of Example 4. An account of the general version of this important perturbation result is given in [3, Section 2.6].

As mentioned earlier, non-negative elements in ℓ^1_μ are taken to be sequences $f = (f_1, f_2, \ldots)$ with $f_j \geq 0$ for all j, in which case we write $f \geq \mathbf{0}$. An operator $T : \ell^1_\mu \supseteq D(T) \to \ell^1_\mu$, is said to be non-negative if $Tf \geq \mathbf{0}$ for all non-negative $f \in D(T)$.

Theorem 6 (See [2, Theorem 2.1] and [4, Corollary 5.17]) *Let the operators $A : \ell^1_\mu \supseteq D(A) \to \ell^1_\mu$ and $B : \ell^1_\mu \supseteq D(B) \to \ell^1_\mu$ have the following properties.*

(i) A is the infinitesimal generator of a semigroup of contractions $\{S_A(t)\}_{t\geq 0}$ on ℓ^1_μ, with $S_A(t) \geq 0$ for all $t \geq 0$.
(ii) B is non-negative and $D(B) \supseteq D(A)$.
(iii) For each non-negative f in $D(A)$,

$$\sum_{j=1}^{\infty} j^{\mu}(Af + Bf)_j \leq 0.$$

Then there exists a strongly continuous semigroup of contractions, $\{S(t)\}_{t\geq 0}$, on ℓ^1_μ satisfying the Duhamel equation

$$S(t)f = S_A(t)f + \int_0^t S(t-s)BS_A(s)f\, ds,\ f \in D(A).$$

Each $S(t)$ is non-negative and the infinitesimal generator of the semigroup is an extension L of $A + B$.

Example 9 We now show that a straightforward application of this perturbation theorem establishes the existence and uniqueness of solutions to the fragmentation equation for a wide class of fragmentation rate coefficients. Once again we work in the state space ℓ^1_1, and we take A and B to be the operators

$$(Af)_n := -a_n f_n,\ n \in \mathbb{N}, \quad D(A) = \{f \in \ell^1_1 : Af \in \ell^1_1\},$$

$$(Bf)_n := \sum_{j=n+1}^{\infty} b_{n,j}\, a_j f_j,\ n \in \mathbb{N}, \quad D(B) = D(A).$$

Then

1. By arguing as in [8, Example 6], it is not difficult to prove that the operator A is the infinitesimal generator of a strongly continuous semigroup of contractions $\{S_A(t)\}_{t\geq 0}$ on ℓ^1_1 given by

$$(S_A(t)f)_n := e^{-a_n t} f_n,\ n \in \mathbb{N}.$$

It is clear that $S_A(t) \geq 0$ for each t.

2. The calculations used in Example 8 can be repeated to show that

$$\begin{aligned}\|Bf\| &\le \sum_{n=1}^{\infty}\sum_{j=n+1}^{\infty} na_j\, b_{n,j}|f_j| \\ &= \sum_{j=2}^{\infty}\left(\sum_{n=1}^{j-1} nb_{n,j}\right) a_j|f_j| \\ &= \sum_{j=2}^{\infty} j\, a_j|f_j| = \|Af\|,\ \forall\, f \in D(A).\end{aligned}$$

Consequently, B is well defined on $D(A)$ and $Bf \ge 0$ for all $f \in D(A)$ with $f \ge 0$.
3. A similar argument shows that

$$\sum_{n=1}^{\infty} n(Af + Bf)_n = 0\ \forall\ f \in D(A) \text{ with } f \ge 0.$$

Consequently, by the Kato–Voigt Perturbation Theorem, there exists a strongly continuous semigroup of contractions $\{S(t)\}_{t\ge 0}$ generated by an extension L of the operator $(A + B, D(A))$, with $S(t)f \ge 0$ for all non-negative $f \in \ell_1^1$.

In this example it is possible to show that L is the closure of $(A + B, D(A))$ and also that

$$\sum_{n=1}^{\infty} n(Lf)_n = 0\ \forall\ f \in D(L) \text{ with } f \ge 0;$$

see [20]. Consequently, the ACP

$$u'(t) = L(u(t)),\ u(0) = \mathring{u} \in D(L),\ \mathring{u} \ge 0,$$

with $L = \overline{A + B}$, has a unique strongly differentiable solution $u : [0,\infty) \to D(L)$ given by $u(t) = S(t)\mathring{u}$.

Other results that can be established for this discrete-size fragmentation equation are:

- If the sequence (a_n) is monotonic increasing, then $S(t) : D(A) \to D(A)$ for all $t \ge 0$ and therefore $u(t) = S(t)\mathring{u}$ is the unique strong solution of the ACP

$$u'(t) = A\,(u(t)) + B\,(u(t)),\ u(0) = \mathring{u} \in D(A),\ \mathring{u} \ge 0.$$

- Suppose that $a_n > 0$ for all $n \geq 2$. Then

$$(A + B)\overline{u} = (0, 0, \ldots) \text{ in } \ell_1^1 \Leftrightarrow \overline{u} = ce_1,$$

where c is a constant and $e_1 = (1, 0, 0, \ldots)$. Moreover, it can be shown that $S(t)u_0 \to M(\mathring{u})e_1$ in ℓ_1^1 as $t \to \infty$, where $M(\mathring{u}) = \sum_{n=1}^{\infty} n\, \mathring{u}_n$. This situation is similar to that observed with the SIR model in that we have infinitely many equilibria, and the equilibrium that any given solution converges to is uniquely determined by the initial data.

See [5] and [20] for further details.

3.2 Semi-linear Infinite-Dimensional Evolution Equations

To conclude, we return to the semi-linear ACP (23). We shall assume that the linear operator L is the infinitesimal generator of a strongly continuous semigroup $\{S(t)\}_{t \geq 0}$ on X. A strong solution on $[0, t_0)$ of this ACP is a function $u : [0, t_0) \to X$ such that

(i) u is strongly continuous on $[0, t_0)$;
(ii) u has a continuous strong derivative on $(0, t_0)$;
(iii) $u(t) \in D(L)$ for $0 \leq t < t_0$;
(iv) $u(t)$ satisfies (23) for $0 \leq t < t_0$.

Suppose that u is a strong solution. Then, under suitable assumptions on N, u will also satisfy the Duhamel equation

$$u(t) = S(t)\mathring{u} + \int_0^t S(t-s)N(u(s))\, ds, \quad 0 \leq t < t_0. \tag{37}$$

This leads to the following definition of a weaker type of solution to the ACP.

Definition 4 A mild solution on $[0, t_0)$ of (23) is a function $u : [0, t_0) \to X$ such that

(i) u is strongly continuous on $[0, t_0)$;
(ii) u satisfies (37) on $[0, t_0)$.

The definitions given earlier for a function on the finite-dimensional space $\mathbb{R}^n$ to be Fréchet differentiable, or to satisfy a local Lipschitz condition, extend to operators on infinite-dimensional spaces. In particular, the nonlinear operator $N : X \to X$ satisfies a local Lipschitz condition on X if, for each $\mathring{u} \in X$, there exists a closed ball $\overline{B}_r(\mathring{u}) := \{f \in X : \|f - \mathring{u}\| \leq r\}$ such that

$$\|N(f) - N(g)\| \leq k\|f - g\|, \quad \forall f, g \in \overline{B}_r(\mathring{u}).$$

Also, N is Fréchet differentiable at $f \in X$ if an operator $N_f \in B(X)$ exists such that

$$N(f+h) = N(f) + N_f(h) + E(f,h),\ h \in X,$$

where

$$\lim_{\|h\|\to 0} \frac{\|E(f,h)\|}{\|h\|} = 0.$$

The operator N_f is the Fréchet derivative of N at f.

Theorem 7 *Let $L \in \mathcal{G}(M,\omega;X)$ and let N satisfy a local Lipschitz condition on X. Then there exists a unique mild solution of the ACP on some interval $[0, t_{max})$. Moreover, if $t_{max} < \infty$, then*

$$\|u(t)\| \to \infty \text{ as } t \to t_{max}^-.$$

Theorem 8 *Let $L \in \mathcal{G}(M,\omega;X)$ and let N be continuously Fréchet differentiable on X. Then the mild solution of the semi-linear ACP, with $\mathring{u} \in D(L)$, is a strong solution.*

For proofs of these results, see [21, Chapter 6] and [7, Chapter 3].

Example 10 We now describe how these results have been applied to the discrete C-F equation in [20,22] and [5]. Having already established that $L = \overline{A+B}$ is the infinitesimal generator of a strongly continuous positive semigroup of contractions on the space ℓ_1^1, we express the full C-F equation as the semi-linear ACP

$$u'(t) = L\left(u(t)\right) + N\left(u(t)\right),\ t > 0, \quad u(0) = \mathring{u} \in D(L),$$

where

$$(Nf)_n := \frac{k}{2}\sum_{j=1}^{n-1} f_{n-j} f_j - k \sum_{j=1}^{\infty} f_n f_j,\ f \in \ell_1^1.$$

We shall show below that $N(f) \in \ell_1^1$ for all $f \in \ell_1^1$. For this it is convenient to introduce the following bilinear operator

$$\tilde{N}(f,g) := \tilde{N}_1(f,g) - \tilde{N}_2(f,g),$$

where, for $f, g \in \ell_1^1$,

$$[\tilde{N}_1(f,g)]_n := \frac{k}{2}\sum_{j=1}^{n-1} f_{n-j} g_j, \quad [\tilde{N}_2(f,g)]_n := k\sum_{j=1}^{\infty} f_n g_j,\ n \in \mathbb{N}.$$

Note that $N(f) = \tilde{N}(f, f)$. Also, it is straightforward to verify that $\tilde{N}(\cdot, \cdot)$ is linear in both left-hand and right-hand arguments. Consequently,

$$\tilde{N}(f + h, f + h) = \tilde{N}(f, f) + \tilde{N}(f, h) + \tilde{N}(h, f) + \tilde{N}(h, h). \tag{38}$$

Now,

$$\begin{aligned}\|\tilde{N}_1(f, g)\| &\le \frac{k}{2}\sum_{n=1}^{\infty}\sum_{j=1}^{n-1} n|f_{n-j}|\,|g_j| \\ &= \frac{k}{2}\sum_{j=1}^{\infty}\sum_{n=j+1}^{\infty} n|f_{n-j}|\,|g_j| \\ &= \frac{k}{2}\sum_{j=1}^{\infty}\sum_{i=1}^{\infty}(i + j)|f_i|\,|g_j| \\ &\le 2\frac{k}{2}\|f\|\,\|g\| = k\|f\|\,\|g\|.\end{aligned}$$

Similarly,

$$\|\tilde{N}_1(f, g)\| \le k\|f\|\,\|g\|.$$

Hence

$$\|\tilde{N}(f, g)\| \le 2k\|f\|\,\|g\| \text{ and } \|N(f)\| \le 2k\|f\|^2.$$

In the case when $N(f) \ge 0$, we can also deduce that

$$\begin{aligned}\|N(f)\| &= \sum_{n=1}^{\infty} n[N(f)]_n \\ &= \frac{k}{2}\left(\sum_{i=1}^{\infty} if_i\right)\left(\sum_{j=1}^{\infty} f_j\right) + \frac{k}{2}\left(\sum_{i=1}^{\infty} f_i\right)\left(\sum_{j=1}^{\infty} jf_j\right) \\ &\quad -k\left(\sum_{i=1}^{\infty} if_i\right)\left(\sum_{j=1}^{\infty} f_j\right) = 0.\end{aligned}$$

The bilinearity of $\tilde{N}(\cdot, \cdot)$ leads immediately to the Fréchet differentiability of N. From (38), we obtain

$$N(f + h) = N(f) + N_f(h) + N(h),$$

where

$$N_f(h) := \tilde{N}(f, h) + \tilde{N}(h, f). \tag{39}$$

For fixed $f \in \ell_1^1$, N_f is a linear operator on ℓ_1^1 and also

$$\|N_f(h)\| \leq 4k\|f\| \, \|h\| \, , \; \forall h \in \ell_1^1 \tag{40}$$

showing that $N_f \in B(X)$. Moreover,

$$\frac{\|N(h)\|}{\|h\|} \leq 2k\|h\| \to 0 \text{ as } \|h\| \to 0.$$

Hence, N is Fréchet differentiable at each $f \in \ell_1^1$, with Fréchet derivative given by (39). Moreover, inequality (40) can be used to establish that N_f is continuous in f. (Note that this also means that N is locally Lipschitz continuous.) We can now apply Theorems 7 and 8 to conclude that the semi-linear ACP has a unique, locally defined (in time) strong solution.

To complete our analysis, we must show that the solution $u(t)$ is non-negative for all t for which it is defined. We would also like to establish that the solution is defined for all $t \geq 0$. It turns out that the latter can be deduced directly from the former since

$$\frac{d}{dt}\|u(t)\| = \|L\,(u(t)) + N\,(u(t))\| = \sum_{n=1}^{\infty} n(L\,(u(t))_n + \sum_{n=1}^{\infty} n(N\,(u(t))_n = 0,$$

showing that $\|u(t)\|$ cannot blow up in finite time. The proof that the solution remains non-negative is the most involved part of the argument and so only some outline details will be supplied here (see [20] for further information). In essence, we use the following trick. The ACP is rewritten as

$$u'(t) = (L(u(t) - \alpha u(t)) + (\alpha u(t) + N(u(t)),$$

where the constant α is chosen so that $(N + \alpha)u(t) \geq 0$ for all t in some interval $[0, t_0]$. The operator $L - \alpha I$ is the infinitesimal generator of the positive semigroup $\{e^{-\alpha t} S(t)\}_{t \geq 0}$ (where $\{S(t)\}_{t \geq 0}$ is the positive semigroup generated by L). The solution u of this modified equation satisfies the integral equation

$$u(t) = e^{-\alpha t} S(t) \mathring{u} + \int_0^t e^{-\alpha(t-s)} S(t-s)(N + \alpha)u(s)\, ds =: T_\alpha(u(t)), \; t \in [0, t_0].$$

The value t_0 is selected so that the operator T_α on the right-hand side of the above equation is a contraction on a suitable Banach space of ℓ_1^1-valued functions and so we can obtain the solution u (the fixed point of this contraction) by means of successive iterations of T_α on the initial state $\mathring{u} > 0$. Since T_α is positivity preserving, it follows that $u(t) \geq 0$ for all $t \in [0, t_0]$. We then repeat this argument, but now with $u(t_0)$ as the initial state, and continue in this manner.

References

1. J.M. Ball, J. Carr, The discrete coagulation–fragmentation equations: existence, uniqueness and density conservation. J. Stat. Phys. **61**, 203–234 (1990)
2. J. Banasiak, On an extension of the Kato–Voigt perturbation theorem for substochastic semigroups and its application. Taiwanese J. Math. **5**, 169–191 (2001)
3. J. Banasiak, Kinetic models in natural sciences, in *Evolutionary Equations with Applications to Natural Sciences*, ed. by J. Banasiak, M. Mokhtar-Kharroubi. Lecture Notes in Mathematics (Springer, Berlin, 2014)
4. J. Banasiak, L. Arlotti, *Perturbations of Positive Semigroups with Applications* (Springer, New York, 2006)
5. J. Banasiak, W. Lamb, The discrete fragmentation equation: semigroups, compactness and asynchronous exponential growth. Kinetic Relat. Model **5**, 223–236 (2012)
6. A. Belleni-Morante, *Applied Semigroups and Evolution Equations* (Clarendon Press, Oxford, 1979)
7. A. Belleni-Morante, A.C. McBride, *Applied Nonlinear Semigroups* (Wiley, Chichester, 1998)
8. A. Bobrowski, Boundary conditions in evolutionary equations in biology, in *Evolutionary Equations with Applications to Natural Sciences*, ed. by J. Banasiak, M. Mokhtar-Kharroubi. Lecture Notes in Mathematics (Springer, Berlin, 2014)
9. N.F. Britton, *Essential Mathematical Biology* (Springer, London, 2003)
10. J.F. Collet, Some modelling issues in the theory of fragmentation–coagulation systems. Commun. Math. Sci. **1**, 35–54 (2004)
11. F.P. da Costa, Existence and uniqueness of density conserving solutions to the coagulation–fragmentation equations with strong fragmentation. J. Math. Anal. Appl. **192**, 892–914 (1995)
12. S.P. Ellner, J. Guckenheimer, *Dynamic Models in Biology* (Princeton University Press, Princeton, 2006)
13. K.-J. Engel, R. Nagel, *One-Parameter Semigroups for Linear Evolution Equations*. Graduate Texts in Mathematics (Springer, New York, 2000)
14. M. Farkas, *Dynamical Models in Biology* (Academic, San Diego, 2001)
15. M.W. Hirsch, S. Smale, *Differential Equations, Dynamical Systems, and Linear Algebra* (Academic, Orlando, 1974)
16. E. Kreyszig, *Introductory Functional Analysis with Applications* (Wiley, New York, 1978)
17. P. Laurençot, The discrete coagulation equations with multiple fragmentation. Proc. Edinburgh Math. Soc. **45**, 67–82 (2002)
18. P. Laurençot, Weak compactness techniques and coagulation equations, in *Evolutionary Equations with Applications to Natural Sciences*, ed. by J. Banasiak, M. Mokhtar-Kharroubi. Lecture Notes in Mathematics (Springer, Berlin, 2014)
19. A.C. McBride, *Semigroups of Linear Operators: An Introduction*. Research Notes in Mathematics (Pitman, Harlow, 1987)
20. A.C. McBride, A.L. Smith, W. Lamb, Strongly differentiable solutions of the discrete coagulation–fragmentation equation. Physica D **239**, 1436–1445 (2010)
21. A. Pazy, *Semigroups of Linear Operators and Applications to Partial Differential Equations* (Springer, Berlin, 1983)
22. L. Smith, W. Lamb, M. Langer, A.C. McBride, Discrete fragmentation with mass loss. J. Evol. Equ. **12**, 181–201 (2012)

23. J.A.D. Wattis, An introduction to mathematical models of coagulation–fragmentation processes; a discrete deterministic mean-field approach. Physica D **222**, 1–20 (2006)
24. S. Wiggins, *Introduction to Applied Nonlinear Dynamical Systems and Chaos* (Springer, New York, 1990)
25. R.M. Ziff, An explicit solution to a discrete fragmentation model. J. Phys. A Math. Gen. **25**, 2569–2576 (1992)

Boundary Conditions in Evolutionary Equations in Biology

Adam Bobrowski

1 A Gentle Introduction to the Theory of Semigroups of Operators

Let me start by quoting Walter Rudin, who in the prologue to his book [43] writes about the exponential function:

> This is the most important function in mathematics. It is defined, for every complex number z, by:
>
> $$\exp(z) = \sum_{n=0}^{\infty} \frac{z^n}{n!}. \tag{1}$$

We will try to extend this definition to the case where z may be replaced by a linear operator: in fact, we will see that for a large class of operators A one may construct a function

$$\mathbb{R}^+ \ni t \mapsto \mathrm{e}^{tA},$$

that will be in many aspects analogous to (1).

A. Bobrowski (✉)
Lublin University of Technology, Nadbystrzycka 18, 20-618 Lublin, Poland
e-mail: a.bobrowski@pollub.pl; bobrowscy@gmail.com

© Springer International Publishing Switzerland 2015

J. Banasiak, M. Mokhtar-Kharroubi (eds.), *Evolutionary Equations with Applications in Natural Sciences*, Lecture Notes in Mathematics 2126,
DOI 10.1007/978-3-319-11322-7_2

1.1 Bounded Linear Operators and Their Exponentials

Roughly speaking, an operator is a function having functions as arguments and values. For example we may think of the following maps:

$$\begin{aligned} f &\mapsto f', \\ f &\mapsto f'', \\ f &\mapsto f^2, \\ f &\mapsto g, \quad g(x) = f(x+a), \qquad x \in \mathbb{R}, \\ (f(1),\ldots,f(n)) &\mapsto (f(1),\ldots,f(n))A_{n\times n}, \end{aligned}$$

where a is a given number and $A_{n\times n}$ is a given square matrix. An operator A is said to be linear if for all scalars α and β we have

$$A(\alpha f + \beta g) = \alpha Af + \beta Ag, \qquad f, g \in D(A),$$

where $D(A)$ denotes the domain of A. From among the four operators listed above only one is not linear. We note that it is customary to use no parentheses for arguments of linear operators: hence, we write Af instead of $A(f)$.

The notion of continuity of operators requires extra structure, which is conveniently provided by Banach spaces, i.e., vector spaces with complete norm. To recall, a function $\|\cdot\|$ mapping a vector space of functions into $\mathbb{R}^+$ is said to be a norm iff for all $f, g \in \mathbb{X}$ and all scalars α,

1. $\|f\| = 0$ iff $f \equiv 0$,
2. $\|\alpha f\| = |\alpha|\,\|f\|$,
3. $\|f + g\| \le \|f\| + \|g\|$,

the third property being commonly referred to as the triangle inequality. Examples include the spaces of continuous functions on compact sets with supremum norm, i.e., with topology of uniform convergence, absolutely integrable functions with respect to a given measure (with norm being the integral of the absolute value of a function), etc. Completeness of the norm simply means that all Cauchy sequences converge, i.e., that the space does not have "holes". An operator $A : \mathbb{B} \to \mathbb{B}$ mapping a Banach space $\mathbb{B}$ into itself is said to be continuous if convergence of its arguments implies convergence of the corresponding values:

$$\|f_n - f\| \to 0 \quad \Rightarrow \quad \|Af_n - Af\| \to 0;$$

where $\|\cdot\|$ is the norm in $\mathbb{B}$. We note that completeness of $\mathbb{B}$ is not required for the definition of continuity, but normed spaces which are not complete do not have good properties and the theory is less satisfactory.

Strikingly, for linear operators continuity is equivalent to boundedness, i.e. to the condition:

$$\exists_{M>0}\forall_{f\in\mathbb{B}} \quad \|Af\| \leq M\|f\|.$$

The smallest constant M with property exhibited above is denoted $\|A\|$ and is a norm in the space of all bounded (continuous) operators in $\mathbb{B}$, denoted $\mathscr{L}(\mathbb{B})$. What makes the theory beautiful (and useful) is that the space of bounded linear operators on $\mathbb{B}$ is again a Banach space when equipped with this norm. (That $\mathscr{L}(\mathbb{B})$ is a linear space, becomes obvious once we define $(\alpha A + \beta B)x := \alpha Ax + \beta Bx$.)

Having recalled the basic definitions, let us turn to the question of how to define an operator exponential for a bounded operator. Of course, if the definition is to be meaningful, this new object must resemble the exponential function for complex numbers. In other words, we somehow need to mimic (1). Luckily, the very reason that makes (1) work is the fact that the space of complex numbers is complete, and the same argument may be applied to operators. To explain we recall that every absolutely convergent series in a Banach space converges. This means that for any sequence $(f_n)_{n\geq 1}$ of elements of a Banach space condition $\sum_{n\geq 1}\|f_n\| < \infty$ implies that the series $\sum_{n\geq 1} f_n$ is well-defined, i.e. the limit $\lim_{n\to\infty}\sum_{k=1}^{n} f_k$ exists. To see this, we note that the triangle inequality implies, for all $N \geq n$,

$$\|\sum_{k=n}^{N} f_k\| \leq \sum_{k=n}^{N}\|f_k\|.$$

Hence, vectors $g_n = \sum_{k=1}^{n} f_k$ form a Cauchy sequence, implying existence of the limit $\lim_{n\to\infty} g_n$.

This allows the following definition. For $A \in \mathscr{L}(\mathbb{B})$ and $t \in \mathbb{R}$ (in fact, if $\mathbb{B}$ is complex we could take complex t):

$$\mathrm{e}^{tA} := \sum_{n=0}^{\infty}\frac{t^n A^n}{n!}, \qquad t \in \mathbb{R}, \tag{2}$$

for we have $\sum_{n=0}^{\infty}\|\frac{t^n A^n}{n!}\| \leq \sum_{n=0}^{\infty}\frac{|t|^n\|A\|^n}{n!} = \mathrm{e}^{|t|\|A\|} < \infty$. In the previous-to-last step we have used the fact, easily established by the definition of the operator norm, that

$$\|AB\| \leq \|A\|\,\|B\|$$

for all $A, B \in \mathscr{L}(\mathbb{B})$.

Such a function has a number of good properties. To begin with, if A and B commute ($AB = BA$), then

$$\mathrm{e}^{A+B} = \mathrm{e}^{A}\mathrm{e}^{B} = \mathrm{e}^{B}\mathrm{e}^{A}. \tag{3}$$

In particular,

$$e^{(t+s)A} = e^{tA}e^{sA}, \qquad e^{-tA} = (e^{tA})^{-1}, \qquad e^{t(A-I)} = e^{-t}e^{tA}; \tag{4}$$

(if, for a given $A \in \mathscr{L}(\mathbb{B})$ there is a $B \in \mathscr{L}(\mathbb{B})$ such that $BA = I_{\mathbb{B}} = AB$, then we write $B = A^{-1}$; here $I_{\mathbb{B}}$ is the identity operator defined by $I_{\mathbb{B}}f = f$). The first equality here is the important **semigroup property** (in fact, here it is a group property, since the equality holds for all $s, t \in \mathbb{R}$).

Example 1 Take $a \geq 0$ and $b \geq 0$ such that $a + b > 0$, and let $\mathbb{B} = \mathbb{R}^2$. The space $\mathscr{L}(\mathbb{B})$ may be identified with the space of 2×2 matrices. For $A = \begin{pmatrix} -a, & a \\ b, & -b \end{pmatrix}$ we have

$$e^{tA} = \frac{1}{a+b}\begin{pmatrix} b + ae^{-(a+b)t}, & a - ae^{-(a+b)t} \\ b - be^{-(a+b)t}, & a + be^{-(a+b)t} \end{pmatrix}.$$

To prove this we note first that $e^{tA} = e^{-(a+b)t}e^{tB}$ where

$$B = A + (a+b)I_{\mathbb{B}} = \begin{pmatrix} b, & a \\ b, & a \end{pmatrix}.$$

Since $B^2 = (a+b)B$, we have, by induction $B^n = (a+b)^{n-1}B$. This shows that

$$e^{tB} = I_{\mathbb{B}} + \frac{1}{a+b}\sum_{n=1}^{\infty}\frac{t^n(a+b)^n}{n!}B = I_{\mathbb{B}} + \frac{1}{a+b}(e^{(a+b)t} - 1)B.$$

Therefore,

$$e^{tA} = e^{-(a+b)t}I_{\mathbb{B}} + \frac{1}{a+b}(1 - e^{-(a+b)t})B = \frac{1}{a+b}(B - e^{-(a+b)t}A),$$

proving the claim.

We note that for $t \geq 0$, the entries of e^{tA} are transition probabilities of a Markov chain with two states. At state 0 the chain waits for an exponential time with parameter a and then jumps to 1, where it waits for an exponential time with parameter b to come back to 0, and so on. The matrix A is an intensity matrix for this chain. Denoting by $X(t)$ the state of the chain at time $t \geq 0$ we obtain for example that the conditional probability of $X(t)$ being equal to 1 given that $X(0) = 1$ is the upper right entry in e^{tA}: $\Pr(X(t) = 1|X(0) = 0) = a - ae^{-(a+b)t}$.

Example 2 Let $\mathbb{B} = C[0,\infty]$ be the space of continuous functions on $[0,\infty]$ with limits at ∞. Equipped with the supremum norm, $\mathbb{B}$ is a Banach space, and for given $a > 0$ the operator $Af(x) = a[f(x+1) - f(x)], x \geq 0$ is bounded, since $\|Af\| \leq 2a\|f\|$. To compute e^{tA} for $t \geq 0$ we use the definition and the fact that

$B := A + aI_{\mathbb{B}}$ is a scalar multiple of the shift operator: $Bf(x) = af(x+1), x \geq 0$ so that $B^n f(x) = a^n f(x+n), x \geq 0$. Therefore,

$$\begin{aligned} e^{tA} f(x) &= e^{-at} e^{t(A+aI)} f(x) = \sum_{n=0}^{\infty} e^{-at} \frac{a^n t^n}{n!} f(x+n) \\ &= \mathbb{E} f(x + N(t)), \qquad t \geq 0, \end{aligned}$$

where $\mathbb{E}$ denotes expected value. In the last line $N(t)$ is a Poisson-distributed random variable with parameter at: $\Pr(N(t) = k) = e^{-at} \frac{(at)^k}{k!}$. In other words, the operator exponential $e^{tA}, t \geq 0$ describes a Poisson process. In this process, if the starting point is x then after time $t \geq 0$ its position is random: with probability $\Pr(N(t) = k)$ the process is at $x + k$.

Example 3 In the same space $\mathbb{B} = C[0, \infty]$, let $Af = f'$ with domain $D(A) = C^1[0, \infty]$. A is unfortunately unbounded: for f_n defined by $f_n(x) = e^{-nx}, x \geq 0$, we have $\|f_n\| = 1$, while $\|Af_n\| = n$; hence, there is no constant M such that $\|Af_n\| \leq M\|f_n\|$. Thus, it is not clear whether the series (2) converges. In fact, its nth term is defined merely for $f \in C^n[0, \infty]$ (n-times continuously differentiable functions with the nth order derivative in $C[0, \infty]$). However, for $f \in C^\infty[0, \infty]$ and $x \geq 0$ we can write:

$$\begin{aligned} e^{tA} f(x) &= \sum_{n=0}^{\infty} \frac{t^n A^n}{n!} f(x) \\ &= f(x) + tf'(x) + \frac{t^2}{2} f''(x) + \frac{t^3}{3!} f'''(x) + \dots \\ &= f(x+t), \end{aligned}$$

with the last equality holding provided f is analytic. This suggests that for all $f \in C[0, \infty]$ we should define

$$e^{tA} f(x) = f(x+t).$$

It is easy to see that with such a definition $e^{tA} e^{sA} = e^{(s+t)A}$, i.e., that the semigroup property holds [since we are dealing with an unbounded operator, the semigroup property is not a consequence of (4)]. The so defined semigroup describes deterministic movement to the right with speed $v = 1$: if the starting point is x then at time t the process is at $x + t$. Note, however, that the operators e^{tA} shift functions in $\mathbb{B}$ to the left.

An important moral to learn from this example is that with unbounded operators we should not expect the exponent to be defined for negative t: here $e^{-tA} = (e^{tA})^{-1}$ is undefined, since for $t > 0$ shifts to the left do not have inverses (part of the shape of the shifted function is for ever lost). This example exemplifies also the fact that deterministic movements are usually described by means of first order differential operators: here $A = \frac{d}{dx}$. Finally, despite the fact that here the exponential formula somewhat worked, we should not expect that it would do its job in general. As we shall see later, for unbounded linear operators (2) should be replaced by, for example:

$$e^{tA} = \lim_{n\to\infty}\left(I - \frac{tA}{n}\right)^{-n}, \qquad t \geq 0.$$

1.2 Exponentials of Unbounded Operators

Before turning to the definition of an operator exponential for unbounded operators A, we will investigate the question of where does the semigroup property come from. A better understanding of this crucial condition, will allow constructing exponentials in a new way. So what is the *real* reason for:

$$e^{tA}e^{sA} = e^{(s+t)A}, \qquad s, t \geq 0?$$

Previously, we have derived this property from (3) which in turn required definition (2) and some manipulations on the power series involved. Here, we will proceed differently. Fixing a bounded A and $f \in \mathbb{B}$, we consider the function $u(t) = e^{tA}f$ with values in $\mathbb{B}$. A simple calculation reveals that u is differentiable with $u'(t) = Ae^{tA}f = Au(t), t \geq 0$. In other words, u is a solution to the differential equation

$$u'(t) = Au(t)$$

with initial condition $u(0) = f$; i.e., a solution to the Cauchy problem related to operator A. On the other hand, using the Banach fixed point theorem and the fact that A is bounded it is easy to check that solutions to this equation are uniquely determined. To continue, we fix $s \geq 0$ and consider $v(t) = e^{(t+s)A}f = u(t+s)$. We have $v'(t) = u'(t+s) = Au(t+s) = Av(t), t \geq 0$. Now, since $v(0) = u(s) = e^{sA}f$, uniqueness of solutions forces

$$e^{(t+s)A}f = v(t) = e^{tA}v(0) = e^{tA}e^{sA}f,$$

which, on account of f being arbitrary, is none other than the semigroup property. In other words, we conclude that the semigroup property reflects uniqueness of solutions of the Cauchy problem. The point is that there are many more operators for which the Cauchy problem is well-posed in the sense provided below than these for which the series (2) converges. In this context, the following theorem is not surprising at all, but quite important.

Theorem 1 *If A is a (not necessarily bounded) operator such that the differential equation (in a Banach space)*

$$\frac{\mathrm{d}u(t)}{\mathrm{d}t} = Au(t), t \geq 0, \qquad u(0) = f$$

has exactly one solution u_f for f in a dense set D, and if this solution depends continuously on f, then the formula

$$\mathrm{e}^{tA} f = u_f(t) \tag{5}$$

defines an exponential function for A (a semigroup of operators).

The proof is quite obvious: for $f \in D$ uniqueness of solutions implies the semigroup property, and the fact that D is dense allows defining $\mathrm{e}^{tA} f$ for $f \in \mathbb{B}$ by continuity.

Back to Example 3 A partial differential equation:

$$\frac{\partial u(t,x)}{\partial t} = \frac{\partial u(t,x)}{\partial x}, t \geq 0, \qquad u(0,x) = f(x), x \in [0,\infty),$$

has exactly one solution for $f \in C^1[0,\infty]$, given by

$$u(t,x) = f(x+t).$$

This equation is identical to the ordinary differential equation

$$\frac{\mathrm{d}u(t)}{\mathrm{d}t} = Au(t), t \geq 0, \qquad u(0) = f \in C^1[0,\infty], \tag{6}$$

in the Banach space $C[0,\infty]$, where $A = \frac{\mathrm{d}}{\mathrm{d}x}$. Clearly, $C^1[0,\infty]$ is dense in $C[0,\infty]$ and solutions depend continuously on initial data f (in the sense that $\lim_{n\to\infty} \| f_n - f\| = 0$ implies $\lim_{n\to\infty} \| f_n(\cdot + t) - f(\cdot + t)\| = 0$). In view of Theorem 1, this gives

$$\mathrm{e}^{t\frac{\mathrm{d}}{\mathrm{d}x}} f(x) = f(x+t),$$

in agreement with our previous guess.

Note that there is a couple of subtle points here: in (6) derivates are taken in the sense of topology in $C[0,\infty]$, i.e., the topology of uniform convergence: $\lim_{h\to 0} \|u'(t) - \frac{u(t+h)-u(t)}{h}\| = 0$, while these in the original equation are point-wise, calculated for each x separately. Hence, apparently, solutions of (6) are solutions of the original equation, but not vice versa. A more thorough analysis, however, reveals that the opposite statement is also true: solutions of the original equation solve (6) as well. Moreover, we note that while for $f \in C^1[0,\infty]$, $u(t,x) = f(x+t)$ solves (6), a similar statement for $f \notin C^1[0,\infty]$ does not make much sense since u is not even differentiable—hence, we speak of generalized, or mild solutions.

Example 4 The PDE:

$$\frac{\partial u(t,x)}{\partial t} = \frac{1}{2}\frac{\partial^2 u(t,x)}{\partial x^2}, t \ge 0, \qquad u(0,x) = f(x), x \in \mathbb{R},$$

has the unique solution for $f \in C^2[-\infty,\infty]$ given by

$$u(t,x) = \frac{1}{\sqrt{2\pi t}} \int_{-\infty}^{\infty} \mathrm{e}^{-\frac{y^2}{2t}} f(x+y)\,\mathrm{d}y.$$

Hence, arguing as above, we obtain

$$\mathrm{e}^{t\frac{1}{2}\frac{\mathrm{d}^2}{\mathrm{d}x^2}} f(x) = \frac{1}{\sqrt{2\pi t}} \int_{-\infty}^{\infty} \mathrm{e}^{-\frac{y^2}{2t}} f(x+y)\,\mathrm{d}y = \mathbb{E}\, f(x+w(t)),$$

where $w(t)$ is a normal variable with expected value 0 and variance t. In other words, $\mathrm{e}^{t\frac{1}{2}\frac{\mathrm{d}^2}{\mathrm{d}x^2}}$ describes position at time t of a random traveler performing a standard Brownian motion on a line.

To recapitulate: Theorem 1 describes a class of linear operators A for which an exponential function $\mathbb{R}^+ \ni t \mapsto \mathrm{e}^{tA}$ may be defined. (Such operators are called **generators** of semigroups of operators.) This class includes the bounded linear operators, but is in fact far larger than the latter. As it is seen from the examples, such exponential functions describe both deterministic and stochastic processes. Interestingly, the entire information on exponential function and hence the information on the whole deterministic or stochastic process is hidden in a single (usually unbounded) operator. For instance, in Example 1 all transition probabilities are hidden in the intensity matrix, and in Example 4, the Gaussian distribution is hidden in the operator of second derivative.

The generators in the spaces of continuous functions are quite often of the following form:

$$Af(x) = a(x)f''(x) + b(x)f'(x) + \text{ integral operator}, \qquad x \in \mathbb{R}$$

where $a > 0$ and b are continuous functions. The related processes are 'compositions' of three simpler ones:

- diffusion with variance $a(x)$ depending on position x,
- deterministic movement along trajectories of the ODE $x'(t) = b(x(t))$, and
- jumps (for example, Poisson process),

However, an operator is more than just a "map". In fact, it is a "map" on a specific domain, and as we shall see later the domain may include information on the behavior of the underlying process on the boundary.

1.3 From Semigroups to Generators

As we have seen, given an unbounded operator A, sometimes we can construct its exponential function $\mathbb{R}^+ \ni t \mapsto e^{tA}$. Denoting $T(t) = e^{tA}$, we then obtain a family of operators such that

$$T(t)T(s) = T(s+t), \qquad t \geq 0. \tag{7}$$

For $f \in D(A)$, we also have (see Theorem 1) $\lim_{t\to 0+} T(t)f = f$, and it transpires that this condition may be extended to all $f \in \mathbb{B}$:

$$\lim_{t\to 0+} T(t)f = T(0)f = f, \qquad f \in \mathbb{B}. \tag{8}$$

Such a family is called a (strongly continuous) semigroup of operators.

A natural question arises of whether given a strongly continuous semigroup of operators $\{T(t), t \geq 0\}$, one can find an operator A such that $T(t) = e^{tA}, t \geq 0$. The answer is in the positive, and Theorem 1 suggests the way to construct A: Af should be the right-hand derivative of $t \mapsto T(t)f$ at $t = 0$:

$$Af = \lim_{t\to 0+} \frac{1}{t}(T(t)f - f), \qquad D(A) = \{f \,|\, \text{the limit } \lim_{t\to 0+} \frac{1}{t}(T(t)f - f) \text{ exists}\}.$$

It is a part of the Hille–Yosida theorem (see below) that $t \mapsto T(t)$ is indeed the exponential function of the so-defined A. To recall, instead of saying that $t \mapsto T(t)$ is the exponential function of A, we also often say then that A **generates** $\{T(t), t \geq 0\}$.

Let us look at an example that will be of importance later.

Example 5 Let $\mathbb{B} = L^1(\mathbb{R}^+)$ be the space of (classes of) absolutely integrable functions on $\mathbb{R}^+$, with the norm $\|f\| = \int_{\mathbb{R}+} |f(x)|\,dx$ and consider the following operators:

$$T(t)f(x) = \begin{cases} f(x-t), & x \geq t, \\ 0, & x < t. \end{cases} \tag{9}$$

It is clear that $T(t)f \in L^1(\mathbb{R}^+)$ and $\|T(t)f\| = \|f\|$ so that all these operators are bounded with

$$\|T(t)\| = 1. \tag{10}$$

A simple calculation shows that the semigroup property (7) holds. For the proof of (8) we consider $e_\lambda, \lambda > 0$ where $e_\lambda(x) = \mathrm{e}^{-\lambda x}, x \ge 0$. Then

$$\|T(t)e_\lambda - e_\lambda\| = \int_0^t \mathrm{e}^{-\lambda x}\,\mathrm{d}x + \int_t^\infty [\mathrm{e}^{-\lambda(x-t)} - \mathrm{e}^{-\lambda x}]\,\mathrm{d}x = 2\int_0^t \mathrm{e}^{-\lambda x}\,\mathrm{d}x,$$

which converges to 0, as $t \to 0$. It follows that (8) holds for all linear combinations of $e_\lambda, \lambda > 0$. On the other hand, since $\{e_\lambda, \lambda > 0\}$ is a total set in $L^1(\mathbb{R}^+)$, a three-epsilon argument based on (10) shows that (8) holds for all $f \in L^1(\mathbb{R}^+)$, establishing that $\{T(t), t \ge 0\}$ is a semigroup of operators.

To find the generator of $\{T(t), t \ge 0\}$ we proceed as follows. Suppose $f \in D(A)$ and $Af = g$. Then, for $x > t > 0$,

$$\begin{aligned}\int_0^x [T(t)f(y) - f(y)]\,\mathrm{d}y &= \int_t^x f(y-t)\,\mathrm{d}y - \int_0^x f(y)\,\mathrm{d}y \\ &= \int_0^{x-t} f(y)\,\mathrm{d}y - \int_0^x f(y)\,\mathrm{d}y \\ &= -\int_{x-t}^x f(y)\,\mathrm{d}y. \end{aligned} \tag{11}$$

Since

$$\int_0^x \left|\frac{T(t)f(y) - f(y)}{t} - g(y)\right| \mathrm{d}y \le \left\|\frac{T(t)f(y) - f(y)}{t} - g(y)\right\| \underset{t\to 0}{\longrightarrow} 0,$$

we have $\lim_{t\to 0+} \int_0^x \frac{T(t)f(y)-f(y)}{t}\,\mathrm{d}y = \int_0^x g(y)\,\mathrm{d}y$ for all $x > 0$. On the other hand, by (11), for almost all $x > 0$ (with respect to the Lebesgue measure),

$$\lim_{t\to 0+} \int_0^x \frac{T(t)f(y) - f(y)}{t}\,\mathrm{d}y = -\lim_{t\to 0+} \frac{1}{t}\int_{x-t}^x f(y)\,\mathrm{d}y = -f(x).$$

Redefining f if necessary on a Lebesgue null set, we obtain

$$f(x) = -\int_0^x g(y)\,\mathrm{d}y. \tag{12}$$

This implies that f is absolutely continuous with $f(0) = 0$ and $f'(x) = -g(x)$. Therefore, $D(A)$ is contained in the set of such functions and on this set $Af = -f'$.

To prove the converse inclusion, we need an auxiliary result: Consider the operators $U(t) \in \mathcal{L}(L^1(\mathbb{R}^+)), t \ge 0$ given by

$$U(t)h(x) = \int_0^1 h(x+ty)\,\mathrm{d}y, \qquad h \in L^1(\mathbb{R}^+), x \ge 0.$$

Clearly,

$$\int_0^\infty |U(t)h(x)|\,\mathrm{d}x \le \int_0^1 \int_0^\infty |h(x+ty)|\,\mathrm{d}x\,\mathrm{d}y \le \int_0^1 \|h\|\,\mathrm{d}y = \|h\|,$$

showing that $U(t)h \in L^1(\mathbb{R}^+)$ provided $h \in L^1(\mathbb{R}^+)$, and that $\|U(t)\| \le 1$. For e_λ defined at the beginning of this example, $U(t)e_\lambda = \int_0^1 \mathrm{e}^{-\lambda(x+ty)}\,\mathrm{d}y = \mathrm{e}^{-\lambda x}\int_0^1 \mathrm{e}^{-\lambda ty}\,\mathrm{d}y$, i.e.,

$$U(t)e_\lambda = \int_0^1 \mathrm{e}^{-\lambda ty}\,\mathrm{d}y\, e_\lambda,$$

implying $\lim_{t\to 0+} U(t)e_\lambda = e_\lambda$. The three-epsilon argument shows as before that

$$\lim_{t\to 0+} U(t)h = h, \qquad \text{for all } h \in L^1(\mathbb{R}^+). \tag{13}$$

A similar (but simpler) reasoning proves that

$$\lim_{t\to 0+} V(t)h = h, \qquad \text{for all } h \in L^1(\mathbb{R}^+), \tag{14}$$

where $V(t)h(x) = h(x+t), x \ge 0, t \ge 0$. (In fact, $\{V(t), t \ge 0\}$ is a semigroup of operators.)

Coming back to the proof of the other inclusion, we consider f of the form (12). Then

$$\left\|\frac{T(t)f - f}{t} - g\right\| = \int_t^\infty \left|\frac{f(x-t)-f(x)}{t} - g(x)\right|\,\mathrm{d}x + \int_0^t \left|\frac{f(x)}{t} + g(x)\right|\mathrm{d}x.$$

Since the second term does not exceed $\frac{1}{t}\int_0^t |f(x)|\,\mathrm{d}x + \int_0^t |g(x)|\,\mathrm{d}x \to 0$, as $t \to 0+$, f being continuous with $f(0) = 0$, we need to show that the first term converges to 0, as $t \to 0+$. Using $f(x-t) - f(x) = \int_{x-t}^x g(y)\,\mathrm{d}y$ and writing $g(x) = \frac{1}{t}\int_{x-t}^x g(x)\,\mathrm{d}y$, we see that this term can be estimated by

$$\begin{aligned}
\int_t^\infty \left|\frac{1}{t}\int_{x-t}^x [g(y)-g(x)]\,\mathrm{d}y\right|\mathrm{d}x &= \int_0^\infty \left|\frac{1}{t}\int_x^{x+t} [g(y)-g(x+t)]\,\mathrm{d}y\right|\mathrm{d}x\\
&= \int_0^\infty \left|\frac{1}{t}\int_0^t [g(x+y)-g(x+t)]\,\mathrm{d}y\right|\mathrm{d}x\\
&= \int_0^\infty \left|\int_0^1 [g(x+ty)-g(x+t)]\,\mathrm{d}y\right|\mathrm{d}x\\
&\le \int_0^\infty \left|\int_0^1 [g(x+ty)-g(x)]\,\mathrm{d}y\right|\mathrm{d}x
\end{aligned}$$

$$+\int_0^\infty |g(x) - g(x+t)|\,\mathrm{d}x \tag{15}$$

$$= \|U(t)g - g\| + \|V(t)g - g\|. \tag{16}$$

This converges to 0 by (13) and (14), completing the proof of the fact that $D(A)$ is composed precisely of f of the form (12) and that $Af = -f'$.

1.4 *The Hille–Yosida Theorem*

In this section, we describe briefly the celebrated Hille–Yosida theorem [8, 21, 22, 29, 32, 41]. To explain, for construction of a semigroup of operators, Theorem 1 requires knowing that a certain Cauchy problem is well-posed, i.e. that its solutions are unique on a dense set and that they depend continuously on initial data. However, checking this in practice is usually very difficult, and there is the need for other conditions guaranteeing existence of the semigroup.

The Hille–Yosida theorem says that an operator A is the generator of a semigroup $\{\mathrm{e}^{tA}, t \ge 0\}$ of contractions (i.e. such that $\|\mathrm{e}^{tA}\| \le 1$) iff it is closed, densely defined and for all $\mu > 0$, $\mu - A$ is invertible with a bounded inverse such that

$$\|\mu(\mu - A)^{-1}\| \le 1.$$

Moreover, if these conditions are satisfied

$$\mathrm{e}^{tA} = \lim_{n\to\infty} \left(1 - \frac{t}{n}A\right)^{-n}. \tag{17}$$

We note that $(1 - \frac{t}{n}A)^{-n} := [(1 - \frac{t}{n}A)^{-1}]^n = [\mu(\mu - A)^{-1}]^n$ where $\mu = \frac{n}{t}$, so that all operators on the right-hand side of (17) are contractions.

To explain the theorem, we need to cover the notion of a closed operator first. An operator A in a Banach space $\mathbb{B}$ is said to be closed iff for any sequence $(f_n)_{n\ge1}$ of elements of its domain $D(A)$ the conditions $\lim_{n\to\infty} f_n = f$ and $\lim_{n\to\infty} Af_n = g$ imply that $f \in D(A)$ and $Af = g$. It is worth stressing that, although the definition somewhat resembles that of continuity, it describes quite a different phenomenon: the assumption that both limits $\lim_{n\to\infty} f_n = f$ and $\lim_{n\to\infty} Af_n = g$ exist is much stronger than the sole requirement that $\lim_{n\to\infty} f_n = f$. For that reason, all continuous operators are closed, but not vice versa. The definition simply means that the graph of A, i.e., the set of points in $\mathbb{B} \times \mathbb{B}$ of the form (f, Af), where $f \in D(A)$ is closed; here, the Cartesian product $\mathbb{B} \times \mathbb{B}$ is equipped with one of the usual norms, e.g. $\|(x, y)\|_{\mathbb{B}\times\mathbb{B}} = \|x\|_{\mathbb{B}} + \|y\|_{\mathbb{B}}$.

Continuation of Example 3 As an instance, we know from Example 3 that $A = \frac{\mathrm{d}}{\mathrm{d}x}$ in $C[0, \infty]$ is not a bounded operator. However, it is closed. To see this, we consider differentiable $f_n \in C[0,\infty]$, such $f_n' \in C[0,\infty]$, $\lim_{n\to\infty} f_n = f$, and $\lim_{n\to\infty} f_n' = g$ for some $g \in C[0,\infty]$. These assumptions imply that for each $x \geq 0$ we have

$$f_n(x) = f_n(0) + \int_0^x f_n'(y)\,\mathrm{d}y, \qquad n \geq 1.$$

Estimating

$$\begin{aligned}\left|\int_0^x f_n'(y)\,\mathrm{d}y - \int_0^x g(y)\,\mathrm{d}y\right| &\leq \int_0^x |f_n'(y) - g(y)|\,\mathrm{d}y \\ &\leq \int_0^x \|f_n' - g\|\,\mathrm{d}y \leq x\|f_n' - g\|,\end{aligned}$$

we deduce that the integrals $\int_0^x f_n'(y)\,\mathrm{d}y$ converge to $\int_0^x g(y)\,\mathrm{d}y$. Since uniform convergence implies pointwise convergence, letting $n \to \infty$ above yields

$$f(x) = f(0) + \int_0^x g(y)\,\mathrm{d}y, \qquad \text{for all } x \geq 0,$$

showing that f is differentiable with $f' = g$. This, however, establishes that $A = \frac{\mathrm{d}}{\mathrm{d}x}$ is closed, as desired.

Example 6 For another, simpler, example, consider the space l^1 of absolutely convergent sequences $f = (f(k))_{k\geq 1}$ equipped with the norm

$$\|f\| = \sum_{k\geq 1} |f(k)| < \infty.$$

Also, let $Af = (-kf(k))_{k\geq 1}$ with domain $D(A) = \{f \mid \sum_{k\geq 1} k|f(k)| < \infty\}$. We note that A is not bounded, because taking $f_n = (0,\ldots,0,1,0,\ldots)$ (1 on the nth coordinate), we obtain $\|f_n\| = 1$ and $Af_n = -nf_n$, so that $\|Af_n\| = n$. However, A is closed. To see this, consider $f_n \in D(A)$ such that $\lim_{n\to\infty} f_n = f$ and $\lim_{n\to\infty} Af_n = g \in l^1$. Since $Af_n = (-kf_n(k))_{k\geq 1}$, this means that

$$\lim_{n\to\infty} \sum_{k\geq 1} |g(k) + kf_n(k)| = 0.$$

In particular, for all k, $\lim_{n\to\infty} kf_n(k) = -g(k)$. Similarly, $\lim_{n\to\infty} f_n(k) = f(k)$, implying $-kf(k) = g(k)$. Therefore, $\sum_{k\geq 1} |kf(k)| = \sum_{k\geq 1} |g(k)| < \infty$, proving that $f \in D(A)$ and $Af = g$.

Coming back to the Hille–Yosida theorem: the idea of formula (17) is that A is unbounded, hence "large". This implies that $(\mu - A)^{-1}$ should be "small", hopefully bounded. More precisely, in the **resolvent equation**

$$\mu f - Af = g$$

where g is given and f is a solution we are searching for, the map $f \mapsto \mu f - Af$ is unbounded, hence g is large as compared to f. But this is the same as saying that f is small as compared to g. The map $g \mapsto f$ should be bounded. This is illustrated in our example.

Continuation of Example 6 The operator A from Example 6 is densely defined (for $f \in l^1$ take $f_n = (f(1), \ldots, f(n-1), f(n), 0, \ldots)$; then $\lim_{n\to\infty} \|f_n - f\| = \lim_{n\to\infty} \sum_{k\geq n+1}^{\infty} |f(k)| = 0$ and $f_n \in D(A)$). The resolvent equation for A takes the form:

$$\mu(f(k))_{k\geq 1} + (kf(k))_{k\geq 1} = (g(k))_{k\geq 1},$$

where μ and g are given, and we are to find $f \in D(A)$. This equation may be written as $\mu f(k) + kf(k) = g(k)$ for all k or, $f(k) = \frac{1}{\mu+k} g(k)$ for all k. In other words, the unique solution to the resolvent equation is

$$(f(k))_{k\geq 1} = (\frac{1}{\mu + k} g(k))_{k\geq 1},$$

without no assumptions on g besides $g \in l^1$. Since we have

$$\mu\|f\| = \sum_{k\geq 1} \frac{\mu}{\mu + k}|g(k)| \leq \sum_{k\geq 1} |g(k)| = \|g\|, \qquad \mu > 0$$

the conditions of the Hille–Yosida theorem are met.

To find the form of e^{tA} we note that

$$\left(1 - \frac{t}{n}A\right)^{-n} g = [\mu(\mu - A)^{-1}]^n g = \left(\left(\frac{\mu}{\mu + k}\right)^n g(k)\right)_{k\geq 1}$$
$$= \left(\frac{1}{(1 + \frac{kt}{n})^n} g(k)\right)_{k\geq 1}$$

where $\mu = \frac{n}{t}$. It follows that $\lim_{n\to\infty} \left(1 - \frac{t}{n}A\right)^{-n} g = (\mathrm{e}^{-kt} g(k))_{k\geq 1}$. It is easy to see that each coordinate of the left-hand side converges to the corresponding coordinate of the right-hand side. The fact that the left-hand side converges to the right-hand side in the l^1 norm may be deduced from the Dominated Convergence Theorem.

2 Boundary Conditions

2.1 *McKendrick–von Foerster Model*

The semigroup of Example 5 may be thought of as an (oversimplified) model of dynamics of an age-structured population. More specifically, we think of a non-negative $f \in L^1(\mathbb{R}^+)$ as an age-profile of the population, so that

$$\int_{a_1}^{a_2} f(a)\,\mathrm{d}a$$

represents the number of individuals of age in the interval $[a_1, a_2]$. If there are neither deaths nor births, and f is an initial population's age-profile, then $T(t)f$ is its age-profile at time $t > 0$. This is because an individual that at time t is of age a, was initially of age $a - t$, provided $a > t$; also, there are no individuals of age $a < t$ because there are no births.

A more realistic model is the one related to the semigroup

$$T_\mu(t)f(a) = \begin{cases} \mathrm{e}^{-\int_{a-t}^{a} \mu(x)\,\mathrm{d}x} f(a-t), & a \geq t \\ 0, & a < t, \end{cases} \qquad f \in L^1(\mathbb{R}^+), \tag{18}$$

where $\mu \geq 0$ is a bounded, integrable function (the assumption of integrability may be disposed of, we introduce it for simplicity merely). We think of μ is a death rate (mortality function), so that $\mathrm{e}^{-\int_0^a \mu(y)\,\mathrm{d}y}$ is the probability that an individual does not die before age $a > 0$. Then

$$\mathrm{e}^{-\int_{a-t}^{a} \mu(x)\,\mathrm{d}x} = \frac{\mathrm{e}^{-\int_0^a \mu(x)\,\mathrm{d}x}}{\mathrm{e}^{-\int_0^{a-t} \mu(x)\,\mathrm{d}x}} \tag{19}$$

is the probability that an individual will reach age a given that we know he/she has reached age $a - t$. Thus $\mathrm{e}^{-\int_{a-t}^{a} \mu(x)\,\mathrm{d}x} f(a-t)$ is the "number" of individuals who were of age $a - t$ at time 0 and survived to age a, i.e. the 'number' of individuals of age a at time t. The second line of the definition of T_μ reflects the fact that the model still does not account for births: at time t there are no individuals of age less than t.

To find the generator of T_μ, we recall the following general scheme: Two semigroups $\left(\mathrm{e}^{tA}\right)_{t\geq 0}$ and $\left(\mathrm{e}^{tB}\right)_{t\geq 0}$ defined in a Banach space $\mathbb{B}$ are said to be isomorphic (or similar) [8, 21, 22] iff there is an isomorphism $I \in \mathcal{L}(\mathbb{B})$ such that

$$I\mathrm{e}^{tA} = \mathrm{e}^{tB} I, \qquad t \geq 0.$$

Then, as it is easy to check, $f \in D(A)$ iff $If \in D(B)$, and we have $IAf = BIf$.

To see that this scheme applies to our situation, we consider $I \in \mathscr{L}(L^1(\mathbb{R}^+))$ given by $(If)(a) = e^{\int_0^a \mu(x)\,dx} f(a), a \geq 0$. The operator I is an isomorphism of $L^1(\mathbb{R}^+)$ with $\|I\| \leq e^{\int_0^\infty \mu(a)\,da}$ (it is here that we use the assumption $\mu \in L^1(\mathbb{R}^+)$) and $(I^{-1}f)(a) = e^{-\int_0^a \mu(x)\,dx} f(a), a \geq 0$ so that $\|I^{-1}\| \leq 1$. A direct calculation shows that

$$T_\mu(t) = I^{-1}T(t)I,$$

i.e. that T_μ is similar to T.

Hence, $D(A_\mu)$ (the domain of the generator A_μ of T_μ) is composed of f of the form $f(a) = e^{\int_0^a \mu(x)\,dx} g(a)$ where $g \in D(A)$. Since the product of two absolutely continuous functions is absolutely continuous, it follows that $D(A_\mu)$ is the set of absolutely continuous functions vanishing at $x = 0$. In other words, $D(A_\mu) = D(A)$, and

$$\begin{aligned} A_\mu f(a) = I^{-1}AIf(a) &= -e^{-\int_0^a \mu(x)\,dx}\left(e^{\int_0^a \mu(x)\,dx} f(a)\right)' \\ &= -e^{-\int_0^a \mu(x)\,dx}\left(\mu(a)e^{\int_0^a \mu(x)\,dx} f(a) + e^{\int_0^a \mu(x)\,dx} f'(a)\right) \\ &= -f'(a) - \mu(a)f(a), \end{aligned}$$

where $'$ denotes derivative with respect to a. To recapitulate: these semigroups have the same domain

$$D(A) = \{f \in L^1(\mathbb{R}^+) |\text{ there is } g \in L^1(\mathbb{R}^+), \text{ such that } f(x) = \int_0^x g(y)\,dy\},$$

and

$$Af = -f' \quad \text{while} \quad A_\mu f = Af - \mu f, \qquad f \in D(A). \tag{20}$$

Therefore, the model taking account of deaths is obtained from a basic one by perturbing the original generator. However, to include births one must resort to another kind of perturbation, the perturbation of the domain of the generator, or the perturbation of the boundary condition. To explain: let us look at the Cauchy problem related to the operator A_μ. Formally, it reads

$$\frac{\partial u(t,a)}{\partial t} = -\frac{\partial u(t,a)}{\partial a} - \mu(a)u(t,a) = [A_\mu u](t,a), \qquad u(0,a) = f(a), \tag{21}$$

where $[A_\mu u](t,a)$ denotes the value of the operator A_μ on $u(t,\cdot)$ evaluated at a. However, there are at least two remarks that have to be made here. First of all, the time-derivative here is not taken pointwise but in the sense of $L^1(\mathbb{R}^+)$, and the equation holds almost everywhere with respect to a. Secondly, the problem is not well-posed and in fact does not model anything unless we specify "boundary conditions". For, as we have already mentioned, for general $f \in L^1(\mathbb{R}^+)$ the function $t \mapsto T_\mu(t)f$ is merely a formal, or mild solution to the Cauchy problem. It is only for $f \in D(A_\mu)$ that it "really" solves (21). In particular, for all $t \geq 0$, $u(t,\cdot)$ is a member of $D(A_\mu)$ (otherwise, the right-hand side would not make sense), hence, an absolutely continuous function with

$$u(t,0) = 0. \tag{22}$$

This is the boundary condition we have alluded to above, and it clearly is visible in the domain of the generator. Its meaning is transparent: there are no births, $u(t,0)$ being the number of individuals of age 0 at time 0. It is only with this boundary condition that (21) describes a population with deaths but no births.

Now, the full McKendrick–von Foerster model involves births that are introduced via a birth rate function b, a bounded measurable function on $\mathbb{R}^+$. Instead of (22) we introduce the boundary condition

$$u(t,0) = \int_0^\infty b(a)u(t,a)\,\mathrm{d}a, \tag{23}$$

which interpreted means that the number of newborns at time $t \geq 0$ depends on the birth rate and population structure at that time: $b(a)u(t,a)$ is the number of individuals born of the members of the population of age a at time t, and we need to integrate over $a \in \mathbb{R}^+$ to obtain the total number of newborns.

Notably, births do not change the way the generator "acts", but rather its domain. Denoting by $A_{\text{McK-F}}$ the generator in the full model, we have thus (comp. [21,22,33])

$$\begin{aligned} A_{\text{McK-F}}f &= -f' - \mu f, \\ D(A_{\text{McK-F}}) &= \{\text{abs. cont. functions with } f(0) = \int_0^\infty b(a)f(a)\,\mathrm{d}a\}. \end{aligned} \tag{24}$$

This is in fact the main point of this section: in modeling an age-structured population's dynamics we are led to a very natural, non-local boundary condition for the involved equation, or to a semigroup of operators with the generator defined on an interesting domain. The equation reads

$$\frac{\partial u(t,a)}{\partial t} = -\frac{\partial u(t,a)}{\partial a} - \mu(a)u(t,a) = [A_{\text{McK-F}}u](t,a), \qquad u(0,a) = f(a), \tag{25}$$

and is formally the same as (21), but features the different boundary condition (23).

This does not happen too often in the semigroup theory, but here we may find a semi-explicit formula for the semigroup generated by $A_{\text{McK-F}}$. To see this, let us recall that by (9), in the no-births no-deaths case, solutions of the Cauchy problem $u(t,a) = T(t)f(a)$ do not change on the lines $a - t = \mathit{const.}$ (see Fig. 1). Above the main diagonal they are determined by the initial condition f "propagated" along

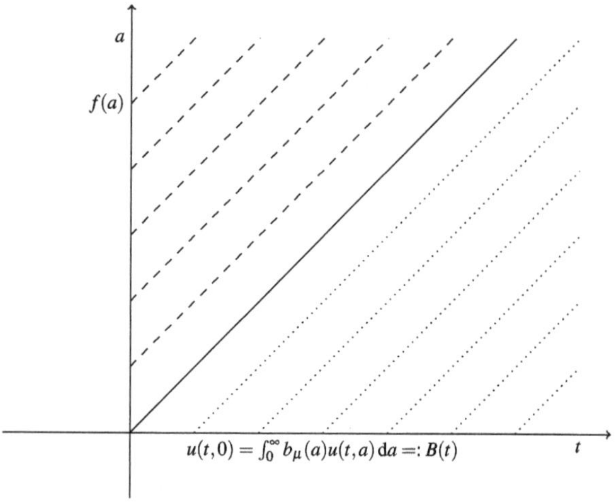

Fig. 1 Solutions to the McKendrick–von Foerster equation

these lines, below the main diagonal they are equal to zero. In the model involving deaths (but still no births) the picture does not change much: above the diagonal the solutions are "tempered" by the survival function (19), below the diagonal they are left intact.

This picture changes substantially when births are incorporated; although this does not influence the form of the solutions above the main diagonal, those below the diagonal change drastically. More specifically, we have

$$T_{\text{McK-F}}(t)f(a) = \begin{cases} e^{-\int_{a-t}^{a}\mu(x)\,dx} f(a-t), & a \ge t \\ e^{-\int_0^a \mu(x)\,dx} B(t-a), & a < t, \end{cases} \qquad f \in L^1(\mathbb{R}^+), \tag{26}$$

where

$$B(t) := T_{\text{McK-F}}(t)f(0)$$

is the total number of newborns at time $t \geq 0$, and the second line may be explained as follows: An individual that at time $t > a$ is of age a was born at time $t-a$ and $e^{-\int_0^a \mu(x)\,dx}$ is the probability of its survival to age a. In other words, $e^{-\int_0^a \mu(x)\,dx} B(t-a)$, is the fraction of individuals born at time $t-a$ that survived to time t. It is thus clear that B, the value of u at the boundary where $a = 0$ determines u below the diagonal (see again Fig. 1).

However, Eq. (26) defines $T_{\text{McK-F}}$ by means of $T_{\text{McK-F}}$ and should be untangled. In fact, it suffices to calculate B. To this end, we consider the following McKendrick renewal equation for B:

$$B(t) = \int_0^\infty b(a+t)e^{-\int_a^{a+t}\mu(x)\,dx} f(a)\,da + \int_0^t b(t-s)e^{-\int_0^{t-s}\mu(x)\,dx} B(s)\,ds.$$

To explain, the two terms in this equation correspond to the number of births from the original population and to the number of births from the descendants of the original population, respectively. More specifically, as in the cases described above,

$$\mathrm{e}^{-\int_a^{a+t} \mu(x)\,\mathrm{d}x} f(a)$$

is the fraction of individuals who at time $t = 0$ were of age a, and survived to age $a+t$. The first integrand is thus the number of babes born of such individuals, $b(a+t)$ being the intensity of giving birth at age $a+t$. As for the second term, it describes individuals born of descendants of the original population: such a descendant must have been born at some time $s < t$ and $\mathrm{e}^{-\int_0^{t-s} \mu(x)\,\mathrm{d}x} B(s)$ is the number of such descendants who survived to time t (i.e., to age $t-s$); $b(t-s)$ is the intensity of child-bearing at age $t-s$, and of course we need to integrate over $s \in [0,t]$.

We note that the first term, denoted $F = F(t)$, depends merely on f, μ and b, and may be treated as given. The other term is a convolution of B with $b_\mu(s) = b(s)\mathrm{e}^{-\int_0^s \mu(x)\,\mathrm{d}x}$ so that the (McKendrick renewal) equation reads:

$$B(t) = F(t) + b_\mu * B(t). \tag{27}$$

An application of Banach's fixed point theorem (see e.g. [20]) in an appropriate space (see [11]) shows that this equation determines B uniquely and that

$$B = F + \sum_{n=1}^{\infty} b_\mu^{n*} * F,$$

with the series converging absolutely; here, b_μ^{n*} denotes the nth convolution power of b_μ. The series has a natural interpretation. F, as we explained above, is the number of direct descendants of the original population, say, daughters. The first term in the series is then the number of grand-daughters, the second is the number of grand-grand-daughters, and so on.

2.2 Feller–Wentzell Boundary Conditions

The partial differential equation of Example 4 models heat conduction in an infinite thin rod (identified with the x-axis $\mathbb{R}$). The similar Cauchy problem on the right half-axis:

$$\frac{\partial u(t,x)}{\partial t} = \frac{1}{2}\frac{\partial^2 u(t,x)}{\partial x^2}, \qquad x \geq 0, t > 0, u(0,x) = f(x),$$

is not well-posed unless boundary conditions are specified: we have encountered such a situation already in discussing the McKendrick–von Foerster model. This is quite understandable physically: if the heat-flow is to be determined, conditions at the rod-end need to be known. Physics suggests three types of such boundary conditions: the first of these, the Neumann boundary condition

$$\frac{\partial u(t,0)}{\partial x} := \frac{\partial u(t,x)}{\partial x}|_{x=0} = 0, \qquad t \geq 0,$$

describes perfectly isolated end, and the Dirichlet boundary condition

$$u(t,0) = 0, \qquad t \geq 0,$$

describes the situation where the end is kept at a specified temperature (here equal to zero). The intermediate Robin boundary condition

$$\frac{\partial u(t,0)}{\partial x} = \gamma u(t,0)$$

models a non perfectly isolated end, where some loss of temperature may occur. And, physically, at least at the beginning of the twentieth century, there seemed to be no other interesting, natural boundary conditions mathematicians may study. (That this is not the case at all, becomes crystal clear after reading excellent G. Goldstein's exposition [30].)

Anyway, this was the situation at the beginning of the twentieth century, when R. A. Fisher and S. Wright came to the stage. They were scientists with mathematical background, working in a field today termed mathematical biology. W. Feller who took up part of their research, in the context of semigroups operators, writes on their work in papers published in the 1950s as follows [25–27]:

> The theory of evolution provides examples of stochastic processes which have not yet been treated systematically.
>
> Existing methods are ... due to R. A. Fisher and S. Wright. They have ... with great ingenuity and admirable resourcefulness ... discovered ... facts of the general theory of stochastic processes.
>
> Essential part of Wright's theory is equivalent to assuming a certain diffusion equation for gene frequency ...
>
> This diffusion equation ... is of a singular type and lead to new types of boundary conditions.

thus acknowledging their role in the development of today's theory of stochastic processes.

To explain the way population genetics has lead to "new types of boundary conditions", we recall the Wright–Fisher model of **genetic drift**. To begin with, we note that in spite of mutations which—given the state of a population—occur independently in each individual, members of (especially: small) populations exhibit striking similarities. This is due to genetic drift, mentioned above, one of the most important forces of population genetics. Simply put, the reason for this phenomenon is that in a population on the one hand new variants are introduced randomly by (neutral) mutations, and on the other many variants are also randomly lost since not all members of the current generation pass their genetic material to the next one.

Wright and Fisher model this phenomenon as follows [8,23,24]. We suppose the population in question to be composed of $2N$ individuals; in doing so we identify individuals with chromosomes (that come in pairs), or even with corresponding loci (places) on these chromosomes. We assume there are only two possible alleles (variants) at this locus: A and a. The size of the population is kept constant all the time, and we consider its evolution in discrete non-overlapping generations formed as follows: an individual in the daughter generation is the same as its parent (reproduction is asexual) and the parent is assumed to be chosen from the parent generation randomly, with all parents being equally probable. In other words, the daughter generation is formed by $2N$ independent draws with replacement from the parent generation. In each draw all parents are equally likely to be chosen and daughters have the same allele as their parents. It should be noted here that such sampling procedure models the genetic drift by allowing some parents not to be selected for reproduction, and hence not contributing to the genetic pool.

Then, the state of the population at time $n \geq 0$ is conveniently described by a single random variable X_n with values in $\{0, \ldots, 2N\}$ being equal to the number of individuals of type A. The sequence $X_n, n \geq 0$ is a time-homogeneous Markov chain with transition probabilities:

$$p_{kl} := \Pr\{X_{n+1} = l | X_n = k\} = \binom{2N}{l} p_k^l (1 - p_k)^{2N-l}, \text{ where } p_k = \frac{k}{2N}. \tag{28}$$

In other words, if $X_n = k$, then X_{n+1} is a binomial random variable with parameter $\frac{k}{2N}$. A typical realization of the Wright–Fisher chain is depicted at Fig. 2, where $N = 5$ and yellow and blue balls represent A and a alleles, respectively. The figure illustrates also the fact that the states 0 and $2N$ are absorbing. This is to say that if for some n, $X_n = 0$ we must have $X_m = 0$ for all $m \geq n$, and a similar statement is true if $X_n = 2N$. Genetically, this expresses the fact that in finite populations, in the absence of other genetic forces, genetic drift (i.e. random change of allele frequencies) reduces variability of population by fixing one of the existing alleles.

Now, imagine that the number $2N$ of individuals is quite large and the individuals are placed on the unit interval $[0, 1]$ with distances between neighboring individuals equal to $\frac{1}{2N}$, so that the kth individual is placed at $\frac{k}{2N}$. Imagine also that the time that elapses from one generation to the other is $\frac{1}{2N}$, so that there are $2N$ generations in a unit interval. Then, the process we observe bears more and more resemblance to a continuous-time continuous-path diffusion process on $[0, 1]$, as depicted at Fig. 3.

Since, conditional on $X_n = k$, the expected single-step displacement $E\,\frac{\Delta X_n}{2N}$ of the approximating process equals $\frac{1}{2N} 2N p_k - \frac{k}{2N} = 0$ with variance $Var\,\frac{\Delta X_n}{2N} = \frac{1}{2N} p_k (1 - p_k)$, in the limit we expect the process starting at $x \in [0, 1]$ to have infinitesimal variance $x(1-x)$ and infinitesimal displacement 0. Indeed, for each N, a point $x \in [0, 1]$ may be identified with $\frac{k}{2N}$ where $k = [2Nx]$, and letting $N \to \infty$

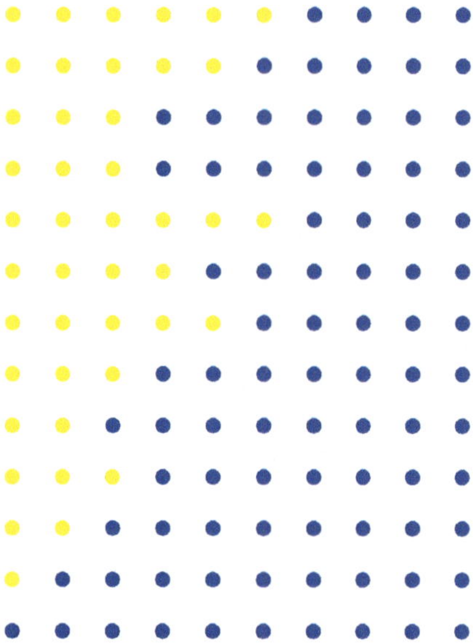

Fig. 2 Wright–Fisher model: each row depicts a generation sampled with replacement from the one lying above it

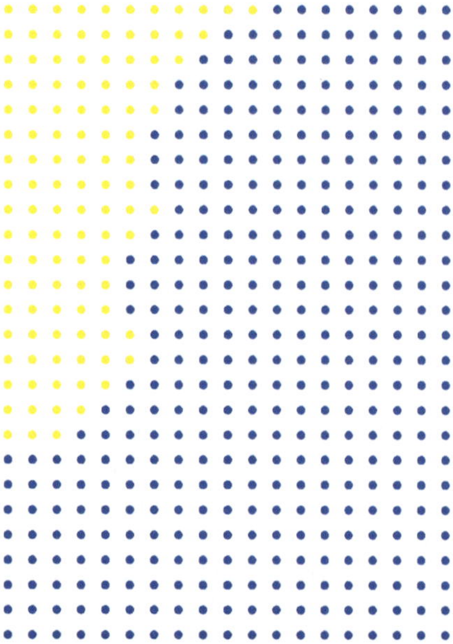

Fig. 3 Diffusion approximation: the border line between the yellow and blue balls resembles a path of a diffusion process

in $\frac{[2Nx]}{2N}(1 - \frac{[2Nx]}{2N})$, which is the infinitesimal variance in a unit time-interval, we obtain $x(1-x)$. Moreover, the limiting process should inherit its boundary behavior from the approximating Markov chains. Therefore, we expect that in the limit we will obtain the process related to the semigroup in $C[0, 1]$, generated by

$$Af(x) = x(1 - x)f''(x) \tag{29}$$

with domain $D(A)$ composed of twice continuously differentiable functions on (0,1) such that $\lim_{x\to 0+} x(1-x)f''(x) = \lim_{x\to 1-} x(1-x)f''(x) = 0$, see e.g. [8, section 8.4.20], [21, pp. 224–226], see also [39, p. 120]. (The requirement that the process is absorbed at $x = 0$ and $x = 1$ is expressed in the fact that $Af(0) = Af(1) = 0$.) The limiting process is often referred to as **Wright's diffusion**, or the **Wright–Fisher diffusion** (without mutations).

A noteworthy variant of the above limit procedure arises when mutations are allowed. Suppose namely that in passing from one generation to the other an individual with allele A may change its state to allele a with probability $\frac{\alpha}{2N}$, and an individual with allele a may change its state to allele A with probability $\frac{\beta}{2N}$, where α and β are non-negative numbers. Then, the number of A alleles is still a Markov chain with transition probabilities of the form (28), but p_k is now changed to

$$p_k = \frac{k}{2N}(1 - \frac{\alpha}{2N}) + (1 - \frac{k}{2N})\frac{\beta}{2N}.$$

Of course, this influences the expected single-step displacement and its variance (conditional on $X_n = k$):

$$E\frac{\Delta X_n}{2N} = p_k - \frac{k}{2N} = \frac{1}{2N}(\beta - (\alpha+\beta)\frac{k}{2N}),$$
$$Var\left(\frac{\Delta X_n}{2N}\right) = \frac{1}{2N}p_k(1-p_k). \tag{30}$$

This suggests that the limit process starting at x will have an infinitesimal unit-time displacement of $\beta - (\alpha+\beta)x$ and the corresponding variance $x(1-x)$, i.e. that the limit process is related to the operator

$$Af(x) = a(x)f''(x) + b(x)f'(x), \qquad x \in (0,1),$$

where $a(x) = x(1-x)$ and $b(x) = \beta - (\alpha+\beta)x$.

This suggests also the following question: can a particle, first of all, reach the boundary? Perhaps it will never reach it and considering boundary conditions is unnecessary? After all, at $x = 0$ the displacement coefficient (usually called 'the drift coefficient' but we will not use this term, to avoid confusion with genetic drift which is expressed in a rather than in b) equals $\beta > 0$ meaning that at the boundary there is a strong tendency to move to the right (this is the force of mutations that causes this!); is diffusion (coefficient a) strong enough to overcome b, making $x=0$ accessible? Can $x = 1$ be reached as well? Or perhaps, $x = 0$ not only allows no particles to reach it, but also is a source of particles constantly entering the interval from the left end? As it transpires, all these situations are possible for a general diffusion process. In other words, a boundary point might be either (see Fig. 4)

Fig. 4 Feller's classification of boundary points

boundary type	accessible?	absorbing?
regular	Y	N
exit	Y	Y
entrance	N	N
natural	N	Y

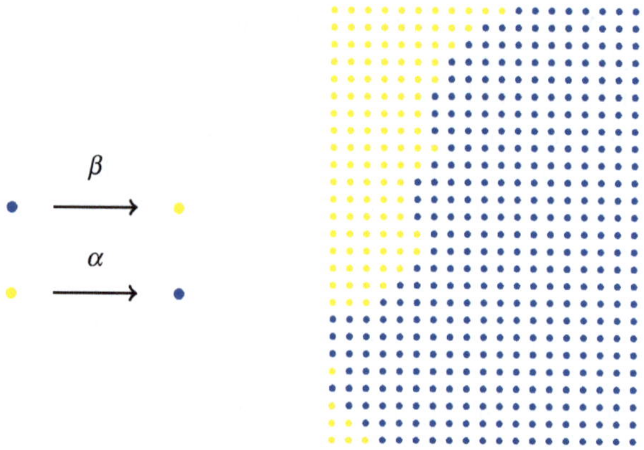

Fig. 5 Diffusion approximation to Wright–Fisher model with mutations

- natural: a particle cannot reach the boundary from the interior, neither can it reach the interior from the boundary,
- entrance: a particle cannot reach the boundary from the interior, but it can reach the interior from the boundary,
- exit: a particle can reach the boundary from the interior, but it cannot reach the interior from the boundary,
- or regular: a particle can reach the boundary from the interior, and it can reach the interior from the boundary.

Before continuing, we note another interesting point. The scaled Wright–Fisher process with mutation, after reaching $\frac{0}{2N}$, will not stay there for ever but for a geometric time τ: $\Pr(\tau \geq n) = (1 - \frac{\beta}{2N})^{2Nn}$ and then resume its motion (see Fig. 5). Hence, general diffusion processes with regular boundary may behave similarly: paths should be able to return to the interior of the unit interval after some random time spent at the boundary.

Hence, we encounter at least two new phenomena, as compared to the interpretation suggested by physicists. In the process modeled by the Neuman boundary condition, heat particles are being reflected from the boundary, and in the Dirichlet boundary condition they are being annihilated there. But here, they may stay at the boundary for some time and then return to the interior. Of course, if we model merely heat flow, the latter behavior is quite impossible (see, however, [30]), but for general diffusion processes, such as the Wright–Fisher diffusion, we cannot rule such a behavior out.

In the case of a regular boundary, a diffusion process may both enter the boundary or leave it, and the full characterization of the process requires description of its behavior there (in the case of natural boundary, on the other hand, no such description is needed and no boundary conditions are imposed). As an example, let us consider the space $C[0,\infty]$ of continuous functions on $\mathbb{R}^+ = [0,\infty)$ with limits at infinity. Also, let A be the operator given by $Af = f''$ with domain composed of twice continuously differentiable functions with $f'' \in C[0,\infty]$, satisfying the boundary condition

$$af''(0) - bf'(0) + cf(0) - d\int_{\mathbb{R}^+_*} f \,\mathrm{d}\mu = 0, \tag{31}$$

where μ is a probability measure on $\mathbb{R}^+_* = (0,\infty)$, and a, b, c and d are given non-negative constants with $c \geq d$ and $a + b > 0$. Note that this boundary condition involves merely $x = 0$, a regular boundary point, but does not touch $x = \infty$, a natural boundary point.

It may be proved that A generates a semigroup of operators in $C[0,\infty]$, and that there is an underlying stochastic process for this semigroup. The form of the generator ($Af = f''$) tells us that, while away from the boundary $x = 0$, the process behaves like a (re-scaled) Brownian motion. To specify the process completely, however, we need to provide rules of its behavior at the boundary, and this is the role of (31). Upon touching the boundary, the particle performing Brownian motion may be stopped there, reflected or killed (i.e. removed from the space); it may also jump somewhere into $\mathbb{R}^+$. The coefficients a, b, c and d may be thought of as describing relative frequencies of such events, and μ is the distribution of particle's position right after the jump. In particular, the case $a = 1, b = c = d = 0$ is the stopped Brownian motion (the particle reaching the boundary stays there for ever), $a = c = d = 0, b = 1$ (Neumann boundary condition) is the reflected Brownian motion whose paths are absolute values of paths of an unrestricted Brownian motion, and $a = b = d = 0, c = 1$ (Dirichlet boundary condition) is the minimal Brownian motion (the particle reaching the boundary disappears).

Consider in more detail the case where $a \neq 0$ and $b = 0$:

$$af''(0) + cf(0) - d\int_{\mathbb{R}^+_*} f \,\mathrm{d}\mu = 0. \tag{32}$$

This is the case of *elementary return Brownian motion*, in which the process after reaching the boundary stays there for a random exponential time T with parameter c/a (see Fig. 6):

$$P(T > t) = \mathrm{e}^{-\frac{c}{a}t}, \quad t \geq 0.$$

At time T, the process either terminates, with probability $1 - \frac{d}{c}$, or jumps, with probability $\frac{d}{c}$, to a random point in $\mathbb{R}^+_*$, the distribution after the jump being μ, and starts its movement afresh.

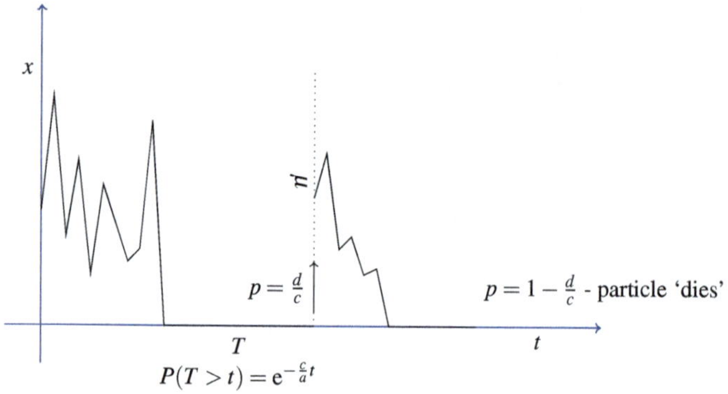

Fig. 6 Elementary return Brownian motion

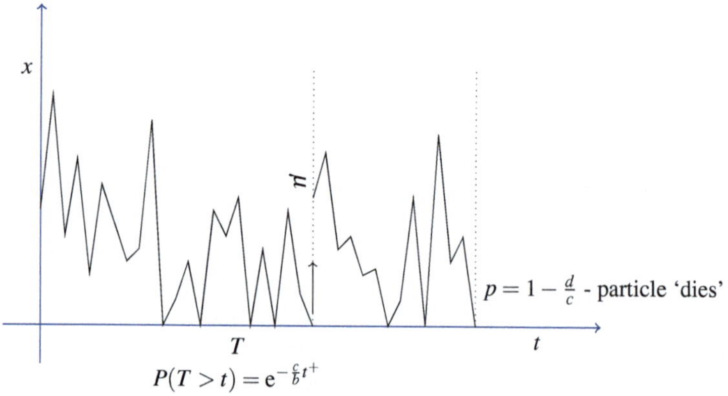

Fig. 7 Elastic barrier

The *elastic barrier* where $a = 0$ and $b \neq 0$:

$$bf'(0) = cf(0) - d\int_{\mathbb{R}_*^+} f\,\mathrm{d}\mu. \tag{33}$$

may be described by analogy (see Fig. 7). Here, after reaching the boundary the trajectory is reflected and the process continues in this fashion for a random time T. The times when the path touches $x = 0$ form a measurable subset of the time axis, and the Lebesgue measure of this set is zero. There is, nevertheless, a way to measure the time spent at the boundary, called the Lévy local time t^+; t^+ itself is a random process (on a separate probability space) increasing only when the Brownian path is at the boundary [34–36,42]. As in the elementary return Brownian motion, at time T distributed according to (see [34, p. 45] or [35, p. 426])

$$P(T > t) = \mathrm{e}^{-\frac{c}{b}t^+}, \quad t \geq 0,$$

the process either terminates, with probability $1 - \frac{d}{c}$, or jumps, with probability $\frac{d}{c}$, to a random point in $\mathbb{R}_*^+$, the distribution after the jump being μ, and starts afresh.

For a more detailed analytic and probabilistic treatment of boundary conditions see e.g. [26, 27, 30, 36, 40] (a very nice insight into boundary conditions may also be gained by considering a characteristic operator—see [19]). A final remark concerns terminology: the study of boundary conditions done by W. Feller for one-dimensional diffusion processes was soon taken up by A.D. Wentzell in the case of multi-dimensional processes [48]. Therefore, in today's literature the boundary conditions for diffusion processes are often called Wentzell boundary conditions, or Feller–Wentzell boundary conditions.

3 Applications, and Recent Developments

3.1 Dealing with Boundary Conditions: Greiner's Approach

Population dynamics with McKendrick-type equations has had a vigorous and long-lasting impact on mathematics and on the theory of semigroups of operators in particular [21, 22, 33, 37, 44, 47]. A similar statement is also true for Feller–Wentzell boundary conditions, the latter forming a core of nowadays theory of stochastic processes [36, 39]; for the influence on semigroups of operators see e.g. [30, 46] and works cited therein.

Within the theory of semigroups of operators, a particularly elegant approach to boundary conditions viewed from the perspective of perturbation theory is due to G. Greiner [31]. To explain, a "classic" perturbation theory deals with the problem of when given a generator A, one may claim that $A+B$ is also a generator: for example, the Phillips perturbation theorem says that this is the case when B is bounded. G. Greiner's paper is concerned with the similar question when it is not the operator itself but its boundary that is perturbed. Here are the details: Let $\mathbb{X}$ and $\mathbb{Y}$ be two Banach spaces, $A : D(A) \to \mathbb{X}$ be a closed operator in $\mathbb{X}$, and $L : D(A) \to \mathbb{Y}$ be a linear operator which is continuous with respect to the graph norm in $D(A)$. (The graph norm is $\|f\|_A = \|f\| + \|Af\|$.) Moreover, assume L to be surjective, and suppose that A_0, defined as the restriction of A to $\ker L$, generates a semigroup of operators in $\mathbb{X}$. Given $F \in \mathscr{L}(\mathbb{X}, \mathbb{Y})$, is the operator A_F defined as the restriction of A to $\ker(L - F)$, the generator as well? This is precisely what is meant when we say that the operator was left intact, but its boundary was "perturbed".

While in general (see [31, Example 1.5]) the answer is in the negative, Greiner's first fundamental theorem [31, Thm 2.1] establishes that A_F is the generator for any F provided there is a constant γ such that for λ larger than some λ_0

$$\|Lf\| > \lambda\gamma\|f\|, \qquad \text{for all } f \in \ker(\lambda - A). \tag{31}$$

To consider a particular example, let us come back to the McKendrick model. Let $W^{1,1}(\mathbb{R}^+)$ be the set of absolutely continuous functions $f \in L^1(\mathbb{R}^+)$ with $f' \in L^1(\mathbb{R}^+)$, i.e., for $f \in W^{1,1}(\mathbb{R}^+)$ we have $f(x) = f(0) + \int_0^x f'(y)\,\mathrm{d}y, x \geq 0$. Also, let

$$A : W^{1,1}(\mathbb{R}^+) \to L^1(\mathbb{R}^+) \qquad Af = -f',$$

and $L : W^{1,1}(\mathbb{R}^+) \to \mathbb{R}$ be given by $Lf = f(0)$. The generator of the semigroup of Example 5 is then A restricted to $\ker L$, the Greiner's operator A_0. In view of (20), the semigroup modeling aging and deaths [see Eq. (18)] is generated by the bounded perturbation of A_0, since the multiplication operator $f \mapsto -\mu f$ is bounded. As we have already stressed, this is not the case with the full McKendrick–von Foerster semigroup: its generator involves a different kind of perturbation, the perturbation of the boundary.

In the set up of Greiner, $\ker(\lambda - A), \lambda > 0$ is spanned here by e_λ where $e_\lambda(x) = \mathrm{e}^{-\lambda x}, x \geq 0$. (For, $f \in \ker(\lambda - A)$ forces $f' = -\lambda f$, and this implies that f' is a continuous function. In other words, f is a "usual" solution to $\lambda f + f' = 0$, i.e, f is a scalar multiple of e_λ.) Since $Le_\lambda = 1$ and $\|e_\lambda\| = \frac{1}{\lambda}$, condition (34) is satisfied with $\gamma = 1$ (in fact, we have equality there). So, Greiner's result (together with the Phillips perturbation theorem) establishes existence of the semigroup generated by the operator of (20) with perturbed domain given in (24).

For a second example, let $C^2[0, \infty]$ be the space of twice continuously differentiable members f of $C[0, \infty]$ with $f'' \in C[0, \infty]$, and let $Af = f''$ and $Lf = f''(0)$ on $C^2[0, \infty]$. It is quite well-known that A restricted to $\ker L$ is the generator of a semigroup. Now, for $\lambda > 0$, all solutions to the differential equation $\lambda f - f'' = 0$ on $\mathbb{R}^+$, are of the form $f(x) = C\mathrm{e}^{-\sqrt{\lambda}x} + D\mathrm{e}^{\sqrt{\lambda}x}$. Since the choice of nonzero D leads out of $C[0, \infty]$, $\ker(\lambda - A)$ is spanned by $e_{\sqrt{\lambda}}$ (see above for this notation). We have $Le_{\sqrt{\lambda}} = \lambda$ and $\|e_{\sqrt{\lambda}}\| = 1$. Hence, condition (34) holds again with $\gamma = 1$. Greiner's theorem thus establishes that for any bounded linear functional $F \in (C[0, \infty])^*)$ the operator A restricted to the kernel of $L - F$ is a generator. In particular, for $Ff = d\int_{\mathbb{R}^+_*} f\,\mathrm{d}\mu - cf(0)$, we obtain existence of the semigroup related to the elementary return Brownian motion [i.e., to boundary condition (32)]. A similar reasoning works for the elastic barrier (33).

3.2 Dealing with Boundary Conditions: Lord Kelvin's Method of Images

Let us start with Feller's construction of the semigroup describing reflected Brownian motion [28, pp. 340–343], which he calls Lord Kelvin's method of images.

The semigroup in question is generated by the operator $Af = \frac{1}{2}f''$ in $C[0, \infty]$ with domain composed of twice continuously differentiable functions with $f'' \in C[0, \infty]$, satisfying the boundary condition (31) with $a = c = d = 0$, i.e., the

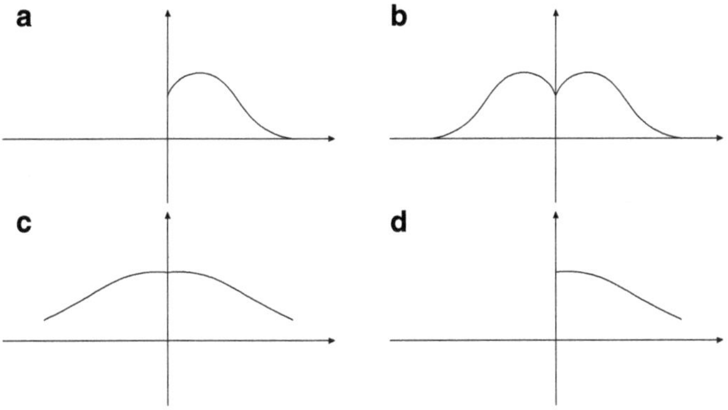

Fig. 8 Method of images: (**a**) a function, (**b**) its even extension, (**c**) the value of the unrestricted Brownian motion semigroup on the even extension, and (**d**) the restriction of the latter to $\mathbb{R}^+$

Neumann condition:

$$f'(0) = 0.$$

As it transpires, this semigroup may be constructed as follows using the semigroup of Example 4. Given $f \in C[0,\infty]$ (Fig. 8a) we consider its even extension (b), apply the semigroup of Example 4 to this extension (c), and then restrict the obtained function to $\mathbb{R}^+$ (d).

That this method works is a miracle in itself, but of course the most intriguing part of it is: *How the Neumann boundary condition is related to even extensions?* A more general problem is

> Can one follow similar lines in constructing semigroups related to the general boundary conditions (31)?

The answer is in positive. But before presenting the solution, let us put method of images in the context of similar or isomorphic semigroups, we have already encountered in Sect. 2.1 in a particular case. Given two strongly continuous semigroups $\{e^{tA}, t \geq 0\}$ and $\{e^{tG}, t \geq 0\}$ in Banach spaces $\mathbb{X}$ and $\mathbb{B}$, respectively, we say that they are similar or isomorphic if there exists an isomorphism $I : \mathbb{X} \to \mathbb{B}$ such that $e^{tA} = I^{-1}e^{tG}I, t \geq 0$. Then, the generators A and G are related by

$$D(A) = \{x \in \mathbb{X}; Ix \in D(G)\}, \qquad Ax = I^{-1}GIx. \tag{35}$$

A particular case of this situation is as follows. Suppose that Λ is a subset of $\mathbb{R}$, $\mathbb{X}$ is a space of real (or complex) functions on Λ, and we are interested in proving that a certain operator A in $\mathbb{X}$ is the generator of a strongly continuous semigroup. Assume also that there exists $\Lambda \subset \Lambda' \subset \mathbb{R}$ and a strongly continuous semigroup $\{T(t), t \geq 0\}$ of operators in a space $\mathbb{B}_0$ of real (or complex) functions on Λ', generated by an operator G_0 resembling A. Usually, A is "G_0 with a boundary condition". One way to approach such a problem is extending functions in $\mathbb{X}$ to functions in $\mathbb{B}_0$ so that the set of these extensions is an invariant subspace $\mathbb{B} \subset \mathbb{B}_0$ for $\{T(t), t \geq 0\}$. Then, the part G of G_0 in $\mathbb{B}$ is the generator of $\{T(t)_{|\mathbb{B}}, t \geq 0\}$. Moreover, quite often, $\mathbb{X}$ and $\mathbb{B}$ are then isomorphic with natural isomorphism I mapping a function on Λ to its extension to Λ', and $R := I^{-1}$ mapping a function on Λ' to its restriction on Λ. Then, there exists the semigroup in $\mathbb{X}$ that is similar to $\{e^{tG}, t \geq 0\}$. Hence, to show that A is the generator, it suffices to show (35). Then, as a bonus, we obtain the explicit form of e^{tA}, referred to as the *abstract Kelvin formula*:

$$e^{tA} f = Re^{tG} If, \qquad f \in \mathbb{X}, t \geq 0. \tag{36}$$

As with the Feller's construction, the non-trivial part in the procedure described above is finding the way of extending functions on Λ to functions on Λ' so that all the remaining steps are valid. To explain the idea of deriving formulae for such extensions, let us recall that any strongly continuous semigroup leaves the domain of its generator invariant. Hence, if members f of $D(A)$ are characterized by a functional equation (a boundary condition), say $F(f) = 0$, then we must have $F(Re^{tG}If) = 0, t \geq 0$. This, when coupled with $(If)_{|\Lambda} = f$, often determines $If, f \in D(A)$, and, by density of $D(A)$, all If.

As we will see, the method of images often leads to the most natural approach to the generation problem [11]. To begin with, let us consider the full McKendrick–von Foerster semigroup generated by (24), and ask the question of whether given $f \in L^1(\mathbb{R}^+)$ we may choose its extension $\tilde{f}$ to the whole real line so that

$$T_{\text{McK-F}}(t) f(a) = e^{-\int_{a-t}^{a} \mu(x)\,dx} \tilde{f}(a-t), \qquad a, t \geq 0, \tag{37}$$

where $\mu(a) = 0$ for $a < 0$. Certainly, it suffices to find $g(a) = \tilde{f}(-a), a \geq 0$. Since for $f \in D(A_{\text{McK-F}})$ the right-hand side of (37) has to belong to $D(A_{\text{McK-F}})$ as well, we calculate [see (24)]:

$$\begin{aligned} g(t) = \tilde{f}(-t) &= \int_0^\infty b(a) e^{-\int_{a-t}^{a} \mu(x)\,dx} \tilde{f}(a-t)\,da \\ &= \int_t^\infty b(a) e^{-\int_{a-t}^{a} \mu(x)\,dx} f(a-t)\,da + \int_0^t b(a) e^{-\int_{a-t}^{a} \mu(x)\,dx} g(t-a)\,da, \\ &= \int_0^\infty b(a+t) e^{-\int_{a}^{a+t} \mu(x)\,dx} f(a)\,da + \int_0^t b(a) e^{-\int_0^{a} \mu(x)\,dx} g(t-a)\,da \end{aligned}$$

where in the last step we used $\mu(a) = 0$ for $a < 0$. In other words, g satisfies the McKendrick renewal equation (27). Since the solution to this equation is unique, we see that $g = B$, the total number of newborns. Of course, we could have obtained this directly by comparing (37) with (26) but the point is that we have derived the renewal equation, and the explicit form for $T_{\text{McK-F}}$ without population dynamics considerations.

Coming back to the Brownian motion semigroups, we will, following [10, 16], provide a construction for the case of the Robin boundary condition (the general case of Feller–Wentzell boundary conditions (31) is treated in detail in [10]). We start by re-writing the semigroup of Example 4 as follows:

$$\begin{aligned} e^{t\frac{1}{2}\frac{d^2}{dx^2}} f(x) &= \frac{1}{\sqrt{2\pi t}} \int_{-\infty}^{\infty} e^{-\frac{u^2}{2t}} \frac{1}{2}[f(x+u) + f(x-u)]\, du \\ &= \frac{1}{\sqrt{2\pi t}} \int_{-\infty}^{\infty} e^{-\frac{u^2}{2t}} C(u) f(x)\, du, \end{aligned} \tag{38}$$

where

$$C(u)f(x) = \frac{1}{2}[f(x+u) + f(x-u)] \tag{39}$$

The family $\{C(t), t \in \mathbb{R}\}$ is an example of an operator cosine family, since it satisfies the functional equation:

$$2C(t)C(s) = C(t+s) + C(t-s).$$

Such families have similar properties to semigroups of operators [29]; a generator of a cosine family is defined as

$$Af = \lim_{t \to 0} 2\frac{C(t)f - f}{t^2},$$

on the domain composed of f for which the limit involved exists. A direct argument shows that the generator of the cosine family (39) is $Af = f''$ with domain equal to the space $C^2[-\infty, \infty]$ (of twice continuously differentiable f with $f'' \in C[-\infty, \infty]$).

As with semigroups, it transpires that cosine families preserve domains of their generators:

$$C(t)[D(A)] \subset D(A).$$

It is also important to note that each generator of a cosine family is the generator of a semigroup (the two families being related by a formula of the type (38), termed the abstract Weierstrass formula).

This suggests the following claim: the operator $Af = f''$ with domain composed of members of $C^2[0, \infty]$ (the space of twice continuously differentiable f with $f'' \in C[0, \infty]$) satisfying $f'(0) = \gamma f(0)$ is not only the generator of a semigroup but also of a cosine family. Moreover, for $f \in C[0, \infty]$ one may find its extension $\tilde{f} \in C[-\infty, \infty]$ such that the cosine family in question is given by

$$C_\gamma(t) f(x) = C(t)\tilde{f}(x), \qquad t \in \mathbb{R}, x \geq 0. \tag{40}$$

Again, we need to find $g(x) = \tilde{f}(-x), x \geq 0$. The requirement that C_γ leaves the domain of its generator invariant, forces

$$\frac{\mathrm{d}}{\mathrm{d}x}\left[\tilde{f}(x+t) + \tilde{f}(x-t)\right]_{|x=0} = \gamma(\tilde{f}(t) + \tilde{f}(-t)), \qquad t \geq 0.$$

This, however, is the same as

$$f'(t) - g'(t) = \gamma(f(t) + g(t)).$$

Using $g(0) = f(0)$ ($\tilde{f}$ is supposed to be continuous!), we find the solution

$$g(t) = f(t) - 2\gamma \int_0^t \mathrm{e}^{-\gamma(t-s)} f(s)\, \mathrm{d}s.$$

It is easy to check that $g \in C[0,\infty]$, i.e., that $\tilde{f} \in C[-\infty,\infty]$ and that if $f'(0) = \gamma f(0)$, then $g'(0) = f'(0) - 2\gamma f(0) = -f'(0)$ and $g''(0) = f''(0) - \gamma(f'(0) + g'(0)) = f''(0)$. These relations allow checking that $\tilde{f} \in C^2[-\infty,\infty]$ provided $f'(0) = \gamma f(0)$, and so (40) indeed defines the cosine family we were looking for. The reader will notice that for $\gamma = 0$, g defined above equals to f, recovering the fact that for Neumann boundary condition the proper extension is the even one.

3.3 Solea Solea *Population*

Many population dynamics studies use the McKendrick equation as a building block [44]: this is the case also in the model of a *solea solea* population with both age and vertical structures, due to O. Arino et al. [1, 45]. In the model, the fish habitat is divided into N spatial patches and the fish densities, or age profiles u_i, in the ith patch satisfy the following system of equations:

$$\frac{\partial u_i(t,a)}{\partial t} + \frac{\partial u_i(t,a)}{\partial a} = -\mu_i(a)u_i(t,a) + \epsilon^{-1}\sum_{j=1}^{N} k_{ij}(a)u_j(t,a), \tag{41}$$

$$u_i(t,0) = \int_0^\infty b_i(a)u_i(t,a)\,\mathrm{d}a, \quad i = 1,\ldots,N,$$

where "t" stands for time, and μ_i and b_i are age-specific and patch-specific mortality and birth rates.

In the absence of the terms $\epsilon^{-1}\sum_{j=1}^{N} k_{ij}(a)u_j(t,a)$, each patch could be treated separately and the population densities there would satisfy the McKendrick equation. The matrix $k(a) = (k_{ij}(a))$ is composed of intensities of movements between patches that occur on a daily basis: the sum of entries in each column of the matrix is zero. The factor ϵ^{-1} (with $\epsilon \ll 1$) corresponds to the fact that the age-related processes and vertical migrations (between the patches) occur at different time scales, a day being the fast time scale as compared to the fish life time.

The main question addressed in [1] is whether in modelling such populations one may disregard the vertical migration to work with a model that has been aggregated, or averaged, over the whole water column. To this end, the authors assume that the matrix k is irreducible and hence possesses the unique normalized right eigenvector $\mathbf{v}(a) = (v_i(a))_{i=1,\ldots,N}$, corresponding to the simple dominant eigenvalue 0. Since this vector describes the stable population distribution among the patches, a heuristic argument makes plausible the *ansatz* that approximately we have

$$\frac{u_i(t,a)}{u(t,a)} = v_i(a), \quad i = 1,\ldots,N, a \geq 0, \tag{42}$$

where $u = \sum_{i=1}^{N} u_i$. In other words, it is assumed that the migrations governed by k occur so fast, as compared to the ageing processes, that the population distribution over the patches reaches the (age-specific) equilibrium long before the ageing process intervenes. This corresponds to letting $\epsilon \to 0$ in (41). In such a simplified, aggregated, model the population density satisfies the McKendrick equation with averaged birth and mortality rates:

$$\begin{aligned} \frac{\partial u(t,a)}{\partial t} + \frac{\partial u(t,a)}{\partial a} &= -\mu_{\mathrm{a}}(a)u(t,a), \\ u(t,0) &= \int_0^\infty b_{\mathrm{a}}(a)u(t,a)\,\mathrm{d}a, \end{aligned} \tag{43}$$

where "a" stands for "aggregated", $\mu_{\mathrm{a}} = \sum_{i=1}^{N} v_i \mu_i$ and $b_{\mathrm{a}} = \sum_{i=1}^{N} v_i b_i$. Here, the weights v_i reflect the underlying, hidden spatial structure. Notably, the resulting boundary condition is a convex combination of the boundary conditions occurring in (41).

This effect is very similar to that observed in [9, 13, 14] where, motivated by a number of biological models, the authors study convex combinations of Feller generators resulting from "averaging" the stochastic processes involved. In fact, these two effects are in a sense dual: under certain regularity conditions on the model's parameters, the predual of the McKendrick semigroup may be constructed in a space of continuous functions [13]. Then, a perturbation of a boundary condition becomes a perturbation of the generator, and the convergence discussed above may be put in the context of [9, 13, 14], see [13] for details.

In [5, 6], the problem of the convergence of solutions of (41) as $\epsilon \to 0$ was fully solved using asymptotic analysis (even in a more general model). However, the authors did not consider the problem as an example of a convex combination of boundary conditions. In [4], the problem is put in the framework of Greiner [31] to deal with abstract boundary conditions, instead of the particular ones of the McKendrick equation. More specifically, the semigroup with the generator's domain equal to $\ker[\Phi_1 \circ \alpha + \Phi_2 \circ (1-\alpha)]$ is approximated by a family of semigroups with the generators' domains involving $\ker \Phi_1$ and $\ker \Phi_2$. We point out that this result allows dealing only with the case where N—the number of patches in the model of *solea solea*—is 2; it is not yet clear how to extend the methods of [4] to deal with the general case. Note that the problem posed here is in a sense converse to the result of [1, 5, 6]: there, a complex model is reduced to a simpler one involving convex combination of the boundary conditions while here, given a convex combination of Feller generators, we construct an approximating sequence of semigroups.

Here are the details: Throughout it is assumed that (34) holds. Also, given a bounded linear operator $\alpha \in \mathscr{L}(\mathbb{X})$, and two operators $F_1, F_2 \in \mathscr{L}(\mathbb{X}, \mathbb{Y})$, we define

$$F_{\mathrm{a}} = F_1\alpha + F_2\beta$$

where $\beta = I_{\mathbb{X}} - \alpha$ ("a" for "average"). By Greiner's theorem, $A_i := A_{F_i}$ and $A_{\mathrm{a}} := A_{F_{\mathrm{a}}}$, are generators with $D(A_i) = \ker \Phi_i$, where $\Phi_i = L - F_i, i = 1, 2$ and $D(A_{\mathrm{a}}) = \ker \Phi_{\mathrm{a}}$, where $\Phi_{\mathrm{a}} = L - F_{\mathrm{a}}$.

Our main goal is to approximate $\left(\mathrm{e}^{tA_{\mathrm{a}}}\right)_{t\geq 0}$ by means of semigroups build from $\left(\mathrm{e}^{tA_1}\right)_{t\geq 0}$ and $\left(\mathrm{e}^{tA_2}\right)_{t\geq 0}$. To this end, we introduce operators $\mathscr{A}_\kappa, \kappa > 0$, in $\mathbb{X} \times \mathbb{X}$ given by

$$D(\mathscr{A}_\kappa) = D(A_1) \times D(A_2) = \ker \Phi_1 \times \ker \Phi_2,$$

$$\mathscr{A}_\kappa = \begin{pmatrix} A_1 & 0 \\ 0 & A_2 \end{pmatrix} + \kappa \begin{pmatrix} -\beta & \alpha \\ \beta & -\alpha \end{pmatrix} =: \mathscr{A}_0 + \kappa \mathscr{Q}.$$

We assume that the semigroup generated by $\mathscr{A}_0$, say $\left(\mathrm{e}^{t\mathscr{A}_0}\right)_{t\geq 0}$ is semicontractive, i.e., it satisfies

$$\|\mathrm{e}^{t\mathscr{A}_0}\| \leq \mathrm{e}^{\omega t}, \qquad t \geq 0, \tag{44}$$

for some $\omega \in \mathbb{R}$ and that

$$\mathscr{P} := \mathscr{Q} + I_{\mathbb{X}\times\mathbb{X}} = \begin{pmatrix} \alpha & \alpha \\ \beta & \beta \end{pmatrix} \tag{45}$$

is a contraction in $\mathbb{X} \times \mathbb{X}$. (The former condition is automatically satisfied if $\left(\mathrm{e}^{tA_0}\right)_{t\geq 0}$ is semicontractive—see the remark on page 215 in [31].) We note that $\mathscr{P}$ is idempotent, hence

$$\begin{aligned} \mathrm{e}^{\kappa t \mathscr{Q}} &= \mathrm{e}^{-\kappa t}\mathrm{e}^{\kappa t \mathscr{P}} = \mathrm{e}^{-\kappa t}\left[I_{\mathbb{X}\times\mathbb{X}} + (\mathrm{e}^{t\kappa} - 1)\mathscr{P}\right] \\ &= \mathrm{e}^{-\kappa t} I_{\mathbb{X}\times\mathbb{X}} + (1 - \mathrm{e}^{-\kappa t})\mathscr{P}. \end{aligned} \tag{46}$$

It follows that $\|e^{t\mathscr{Q}}\| \le 1$ and for the semigroups generated by $\mathscr{A}_\kappa$ (which exist by the Phillips perturbation theorem) we have, by the Trotter product formula,

$$\|e^{t\mathscr{A}_\kappa}x\| \le \lim_{n\to\infty} \left\| \left[e^{\frac{t}{n}\mathscr{A}_0} e^{\frac{\kappa t}{n}\mathscr{Q}} \right]^n x \right\| \le e^{\omega t}\|x\|, \qquad x \in \mathbb{X}\times\mathbb{X},$$

so that

$$\|e^{t\mathscr{A}_\kappa}\| \le e^{\omega t}, \qquad \kappa > 0, t \ge 0. \tag{47}$$

Operator $\mathscr{P}$ is a projection on the subspace $\mathbb{X}' \subset \mathbb{X}\times\mathbb{X}$ of vectors of the form $\binom{\alpha x}{\beta x}$; the latter space is isomorphic to $\mathbb{X}$ with isomorphism $\mathscr{I} : \mathbb{X} \to \mathbb{X}'$ given by $\mathscr{I}x = \binom{\alpha x}{\beta x}$.

Theorem 3.1 *In the above setup, assume that α leaves $D(A)$ invariant. Then,*

$$\lim_{\kappa\to+\infty} e^{t\mathscr{A}_\kappa}\begin{pmatrix} x_1 \\ x_2 \end{pmatrix} = \mathscr{I}e^{tA_a}\mathscr{I}^{-1}\mathscr{P}\begin{pmatrix} x_1 \\ x_2 \end{pmatrix} = \begin{pmatrix} \alpha e^{t\mathscr{A}_a}(x_1+x_2) \\ \beta e^{t\mathscr{A}_a}(x_1+x_2) \end{pmatrix}, \qquad t > 0, x_1, x_2 \in \mathbb{X}. \tag{48}$$

For $\binom{x_1}{x_2} \in \mathbb{X}'$ the same is true for $t = 0$ as well, and the limit is almost uniform in $t \in [0,\infty)$; for other $\binom{x_1}{x_2}$ the limit is almost uniform in $t \in (0,\infty)$.

Intuitively, this result may be explained as follows. The components of the semigroup $\left(e^{t\mathscr{A}_0}\right)_{t\ge0}$ are uncoupled, while in $\left(e^{t\mathscr{A}_\kappa}\right)_{t\ge0}$ the coupling is realised by the operator $\mathscr{Q}$ which may be thought of as describing a Markov chain switching one dynamics into the other (the jumps' intensities are state-dependent, see examples given later). As $\kappa \to \infty$, the Markov chain reaches its statistical equilibrium, so that with "probability" α it chooses the first dynamics, and with "probability" β, it chooses the second dynamics. This results in a convex combination of boundary conditions in the limit semigroup. (Compare the main theorem in [9], see also [13].)

Theorem 3.2 *Under conditions of the previous theorem, let $\mathscr{B} = \begin{pmatrix} B_1 & 0 \\ 0 & B_2 \end{pmatrix}$, where B_1 and B_2 are bounded linear operators. Then,*

$$\lim_{\kappa\to+\infty} e^{t(\mathscr{A}_\kappa+\mathscr{B})}\begin{pmatrix} x_1 \\ x_2 \end{pmatrix} = \mathscr{I}e^{t(A_a+B_1\alpha+B_2\beta)}\mathscr{I}^{-1}\mathscr{P}\begin{pmatrix} x_1 \\ x_2 \end{pmatrix}, \qquad t > 0, x_1, x_2 \in \mathbb{X}. \tag{49}$$

For $\binom{x_1}{x_2} \in \mathbb{X}'$ the same is true for $t = 0$ as well, and the limit is almost uniform in $t \in [0,\infty)$; for other $\binom{x_1}{x_2}$ the limit is almost uniform in $t \in (0,\infty)$.

Remark 1 For Theorems 3.1 and 3.2, besides (34) and (44), we assume that $\mathscr{P}$, defined in (45), is a contraction in $\mathbb{X}\times\mathbb{X}$ and α leaves $D(A)$ invariant. While the nature of the first and the last conditions is transparent, the other two require a comment. As already mentioned, together they imply stability condition (47) (which is a common assumption in convergence theorems), and in fact our theorems remain true if we simply assume (47). However, for the sake of applications it is more convenient to assume the two conditions discussed above. Out of these two, the one requiring $\mathscr{P}$ to be a contraction seems to be most restrictive, apparently excluding spaces with supremum norm. On the other hand, this assumption is often satisfied in L^1-type spaces. (Similarly, the "dual" theorem in [9, 13] is designed for spaces of continuous functions.) In particular, if $\mathbb{X}$ is an AL-space, i.e. a Banach lattice such that

$$\|x + y\| = \|x\| + \|y\|, \qquad x, y \geq 0,$$

and $\mathbb{X} \times \mathbb{X}$ is equipped with the order "$\binom{x}{y} \geq 0$ iff $x \geq 0$ and $y \geq 0$" and the norm $\left\|\binom{x}{y}\right\| = \|x\| + \|y\|$, then $\mathscr{P}$ is a contraction provided α and β are positive operators. For, in such a case,

$$\begin{aligned} \left\|\mathscr{P}\begin{pmatrix} x \\ y \end{pmatrix}\right\| &= \left\|\begin{pmatrix} \alpha(x+y) \\ \beta(x+y) \end{pmatrix}\right\| = \|\alpha(x+y)\| + \|\beta(x+y)\| \\ &= \|x+y\| \leq \left\|\begin{pmatrix} x \\ y \end{pmatrix}\right\|, \qquad x, y \geq 0, \end{aligned}$$

and $\mathscr{P}$ is positive. Hence, [see e.g. [3, Proposition 2.67]]

$$\|\mathscr{P}\| = \sup_{\left\|\binom{x}{y}\right\|=1, \binom{x}{y}\geq 0} \left\|\mathscr{P}\begin{pmatrix} x \\ y \end{pmatrix}\right\| \leq 1.$$

Example 7 As we have already seen, in the motivating example of dynamics of *sola sola*,

$$A : W^{1,1}(\mathbb{R}^+) \to L^1(\mathbb{R}^+) \qquad Af = -f',$$

and $L : W^{1,1}(\mathbb{R}^+) \to \mathbb{R}$ is given by $Lf = f(0)$, and condition (34) is satisfied with $\gamma = 1$.

For $b_i \in L^\infty(\mathbb{R}^+), i = 1, 2$, the functionals $F_i f = \int_0^\infty b_i(a) f(a)\,da$ are linear and bounded. Hence, A_{F_i} generates a semigroup of operators and so does $A_{F_i} + B_i$, where given $\mu_i \in L^\infty(\mathbb{R}^+)$, B_i is a (bounded) multiplication operator $f \mapsto -\mu_i f$. It is well-known [see e.g. [11, 21, 33]] that there is ω such that $\|e^{t(A_{F_i}+B_i)}\| \leq e^{\omega t}, i = 1, 2$, implying (44).

Let $\alpha \in W^{1,\infty}(\mathbb{R}^+)$ satisfy $0 \leq \alpha \leq 1$. Then the related multiplication operator (denoted by the same letter) is bounded in $L^1(\mathbb{R}^+)$, and leaves $D(A) = W^{1,1}(\mathbb{R}^+)$ invariant. Moreover, the related operator $\mathscr{P}$ [see (45)] in $L^1(\mathbb{R}^+) \times L^1(\mathbb{R}^+)$, equipped with the norm $\left\|\binom{f_1}{f_2}\right\| = \|f_1\|_{L^1(\mathbb{R}^+)} + \|f_2\|_{L^1(\mathbb{R}^+)}$, is a contraction (see Remark 1). Hence, all assumptions of Theorem 3.2 are satisfied. This again establishes that the general model (41) (with $N = 2$ and normalized matrix k) may be approximated by the averaged one (43).

3.4 Modelling Neurotransmitters

In our last section we discuss the results of [15], where a link between two recent models of dynamics of synaptic depression was provided. This involves Feller–Wentzell-type transmission conditions, as we explain in what follows.

In an attempt to understand phenomena behind synaptic depression, Aristizabal and Glavinovič introduced a simple ODE model of dynamics of levels of neurotransmitters [2]. They adopted the following widely accepted, simplified but sufficiently accurate description (see papers cited in [2]): neurotransmitters are localized in three compartments, or pools: the large pool, where also their synthesis takes place, the small intermediate pool, and the immediately available pool, from which they are released during stimulus. Moreover, they assumed that the dynamics of levels $U_i, i = 1, 2, 3$ of vesicles with neurotransmitters in the pools is analogous to that of voltages across the capacitors in the electric circuit reproduced (with minor changes) as our Fig. 9.

This results in the following system of ODEs for U_i:

$$\begin{pmatrix} U_1' \\ U_2' \\ U_3' \end{pmatrix} = Q \begin{pmatrix} U_1 \\ U_2 \\ U_3 \end{pmatrix} + \begin{pmatrix} 0 \\ 0 \\ \frac{1}{R_3C_3}(E - U_3) \end{pmatrix}, \tag{50}$$

where

$$Q = \begin{pmatrix} -\frac{1}{R_0C_1} - \frac{1}{R_1C_1} & \frac{1}{R_1C_1} & 0 \\ \frac{1}{R_1C_2} & -\frac{1}{R_2C_2} - \frac{1}{R_1C_2} & \frac{1}{R_2C_2} \\ 0 & \frac{1}{R_2C_3} & -\frac{1}{R_2C_3} \end{pmatrix}.$$

For the electric circuit, E denotes the electromotive source, the constants C_is are capacitors' sizes, while R_is characterize the resistors. Biologically, E represents

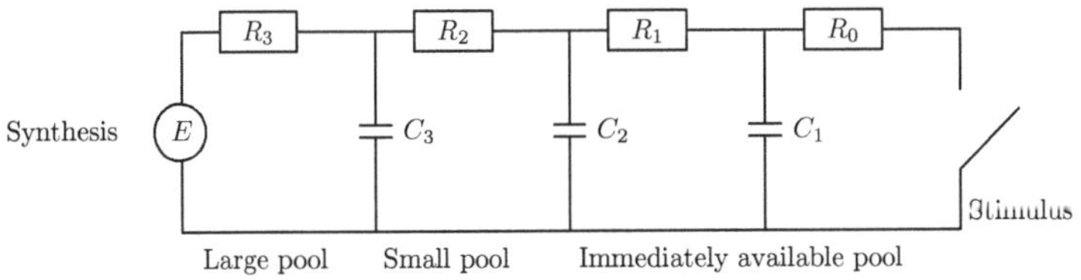

Fig. 9 The ODE model of Aristizabal and Glavinovič

synthesis and C_is are the capacities to store vesicles, but R_is do not have a clear meaning. Merely the compounds $\frac{1}{R_i C_j}$ are interpreted as the pools' replenishment rates.

A more recent PDE model of Bielecki and Kalita [7] zooms in on vesicles with neurotransmitters, and assumes that they move according to a diffusion process in a three dimensional domain Ω. As a result, in the linear version of the model, the (unknown) concentration ρ of vesicles in the cytoplasm satisfies a Fokker-Planck-type equation

$$\frac{\partial \rho}{\partial t} = A\rho + \beta(\overline{\rho} - \rho) \tag{51}$$

where A is a second order, elliptic partial differential operator, $\beta : \Omega \to \mathbb{R}$ is the rate of neurotransmitters' production, and $\overline{\rho}$ is a balance concentration of vesicles.

We are looking for a connection between these models. To this end, first we note that equations (50) and (51) are of quite a different nature: while (in the absence of stimulus and production) the latter is conservative, the former is not. However, the dual to the matrix Q in (50) is an intensity matrix (it has non-negative off-diagonal entries, and the sums of its entries in each column are zero), and the dual equation to (50) is conservative. This suggests that to find a link between the two models, one must first pass from the description of dynamics of densities (concentrations), to that of expected values, i.e. instead of considering the Fokker-Planck-type equation (51) one should pass to the dual Kolmogorov backward equation. Moreover, to find such a link, one needs to specify the way vesicles move from one pool to another, i.e. to specify the transmission conditions [17], which are missing in Bielecki and Kalita's model. (In proving formula (52) below, instead of introducing appropriate transmission conditions, Bielecki and Kalita use what they call "technical conditions", without showing that these technical conditions are satisfied.) These transmission conditions, describing communication between pools, are of crucial importance in the analysis.

In [15] we introduced a one-dimensional version of the Bielecki and Kalita model, where vesicles perform a Brownian motion on three adjacent intervals (corresponding to pools) with diffusion coefficients varying from pool to pool, and where the mechanism of passing from one pool to another is specified by means of transmission conditions. Our main result says that as the diffusion coefficients in the model tend to infinity and the boundary and transmission conditions are scaled in an appropriate way, the solutions to the related Cauchy problems converge to those of the model of Aristizabal and Glavinovič. Roughly speaking, if diffusion in three separate pools is large and communication between pools is slow, the ODE model is a good approximation of the PDE model.

We note that to show a connection between the two models, Bielecki and Kalita also divide Ω into three subregions Ω_3, Ω_2 and Ω_1, corresponding to the three pools. They assume that the diffusion process the vesicles perform is a three-dimensional

Brownian motion and the diffusion coefficients, say $\sigma_1, \sigma_2, \sigma_3$ vary from region to region, and suggest (see [7, Thm 2]) that the quantities

$$U_i = \frac{\int_{\Omega_i} \rho}{\text{volume}\Omega_i}, \quad i = 1, 2, 3, \tag{52}$$

satisfy the ODE system (50) of Aristizabal and Glavinovič with $C_i = \frac{\text{volume}\Omega_i}{\sigma_i}$. However, this formula is at least doubtful: the proof of (52) given in [7] contains a number of errors. Moreover, in the absence of stimulus and neurotransmitter's production, the total number of vesicles should remain constant. Hence, $U_1 + U_2 + U_3 = const$, provided volume$\Omega_i = 1, i = 1, 2, 3$. However, system (50) is not conservative, i.e. $(U_1 + U_2 + U_3)' \neq 0$ unless all C_i' are the same [no stimulus case is obtained by letting $R_0 \to \infty$, and no production results in removing the second summand in (50)]. Hence, formula (52) cannot hold unless all σ_i' are the same.

In our model we imagine the three pools as three adjacent intervals $[0, r_3]$, $[r_3, r_2]$ and $[r_2, r_1]$ of the real line, corresponding to the large, the small and the immediately available pools, respectively. As in the model of Bielecki and Kalita, in each of those intervals, vesicles perform Brownian motions with respective diffusion coefficients σ_3, σ_2 and σ_1. The pools are separated by semi-permeable membranes located at $x = r_3$ and $x = r_2$. (We note that no such membranes exist physically; they are merely imaginary. Alternatively, instead of membranes and their permeability we may speak of intensities of passing from one pool to the other.) Therefore, it is convenient to think of the actual state space Ω of the process performed by the vesicles as the union of three intervals:

$$\Omega := \Omega_3 \cup \Omega_2 \cup \Omega_1 := [0, r_3^-] \cup [r_3^+, r_2^-] \cup [r_2^+, r_1].$$

(In order to keep our notations consistent with those of [2] and [7], the intervals are numbered "from the right to the left".) Note that r_3 is now split into two points: r_3^- and r_3^+, representing positions to the immediate left and to the immediate right from the first membrane; a similar remark concerns r_2. Vesicles in all pools may permeate through the imaginary membrane(s) to the adjacent pool(s), and their ability to filter from the ith into the jth pool is characterized by permeability coefficients $k_{ij} \geq 0$, $i, j = 1, 2, 3, |i - j| = 1$. The left end-point $x = 0$ is a reflecting boundary for the process, and the right end-point $x = r_1$ is an elastic boundary with elasticity coefficient $k_{10} \geq 0$. The case $k_{10} > 0$ characterizes the boundary during stimulus, and $k_{10} = 0$ describes it in between stimuli (i.e., when there is no stimulus, $x = r_1$ is a reflecting boundary). Hence, k_{10} characterizes vesicles' ability to be released from the terminal bouton.

To describe our model more formally, we note that Ω is a (disconnected) compact space and the function σ defined on Ω by

$$\sigma(x) = \sigma_i, \quad x \in \Omega_i, i = 1, 2, 3,$$

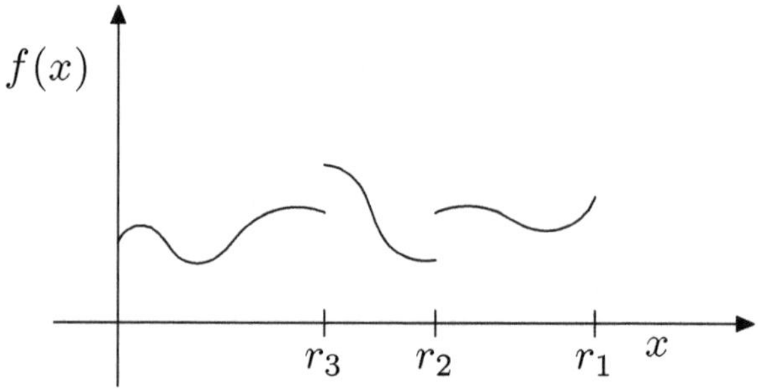

Fig. 10 A typical member of $C(\Omega)$

is continuous. A typical member of the Banach space

$$\mathbb{B} = C(\Omega)$$

of complex continuous functions on Ω is depicted at Fig. 10. We note that the space is isometrically isomorphic to the Cartesian product $C(\Omega_1) \times C(\Omega_2) \times C(\Omega_3)$ of the spaces of continuous functions on the three intervals. In other words, a member of $C(\Omega)$ may be identified with three continuous functions f_1, f_2, f_3 being restrictions of f to the three intervals $\Omega_1, \Omega_2, \Omega_3$, respectively; we have $\|f\|_{C(\Omega)} = \max_{i=1,2,3} \|f_i\|_{C(\Omega_i)}$.

The first result in [15] says that the operator A in $C(\Omega)$ defined as

$$Af = \sigma f'' \tag{53}$$

for twice continuously differentiable functions f on Ω satisfying the conditions:

$$\begin{aligned} f'(0) &= 0, & f'(r_3^-) &= k_{32}[f(r_3^+) - f(r_3^-)], \\ f'(r_3^+) &= k_{23}[f(r_3^+) - f(r_3^-)], & f'(r_2^-) &= k_{21}[f(r_2^+) - f(r_2^-)], \\ f'(r_2^+) &= k_{12}[f(r_2^+) - f(r_2^-)], & f'(r_1) &= -k_{10} f(r_1), \end{aligned} \tag{54}$$

generates a semigroup $\{e^{tA}, t \geq 0\}$ in $C(\Omega)$. The semigroup describes the dynamics of expected values of neurotransmitters' numbers in the three pools.

The results presented in Sect. 2.2 allow interpreting the boundary and transition conditions (54) as follows. Turning, for example, to the third condition in (54), we note that it has the form of the elastic barrier condition (33) with 0 replaced by r_3^+, $b = 1, c = d = k_{23}$ and μ equal to the Dirac measure at r_3^-. Hence, it describes the process in which the vesicles in Ω_2 bounce from the imaginary membrane separating it from Ω_3 to filter into the latter interval at a random time T with distribution:

$$P(T > t) = e^{-k_{23} t^+}, \quad t \geq 0.$$

(Stochastic analysis of the related *snapping out Brownian motion* may be found in [38].) In particular, the larger k_{23} is, the shorter is the time needed for the vesicle to filter through the imaginary membrane. Hence, k_{23} is truly a permeability coefficient for passing from Ω_2 to Ω_3. On the other hand, as indicated previously, dividing k_{23} by κ_n, where $(\kappa_n)_{n\geq 1}$ is a sequence of positive numbers tending to infinity, and letting $n \to \infty$ we obtain $P(T > t) = 1$, i.e. the time to filter through the membrane is infinite and the boundary is reflecting. The interpretation of the other boundary and transmission conditions in (54) is analogous.

The main point is to study the limit of the semigroups generated by the operators $A_{\kappa_n}, n \geq 1$ defined by (53) with σ replaced by $\kappa_n \sigma$ and all permeability coefficients in (54) divided by κ_n. More specifically, it is proved in [15] that

$$\lim_{n\to\infty} \mathrm{e}^{tA_{\kappa_n}} f = \mathrm{e}^{tQ} Pf, \qquad t > 0, f \in C(\Omega) \tag{55}$$

where Q is given by

$$Q = \begin{pmatrix} -k'_{10} - k'_{12} & k'_{12} & 0 \\ k'_{21} & -k'_{21} - k'_{23} & k'_{23} \\ 0 & k'_{32} & -k'_{32} \end{pmatrix}, \qquad k'_{ij} = \frac{\sigma_i k_{ij}}{|\Omega_i|}, \tag{56}$$

where $|\Omega_i|$ is the length of the ith interval, and the operator P given by

$$Pf = (|\Omega_i|^{-1} \int_{\Omega_i} f)_{i=1,2,3}$$

is a projection on the subspace $\mathbb{B}_0$ of $C(\Omega)$ of functions that are constant on each of the three subintervals separately; the subspace may be identified with $\mathbb{R}^3$, and its members may be identified with triples of real numbers.

Intuitively, as the diffusion coefficients increase, the transition probabilities between points in each interval separately tend to 1. As a result, the points become indistinguishable and may be lumped together. Points from different intervals may not be lumped together since as $n \to \infty$ the permeability coefficients $\frac{k_{ij}}{\kappa_n}$ tend to zero and in the limit the membranes become reflecting boundaries, separating the intervals. However, because of the intimate relation between diffusion and permeability coefficients, the three states of the limit process, i.e. the three intervals contracted to three separate points, communicate as the states of a Markov chain with intensity matrix Q (see Fig. 11).

Comparing intensity matrices of (50) and (56), we obtain the following relations between parameters in the two models:

$$\frac{1}{R_0 C_1} = \frac{\sigma_1 k_{10}}{|\Omega_1|}, \frac{1}{R_1 C_1} = \frac{\sigma_1 k_{12}}{|\Omega_1|}, \frac{1}{R_1 C_2} = \frac{\sigma_2 k_{21}}{|\Omega_2|},$$
$$\frac{1}{R_2 C_2} = \frac{\sigma_2 k_{23}}{|\Omega_2|}, \frac{1}{R_2 C_3} = \frac{\sigma_3 k_{32}}{|\Omega_3|}. \tag{57}$$

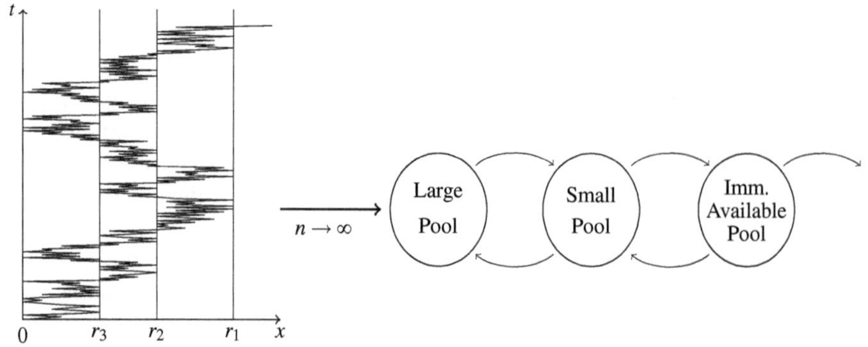

Fig. 11 Approximating a Markov chain by diffusion processes

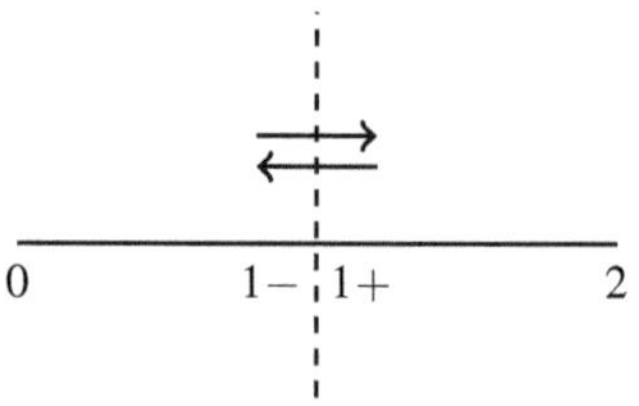

Fig. 12 Two pools

These relations agree with the intuition that the replenishment rate from the ith to the jth pool is directly proportional to permeability of the imaginary membrane separating them and to the speed of diffusion in the ith interval, and inversely proportional to the length of this interval.

Perhaps this mechanism will become yet more clear if we consider the case of two pools. For simplicity, we assume, as at our Fig. 12, that two intervals representing pools are of the same unit length, and consider first the operator $Af = f''$ defined for twice continuously differentiable functions in $\mathbb{B} = C[0,1] \times C[1,2]$ (each pair of functions in $\mathbb{B}$ is identified with a single function on $[0,2]$, continuous in this interval except perhaps at $x = 1$ where it has limits from the left and from the right), satisfying boundary and transmission conditions

$$f'(0) = f'(2) = 0,$$
$$f'(1-) = \alpha[f(1+) - f(1-)],$$
$$f'(1+) = \beta[f(1+) - f(1-)],$$

where α and β play the role of permeability coefficients of the membrane located at $x = 1$ (see Fig. 12). Now, if we replace $Af = f''$ by $A_n f = nf''$ and divide the permeability coefficients by n, the related semigroups e^{tA_n} will converge, as $n \to \infty$, to $e^{tQ}P$ where

$$Pf = \left(\int_0^1 f, \int_1^2 f \right)$$

is a map from $\mathbb{X}$ to $\mathbb{R}^2$ (identified with the subspace $\subset \mathbb{X}$ of functions that are constant on each of the intervals) and

$$Q = \begin{pmatrix} -\alpha & \alpha \\ \beta & -\beta \end{pmatrix},$$

is the intensity matrix of the simplest Markov chain.

But, on the other hand, the principle discovered here is more general than it appears [12]. In fact, we may forget about adjacent intervals and consider diffusions on graphs. More specifically, imagine a finite graph $\mathcal{G}$ without loops, and a Markov process on $\mathcal{G}$ obeying the following informal rules.

- While on the ith edge, imagined as a C^1 curve in $\mathbb{R}^3$, the process behaves like a one-dimensional Brownian motion with variance $\sigma_i > 0$.
- Graph's vertices are semi-permeable membranes, allowing communication between the edges; permeability coefficients p_{ij}, describing the possibility to filter through the membrane from the ith to the jth edge, depend on the edges. In particular, p_{ij} is in general different from p_{ji}. At each vertice, the process may also be killed and removed from the state-space.

Now, suppose the diffusion's speed increases while membranes' permeability decreases (i.e. $\sigma_i \to \infty$ and $p_{ij} \to 0$). As a result, points in each edge communicate almost immediately and in the limit are lumped together, but the membranes prevent lumping of points from different edges. It transpires, nevertheless, that the assumption that the rate with which permeability coefficients tend to zero is the same as the rate with which the diffusion coefficients tend to infinity, leads to a limit process in which communication between lumped edges is possible. The lumped edges form then the vertices in the so-called *line graph of* $\mathcal{G}$ (see [18]) and communicate as the states of a Markov chain with jumps' intensities directly proportional to permeability coefficients p_{ij} and the diffusion coefficients σ_i, and inversely proportional to the edges' lengths (see Fig. 13). The assumption on the rate is important: if diffusion coefficients tend to infinity slower than the permeability coefficients tend to zero, there is no communication between the vertices in the limit line graph, and in the opposite case all points of the original graph are lumped together, and nothing interesting happens.

This procedure may also be reversed: given a finite-state Markov chain, we may find a graph $\mathcal{G}$ and construct a fast diffusion on $\mathcal{G}$ approximating the chain. See [12] for details.

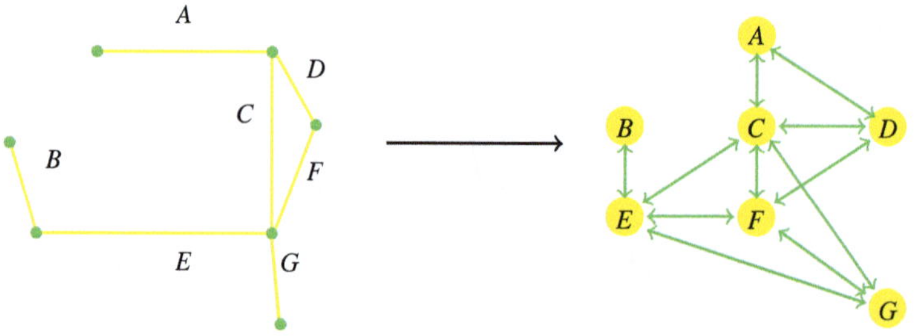

Fig. 13 From diffusion on $\mathcal{G}$ to a Markov chain on the vertices of the line graph of $\mathcal{G}$; edges "shrink" to vertices, vertices "split" into edges

Acknowledgements This research was partially supported by the Polish Government under grant 6081/B/H03/2011/40.

References

1. O. Arino, E. Sanchez, R. Bravo de la Parra, P. Auger, A singular perturbation in an age-structured population. SIAM J. Appl. Math. **60**(2), 408–436 (1999–2000)
2. F. Aristizabal, M.I. Glavinovic, Simulation and parameter estimation of dynamics of synaptic depression. Biol. Cybern. **90**, 3–18 (2004)
3. J. Banasiak, L. Arlotti, *Perturbations of Positive Semigroups with Applications* (Springer, London, 2006)
4. J. Banasiak, A. Bobrowski, A semigroup related to a convex combination of boundary conditions obtained as a result of averaging other semigroups. J. Evol. Equ. (2014). doi: 10.1007/s00028-014-0257-z
5. J. Banasiak, A. Goswami, Singularly perturbed population models with reducible migration matrix. 1. Sova-Kurtz theorem and the convergence to the aggregated model. DCDS-A **35**, 617–635 (2015)
6. J. Banasiak, A. Goswami, S. Shindin, Aggregation in age and space structured population models: an asymptotic analysis approach. J. Evol. Equ. **11**, 121–154 (2011)
7. A. Bielecki, P. Kalita, Model of neurotransmitter fast transport in axon terminal of presynaptic neuron. J. Math. Biol **56**, 559–576 (2008)
8. A. Bobrowski, *Functional Analysis for Probability and Stochastic Processes* (Cambridge University Press, Cambridge, 2005)
9. A. Bobrowski, On a semigroup generated by a convex combination of two Feller generators. J. Evol. Equ. **7**(3), 555–565 (2007)
10. A. Bobrowski, Generation of cosine families via Lord Kelvin's method of images. J. Evol. Equ. **10**(3), 663–675 (2010)
11. A. Bobrowski, Lord Kelvin's method of images in the semigroup theory. Semigroup Forum **81**, 435–445 (2010)
12. A. Bobrowski, From diffusions on graphs to Markov chains via asymptotic state lumping. Ann. Henri Poincare **13**, 1501–1510 (2012)
13. A. Bobrowski, R. Bogucki, Semigroups generated by convex combinations of several Feller generators in models of mathematical biology. Stud. Math. **189**, 287–300 (2008)
14. A. Bobrowski, R. Bogucki, Two theorems on singularly perturbed semigroups with applications to models of applied mathematics. Discrete Continuous Dyn. Syst. B **17**(3), 735–757 (2012)

15. A. Bobrowski, K. Morawska, From a PDE model to an ODE model of dynamics of synaptic depression. Discrete Continuous Dyn. Syst. B **17**(7), 2313–2327 (2012)
16. R. Chill, V. Keyantuo, M. Warma, Generation of cosine families on $L^p(0,1)$ by elliptic operators with Robin boundary conditions, in *Functional Analysis and Evolution Equations. The Günter Lumer Volume*, ed. by H. Amann (Birkhauser, Basel, 2007), pp. 113–130
17. R. Dautray, J.-L. Lions, *Mathematical Analysis and Numerical Methods for Science and Technology*, vol. 3 (Springer, Berlin, 1990). Spectral theory and applications, With the collaboration of Michel Artola and Michel Cessenat, Translated from the French by John C. Amson
18. N. Deo, *Graph Theory with Applications to Engineering and Computer Science* (Prentice-Hall, Englewood Cliffs, 1974)
19. E.B. Dynkin, A.A. Yushkevich, *Markov Processes. Theorems and Problems* (Plenum Press, New York, 1969)
20. R.E. Edwards, *Functional Analysis. Theory and Applications* (Dover Publications, New York, 1995)
21. K.-J. Engel, R. Nagel, *One-Parameter Semigroups for Linear Evolution Equations* (Springer, New York, 2000)
22. K.-J. Engel, R. Nagel, *A Short Course on Operator Semigroups* (Springer, New York, 2006)
23. W.J. Ewens, *Mathematical Population Genetics*, 2nd edn. (Springer, New York, 2004)
24. W. Feller, *An Introduction to Probability Theory and Its Applications*, vol. 1 (Wiley, New York, 1950) [3rd edn., 1970]
25. W. Feller, Diffusion processes in genetics, in *Proceedings of the Second Berkeley Symposium on Mathematical Statistics and Probability, 1950* (University of California Press, Berkeley/Los Angeles, 1951), pp. 227–246
26. W. Feller, Diffusion processes in one dimension. Trans. Am. Math. Soc. **77**(1), 468–519 (1952)
27. W. Feller, The parabolic differential equations and the associated semi-groups of transformations. Ann. Math. **55**, 468–519 (1952)
28. W. Feller, *An Introduction to Probability Theory and Its Applications*, vol. 2 (Wiley, New York, 1966) [2nd edn., 1971]
29. J.A. Goldstein, *Semigroups of Linear Operators and Applications* (Oxford University Press, New York, 1985)
30. G.R. Goldstein, Derivation and physical interpretation of general boundary conditions. Adv. Differ. Equ. **11**(4), 457–480 (2006)
31. G. Greiner, Perturbing the boundary conditions of a generator. Houston J. Math. **13**(2), 213–229 (1987)
32. E. Hille, R.S. Phillips, *Functional Analysis and Semi-Groups*. American Mathematical Society Colloquium Publications, vol. 31 (American Mathematical Society, Providence, 1957)
33. M. Iannelli, *Mathematical Theory of Age-Structured Population Dynamics*. Applied Mathematical Monographs, vol. 7 (Giardini Editori E Stampatori, Pisa, 1995)
34. K. Itô, H.P. McKean Jr., *Diffusion Processes and Their Sample Paths* (Springer, Berlin, 1996). Repr. of the 1974 ed.
35. I. Karatzas, S.E. Shreve, *Brownian Motion and Stochastic Calculus* (Springer, New York, 1991)
36. S. Karlin, H.M. Taylor, *A Second Course in Stochastic Processes* (Academic [Harcourt Brace Jovanovich Publishers], New York, 1981)
37. A. Lasota, M.C. Mackey, *Chaos, Fractals, and Noise. Stochastic Aspects of Dynamics* (Springer, Berlin, 1994)
38. A. Lejay, The snapping out Brownian motion. hal-00781447, December 2012
39. T.M. Liggett, *Continuous Time Markov Processes. An Introduction* (American Mathematical Society, Providence, 2010)
40. P. Mandl, *Analytical Treatment of One-Dimensional Markov Processes* (Springer, New York, 1968)
41. A. Pazy, *Semigroups of Linear Operators and Applications to Partial Differential Equations* (Springer, New York, 1983)

42. D. Revuz, M. Yor, *Continuous Martingales and Brownian Motion*, 3rd edn. (Springer, Berlin, 1999)
43. W. Rudin, *Real and Complex Analysis*, 3rd edn. (McGraw-Hill Book, New York, 1987)
44. R. Rudnicki (ed.), *Mathematical Modelling of Population Dynamics*. Banach Center Publications, vol. 63 (Polish Academy of Sciences Institute of Mathematics, Warsaw, 2004). Papers from the conference held in Będlewo, June 24–28, 2002
45. E. Sanchez, R. Bravo de la Parra, P. Auger, P. Gomez-Mourelo, Time scales in linear delayed differential equations. J. Math. Anal. Appl. **323**, 680–699 (2006)
46. K. Taira, *Semigroups, Boundary Value Problems and Markov Processes*. Springer Monographs in Mathematics (Springer, Berlin, 2004)
47. G.F. Webb, *Theory of Nonlinear Age-Dependent Population Dynamics*. Monographs and Textbooks in Pure and Applied Mathematics, vol. 89 (Marcel Dekker, New York, 1985)
48. A.D. Wentzell, On lateral conditions for multidimensional diffusion processes. Teor. Veroyatnost. i Primenen. **4**, 172–185 (1959). English translation: Theory Prob. Appl. **4**, 164–177 (1959)

Introduction to Complex Networks: Structure and Dynamics

Ernesto Estrada

1 Introduction

1.1 Motivations

This chapter is written with graduate students in mind. During the very encouraging meeting at the African Institute for Mathematical Sciences (AIMS) for the CIMPA-UNESCO-MESR-MINECO-South Africa Research School on "*Evolutionary Equations with Applications in Natural Sciences*" I noticed a great interest of graduate and postgraduate students in the field of complex networks. This chapter is then an elementary introduction to the field of complex networks, not only about the dynamical processes taking place on them, as originally planned, but also about the structural concepts needed to understand such dynamical processes. At the end of this chapter I will provide some basic material for the further study of the topics covered here, apart from the references cited in the main text. This is aimed to help students to navigate the vast literature that has been generated in the last 15 years of studying complex networks from an interdisciplinary point of view.

The study of complex networks has become a major topic of interdisciplinary research in the twentyfirst century. Complex systems are ubiquitous in nature and made-made systems, and because complex networks can be considered as the skeleton of complex systems they appear in a wide range of scenarios ranging from social and ecological to biological and technological systems. The concept of "complexity" may well refer to a quality of the system or to a quantitative characterisation of that system [40, 44]. As a quality of the system it refers to

E. Estrada (✉)
Department of Mathematics & Statistics, University of Strathclyde, Glasgow G1 1XQ, UK
e-mail: ernesto.estrada@strath.ac.uk

© Springer International Publishing Switzerland 2015

J. Banasiak, M. Mokhtar-Kharroubi (eds.), *Evolutionary Equations with Applications in Natural Sciences*, Lecture Notes in Mathematics 2126,
DOI 10.1007/978-3-319-11322-7_3

what makes the system complex. In this case complexity refers to the presence of emergent properties in the system. That is, to the properties which emerge as a consequence of the interactions of the parts in the system. In its second meaning, complexity refers to the amount of information needed to specify the system.

In the so-called complex networks there are many properties that emerge as a consequence of the global organisational structure of the network. For instance, a phenomenon known as "small-worldness" is characterised by the presence of relatively small average path length (see further for definitions) and a relatively high number of triangles in the network. While the first property appears in randomly generated networks, the second "emerges" as a consequence of a characteristic feature of many complex systems in which relations display a high level of transitivity. This second property is not captured by a random generation of the network.

By considering complexity in its quantitative edge we may attempt to characterise complex networks by giving the minimum amount of information needed to describe them. For the sake of comparison let us also consider a regular and a random graph of the same size of the real-world network we want to describe. For the case of a regular graph we only need to specify the number of nodes and the degree of the nodes (recall that every node has the same degree). With this information many non-isomorphic graphs can be constructed, but many of their topological and combinatorial properties are determined by the information provided. In the case of the random network we need to specify the number of nodes and the probability for joining pairs of nodes. As we will see in a further section, most of the structural properties of these networks are determined by this information only. In contrast, to describe the structure of one of the networks representing a real-world system we need an awful amount of information, such as: number of nodes and links, degree distribution, degree-degree correlation, diameter, clustering, presence of communities, patterns of communicability, and other properties that we will study in this chapter. However, even in this case a complete description of the system is still far away. Thus, the network representation of these systems deserves the title of complex networks because:

1. there are properties that emerge as a consequence of the global topological organisation of the system,
2. their topological structures cannot be trivially described like in the cases of random or regular graphs.

Complex networks can be classified according to the nature of the interactions among the entities forming the nodes of the network. Some examples of these classes are:

- ***Physical linking***: pairs of nodes are physically connected by a *tangible link*, such as a cable, a road, a vein, etc. Examples are: Internet, urban street networks, road networks, vascular networks, etc.

- ***Physical interactions***: links between pairs of nodes represents ***interactions*** which are determined by a *physical force*. Examples are: protein residue networks, protein-protein interaction networks, etc.
- ***"Ethereal" connections***: links between pairs of nodes are *intangible*, such that information sent from one node is received at another irrespective of the "physical" trajectory. Examples are: WWW, airports network.
- ***Geographic closeness***: nodes represent regions of a surface and their connections are determined by their *geographic proximity*. Examples are: countries in a map, landscape networks, etc.
- ***Mass/energy exchange***: links connecting pairs of nodes indicate that some ***energy or mass*** has been *transferred* from one node to another. Examples are: reaction networks, metabolic networks, food webs, trade networks, etc.
- ***Social connections***: links represent any kind of *social relationship* between nodes. Examples are: friendship, collaboration, etc.
- ***Conceptual linking***: links indicate *conceptual relationships* between pairs of nodes. Examples are: dictionaries, citation networks, etc.

1.2 General Concepts of Networks

Here we define a network as the triple $G = (V, E, f)$, where V is a finite set of nodes , $E \subseteq V \otimes V = \{e_1, e_2, \cdots, e_m\}$ is a set of links and f is a *mapping* which associates some elements of E to a pair of elements of V, such as that if $v_i \in V$ and $v_j \in V$ then $f : e_p \to [v_i, v_j]$ and $f : e_q \to [v_j, v_i]$ [14, 23]. A *weighted network* is defined by replacing the set of links E by a set of link weights $W = \{w_1, w_2, \cdots, w_m\}$, such that $w_i \in \Re$. Then, a weighted network is defined by $G = (V, W, f)$. Two nodes u and v in a network are said to be *adjacent* if they are joined by a link $e = \{u, v\}$. Nodes u and v are *incident* with the link e, and the link e is *incident* with the nodes u and v. The *node degree* is the number of links which are incident with a given node. In directed networks, those where each edge has an arrow pointing from one node to another, the node u is *adjacent* to node v if there is a directed link from u to v $e = (u, v)$. A link from u to v is *incident from* u and *incident to* v; u is *incident to* e and v is *incident from* e. The *in-degree* of a node is the number of links incident to it and its *out-degree* is the number of links incident from it. The graph $S = (V', E')$ is called a subgraph of a network $G = (V, E)$ if and only if $V' \subseteq V$ and $E' \subseteq E$.

An important concept in the analysis of networks is that of walk. A (directed) walk of length l is any sequence of (not necessarily different) nodes $v_1, v_2, \cdots, v_l, v_{l+1}$ such that for each $i = 1, 2, \cdots, l$ there is link from v_i to v_{i+1}. This walk is referred to as a walk from v_1 to v_{l+1}. A *closed walk* (CW) of length l is a walk $v_1, v_2, \cdots, v_l, v_{l+1}$ in which $v_{l+1} = v_1$. A walk of length l in which all the nodes (and all the links) are distinct is called a *path*, and a closed walk in which all the links and all the nodes (except the first and last) are distinct is a *cycle*. If there is a path between each pair of nodes in a network, the network is said to be connected.

Every connected subgraph is a *connected component* of the network. The analogous concept for the directed network is that of strongly connected network. A directed network is *strongly connected* if there is a path between each pair of nodes. The *strongly connected components* of a directed network are its maximal strongly connected subgraphs.

A common representation of the topology of a network G is through the adjacency matrix. It is a square matrix **A** whose entries are defined by

$$A_{ij} = \begin{cases} 1 \text{ if } i, j \in E \\ 0 \text{ otherwise} \end{cases}. \tag{1}$$

A is symmetric for undirected networks and possibly un-symmetric for directed ones.

Another important matrix representation of a network is through its Laplacian matrix **L**, which is the discrete analogous of the Laplacian operator [32]. The entries of this matrix are defined by

$$L_{uv} = \begin{cases} -1 \text{ if } uv \in E, \\ k_u \quad \text{if } u = v, \\ 0 \text{ otherwise.} \end{cases} \tag{2}$$

Let us designate by ∇ the incidence matrix of the network, which is an $n \times m$ matrix whose rows and columns represents the nodes and edges of the network, respectively, such that

$$\nabla_{ue} = \begin{cases} +1 \text{ if} e \in E \text{is incoming to node } u, \\ -1 \text{ if} e \in E \text{is outcoming from node } u, \\ 0 \quad \text{otherwise.} \end{cases} \tag{3}$$

Then,

$$\mathbf{L} = \nabla \nabla^T. \tag{4}$$

If we designate by **K** the diagonal matrix of node degrees in the network, the Laplacian and adjacency matrices of a network are related as follows: $\mathbf{L} = \mathbf{K} - \mathbf{A}$.

The spectrum of the adjacency matrix of a network can be written as

$$Sp\mathbf{A} = \begin{pmatrix} \lambda_1(\mathbf{A}) & \lambda_2(\mathbf{A}) & \cdots & \lambda_n(\mathbf{A}) \\ m(\lambda_1(\mathbf{A})) & m(\lambda_2(\mathbf{A})) & \cdots & m(\lambda_n(\mathbf{A})) \end{pmatrix}, \tag{5}$$

where $\lambda_1(\mathbf{A}) \geq \lambda_2(\mathbf{A}) \geq \cdots \geq \lambda_n(\mathbf{A})$ are the distinct eigenvalues of **A** and $m(\lambda_1(\mathbf{A})), m(\lambda_2(\mathbf{A})), \cdots, m(\lambda_n(\mathbf{A}))$ are their multiplicity.

In the case of the Laplacian matrix the spectrum can be written in a similar way:

$$SpL = \begin{pmatrix} \mu_1(\mathbf{L}) & \mu_2\mathbf{L} & \cdots & \mu_n(\mathbf{L}) \\ m(\mu_1(\mathbf{L})) & m(\mu_2(\mathbf{L})) & \cdots & m(\mu_n(\mathbf{L})) \end{pmatrix}. \tag{6}$$

The Laplacian matrix is positive semi-definite and the multiplicity of 0 as an eigenvalue is equal to the number of connected components in the network. Then, the second smallest eigenvalue of **L**, $\mu_2(\mathbf{L})$, is usually called the algebraic connectivity of the network [19].

An important property for the study of complex networks is its degree distribution. Let $p(k) = n(k)/n$, where $n(k)$ is the number of nodes having degree k in the network of size n [2]. That is, $p(k)$ represents the probability that a node selected uniformly at random has degree k. The histogram of $p(k)$ versus k represents the degree distribution of the network. Determining the degree distribution of a network is a complicated task. Among the difficulties usually found we can mention the fact that sometimes the number of data points used to fit the distribution is too small and sometimes the data are very noisy. For instance, in fitting power-law distributions, the tail of the distribution, the part which corresponds to high-degrees, is usually very noisy. There are two main approaches in use for reducing this noise effect in the tail of probability distributions. One is the binning procedure, which consists in building a histogram using bin sizes which increase exponentially with degree. The other approach is to consider the cumulative distribution function (CDF) [11].

There are many local properties which are used to characterise the nodes and links of complex networks. One of the most important ones is the so-called clustering coefficient introduced by Watts and Strogatz in 1998 [47]. For a given node i the clustering coefficient is the number of triangles connected to this node $|C_3(i)|$ divided by the number of triples centred on it

$$C_i = \frac{2|C_3(i)|}{k_i(k_i - 1)}, \tag{7}$$

where k_i is the degree of the node. The average value of the clustering for all nodes in a network $\bar{C}$ has been extensively used in the analysis of complex networks

$$\bar{C} = \frac{1}{n}\sum_{i=1}^{n} C_i. \tag{8}$$

Another group of local measures for the nodes of a network are the centrality measures [17, 22, 25, 29]. These measures try to capture the notion of "importance" of nodes in networks by quantifying the ability of a node to communicate directly with other nodes, or its closeness to many other nodes or the number of pairs of nodes which need a specific node as intermediary in their communications. The simplest example of these measures is the degree of a node. A generalisation of this concept can be seen through the use of the eigenvector associated with the largest eigenvalue of the adjacency matrix of the network. This centrality, known as the eigenvector centrality, captures the influence not only of nearest neighbours but also of more distant nodes in a network [8, 9]. It can be formally defined as

$$\varphi(i) = \left(\frac{1}{\lambda_1}\mathbf{A}\varphi_1\right)_i. \tag{9}$$

The *closeness centrality* measures how close a node is from the rest of the nodes in the network and [22] is expressed mathematically as follows

$$CC(u) = \frac{n-1}{S(u)}, \tag{10}$$

where the distance sum $s(u)$ is

$$S(u) = \sum_{v \in V(G)} d(u, v). \tag{11}$$

The *betweenness centrality* quantifies the importance of a node in the communication between other pairs of nodes in the network [22]. It measures the proportion of information that passes through a given node in communications between other pairs of nodes in the network and it is defined as:

$$BC(k) = \sum_i \sum_j \frac{\rho(i,k,j)}{\rho(i,j)}, \quad i \neq j \neq k. \tag{12}$$

where $\rho(i, j)$ is the number of shortest paths from node i to node j, and $\rho(i, k, j)$ is the number of these shortest paths that pass through the node k in the network.

The *subgraph centrality* counts the number of closed walks starting and ending at a given node, which are mathematically given by the diagonal entries of $\mathbf{A}^k$. A penalisation is used for longer walks, such that they contribute less to the centrality than the shortest walks [15–18]. It is defined as:

$$EE(i) = \left(\sum_{l=0}^{\infty} \frac{\mathbf{A}^l}{l!}\right)_{ii} = (e^{\mathrm{A}})_{ii}. \tag{13}$$

Another characteristic feature of complex networks is the presence of communities of nodes which are more tightly connected among them than with the rest of the nodes in the network. In general it is considered that a community is a subset of nodes in a network for which the density of connections is significantly larger than the density of connections between them and the rest of the nodes. The reader is directed to the specialised literature to obtain information about the many methods available for detecting communities in networks [20].

The quality of a partition of a network into several communities can be measured by mean of a few indices. The most popular among these quality criteria is the so-called *modularity index*. In a network consisting of n_V partitions, $V_1, V_2, \ldots, V_{nC}$, the modularity is the sum over all partitions of the difference between the fraction of links inside each partition and the expected fraction by considering a random network with the same degree for each node [36]:

$$Q = \sum_{k=1}^{nc} \left[\frac{|E_k|}{m} - \left(\frac{\sum_{j \in V_k} k_j}{2m} \right)^2 \right], \tag{14}$$

where $|E_k|$ is the number of links between nodes in the kth partition of the network. Modularity is interpreted in the following way. If $Q = 0$, the number of intra-cluster links is not bigger than the expected value for a random network. Otherwise, $Q = 1$ means that there is a strong community structure in the network given by the partition analysed.

2 Models of Networks

It is useful when studying complex networks to use random networks as models to compare the properties we are studying. This allows us to understand whether such property is the result of any natural evolutionary process or simply a randomly appearing artefact of network generation. In a random network, a given set of nodes are connected in a random way.

2.1 The Erdös-Rényi model

The simplest model of random network was introduced by Erdös and Rényi [13] in which we start by considering n isolated nodes and with probability $p > 0$ a pair of nodes is connected by an edge. Consequently, the network is determined only by the number of nodes and edges and it can be written as $G(n, m)$ or $G(n, p)$. In Fig. 1 we illustrate some examples of Erdös-Rényi random graphs with the same number of nodes and different linking probabilities.

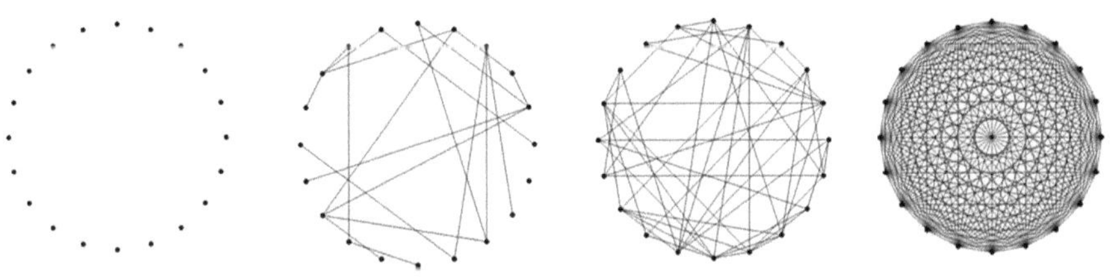

Fig. 1 Illustration of the changes of an Erdös-Rényi random network with 20 nodes and probabilities that increases from zero (*left*) to one (*right*)

A few properties of Erdös-Rényi (ER) random networks are summarised below [5, 24, 37, 48]:

(i) The expected number of edges per node:

$$\bar{m} = \frac{n(n-1)p}{2}. \quad (15)$$

(ii) The expected node degree:

$$\bar{k} = (n-1)p. \quad (16)$$

(iii) The degrees follow a Poisson distribution of the form

$$p(k) = \frac{e^{-\bar{k}}\bar{k}^k}{k!}, \quad (17)$$

as illustrated in Fig. 2:

(iv) The average path length for large n:

$$\bar{l}(H) = \frac{\ln n - \gamma}{\ln(pn)} + \frac{1}{2}, \quad (18)$$

where $\gamma \approx 0.577$ is the Euler-Mascheroni constant.

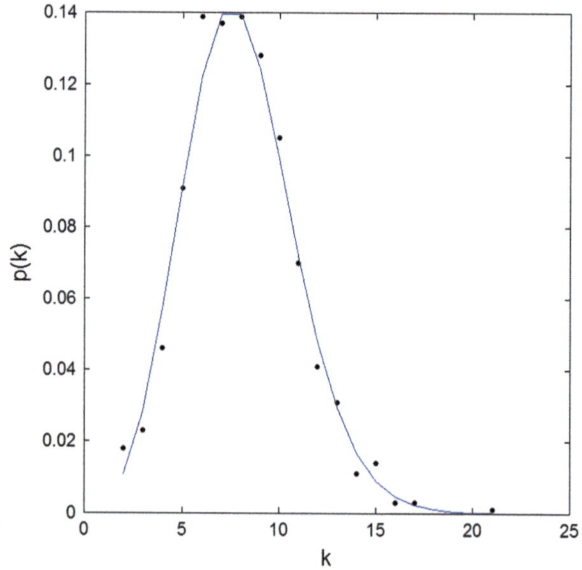

Fig. 2 Illustration of the degree distribution of an Erdös-Rényi (ER) random network with 1,000 nodes and 4,000 links. The *solid line* is the expected distribution and the *dots* represents the values for the average of 100 realizations

(v) The average clustering coefficient:

$$\bar{C} = p = \delta(G). \tag{19}$$

(vi) When increasing p, most nodes tends to be clustered in one giant component, while the rest of nodes are isolated in very small components

(vii) The structure of $G_{ER}(n,p)$ changes as a function of $p = \bar{k}/(n-1)$ giving rise to the following three stages:

(a) *Subcritical* $\bar{k} < 1$, where all components are simple and very small. The size of the largest component is $S = O(\ln n)$.

(b) *Critical* $\bar{k} = 1$, where the size of the largest component is $S = \Theta(n^{2/3})$.

(c) *Supercritical* $\bar{k} > 1$, where the probability that $(f - \varepsilon)n < S < (f + \varepsilon)n$ is 1 when $n \to \infty$ $\varepsilon > 0$ for, where $f = f(\bar{k})$ is the positive solution of the equation: $e^{-\bar{k}f} = 1 - f$. The rest of the components are very small, with the second largest having size about $\ln n$.

(viii) The largest eigenvalue of the adjacency matrix in an ER network grows proportionally to $n : \lim_{n\to\infty}(\lambda_1(\mathbf{A})/n) = p$.

(ix) The second largest eigenvalue grows more slowly than $\lambda_1 : \lim_{n\to\infty}(\lambda_2(\mathbf{A})/n^{\varepsilon}) = 0$ for every $\varepsilon > 0.5$.

(x) The smallest eigenvalue also grows with a similar relation to $\lambda_2(\mathbf{A})$: $\lim_{n\to\infty}(\lambda_n(\mathbf{A})/n^{\varepsilon}) = 0$ for every $\varepsilon > 0.5$.

(xi) The spectral density of an ER random network follows the Wigner's semicircle law, which is simply written as:

$$\rho(\lambda) = \begin{cases} \frac{\sqrt{4-\lambda^2}}{2\pi} & -2 \leq \lambda/r \leq 2,\ r = \sqrt{np(1-p)} \\ 0 & \text{otherwise.} \end{cases} \tag{20}$$

2.2 *Small-World Networks*

Despite the great usability of ER random networks as null models for studying complex networks it has been observed empirically that they do not reproduce some important properties of real-world networks. These empirical evidences can be traced back to the famous experiment carried out by Stanley Milgram in 1967 [31]. Milgram asked some randomly selected people in the U.S. cities of Omaha (Nebraska) and Wichita (Kansas) to send a letter to a target person who lives in Boston (Massachusetts) on the East Coast. The rules stipulated that the letter should be sent to somebody the sender knows personally. Although the senders and the target were separated by about 2,000 km the results obtained by Milgram were surprising because:

1. The average number of steps needed for the letters to arrive to its target was around 6.

2. There was a large group inbreeding, which resulted in acquaintances of one individual feedback into his/her own circle, thus usually eliminating new contacts.

These results transcended to the popular culture as the "small-world" phenomenon or the fact that every pair of people in the World are at "six degrees of separation" only. Practically in every language and culture we have a phrase saying that the World is small enough so that a randomly chose person has a connection with some of our friends.

The Erdös-Rényi random network reproduces very well the observation concerning the relatively small average path length, but it fails in reproducing the large group inbreeding observed. That is, the number of triangles and the clustering coefficient in the ER network are very small in comparison with those observed in real-world systems. In 1998 Watts and Strogatz [47] proposed a model that reproduces the two properties mentioned before in a simple way. Let n be the number of nodes and k be an even number, the Watt-Strogatz model starts by using the following construction (see Fig. 3):

1. Place all nodes in a circle;
2. Connect every node to its first $k/2$ clockwise nearest neighbours as well as to its $k/2$ counter-clockwise nearest neighbours;
3. With probability p rewire some of the links in the circulant graph obtained before.

The network constructed in the steps (i) and (ii) is a ring (a circulant graph), which for $k > 2$ is full of triangles and consequently has a large clustering coefficient. The average clustering coefficient for these networks is given by [4]

$$\bar{C} = \frac{3(k-2)}{4(k-1)}, \tag{21}$$

which means that $\bar{C} = 0.75$ for very large values of k.

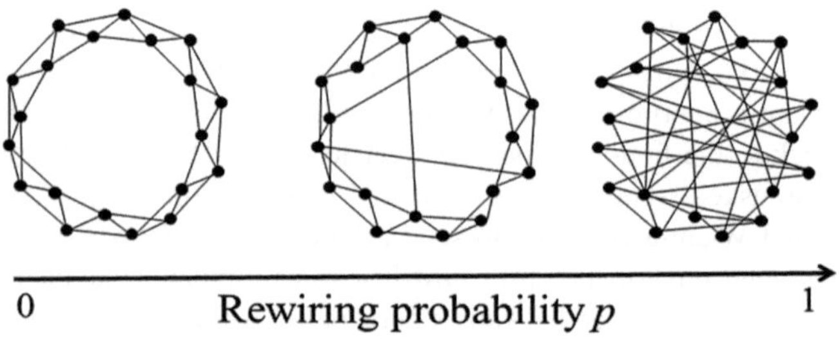

Fig. 3 Schematic representation of the evolution of the rewiring process in the Watts-Strogatz model

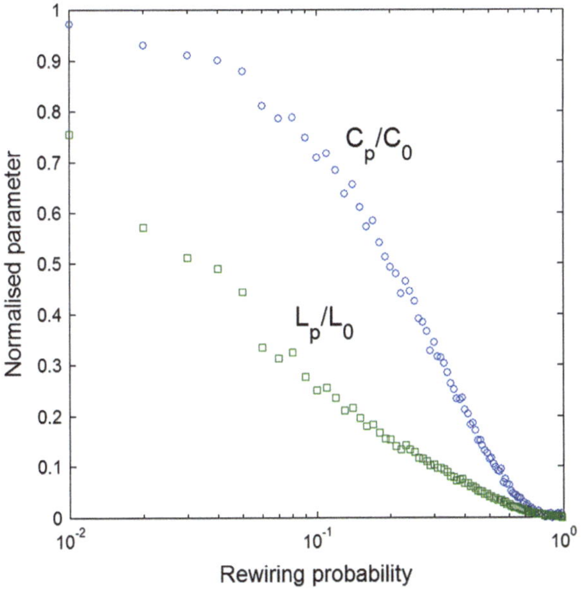

Fig. 4 Representation of the variation in the average path length and clustering coefficient with the change of the rewiring probability in the Watts-Strogatz model with 100 nodes and 5,250 links

As can be seen in Fig. 3 (left) the shortest path distance between any pair of nodes which are opposite to each other in the network is relatively large. This distance is, in fact, equal to $\lceil \frac{n}{k} \rceil$. Then

$$\bar{l} \approx \frac{(n-1)(n+k-1)}{2kn}. \tag{22}$$

This relatively large average path length is far from that of the Milgram experiment. In order to produce a model with small average path length and still having relatively large clustering, Watts and Strogatz consider the step (iii) for rewiring the links in that ring. This rewiring makes that the average path length decreases very fast while the clustering coefficient still remains high. In Fig. 4 we illustrate what happens to the clustering and average path length as the rewiring probability change from 0 to 1 in a network.

2.3 "Scale-Free" Networks

The availability of empirical data about real-world complex networks allowed to determine some of their topological characteristics. It was observed in particular that one of these characteristics deviate dramatically from what is expected from a random evolution of the system in the form of ER or WS models. This characteristic is the observed degree distribution of real-world networks. It was observed [2] that many real-world networks display some kind of fat-tailed degree distribution [21], which in many cases followed power-law fits, in contrast with the expected Poisson-like distributions of ER and WS networks. In 1999 Barabási and Albert [2] proposed a model to reproduce this important characteristic of real-world complex networks.

In the Barabási-Albert (BA) model a network is created by using the following procedure. Start from a small number m_0 of nodes. At each step add a new node u to the network and connect it to $m \leq m_0$ of the existing nodes $v \in V$ with probability

$$p_u = \frac{k_u}{\sum_w k_w}. \tag{23}$$

We can assume that we start from a connected random network of the Erdös-Rényi type with m_0 nodes, $G_{ER} = (V, E)$. In this case the BA process can be understood as a process in which small inhomogeneities in the degree distribution of the ER network growths in time. Another option is the one developed by Bollobás and Riordan [7] in which it is first assumed that $d = 1$ and that the ith node is attached to the jth one with probability:

$$p_i = \begin{cases} \frac{k_j}{1+\sum\limits_{j=0}^{i-1}} \text{if } j < i \\ \frac{1}{1+\sum\limits_{j=0}^{i-1}} \text{if } j = i \end{cases}. \tag{24}$$

Then, for $d > 1$ the network grows as if $d = 1$ until nd nodes have been created and the size is reduced to n by contracting groups of d consecutive nodes into one. The network is now specified by two parameters and we denote it by $BA(n, d)$. Multiple links and self-loops are created during this process and they can be simply eliminated if we need a simple network.

A characteristic of BA networks is that the probability that a node has degree $k \geq d$ is given by:

$$p(k) = \frac{2d(d-1)}{k(k+1)(k+2)} \sim k^{-3}, \tag{25}$$

as illustrated in Fig. 5.

This model has been generalised to consider general power-law distributions where the probability of finding a node with degree k decays as a negative power of the degree: $p(k) \sim k^{-\gamma}$. This means that the probability of finding a high-degree node is relatively small in comparison with the high probability of finding low-degree nodes. These networks are usually referred to as "scale-free" networks.

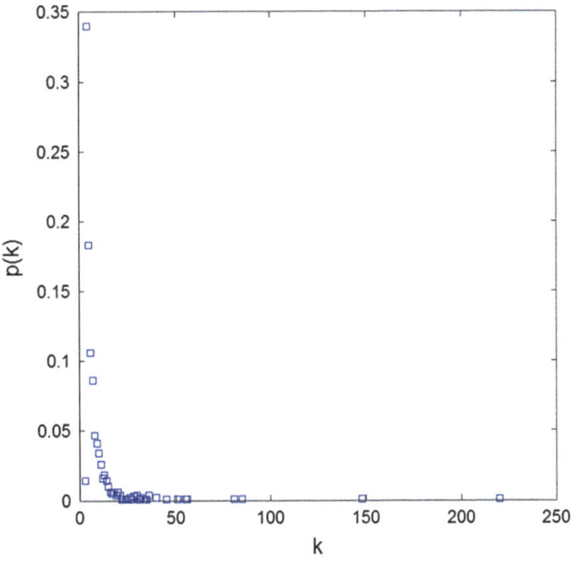

Fig. 5 Illustration of the characteristic power-law degree distribution of a network generated with the BA model

The term scaling describes the existence of a power-law relationship between the probability and the node degree: $p(k) = Ak^{-\gamma}$: multiplying the degree by a constant factor c, only produces a proportionate scaling of the probability:

$$p(k, c) = A(ck)^{-\gamma} = Ac^{-\gamma} \cdot p(k). \tag{26}$$

Power-law relations are usually represented in a logarithmic scale, leading to a straight line, $\ln p\,(k) = -\gamma \ln k + \ln A$, where $-\gamma$ is the slope and $\ln A$ the intercept of the function. Scaling by a constant factor c means that only the intercept of the straight line changes but the slope is exactly the same as before: $\ln p\,(k, c) = -\gamma \ln k - \gamma Ac$.

In the case of BA networks, Bollobás [6] has proved that for fixed values $d \geq 1$, the expected value for the clustering coefficient $\bar{C}$ is given by

$$\bar{C} \sim \frac{d-1}{8} \frac{\log^2 n}{n}, \tag{27}$$

for $n \to \infty$, which is very different from the value $\bar{C} \sim n^{-0.75}$ reported by Barabási and Albert [2] for $d = 2$.

On the other hand, the average path length has been estimated for the BA networks to be as follows [7]:

$$\bar{l} = \frac{\ln n - \ln(d/2) - 1 - \nu}{\ln \ln n + \ln(d/2)} + \frac{3}{2}, \tag{28}$$

where ν is the Euler-Mascheroni constant. This means that for the same number of nodes and average degree, BA networks have smaller average path length than their ER analogues. Other alternative models for obtaining power-law degree distributions with different exponents γ can be found in the literature [12]. In closing, using this preferential attachment algorithm we can generate random networks which are different from those obtained by using the ER method in many important aspects including their degree distributions, average clustering and average path length.

3 Dynamical Processes on Networks

Due to the fact that complex networks represent the topological skeleton of complex systems there are many dynamical processes that can take place on the nodes and links of these networks. We concentrate here in those cases where the topology of the network is static, i.e., the nodes and links do not change in time. Then, we study processes such as the consensus and synchronisation among the nodes in a network, the diffusion of epidemics through the links of a network and the propagation of beliefs by means of replication and mutation processes.

3.1 Consensus

The consensus is a dynamical process in which pairs of connected nodes try to reach agreement regarding a certain quantity of interest [39]. Then, eventually the network as a whole collapses into a consensus state, which is the state in which the differences of the quantity of interest vanish for all pairs of nodes in the system. This process is of great importance in social and engineering sciences where it models situations ranging from social consensus to spatial rendezvous and alignment of autonomous robots [39].

Let $n = |V|$ be the number of agents forming a network, the collective dynamics of the group of agents is represented by the following equations for the continuous-time case:

$$\dot{u}_i(t) = \sum_{j \sim i} \left[u_j(t) - u_i(j) \right], i = 1, \cdots, n \tag{29}$$

$$u_i(0) = z_i, z_i \in \Re$$

which in matrix form are written as

$$\dot{\mathbf{u}}(t) = -\mathbf{L}\mathbf{u}(t), \tag{30}$$

$$\mathbf{u}(0) = \mathbf{u}_0, \tag{31}$$

where $\mathbf{u}_0$ is the original distribution which may represent opinions, positions in space or other quantities with respect to which the agents should reach a consensus. The reader surely already recognized that Eqs. (30)–(31) are identical to the heat equation,

$$\frac{\partial u}{\partial t} = h\nabla^2 u, \tag{32}$$

where h is a positive constant and $\nabla^2 = -\mathbf{L}$ is the Laplace operator. In general this equation is used to model the diffusion of "information" in a physical system, where by information we can understand heat, a chemical substance or opinions in a social network.

A consensus is reached if, for all $u_i(0)$ and all $i, j=1,\ldots,n$, $|u_i(t) - u_j(t)| \to 0$ as $t \to 0$. In other words, the consensus set $A \subseteq \Re^n$ is the subspace $span\{\mathbf{1}\}$, that is

$$A = \{u \in \Re^n | u_i = u_j, \forall i, j\}. \tag{33}$$

A necessary and sufficient condition for the consensus model to converge to the consensus subspace from an arbitrary initial condition is that the network is connected.

The discrete-time version of the consensus model has the form

$$\mathbf{u}_i(t+1) = \mathbf{u}_i(t) + \varepsilon \sum_{(i,j)\in E} \mathbf{A}_{ij}\left[\mathbf{u}_j(t) - \mathbf{u}_i(t)\right] \tag{34}$$

$$\mathbf{u}(0) = \mathbf{u}_0, \tag{35}$$

where $\mathbf{u}_i(t)$ is the value of a quantitative measure on node i and $\varepsilon > 0$ is the step-size. It has been proved that the consensus is asymptotically reached in a connected graph for all initial states if $0 < \varepsilon < 1/k_{\max}$, where $k_{\max}$ is the maximum degree of the graph. The discrete-time collective dynamics of the network can be written in matrix form as [39] as

$$\mathbf{u}(t+1) = \mathbf{P}\mathbf{u}(t), \tag{36}$$

$$\mathbf{u}(0) = \mathbf{u}_0, \tag{37}$$

where $\mathbf{P} = \mathbf{I} - \varepsilon\mathbf{L}$, and $\mathbf{I}$ is the $n \times n$ identity matrix. The matrix $\mathbf{P}$ is the Perron matrix of the network with parameter $0 < \varepsilon < 1/k_{\max}$. For any connected undirected graph the matrix $\mathbf{P}$ is an irreducible, doubly stochastic matrix with all eigenvalues μ_j in the interval $[-1, 1]$ and a trivial eigenvalue of 1. The reader can find the previously mentioned concepts in any book on elementary linear algebra. The relation between the Laplacian and Perron eigenvalues is given by: $\mu_j = 1 - \varepsilon\lambda_j$.

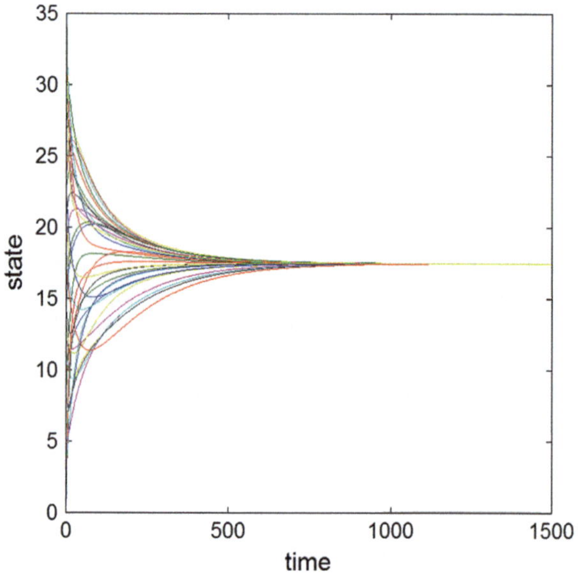

Fig. 6 Time evolution of consensus dynamics in a real-world social network with random initial states for the nodes

In Fig. 6 we illustrate the consensus process in a real-world social network having 34 nodes and 78 edges.

The solution for the consensus dynamics problem is given by

$$\mathbf{u}\left(t\right) = e^{-t\mathbf{L}}\mathbf{u}_0, \tag{38}$$

where

$$\mathbf{u}(t) = e^{-t\mu_1}\left(\varphi_1^T\mathbf{u}_0\right)\varphi_1 + e^{-t\mu_2}\left(\varphi_2^T\mathbf{u}_0\right)\varphi_2 + \cdots + e^{-t\mu_n}\left(\varphi_n^T\mathbf{u}_0\right)\varphi_n. \tag{39}$$

We remind the reader that

$$\begin{aligned} e^{-t\mathbf{L}} &= e^{-t\left(\mathrm{U}\Lambda U^T\right)} = \mathbf{U}e^{-t\boldsymbol{\Lambda}}\mathbf{U}^T \\ &= e^{-t\mu_1}\boldsymbol{\varphi}_1\boldsymbol{\varphi}_1^T + e^{-t\mu_2}\boldsymbol{\varphi}_2\boldsymbol{\varphi}_2^T + \cdots + e^{-t\mu_n}\boldsymbol{\varphi}_n\boldsymbol{\varphi}_n^T, \end{aligned} \tag{40}$$

where

$$\mathbf{U} = \begin{pmatrix} \varphi_1\left(1\right) & \varphi_2\left(1\right) & \cdots & \varphi_n\left(1\right) \\ \varphi_1\left(2\right) & \varphi_2\left(2\right) & \cdots & \varphi_n\left(2\right) \\ \vdots & \vdots & \ddots & \vdots \\ \varphi_1\left(n\right) & \varphi_2\left(n\right) & \cdots & \varphi_n\left(n\right) \end{pmatrix}, \tag{41}$$

and

$$\boldsymbol{\Lambda} = \begin{pmatrix} \mu_1 & 0 & \cdots & 0 \\ 0 & \mu_2 & \cdots & 0 \\ \vdots & \vdots & \ddots & \vdots \\ 0 & 0 & \cdots & \mu_n \end{pmatrix}, \tag{42}$$

such that $0 = \mu_1 < \mu_2 \leq \cdots \leq \mu_n$.

It is know that in a connected undirected network $\mu_1 = 0$and $\mu_j > 0, \forall j \neq 1$. Thus

$$\mathbf{u}(t) \to \left(\varphi_1^T \mathbf{u}_0\right) \varphi_1 = \frac{\mathbf{1}^T \mathbf{u}_0}{n} \mathbf{1} \qquad \text{as} \qquad t \to \infty. \tag{43}$$

Hence $\mathbf{u}(t) \to A$ as $t \to \infty$. In other words, there is a global consensus.

As μ_2 is the smallest positive eigenvalue of the network Laplacian, it dictates the slowest mode of convergence in the above equation. In other words, the consensus model converges to the consensus set in an undirected connected network with a rate of convergence that is dictated by μ_2.

As the states of the nodes evolve toward the consensus set, one has

$$\frac{d}{dt}(\mathbf{1}^T \mathbf{u}(t)) = \mathbf{1}^T \left(-\mathbf{L}\mathbf{u}(t)\right) = -\mathbf{u}(t)^T \mathbf{L}\mathbf{1} = 0, \tag{44}$$

Then

$$\mathbf{1}^T \mathbf{u}(t) = \sum_i u_i(t), \tag{45}$$

is a *constant of motion* for the consensus dynamics. Furthermore, the state trajectory generated by the consensus model converges to the projection of its initial state, in the Euclidean norm, onto the consensus space, since

$$\arg\min_{\mathbf{u} \in A} \|\mathbf{u} - \mathbf{u}_0\| = \frac{\mathbf{1}^T \mathbf{u}_0}{\mathbf{1}^T \mathbf{1}} \mathbf{1} = \frac{\mathbf{1}^T \mathbf{u}_0}{n} \mathbf{1}. \tag{46}$$

As can be seen in Fig. 7 the trajectory of the consensus model retains the centroid of the node's states as its constant of motion.

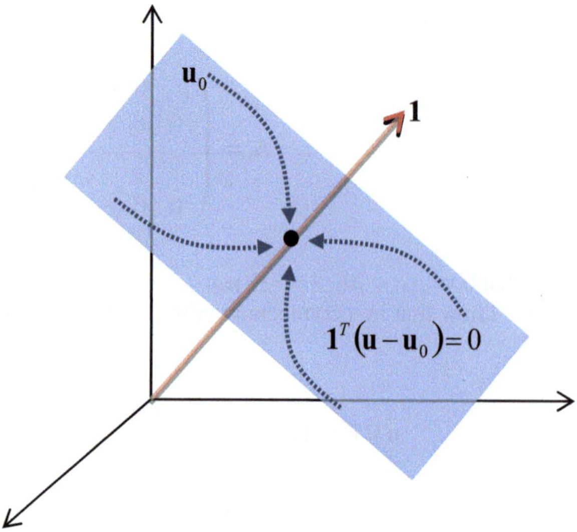

Fig. 7 Illustration of the trajectory of the consensus model (adapted from [30])

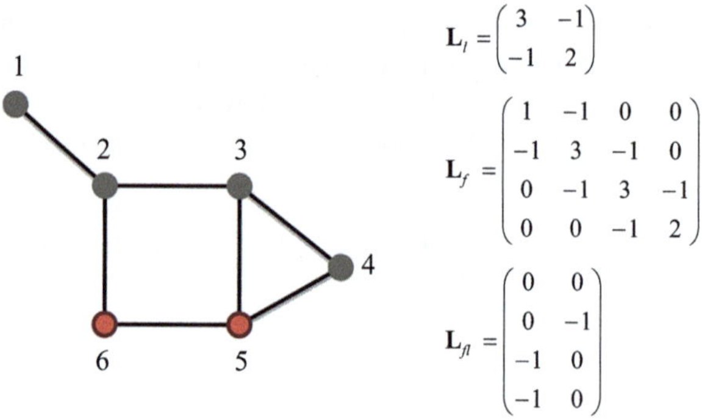

Fig. 8 Illustration of the different Laplacian matrices for a simple network with two leaders (nodes 5 and 6) and four followers (nodes 1–4)

3.1.1 Consensus with Leaders

In many real-world scenarios a group of nodes in the network act as leaders that drive the dynamics of the system. The set of nodes is then divided into leaders and followers. The state of the leaders does not change during the consensus process and the quantity of interest in the process for the followers converges to the convex hull formed by the leaders. In this case the consensus dynamics can be written as

$$\begin{bmatrix} \dot{\mathbf{u}}_f \\ \dot{\mathbf{u}}_l \end{bmatrix} = - \begin{bmatrix} \mathbf{L}_f & \mathbf{L}_{fl} \\ \mathbf{0} & \mathbf{0} \end{bmatrix} \begin{bmatrix} \mathbf{u}_f \\ \mathbf{u}_l \end{bmatrix} - \begin{bmatrix} \mathbf{0} \\ \mathbf{I} \end{bmatrix} \mathbf{u}, \tag{47}$$

where the vector $\mathbf{u}$ and the Laplacian matrix have been split into their parts corresponding to the leaders and followers. The Laplacian $\mathbf{L}_{fl}$ corresponds to the interaction between leaders and followers in the network (see Fig. 8). The dynamics of the followers can then be expressed as:

$$\dot{\mathbf{u}}_f = -\mathbf{L}_f \mathbf{u}_f - \mathbf{L}_{fl} \mathbf{u}_l, \tag{48}$$

We know that $\mathbf{L}$ is positive semi-definite and if the network is connected we have that $N(\mathbf{L}) = \text{span}\{\mathbf{1}\}$. Now, since

$$\mathbf{u}_f^T \mathbf{L}_f \mathbf{u}_f = \begin{bmatrix} \mathbf{u}_f^T & 0 \end{bmatrix} \mathbf{L}_f \begin{bmatrix} \mathbf{u}_f \\ 0 \end{bmatrix}, \tag{49}$$

and

$$\begin{bmatrix} \mathbf{u}_f^T & 0 \end{bmatrix} \notin N(\mathbf{L}), \tag{50}$$

we have that

$$\begin{bmatrix} \mathbf{u}_f^T & 0 \end{bmatrix} \mathbf{L}_f \begin{bmatrix} \mathbf{u}_f \\ 0 \end{bmatrix} > 0, \forall \mathbf{u}_f \in \Re^{n_f}. \tag{51}$$

That is, if the network G is connected then $\mathbf{L}_f$ is positive definite, $\mathbf{L}_f \succ 0$. This means that $\mathbf{L}_f^{-1}$ exists and $\mathbf{u}_f = -\mathbf{L}_f^{-1} \mathbf{L}_{fl} \mathbf{u}_l$ is well defined. All of this implies that given fixed leader opinions $\mathbf{u}_l$, the equilibrium point under the leader-follower dynamics is

$$\mathbf{u}_f = -\mathbf{L}_f^{-1} \mathbf{L}_{fl} \mathbf{u}_l, \tag{52}$$

which is globally asymptotically stable.

An example of a consensus with leaders-followers dynamics in a network is illustrated in Fig. 9.

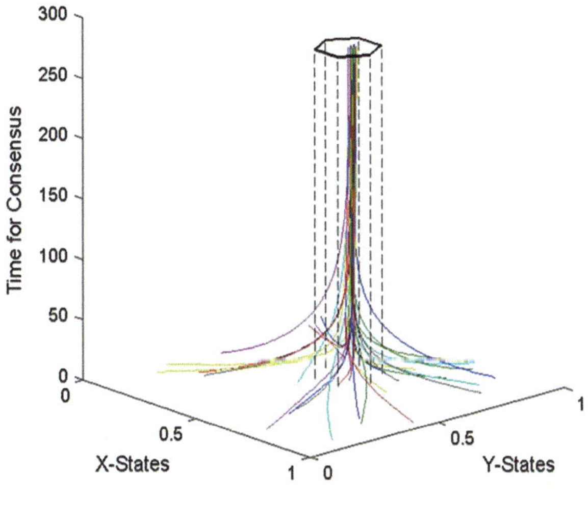

Fig. 9 Illustration of a consensus dynamics in a random network. Six nodes are selected randomly as leaders and the rest are followers which are represented as an hexagon. Two quantities are designated here as of interest for the consensus, which are designated as X- and Y-states

3.1.2 Consensus in Directed Networks

Let us now consider the case of a weighted directed network, like the one illustrated in Fig. 10.

In this case the equations governing the dynamical process can be written as

$$\begin{aligned}\dot{u}_1(t) &= 0,\\ \dot{u}_2(t) &= w_{21}(u_1(t) - u_2(t)),\\ \dot{u}_3(t) &= w_{32}(u_2(t) - x_3(t)) + w_{34}(u_4(t) - u_3(t)),\\ \dot{u}_4(t) &= w_{42}(u_2(t) - u_4(t)) + w_{43}(u_3(t) - u_4(t)).\end{aligned} \tag{53}$$

In matrix form they are

$$\dot{\mathbf{u}}(t) = \begin{pmatrix} 0 & 0 & 0 & 0 \\ -w_{21} & w_{21} & 0 & 0 \\ 0 & -w_{32} & w_{32} + w_{34} & -w_{34} \\ 0 & -w_{42} & -w_{43} & w_{42} + w_{43} \end{pmatrix} \mathbf{u}(t). \tag{54}$$

This equation is similar to the consensus dynamics model that we have considered before and can be written as [30]

$$\dot{\mathbf{u}}(t) = -\mathbf{L}(D)\mathbf{u}(t), \mathbf{u}(0) = \mathbf{u}_0, \tag{55}$$

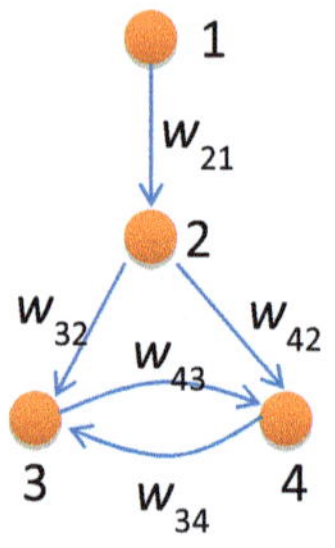

Fig. 10 Illustration of a weighted directed network

where

$$\mathbf{L}(D) = \mathbf{Diag}\left(\mathbf{A}^T\mathbf{1}\right) - \mathbf{A}^T = \begin{pmatrix} 0 & 0 & 0 & 0 \\ -w_{21} & w_{21} & 0 & 0 \\ 0 & -w_{32} & w_{32}+w_{34} & -w_{34} \\ 0 & -w_{42} & -w_{43} & w_{42}+w_{43} \end{pmatrix}, \quad (56)$$

and

$$\mathbf{A} = \begin{pmatrix} 0 & w_{21} & 0 & 0 \\ 0 & 0 & w_{32} & w_{42} \\ 0 & 0 & 0 & w_{43} \\ 0 & 0 & w_{34} & 0 \end{pmatrix}. \quad (57)$$

Let us now introduce a few definitions and results which will help to understand when the system represented by a directed network converges to the consensus set, i.e., when there is a global consensus in a directed network.

We start by introducing the concept of rooted out-branching subgraph [30]. A rooted out-branching subgraph (ROS) is a directed subgraph such that

1. It does not contain a directed cycle and
2. It has a vertex v_r (root) such that for every other vertex v there is a directed path from v_r to v.

An example of ROS is illustrated in Fig. 11 for a small directed network.

A directed network contains a ROS if and only if $\mathbf{rank}\left(\mathbf{L}(D)\right) = n - 1$. In that case, the nullity of the Laplacian $N(\mathbf{L}(D))$ is spanned by the all-ones vector. It is

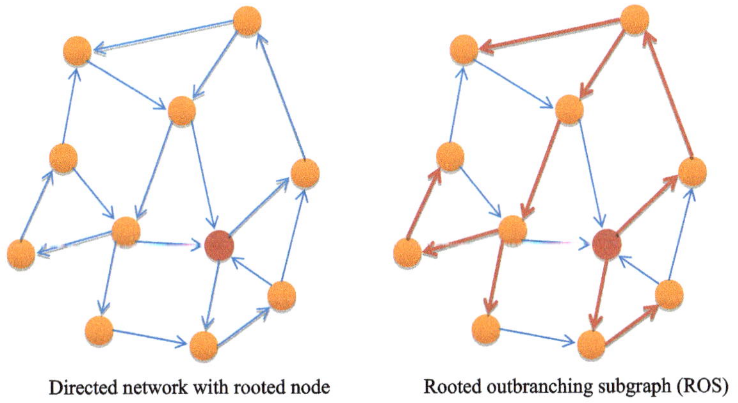

Fig. 11 Illustration of a ROS for a directed network in which a node has been marked in *red*. The ROS, represented in the *left part* of the figure, is constructed for the marked node. The figure has been adapted from [30]

known that for a directed network with n nodes the spectrum of $\mathbf{L}(D)$ lies in the region:

$$\left\{z \in C \mid \left|z - \hat{k}_{in}\right| \leq \hat{k}_{in}\right\}, \tag{58}$$

where $\hat{k}_{in}$ is the maximum (weighted) in-degree in D. That is, for every directed network, the eigenvalues of $\mathbf{L}(D)$ have non-negative real parts. That is, the eigenvalues of $\mathbf{L}(D)$ are contained in the Geršgorin disk of radius $\hat{k}_{in}$ centred at $\hat{k}_{in}$.

Let $\mathbf{L}(D) = \mathbf{PJ}(\boldsymbol{\Lambda})\mathbf{P}^{-1}$ be the Jordan decomposition of $\mathbf{L}(D)$. When D contains a ROS, the non-singular matrix $\mathbf{P}$ can be chosen such that

$$\mathbf{J}(\boldsymbol{\Lambda}) = \begin{pmatrix} 0 & 0 & \cdots & 0 \\ 0 & J(\mu_2) & \cdots & 0 \\ 0 & 0 & \cdots & 0 \\ \vdots & \vdots & \vdots & \vdots \\ 0 & \cdots & 0 & J(\mu_n) \end{pmatrix}, \tag{59}$$

where μ_i $(i = 2, \ldots, n)$ have positive real parts, and $J(\mu_i)$ is the Jordan block associated with eigenvalue μ_i. Consequently,

$$\lim_{t \to \infty} e^{-t\mathbf{J}(\boldsymbol{\Lambda})} = \begin{pmatrix} 1 & 0 & \cdots & 0 \\ 0 & 0 & \cdots & 0 \\ 0 & 0 & \cdots & 0 \\ \vdots & \vdots & \vdots & \vdots \\ 0 & \cdots & 0 & 0 \end{pmatrix}, \tag{60}$$

and $\lim_{t \to \infty} e^{-t\mathbf{L}} = \mathbf{p}_1\mathbf{q}_1^T$, where $\mathbf{p}_1$ and $\mathbf{q}_1^T$ are, respectively, the first column of $\mathbf{P}$ and the first row of $\mathbf{P}^{-1}$, that is, where $\mathbf{p}_1\mathbf{q}_1^T = 1$.

Then, finally we have that for a directed network D containing a ROS, the state trajectory generated by the consensus dynamic model, initialized from $\mathbf{u}_0$, satisfies $\lim_{t \to \infty} \mathbf{u}(t) = \left(\mathbf{p}_1\mathbf{q}_1^T\right)\mathbf{u}_0$, where $\mathbf{p}_1$ and $\mathbf{q}_1^T$, are, respectively, the right and left eigenvectors associated with the zero eigenvalue of $\mathbf{L}(D)$, normalized such that $\mathbf{p}_1\mathbf{q}_1^T = 1$. *As a result, one has* $\mathbf{u}(t) \to A$, i.e., there is a global consensus, *for all initial conditions if and only if D contains a rooted out-branching.*

Two examples from real-world directed networks are illustrated in Figs. 12 and 13.

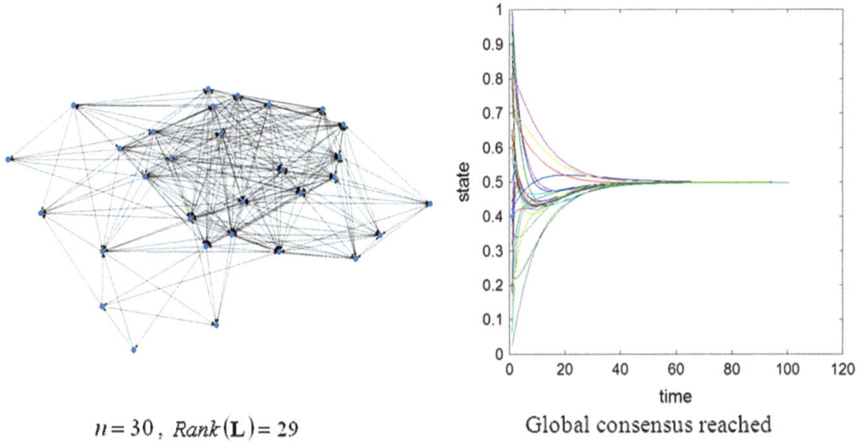

Fig. 12 Illustration of the consensus dynamics in the directed network representing the food web of Coechella Valley, consisting of 30 species and their directed trophic relations. The rank of the Laplacian matrix is $n - 1$ and a global consensus is reached in the system

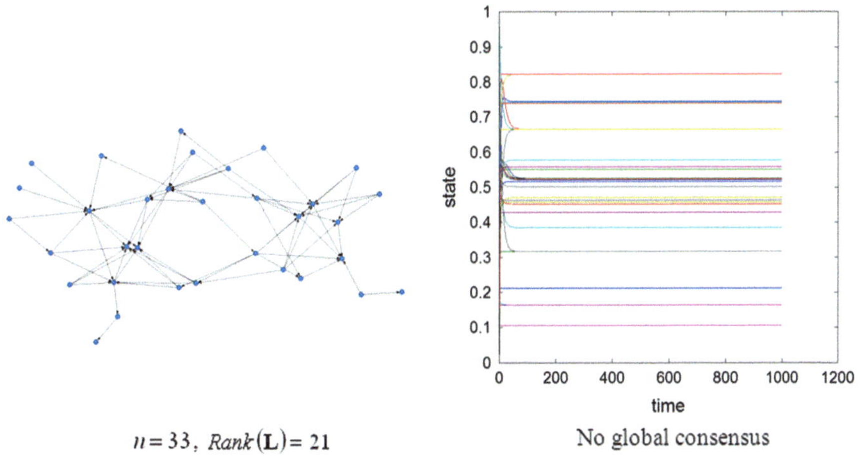

Fig. 13 Illustration of the consensus dynamics in the directed network representing the food web of Chesapeake Bay, consisting of 33 species and their directed trophic relations. The rank of the Laplacian matrix is not $n - 1$ and a global consensus is not reached in the system

3.2 Synchronization in Networks

Synchronization is a phenomenon that appears very frequently in many natural and man-made systems in which a collection of oscillators coupled to each other [1, 10]. They include animal and social behaviour, neurons, cardiac pacemaker cells, among others. In this case the network $G = (V, E)$ represents the couple oscillators, where each node is an n-dimensional dynamical systems described by

$$\dot{x}_i = f(x_i) + c\sum_{j=1}^{n} L_{ij}H(t)x_j, \quad i = 1, \ldots, n, \tag{61}$$

where $x_i = (x_{i1}, x_{i2}, \ldots, x_{iN}) \in \Re^n$ is the state vector of the node i, c is a constant representing the coupling strength, $f(\cdot) : \Re^n \to \Re^n$ is a smooth vector valued function which defines the dynamics, $H(\cdot) : \Re^n \to \Re^n$ is a fixed output function also known as outer coupling matrix, t is the time and L_{ij} are the elements of the Laplacian matrix of the network (sometimes the negative of the Laplacian matrix is taken here). The synchronised state of the network is achieved if

$$x_1(t) = x_2(t) = \cdots = x_n(t) \to s(t), \quad \text{as} \quad t \to \infty. \tag{62}$$

Now, we consider that each entry of the state vector of the network is perturbed by a small perturbation ξ_i, such that we can write $x_i = s + \xi_i$ ($\xi_i << s$). In order to analyse the stability of the synchronised manifold $x_1 = x_2 = \ldots = x_n$, we expand the terms in (61) as

$$f(x_i) \approx f(s) + \xi_i f'(s), \tag{63}$$

$$H(x_i) \approx H(s) + \xi_i H'(s), \tag{64}$$

where the primes refers to the derivatives respect to s. In this way, the evolution of the perturbations is determined by:

$$\dot{\xi}_i = f'(s)\,\xi_i + c\sum_j \left[L_{ij}H'(s)\right]\xi_j. \tag{65}$$

The eigenvectors of the Laplacian matrix are an appropriate set of linear combinations of the perturbations and we can decouple the system of equations for the perturbations by using such eigenvectors. Let ϕ_j be an eigenvector of the Laplacian matrix of the network associated with the eigenvalue μ_j. Then

$$\dot{\phi}_i = \left[f'(s) + c\mu_i H'(s)\right]\phi_i. \tag{66}$$

The solution of these decoupled equations can be obtained by considering that at short times the variations of s are small enough. In this case we have

$$\dot{\phi}_i(t) = \phi_i^0 \exp\left\{\left[f'(s) + c\mu_i H'(s)\right]t\right\}, \tag{67}$$

where ϕ_i^0 is the initially imposed perturbation.

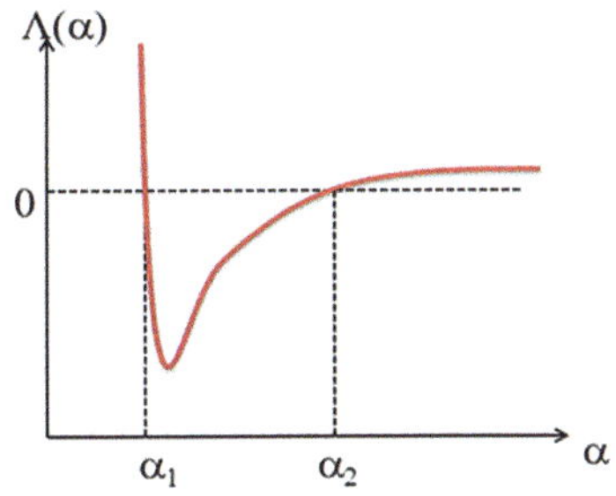

Fig. 14 Schematic representation of the typical behaviour of the master stability function

We now consider the term in the exponential of (67), $\Lambda_i = f'(s) + c\mu_i H'(s)$. If $f'(s) > c\mu_i H'(s)$, the perturbations will increase exponentially, while if $f'(s) < c\mu_i H'(s)$ they will decrease exponentially. So, the behaviour of the perturbations in time is controlled by the magnitude of μ_i. Then, the stability of the synchronised state is determined by the *master stability function*:

$$\Lambda(\alpha) \equiv \max_s \left[f'(s) + \alpha H'(s) \right], \tag{68}$$

which corresponds to a large number of functions f and H is represented in Fig. 14.

As can be seen the necessary condition for stability of the synchronous state is that $c\mu_i$ is between $\alpha_1$and α_2, which is the region where $\Lambda(\alpha) < 0$. Then, the condition for synchronization is [3]:

$$Q := \frac{\mu_N}{\mu_2} < \frac{\alpha_2}{\alpha_1}. \tag{69}$$

That is, the synchronisability of a network is favoured by a small eigenratio Q which indeed depends only on the topology of the network. For instance, in the WS model with rewiring probability $p = 0$, i.e., a circulant network of n nodes with degree $k = 2r$, $r >> 1$, $r << n$, the largest and second smallest eigenvalues of the Laplacian matrix are

$$\mu_N \sim (2r + 1)(1 + 2/3\pi), \tag{70}$$

$$\mu_2 \sim 2\pi^2 r (r + 1)(2r + 1) / \left(3n^2\right), \tag{71}$$

such that the eigenratio is given by

$$Q := \frac{\mu_1}{\mu_{n-1}} \sim \frac{n^2}{r(r + 1)}. \tag{72}$$

This means that the synchronizability of this network is very bad as for a fixed r, $Q \to \infty$ as $n \to \infty$.

However, in the small-world regime the eigenratio is given by [33]

$$Q \cong \frac{np + \sqrt{2p\,(1-p)\,n\log n}}{np - \sqrt{2p\,(1-p)\,n\log n}}, \tag{73}$$

which means that $Q \to 1$ as $n \to \infty$. Indeed, small-world networks are expected to show excellent synchronisability according to their low values of the eigenratio.

The effect of the degree distribution can also be analysed by writing [35]

$$\dot{x}_i = F\,(x_i) + \frac{\sigma}{k_i^{\beta}} \sum_j L_{ij} H\,(x_i), \tag{74}$$

where β is a parameter and k_i is the degree of the corresponding node. Then the coupling matrix can be written as

$$\mathbf{C} = \mathbf{K}^{-\beta}\mathbf{L}, \tag{75}$$

which allow us to write

$$\det\left(\mathbf{K}^{-\beta}\mathbf{L} - \lambda\mathbf{I}\right) = \det\left(\mathbf{K}^{-\beta/2}\mathbf{L}\mathbf{K}^{-\beta/2} - \lambda\mathbf{I}\right), \tag{76}$$

indicating that the coupling matrix is real and nonnegative.

Now, let us write

$$\begin{aligned} \dot{x}_i &= F\,(x_i) + \frac{\sigma}{k_i^{\beta}}\left[k_i H\,(x_i) - \sum_{j\sim i} H\left(x_j\right)\right] \\ &= F\,(x_i) - \sigma k_i^{1-\beta}\left[\bar{H}_i - H\,(x_i)\right], \end{aligned} \tag{77}$$

where

$$\bar{H}_i = \sum_{j\sim i} H\left(x_j\right)/k_i. \tag{78}$$

Let us now consider that the network is random and that the system is close to the synchronised state s. In this case, $\bar{H}_i \approx H(s)$ and we can write

$$\dot{x}_i = F\,(x_i) - \sigma k_i^{1-\beta}\left[H\,(s) - H\,(x_i)\right], \tag{79}$$

which implies that the condition for synchronisability is

$$\alpha_1 < \sigma k_i^{1-\beta} < \alpha_2, \forall i. \tag{80}$$

That is, as soon as one node has degree different from the others the network is more difficult to synchronise. In other words, as soon as a network departs from regularity it is more difficult to synchronise. In fact, the eigenratio in this case is given by

$$Q = \begin{cases} \left(\frac{k_{\max}}{k_{\min}}\right)^{1-\beta} \text{if} \beta \leq 1, \\ \left(\frac{k_{\min}}{k_{\max}}\right)^{1-\beta} \text{if} \beta \geq 1. \end{cases} \tag{81}$$

Consequently, the minimum value of the eigenratio is obtained for $\beta = 1$.

3.2.1 The Kuramoto Model

In real-world situations it is frequent to find that the oscillators are not identical. The Kuramoto model simulates this kind of situations [28]. For describing this model let us consider n planar rotors with angular phase x_i and natural frequency ω_i coupled with strength K and evolving according to:

$$\dot{x}_i = \omega_i + K \sum_{j \sim i} \sin\left(x_i - x_j\right), \tag{82}$$

where $\sum_{j \sim i}$ represents the sum over pairs of adjacent nodes. Using the incidence matrix ∇ (3) we can write (82) in matrix-vector form as follows

$$\dot{\mathbf{x}} = \omega + K \nabla \sin\left(\nabla^T \mathbf{x}\right). \tag{83}$$

The level of synchronisation is quantified by the order parameter:

$$r(t)\, e^{i\psi(t)} = \frac{1}{n} \sum_{j=1}^{n} e^{ix_j(t)}. \tag{84}$$

It represents the collective motion of the group of planar oscillators in the complex plane. If we represent each oscillator phase x_j as a point moving around a unit circle in the complex plane, then the radius measures the coherence and $\psi(t)$ is the average phase of the rotors (see Fig. 15). The synchronisation is then observed when $r(t)$ is non-zero for a group of oscillators and $r(t) \to 1$ indicates a collective movement of all the oscillators with almost identical phases.

The Kuramoto model is frequently solved by considering a complete network of oscillators, i.e., each pair of oscillators is connected, coupled with the same strength

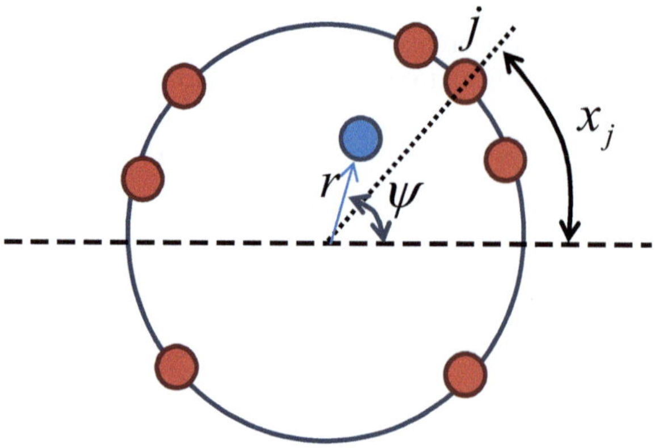

Fig. 15 Representation of the Kuramoto model where the phases correspond to points moving around a unit circle in the complex plane. The *blue circle* represents the centre of mass of these points, which is defined by the order parameter

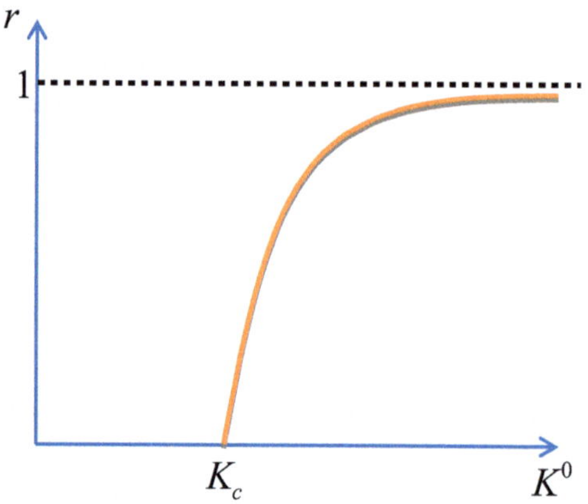

Fig. 16 Mean-field results of the Kuramoto model

$K = K^0/n$, with finite K^0. Then, by multiplying both sides of the order parameter by e^{-ix_l} and taking imaginary parts we obtain [45]:

$$r \sin(\psi - x_l) = \frac{1}{n}\sum_{j=1}^{n} \sin\left(x_j - x_l\right). \tag{85}$$

So that

$$\dot{x}_l = \omega_l + K^0 r \sin(x_l - \psi), \tag{86}$$

in which the interaction term is given by a coupling with the mean phase ψ and the intensity is proportional to the coherence r.

The results can be graphically analysed by considering Fig. 16. It indicates the existence of a critical point K_c, below which there is no synchronization, i.e., $r = 0$. Above this critical point a finite fraction of the oscillators synchronise, and when

$n \to \infty$ and $t \to \infty$ becomes $(K^0 - K_c)^\beta$ with K_c depending on the natural frequencies and $\beta = 1/2$. Thus, in general

$$r = \begin{cases} 0 & K^0 < K_c, \\ (K^0 - K_c)^\beta & K^0 \geq K_c. \end{cases} \tag{87}$$

In small-world networks it has been empirically found that the order parameter is scaled as

$$r(n, K) = n^{-\beta/\nu} F\left[(K - K_c)\, n^{1/\nu}\right], \tag{88}$$

where F is a scaling function and ν describes the divergence of the typical correlation size $(K - K_c)^{-\nu}$. These results indicate that in the Watts-Strogatz model there is always a rewiring probability p for which a finite K_c exists and the network can be fully synchronised as illustrated in Fig. 17. It has been empirically found that the values of β and ν are fully compatible with those expected from the mean field model.

In the case of oscillators with varying degree it has been found that the critical coupling parameter scales as

$$K_c = c\frac{\langle k \rangle}{\langle k^2 \rangle}, \tag{89}$$

where $\langle \cdot \rangle$ indicates average and c is a constant that depends on the distribution of individual frequencies. Because the quantity $\langle k^2 \rangle / \langle k \rangle$ indicates the level of heterogeneity of the degree distribution of the network, it is then clear that if the network has very large degree heterogeneity the synchronisation threshold goes to zero as $n \to \infty$.

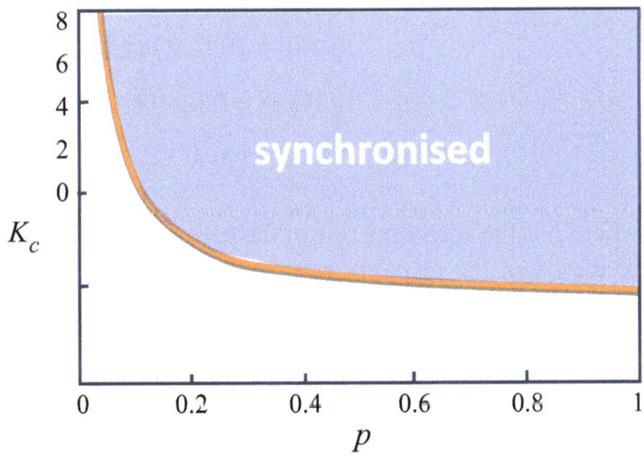

Fig. 17 Relation of the critical coupling parameter K_c with the rewiring probability p in the Watts-Strogatz model of small-world networks

3.3 Epidemics on Networks

The study of epidemics in complex networks has become a major area of research at the intersection of network theory and epidemiology. In general, these models are extensions of the classical models used in epidemiology which consider the influence of the topology of a network on the propagation of an epidemic [26]. The simplest model assumes that an individual who is susceptible (S) to an infection could become infected (I). In a second model the infected individual can also recover (R) from infection. The first model is known as a SI model, while the second is known as a SIR model. In a third model, known as SIS, an individual can be re-infected, so that infections do not confer immunity on an infected individual. Finally, a model known as SIRS allows for recovery and reinfection as an attempt to model the temporal immunity conferred by certain infections. We describe here two of the most frequently analysed epidemic models.

Let us start by considering a random uncorrelated network $G = (V, E)$ with degree distribution $P(k)$, where a group of nodes $S \subseteq V$ are considered susceptible and they can be infected by direct contact with infected individuals. By uncorrelated we mean that the probability that a node with degree k is connected to a node of degree k' is independent of k. Let n_k be the number of nodes with degree $k$in the network, and S_k, I_k, R_k be the number of such nodes which are susceptible, infected and recovered, respectively. Then, $s_k = \frac{S_k}{n_k}$, $x_k = \frac{I_k}{n_k}$ and $r_k = \frac{R_k}{n_k}$ represent the densities of susceptible, infected and recovered individuals with degree k in the network, respectively. The global averages are then given by

$$s = \sum_k s_k P(k), \quad x = \sum_k x_k P(k), \quad r = \sum_k r_k P(k). \tag{90}$$

The evolution of these probabilities in time is governed by the following equations that define the model:

$$\dot{x}_k(t) = \beta k s_k(t) \rho_k(t) - \nu x_k(t), \tag{91}$$

$$\dot{s}_k(t) = -\beta k s_k(t) \rho_k(t), \tag{92}$$

$$\dot{r}_k(t) = \nu x_k(t), \tag{93}$$

where β is the spreading rate of the pathogen, $\nu$is the probability that a node recovers or dies, i.e., the recovery rate, and $\rho_k(t)$ is the density of infected neighbours of nodes with degree k. The initial conditions for the model are: $x_k(0) = 0$, $r_k(0) = 0$ and $s_k(0) = 1 - x_k(0)$. This model is known as the **SIR** (*susceptible-infected-recovered*) model, where there are three compartments as sketched in Fig. 18:

Fig. 18 Diagrammatic representation of a SIR model

When the initial distribution of infected nodes is homogeneous and very small, i.e.,$x_k(0) = x_0 \to 0$, and $s_k(0) \approx 1$ the solution of the model is given by

$$s_k(t) = e^{-\beta k \phi(t)}, \tag{94}$$

$$r_k(t) = \nu \int_0^t x_k(\tau) d\tau, \tag{95}$$

where

$$\phi(t) = \frac{1}{\nu \langle k \rangle} \sum_k (k-1) P(k) r_k(t). \tag{96}$$

The evolution of $\phi(t)$ in time is given by

$$\dot{\phi}(t) = 1 - \langle k \rangle^{-1} - \nu \phi(t) - \langle k \rangle^{-1} \sum_k (k-1) P(k) s_k(t), \tag{97}$$

which allows us to obtain the total epidemic prevalence $r_\infty = \sum_k P(k) r_k(t \to \infty)$ as

$$r_\infty = \sum\nolimits_k P(k)(1 - s_k(t \to \infty)). \tag{98}$$

Using geometric arguments it can be shown that the prevalence of the infection at an infinite time is larger than zero, i.e., $r_\infty > 0$ if

$$\beta \langle k \rangle^{-1} \sum_k k(k-1) \geq \nu, \tag{99}$$

which defines the following *epidemic threshold* [34]

$$\frac{\beta}{\nu} = \frac{\langle k \rangle}{\langle k^2 \rangle - \langle k \rangle}. \tag{100}$$

Below this threshold the epidemic dies out, $r_\infty = 0$, and above it there is a finite prevalence $r_\infty > 0$

The second model that we consider here is the **SIS** (susceptible-infected-susceptible) model, in which the general flow chart of the infection can be represented as in Fig. 19.

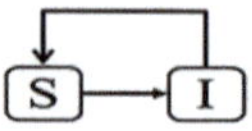

Fig. 19 Diagrammatic representation of a SI model

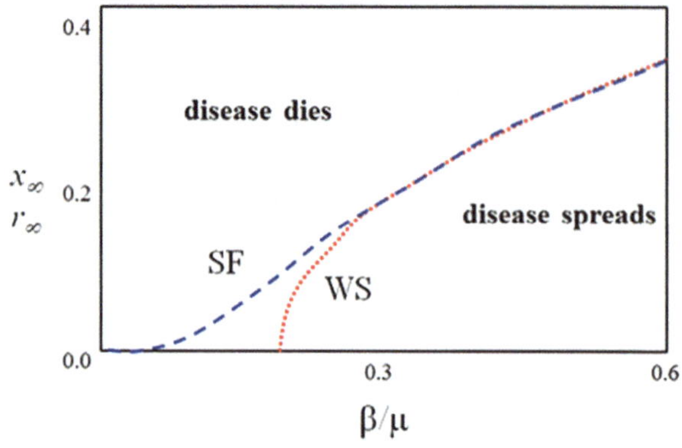

Fig. 20 Illustration of the total prevalence of disease in both the SIR and SIS models in networks with homogeneous (WS) and heterogeneous (SF) degree distributions

The SIS model can be written mathematically as follow

$$\dot{x}_k(t) = \beta k [1 - x_k(t)] \rho_k(t) - \nu x_k(t), \tag{101}$$

where the term ν represents here the rate at which individuals recover from infection and become susceptible again.

The number of infected individuals at the infinite time limit can be obtained by imposing the stationary condition $\dot{x}_k(t) = 0$. Then,

$$x_k = \frac{\beta k \rho_k}{\mu_k + \beta k \rho_k}. \tag{102}$$

Following a similar procedure as for the SIR model, the epidemic threshold for the SIS model is obtained as [43]

$$\frac{\beta}{\nu} = \frac{\langle k \rangle}{\langle k^2 \rangle}. \tag{103}$$

A schematic illustration of the phase diagram for both SIR and SIS models is given in Fig. 20. As can be seen for networks with heterogeneous degree distributions, e.g., Scale Free(SF) networks, there is practically no epidemic threshold because $\frac{\beta}{\nu}$ is always larger than x_∞ for SIS or r_∞ for SIR. This contrasts with the existence of an epidemic threshold for networks with more homogeneous degree distributions like the ones generated with the Watts-Strogatz (WS) model.

4 Replicator-Mutator Dynamics

We turn here our attention to the so-called *replicator-mutator model*, which describes the dynamics of complex adaptive systems, such as in population genetics, autocatalytic reaction networks and the evolution of languages [27, 38, 41, 42]. Let us consider a series $V = \{v_1, v_2, \ldots, v_n\}$ of n agents such that each agent plays one of the n behaviours or strategies available at $\mathbf{b} = \{b_1, b_2, \ldots, b_n\}$. Let $\mathbf{x} = \{x_1, x_2, \ldots, x_n\}$ be a vector such that $0 \leq x_i \leq 1$ is the fraction of individuals using the ith behaviour. Furthermore, we assume that $\mathbf{x}^T\mathbf{1} = 1$, where $\mathbf{1}$ is an $n \times 1$ all-ones vector. Let us assume that if an individual change from behaviour b_j to behaviour b_i she is rewarded with a payoff $A_{ij} \geq 0$. Therefore, we can consider that two individuals v_i and v_j that can change from one behaviour to another are connected to each other in such a way that the connection has a weight equal to $A_{ij} > 0$. The set of connections is $E \subseteq V \times V$, such that we can define a graph $G = (V, E, W, \phi)$, where W is the set of rewards, and $\phi : W \rightarrow E$ is a surjective mapping of the set of rewards onto the set of connections. The rewards between pairs of individuals are then defined by the weighted adjacency matrix $\mathbf{A}$ of the graph. In this case we consider that every node has a self-loop such that $A_{ii} = 1, \forall i \in V$.

The fitness f_i of behaviour b_i is usually assumed to have the following form

$$f_i = f_0 + \sum_{j=1}^{n} A_{ij} x_j, \tag{104}$$

where f_0 is the base fitness, here assumed to be zero. The fitness vector can then be obtained as

$$\mathbf{f} = \mathbf{A}\mathbf{x}, \tag{105}$$

and the average fitness of the population is defined to be

$$\phi = \sum_{j=1}^{n} f_i x_j = \mathbf{x}^T \mathbf{A}\mathbf{x}. \tag{106}$$

Let us now introduce the probability q_{ij} that an individual having behaviour b_i ends up with behaviour b_j. Thus, q_{ii} gives the reliability for an individual with strategy b_i to remain with it. Such probabilities can be represented through the row-stochastic matrix $\mathbf{Q}$ such that

$$\sum_{j=1}^{n} q_{ij} = 1. \tag{107}$$

The rate of change of the fraction of individuals using the ith behaviour with respect to the time is modelled by the *replicator-mutator equation*

$$\dot{x}_i = \sum_{j=1}^{n} x_j f_j q_{ji} - x_i \phi$$
$$= x_i \left(f_i q_{ji} - \phi\right) + \sum_{j \neq i} f_j x_j q_{ji}, \tag{108}$$
$$i = 1, 2, \ldots, n.$$

This is a generalization of the replicator dynamics widely studied in game theory and population dynamics, where $\dot{x}_i$ is assumed to be proportional to x_i and to how far the fitness of that individual exceeds the average fitness of the whole population:

$$\dot{x}_i = x_i \left(f_i - \varphi\right), i = 1, 2, \ldots, n. \tag{109}$$

In matrix form Eqs. (108) and (109) are expressed respectively as

$$\dot{\mathbf{x}} = \mathbf{Q}^T \mathbf{F}\mathbf{x} - \varphi\mathbf{x}, \tag{110}$$

$$\dot{\mathbf{x}} = \mathbf{F}\mathbf{x} - \varphi\mathbf{x}, \tag{111}$$

where $\mathbf{F} = diag\left(\mathbf{f}\right)$. Then, it is obvious that (111) is the particular case in which $\mathbf{Q} = \mathbf{I}$, where $\mathbf{I}$ is the corresponding identity matrix.

The entries of the matrix $\mathbf{Q}$ have been defined in different ways in the literature [27, 38, 41, 42] and the approach to be developed in the current work is compatible with all of them. However, for its simplicity and mathematical elegance we will follow here the approach developed by Olfati-Saber in [38], where the entries of $\mathbf{Q}$ are defined as:

$$q_{ij} = \begin{cases} \mu w_{ij}, & i \neq j \\ 1 - \mu\left(1 - w_{ii}\right), & i = j, \end{cases} \tag{112}$$

where μ is the mutation rate and

$$w_{ij} = \frac{A_{ij}}{\sum_j A_{ij}}, \tag{113}$$

such that the matrix $\mathbf{W} = [w_{ij}]$ is the weighted adjacency matrix of the graph.

Let us define the following operator

$$\Im = \mathbf{I} - \mathbf{K}^{-1}\mathbf{A}, \tag{114}$$

where

$$\mathbf{K} = diag\left(\mathbf{1}^T \mathbf{A}\right). \tag{115}$$

Then, the matrix $\mathbf{Q}$ can be obtained as

$$\begin{aligned}\mathbf{Q} &= \mathbf{I} - \mu\Im \\ &= (1-\mu)\,\mathbf{I} + \mu\mathbf{W} \\ &= (1-\mu)\,\mathbf{I} + \mu\mathbf{K}^{-1}\mathbf{A}.\end{aligned} \tag{116}$$

Let us say a few words about the operator $\Im$. It is known that in a Markov process defined on a graph $G = (V, E)$ with a suitable probability measure, the semigroup transition function $P(\mathbf{x}, t, G), t \in [0, \infty)$ evolves according to the diffusion equation:

$$\frac{dP(\mathbf{x}, t, G)}{dt} = -\Im P(\mathbf{x}, t, G), \tag{117}$$

where

$$\Im = \frac{d}{d\mathbf{x}}\{a(x)\,dx + b(x)\}, \tag{118}$$

and $P(\mathbf{x}, t, G) = \exp(-t\Im)$ is a solution of (118). Therefore, $\Im = \mathbf{I} - \mathbf{K}^{-1}\mathbf{A}$ is the discrete Fokker-Planck operator defined for a Markov process on a graph [46]. Consequently, we can write (117) as

$$\dot{\mathbf{x}} = (\mathbf{I} - \mu\Im)^T\,\mathbf{F}\mathbf{x} - \varphi\mathbf{x}, \tag{119}$$

and the discrete-time version of it is written as

$$\mathbf{x}(t+1) = \mathbf{x}(t) + \varepsilon\left[\left(\mathbf{I} - \mu\Im^T\right)\mathbf{F}\mathbf{x}(t) - \varphi\mathbf{x}(t)\right], \tag{120}$$

where $\varepsilon > 0$ is the time-step of integration/discretization.

At a given time step of the dynamical evolution of the system, the number of individuals in steady-state is accounted for by the *diversity* $n_e(x)$ of the system, which has been defined as [38]

$$n_e(x) = \left(\sum_i x_i^2\right)^{-1}. \tag{121}$$

An order parameter was also introduced in [38] as

$$\rho = \left|\sum_{j=1}^{n} x_j e^{i\theta_j}\right|, \tag{122}$$

where $\theta_j = 2\pi j/n$, $j = 1, \ldots, n$ and $i = \sqrt{-1}$.

Using the diversity measure the following *phases of evolution* have been identified on the basis of the values of the diversity at a very large time, i.e., $t \to \infty$ [38]:

1. *Behavioural flocking*: $n_e = 1$, which indicates that a single dominant behaviour emerges.
2. *Cohesion*: $1 < n_e << n$, in which a few dominant behaviours emerge.
3. *Collapse*: $1 << n_e < n$, where many dominant behaviours emerge.
4. *Complete collapse*: $n_e = n$, where no dominant behaviour emerges.

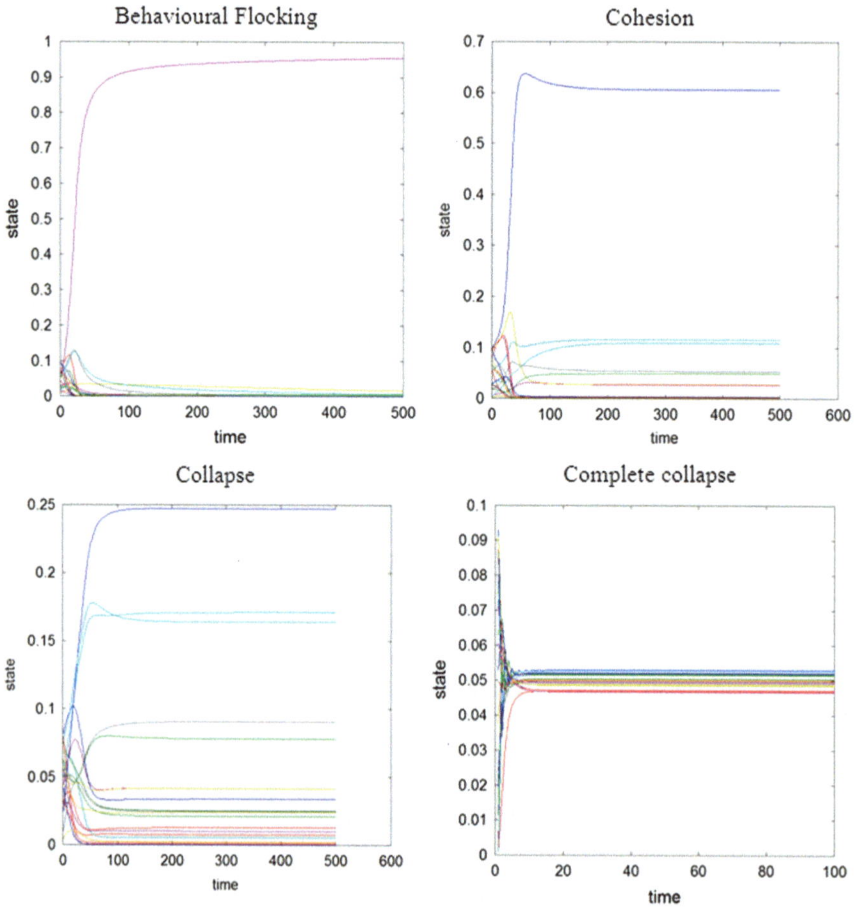

Fig. 21 Illustration of the results obtained with the replicator-mutator model for an ER network with 20 nodes and 40 links. The values of the mutation parameter are 0.001; 0.15; 0.5; 10. The corresponding values of the diversity are 1.04; 2.5; 7.2; 20.0. The order parameters are 0.99; 0.51; 0.24; 0.05

The four phases of the evolution of the replicator-mutator dynamic in a random graph with Erdös-Rényi topology with 20 nodes and 40 links are illustrated in Fig. 21. Notice that in the behavioural flocking phase, the order parameter is close to unity, while in the total collapse phase it goes close to zero.

References

1. A. Arenas, A. Diaz-Guilera, C.J. Pérez-Vicente, Synchronization processes in complex networks. Physica D **224**, 27–34 (2006)
2. A.-L. Barabási, R. Albert, Emergence of scaling in random networks. Science **286**, 509–512 (1999)
3. M. Barahona, L.M. Pecora, Synchronization in small-world systems. Phys. Rev. Lett. **89**, 054101 (2002)
4. A. Barrat, M. Weigt, On the properties of small-world network models. Eur. Phys. J. B **13**, 547–560 (2000)
5. B. Bollobás, *Modern Graph Theory* (Springer, New York, 1998)
6. B. Bollobás, O. Riordan, Mathematical results on scale-free random graphs, in *Handbook of Graph and Networks: From the genome to the Internet*, ed. by S. Bornholdt, H. G. Schuster (Wiley, Weinheim, 2003), pp. 1–32
7. B. Bollobás, O. Riordan, The diameter of a scale-free random graph. Combinatorica **24**, 5–34 (2004)
8. P. Bonacich, Factoring and weighting approaches to status scores and clique identification. J. Math. Sociol. **2**, 113–120 (1972)
9. P. Bonacich, Power and centrality: a family of measures. Am. J. Soc. **92**, 1170–1182 (1987)
10. G. Chen, X. Wang, X. Li, J. Lü, Some recent advances in complex networks synchronization, in *Recent Advances in Nonlinear Dynamics and Synchronization*, ed. by K. Kyamakya (Springer, Berlin, 2009), pp. 3–16
11. A. Clauset, C. RohillaShalizi, M.E.J. Newman, Power-law distributions in empirical data. SIAM Rev. **51**, 661–703 (2010)
12. S.N. Dorogovtsev, J.F.F. Mendes, *Evolution of Networks: From Biological Nets to the Internet and WWW* (Oxford University Press, Oxford, 2003)
13. P. Erdös, A. Rényi, On random graphs. Publ. Math. Debrecen **5**, 290–297 (1959)
14. E. Estrada, *The Structure of Complex Networks. Theory and Applications* (Oxford University Press, Oxford, 2011)
15. E. Estrada, N. Hatano, Statistical-mechanical approach to subgraph centrality in complex networks. Chem. Phys. Let. **439**, 247–251 (2007)
16. E. Estrada, N. Hatano, Communicability in complex networks. Phys. Rev. E **77**, 036111 (2008)
17. E. Estrada, J.A. Rodríguez-Velázquez, Subgraph centrality in complex networks. Phys. Rev. E **71**, 056103 (2005)
18. E. Estrada, N. Hatano, M.Benzi, The physics of communicability in complex networks. Phys. Rep. **514**, 89–119 (2012)
19. M. Fiedler, Algebraic connectivity of graphs. Czech. Math. J. **23**, 298–305 (1973)
20. S. Fortunato, Community detection in graphs. Phys. Rep. **486**, 75–174 (2010)
21. S. Foss, D. Korshunov, S. Zachary, *An Introduction to Heavy-Tailed and Subexponential Distributions* (Springer, Berlin, 2011)
22. L.C. Freeman, Centrality in social networks conceptual clarification. Soc. Netw. **1**, 215–239 (1979)
23. F. Harary, *Graph Theory* (Addison-Wesley, Reading, 1969)
24. S. Janson, The first eigenvalue of random graphs. Comb. Probab. Comput. **14**, 815–828 (2005)

25. L. Katz, A new status index derived from sociometric analysis. Psychometrica **18**, 39–43 (1953)
26. M.J. Keeling, K.T. Eames, Networks and epidemic models. J. Roy. Soc. Interface **2**, 295–307 (2005)
27. N.L. Komarova, Replicator–mutator equation, universality property and population dynamics of learning. J. Theor. Biol. **230**, 227–239 (2004)
28. Y. Kuramoto, *Chemical Oscillations, Waves, and Turbulence* (Springer, Berlin, 1984)
29. A.N. Langville, C.D. Meyer, *Google's PageRank and Beyond. The Science of Search Engine Rankings* (Princeton University Press, Princeton, 2006)
30. M. Meshbahi, M. Egerstedt, *Graph Theoretic Methods in Multiagent Networks* (Princeton University Press, Princeton, 2010), pp. 42–71
31. S. Milgram, The small world problem. Psychol. Today **2**, 60–67 (1967)
32. B. Mohar, The Laplacian spectrum of graphs, in *Graph Theory, Combinatorics, and Applications*, vol. 2, ed. by Y. Alavi, G. Chartrand, O.R. Oellermann, A.J. Schwenk (Wiley, New York, 1991), pp. 871–898
33. B. Mohar, Some applications of Laplace eigenvalues of graphs, in *Graph Symmetry: Algebraic Methods and Applications*, ed. by G. Hahn, G. Sabidussi. NATO ASI Series C (Kluwer, Dordrecht, 1997), pp. 227–275
34. Y. Moreno, R. Pastor-Satorras, A. Vespignani, Epidemic outbreaks in complex heterogeneous networks. Eur. Phys. J. B **26**, 521–529 (2002)
35. A.E. Motter, C. Zhou, J. Kurth, Network synchronization, diffusion, and the paradox of heterogeneity. Phys. Rev. E **71**, 016116 (2005)
36. M.E.J. Newman, Modularity and community structure in networks. Proc. Natl. Acad. Sci. USA **103**, 8577–8582 (2006)
37. M.E.J. Newman, S.H. Strogatz, D.J. Watts, Random graphs with arbitrary degree distributions and their applications. Phys. Rev. E **64**, 026118 (2001)
38. R. Olfati-Saber, Evolutionary dynamics of behavior in social networks, in *46th IEEE Conference on Decision and Control, 2007* (IEEE, Piscataway, NJ, 2007), pp. 4051–4056
39. R. Olfati-Saber, J.A. Fax, R.M. Murray, Consensus and cooperation in networked multi-agent systems. Proc. IEEE **95**, 215–233 (2007)
40. J.M. Ottino, Complex systems. AIChE J. **49**, 292–299 (2003)
41. D. Pais, N.E. Leonard, Limit cycles in replicator-mutator network dynamics, in *50th IEEE Conference on Decision and Control and European Control Conference (CDC-ECC), 2011* (IEEE, Piscataway, NJ, 2011), pp. 3922–3927
42. D. Pais, C.H. Caicedo-Núñez, N.E. Leonard, Hopf bifurcations and limit cycles in evolutionary network dynamics. SIAM J. Appl. Dyn. Syst. **11**, 1754–1784 (2012)
43. R. Pastor-Satorras, A. Vespignani, Epidemic spreading in scale-free networks. Phys. Rev. Lett. **86**, 3200–3203 (2001)
44. R.K. Standish, Concept and definition of complexity, in *Intelligent Complex Adaptive Systems*, ed. by A. Yang, Y. Shan (IGI Global, Hershey, 2008), pp. 105–124. arXiv:0805.0685
45. S.H. Strogatz, R.E. Mirollo, Phase-locking and critical phenomena in lattices of coupled nonlinear oscillators with random intrinsic frequencies. Physica D **143**, 143–168 (1988)
46. H.-F. Wang, E.R. Hancock, Probabilistic relaxation labeling by Fokker-Planck diffusion on a graph, in *Graph-Based Representations in Pattern Recognition* (Springer, Berlin/Heidelberg, 2007), pp. 204–214
47. D.J. Watts, S.H. Strogatz, Collective dynamics of "small-world" networks. Nature **393**, 440–442 (1998)
48. E.P. Wigner, Characteristic vectors of bordered matrices with infinite dimensions. Ann. Math. **62**, 548–564 (1955)

Further Reading

A. Barrat, M. Barthélemy, A. Vespignani, *Dynamical Processes on Complex Networks* (Cambridge University Press, Cambridge, 2008)

G. Caldarelli, *Scale-Free Networks. Complex Webs in Nature and Technology* (Oxford University Press, Oxford, 2007)

E. Estrada, *The Structure of Complex Networks. Theory and Applications* (Oxford University Press, Oxford, 2011)

M.A. Nowak, *Evolutionary Dynamics: Exploring the Equations of Life* (Harvard University Press, Cambridge, 2006)

M. Mesbahi, M. Egerstedt, *Graph Theoretic Methods in Multiagent Networks* (Princeton University Press, Princeton, 2010)

Kinetic Models in Natural Sciences

Jacek Banasiak

1 Introduction

1.1 Preliminaries

In our terminology, a kinetic type equation describes an evolution of a population of objects, depending on attributes from a certain set Ω, subject to a given set of conservation laws. Equations of this type also are referred to as Master Equations. One of the natural ways to describe such a population is by providing the density of the objects with respect to the attributes and investigate how it changes in time. The density, say $u(x)$, is either the number of elements with an attribute x (if the number of possible attributes is finite or countable), or a gives the quantity of elements with attributes in a set A, according to the formula

$$\int_A u(x)d\mu, \tag{1}$$

if x is a continuous variable.

J. Banasiak (✉)
School of Mathematics, Statistics and Computer Science, University of KwaZulu-Natal, Durban, South Africa

Institute of Mathematics, Technical University of Łódź, Łódź, Poland
e-mail: banasiak@ukzn.ac.za

© Springer International Publishing Switzerland 2015

J. Banasiak, M. Mokhtar-Kharroubi (eds.), *Evolutionary Equations with Applications in Natural Sciences*, Lecture Notes in Mathematics 2126,
DOI 10.1007/978-3-319-11322-7_4

In many cases we are interested in tracking the total number of elements of the population which, for a given time t, is given by

$$\sum_{x \in \Omega} u(x,t), \tag{2}$$

if Ω is countable, and by

$$\int_{\Omega} u(x,t)\mathrm{d}x, \tag{3}$$

if Ω is a continuum.

A kinetic equation for u usually is built in the following way. Let $u(x,t)$ be the density of individuals with respect to the attribute(s) $x \in \Omega$ at a time t. Then we balance, for any subset $A \subset \Omega$,

1. the loss of individuals with attributes in A due to changes of the attributes to the ones outside A;
2. the gain of individuals who changed the attributes from outside A to the ones in A;
3. transport through A.

This results in the Master, or Kinetic, Equation of the process,

$$\partial_t u(x,t) = [\mathcal{K}u](x,t) := [\mathcal{T}u](x,t) + [\mathcal{A}u](x,t) + [\mathcal{B}u](x,t), \tag{4}$$

where $\mathcal{A}$ is the loss operator, $\mathcal{B}$ is the gain operator, while $\mathcal{T}$ describes the transport phenomena. Equation (4) is supplemented with the initial state of the system

$$u(x,0) = \mathring{u}(x), \quad x \in \Omega. \tag{5}$$

Only in exceptional circumstances can the problem (4), (5) be solved. Usually we have to prove the existence, uniqueness and other relevant properties of the solution u without knowing its explicit form. There are various ways of doing this. We shall follow the dynamical systems approach. Here, the evolution of the system is described using a family of operators $(\mathcal{G}(t))_{t\geq 0}$, parameterised by time, that map an initial state $\mathring{u}$ of the system to all subsequent states in the evolution; that is, the solution is represented as

$$u(t) = \mathcal{G}(t)\mathring{u}. \tag{6}$$

The solutions of (4); that is, the states of the system, belong to some appropriate state space which is chosen partly due to its relevance to the problem but also for the mathematical convenience. By no means is this choice unique: it is a mathematical intervention into the model.

In the processes discussed in these lectures, an appropriately defined integral of the density over the space of attributes is the total amount of individuals, or the total mass of the system. Due to the conservation laws used to construct the equation, this integral is constant, or changes in some pre-defined way.

From this point of view it is natural to consider such processes as evolutions of densities in an appropriate L_1 space; that is, in the space

$$L_1(\Omega, \mu) = \left\{ u;\ \|u\| = \int_{\Omega} |u| d\mu < +\infty \right\},$$

where μ is an appropriate measure relevant to the process. Such a space will be our state space; that is, the state of the system will be described by a density with finite integral over Ω, which we often will be calling the total mass (irrespective of its real interpretation).

However, we can try to control the process using some other gauge function. For instance, if we were interested in controlling the maximal concentration of the particles, a more proper choice would be to use the functional

$$\sup_{x \in \Omega} |u(x)|$$

as the gauge function.

This approach leads in a natural way to a class of abstract spaces called the Banach spaces.

1.1.1 Interlude: Banach Spaces and Linear Operators

In what follows we shall restrict our attention to the state spaces which are Banach spaces, though more general state spaces such metric or topological vector spaces are also possible, see e.g. [36]. For an in-depth information on the topics discussed here the reader is referred to [23, 26, 47]. To recall, a Banach space is a vector space X, equipped with a finite gauge function $\|\cdot\|$, called norm, satisfying $\|x\| = 0$ if and only if $x = 0$, $\|\alpha x\| = |\alpha|\|x\|$ for each scalar α and $\|x + y\| \leq \|x\| + \|y\|$, $x, y \in X$ and which is complete with respect to the convergence defined by the norm (a space is complete if it contains limits of all its Cauchy sequences).

Example 1 We introduce a class of Banach spaces which will play crucial role in the theory of differential equations which will be used throughout the lectures: the Sobolev spaces.

In general considerations, when dealing with partial derivatives of functions, often only the order of the derivative is important. In such cases, to shorten calculations, we introduce the following notation. Let $\alpha = (\alpha_1, \ldots, \alpha_n)$, $\alpha_i \in \mathbb{N}_0, i = 1, \ldots, n$, be a multi-index and denote $|\alpha| = \alpha_1 + \cdots + \alpha_n$. Then, for a given (locally integrable) function u, we denote any generalized (distributional) derivative of u of order $|\alpha|$ by

$$\partial^\alpha u = \frac{\partial^{|\alpha|}}{\partial x_1^{\alpha_1} \cdots \partial x_n^{\alpha_n}} u.$$

The Sobolev spaces $W_1^m(\Omega)$ are defined as

$$W_1^m(\Omega) := \{u \in L_1(\Omega);\ \partial^\alpha u \in L_1(\Omega),\ |\alpha| \leq m\}.$$

In the same way, starting from the space $L_p(\Omega)$ of functions integrable with a power $p \in [1, \infty[$, we can define Sobolev spaces $W_p^m(\Omega)$, $p \in [1, \infty[$. For $p = \infty$ the corresponding space $L_\infty(\Omega)$ is the space of functions which are bounded almost everywhere on Ω, and $W_\infty^m(\Omega)$ is the space in which all generalized derivatives up to the order m have this property as well.

An object intimately related with a Banach space is a linear operator. A *(linear) operator* from X to Y is a linear function $A : D(A) \to Y$, where $D(A)$ is a linear subspace of X, called the *domain* of A. We often use the notation $(A, D(A))$ to denote the operator A with domain $D(A)$; if the domain is obvious, we simply write A. By $\mathcal{L}(X, Y)$ we denote the space of all bounded operators between X and Y; that is, the operators for which

$$\|A\| := \sup_{\|x\| \leq 1} \|Ax\| = \sup_{\|x\|=1} \|Ax\| < +\infty. \tag{7}$$

The space $\mathcal{L}(X, X)$ is abbreviated as $\mathcal{L}(X)$. We further define the *kernel* (or the *null-space*) of A by

$$\text{Ker}\, A = \{x \in D(A);\ Ax = 0\}$$

and the *range* of A by

$$\text{Ran}\, A = \{y \in Y;\ Ax = y \text{ for some } x \in D(A)\}$$

Furthermore, the *graph* of A is defined as the set $\{(x, y) \in X \times Y;\ x \in D(A), y = Ax\}$. We say that the operator A is *closed* if its graph is a closed subspace of $X \times Y$. Equivalently, A is closed if and only if for any sequence $(x_n)_{n\in\mathbb{N}} \subset D(A)$, if $\lim_{n\to\infty} x_n = x$ in X and $\lim_{n\to\infty} Ax_n = y$ in Y, then $x \in D(A)$ and $y = Ax$.

An operator A in X is *closable* if the closure of its graph is itself a graph of an operator. In such a case the operator whose graph is $\overline{G(A)}$ is called the *closure* of A and denoted by $\overline{A}$.

Example 2 Consider the operator $Af = f'$ in $C([0, 1])$ and $L_1([0, 1])$. Then, [13, Example 2.3], A is unbounded in both spaces, closed in $C([0, 1])$ and not closed, but closable, in $L_1([0, 1])$.

In this way, (4) can be written as the Cauchy problem for an ordinary differential equation in an appropriate Banach space X: find $\mathbb{R}_+ \ni t \to u(t) \in X$ such that

$$\partial_t u = Ku, \quad t > 0, \qquad u(0) = \overset{\circ}{u} \in X, \tag{8}$$

where $K : D(K) \to X$ is a realization of the expression $\mathcal{K}$, defined on some subset $D(K)$ of the chosen state space X. It is clear that a minimum requirement for $D(K)$ is that $[\mathcal{K}u](\cdot) \in X$ for $u \in D(K)$. It is important to remember that the expression $\mathcal{K}$ usually has multiple realizations and finding an appropriate one, such that with $(K, D(K))$ the problem (8) is well posed (often called the generator of the process), is a very difficult task.

We mention the so-called maximal realization of the expression $\mathcal{K}$, $K_{\max}$ defined as the restriction of $\mathcal{K}$ to

$$D(K_{\max}) = \{u \in X;\ x \to [\mathcal{K}u](x) \in X\}.$$

The generator may be, or may be not, equal to $K_{\max}$. In the former case, typically (4) is uniquely solvable in X.

1.2 The Models

Here we introduce the examples which will be discussed in the course.

1.2.1 Transport on Networks

Let us consider a network with some substance flowing along the edges and being redistributed in the nodes. The process of the redistribution of the flow is the loss-gain process governed by the Kirchoff's law (flow-in = flow-out). Thus, though this model does not exactly fit into the framework of (4), it is an example of a kinetic process as defined above.

The network under consideration is represented by a simple directed graph $G = (V(G), E(G)) = (\{v_1, \dots, v_n\}, \{e_1, \dots, e_m\})$ with n vertices $v_1, \dots, v_n$ and m edges (arcs), $e_1, \dots, e_m$. We suppose that G is connected but not necessarily strongly connected, see e.g. [21, 28]. Each edge is normalized so as to be identified with $[0, 1]$, with the head at 0 and the tail at 1. The outgoing incidence matrix, $\Phi^- = (\phi^-_{ij})_{1 \le i \le n, 1 \le j \le m}$, and the incoming incidence matrix, $\Phi^+ = (\phi^+_{ij})_{1 \le i \le n, 1 \le j \le m}$, of this graph are defined, respectively as

$$\phi^-_{ij} = \begin{cases} 1 \text{ if } v_i \overset{e_j}{\to} \\ 0 \text{ otherwise,} \end{cases} \qquad \phi^+_{ij} = \begin{cases} 1 \text{ if } \overset{e_j}{\to} v_i \\ 0 \text{ otherwise.} \end{cases}$$

If the vertex v_i has more than one outgoing edge, we place a non negative weight w_{ij} on the outgoing edge e_j such that for this vertex v_i,

$$\sum_{j \in E_i} w_{ij} = 1,$$

where E_i is defined by saying that $j \in E_i$ if the edge e_j is outgoing from v_i. Naturally, $w_{ij} = 1$ if $E_i = \{j\}$ and, to shorten notation, we adopt the convention that $w_{ij} = 1$ for any j if $E_i = \emptyset$. Then the weighted outgoing incidence matrix, Φ_w^-, is obtained from Φ^- by replacing each nonzero ϕ_{ij}^- entry by w_{ij}. If each vertex has an outgoing edge, then Φ_w^- is row stochastic, hence $\Phi^-\left(\Phi_w^-\right)^T = I_n$ (where the superscript T denotes the transpose). The (weighted) adjacency matrix $\mathbb{A} = (a_{ij})_{1\le i,j\le n}$ of the graph is defined by taking $a_{ij} = w_{jk}$ if there is e_k such that $v_j \stackrel{e_k}{\to} v_i$ and 0 otherwise; that is, $\mathbb{A} = \Phi^+(\Phi_w^-)^T$. An important role is played by the line graph Q of G. To recall, $Q = (V(Q), E(Q)) = (E(G), E(Q))$, where

$$\begin{aligned} E(Q) &= \{uv;\ u, v \in E(G), \text{the head of } u \text{ coincides with the tail of } v\} \\ &= \{\varepsilon_j\}_{1\le j\le k}. \end{aligned}$$

By $\mathbb{B}$ we denote the weighted adjacency matrix for the line graph; that is,

$$\mathbb{B} := (\Phi_w^-)^T \Phi^+. \tag{9}$$

If there is an outgoing edge at each vertex then, from the definition of $\mathbb{B}$, we see that it is column stochastic. A vertex v will be called a source if it has no incoming edge and a sink if there are no outgoing edges.

Example 1.1 Consider the following graph.

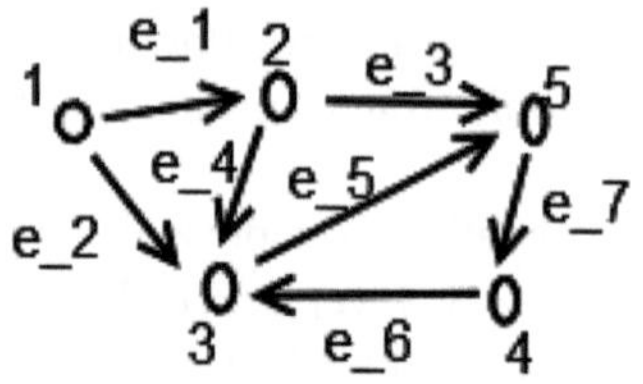

For this graph, the matrices Φ^-, Φ^+ are given below.

$$\Phi^- = \begin{pmatrix} 1\,1\,0\,0\,0\,0\,0 \\ 0\,0\,1\,1\,0\,0\,0 \\ 0\,0\,0\,0\,1\,0\,0 \\ 0\,0\,0\,0\,0\,1\,0 \\ 0\,0\,0\,0\,0\,0\,1 \end{pmatrix}, \quad \Phi^+ = \begin{pmatrix} 0\,0\,0\,0\,0\,0\,0 \\ 1\,0\,0\,0\,0\,0\,0 \\ 0\,1\,0\,1\,0\,1\,0 \\ 0\,0\,0\,0\,0\,0\,1 \\ 0\,0\,1\,0\,1\,0\,0 \end{pmatrix},$$

while the adjacency matrix is given by

$$\begin{pmatrix} 0\,0\,0\,0\,0 \\ 1\,0\,0\,0\,0 \\ 1\,1\,0\,1\,0 \\ 0\,0\,0\,0\,1 \\ 0\,1\,1\,0\,0 \end{pmatrix}.$$

We are interested in a flow on a closed network G. Then the standard assumption is that the flow satisfies the Kirchoff law at the vertices

$$\sum_{j=1}^{m} \phi_{ij}^{-} c_j u_j(1,t) = w_{ij} \sum_{j=1}^{m} \phi_{ij}^{+} c_j u_j(0,t), \quad t > 0, i \in 1, \ldots, n,$$

which, in this context, is the conservation of mass law: the total inflow of mass per unit time equals the total outflow at each node (vertex) of the network.

Let $u_j(x,t)$ be the density of particles at position x and at time $t \geq 0$ flowing along edge e_j for $x \in [0, 1]$. The particles on e_j are assumed to move with velocity $c_j > 0$ which is constant for each j. We consider a generalization of Kirchoff's law by allowing for a decrease/amplification of the flow at the entrances and exits of each vertex. Then the flow is described by

$$\begin{cases} \partial_t u_j(x,t) &= c_j \partial_x u_j(x,t), \quad x \in (0,1), \quad t \geq 0, \\ u_j(x,0) &= f_j(x), \\ \phi_{ij}^{-} \xi_j c_j u_j(1,t) &= w_{ij} \sum\limits_{k=1}^{m} \phi_{ik}^{+} (\gamma_k c_k u_k(0,t)), \end{cases} \tag{10}$$

where $\gamma_j > 0$ and $\xi_j > 0$ are the absorption/amplification coefficients at, respectively, the head and the tail of the edge e_j. If $\gamma_j = \xi_j = 1$ for all $j = 1, \cdots, m$, then we recover the Kirchoff law at the vertices.

Remark 1 We observe that the boundary condition in (10) takes a special form if v_i is either a sink or a source. If it is a sink, then $E_i = \emptyset$ and, by the convention above,

$$0 = \sum_{k=1}^{m} \phi_{ik}^{+} (\gamma_k c_k u_k(0,t)), \quad t > 0, \tag{11}$$

and

$$\phi_{ij}^{-} \xi_j c_j u_j(1,t) = 0, \quad t > 0, j = 1, \ldots, m, \tag{12}$$

if it is a source. The last condition is nontrivial only if $j \in E_i$ as then $\phi_{ij}^{-} \neq 0$.

We consider (10) as an abstract Cauchy problem

$$\mathbf{u}_t = A\mathbf{u}, \qquad \mathbf{u}(0) = \mathbf{f}, \tag{13}$$

in $X = (L_1([0,1]))^m$, where A is the realization of the expression $\mathsf{A} = (c_j \partial_x)_{1\leq j\leq m}$ on the domain

$$D(A) = \{\mathbf{u} \in (W_1^1([0,1]))^m;\ \mathbf{u} \text{ satisfies the b. c. in (10)}\}. \tag{14}$$

We denote $\mathbb{C} = \operatorname{diag}(c_j)_{1\leq j\leq m}, \mathbb{K} = \operatorname{diag}(\xi_j)_{1\leq j\leq m}$ and $\mathbb{G} = \operatorname{diag}(\gamma_j)_{1\leq j\leq m}$. It can be proved, [18], that

$$D(A) = \{\mathbf{u} \in (W_1^1([0,1]))^m;\ \mathbf{u}(1) = \mathbb{K}^{-1}\mathbb{C}^{-1}\mathbb{B}\mathbb{G}\mathbb{C}\mathbf{u}(0)\}. \tag{15}$$

We note that a numerical analysis of a related problem for a nonlinear transport equation is presented in [20].

1.2.2 Epidemiological Models with Age Structure

The gain and loss parts in this model are given by the SIRS system

$$\begin{aligned} S' &= -\Lambda(I)S + \delta I, \\ I' &= \Lambda(I)S - (\delta + \gamma)I, \\ R' &= \gamma I, \end{aligned} \tag{16}$$

where S, I, R are, respectively, the number of susceptibles, infectives and recovered (with immunity) and γ, δ are recovery rates with and without immunity. In other words, the loss and gain operators of (4) are given by

$$\mathcal{A}\begin{pmatrix} S \\ I \\ R \end{pmatrix} = -\begin{pmatrix} \Lambda(I)S \\ (\delta+\gamma)I \\ 0 \end{pmatrix}; \qquad \mathcal{B}\begin{pmatrix} S \\ I \\ R \end{pmatrix} = \begin{pmatrix} \delta I \\ \Lambda(I)S \\ \gamma I \end{pmatrix}. \tag{17}$$

For many diseases the rates of infection and recovery significantly vary with age. Thus the vital dynamics of the population and the infection mechanism can interact to produce a nontrivial dynamics. To model it, we assume that the total population in the absence of disease can be modelled by the linear McKendrick model describing the evolution in time of the density of the population with respect to age $a \in [0, \omega]$, $\omega < \infty$, denoted by $n(a,t)$. The evolution is driven by the processes of death and birth with vital rates $\mu(a)$ and $\beta(a)$, respectively. Due to the epidemics, we split the population into susceptibles, infectives and recovered,

$$n(a,t) = s(a,t) + i(a,t) + r(a,t),$$

so that the scalar McKendrick equation for n splits, according to (16), into the system

$$\begin{aligned}\partial_t s(a,t) + \partial_a s(a,t) + \mu(a)s(a,t) &= -\Lambda(a,i(\cdot,t))s(a,t) + \delta(a)i(a,t),\\ \partial_t i(a,t) + \partial_a i(a,t) + \mu(a)i(a,t) &= \Lambda(a,i(\cdot,t))s(a,t) - (\delta(a)+\gamma(a))i(a,t),\\ \partial_t r(a,t) + \partial_a r(a,t) + \mu(a)r(a,t) &= \gamma(a)i(a,t), \end{aligned}\tag{18}$$

where now the rates are age specific, see [31]. The function Λ is the infection rate (or the force of infection). In the so-called intercohort model, which will be analysed later in these lectures, we use

$$\Lambda(a,i(\cdot,t)) = \int_0^{\omega} K(a,a')i(a',t)da', \tag{19}$$

where K is a nonnegative bounded function which accounts for the age dependence of the infections. For instance, for a typical childhood disease, K should be large for small a, a' and close to zero for large a or a' (not necessarily 0, as usually adults can contract it). System (18) is supplemented by the boundary conditions

$$\begin{aligned} s(0,t) &= \int_0^{\omega} \beta(a)(s(a,t) + (1-p)i(a,t) + (1-q)r(a,t))da,\\ i(0,t) &= p\int_0^{\omega} \beta(a)i(a,t)da, \qquad r(0,t) = q\int_0^{\omega} \beta(a)r(a,t)da, \end{aligned}\tag{20}$$

where $p, q \in [0,1]$ are the vertical transmission parameters of infectiveness and immunity, respectively. Finally, we prescribe the initial conditions

$$s(a,0) = \overset{\circ}{s}(a), \quad i(a,0) = \overset{\circ}{i}(a), \quad r(a,0) = \overset{\circ}{r}(a). \tag{21}$$

1.2.3 Fragmentation–Coagulation Processes

The name may seem very specific, but such processes occur in a wide range of applications, see [34]. Also, they possibly are the most rewarding kinetic processes to study from the analytical point of view.

Fragmentation and coagulation may be discrete, when we assume that there is a minimal size of interacting particles and all clusters are finite ensembles of such fundamental building blocks, and continuous with the matter assumed to be a continuum. The first model is described in the lecture [34]; for the latter see e.g. [35, 36]. Here we only consider the continuous model. In the case of pure fragmentation a standard modelling process leads to the following equation:

$$\partial_t u(x,t) = -a(x)u(x,t) + \int_x^\infty a(y)b(x|y)u(y,t)dy, \tag{22}$$

u is the density of particles of mass x, a is the fragmentation rate and b describes the distribution of particle masses x spawned by the fragmentation of a particle of mass y; that is, the expected number of particles of mass x resulting from fragmentation of a particle of mass y. Further

$$M(t) = \int_0^\infty xu(x,t)dx \tag{23}$$

is the total mass at time t. Local conservation principle requires

$$\int_0^y xb(x|y)dx = y, \tag{24}$$

while the expected number of particles produced by a particle of mass y is given by $n_0(y) = \int_0^y b(x|y)dx$.

Fragmentation can be supplemented by growth/decay, transport or diffusion processes, [11–13, 15], but we will not discuss them here.

If we combine the fragmentation process with coagulation, we will get

$$\partial_t u(x,t) = -a(x)u(x,t) + \int_x^\infty a(y)b(x|y)u(y,t)dy \tag{25}$$
$$-u(x,t)\int_0^\infty k(x,y)u(y,t)dy + \frac{1}{2}\int_0^x k(x-y,y)u(x-y,t)u(y,t)dy.$$

The coagulation kernel $k(x,y)$ represents the likelihood of a particle of size x attaching itself to a particle of size y and, for a moment, we assume that it is a symmetric nonnegative positive function.

Since the fragmentation and coagulation processes just rearrange the mass among the clusters, (23) implies that the natural space to analyse them is

$$X_1 = L_1(\mathbb{R}_+, xdx) = \left\{u;\ \|u\|_1 = \int_0^\infty |u(x)|xdx < +\infty\right\}.$$

However, for the coagulation processes it is important to control also the number of particles, or even some higher moments of the density. The best results are obtained in the scale of spaces $X_{1,m}$, $m \geq 1$, where

$$X_{1,m} = L_1(\mathbb{R}_+, (1+x^m)dx) = \left\{ u;\ \|u\|_{0,m} = \int_0^\infty |u|(1+x^m)dx < +\infty \right\}.$$

2 The Main Tool Box

2.1 Basic Positivity Concepts

The common feature of the introduced models is that the solution originating from a nonnegative density should stay nonnegative; that is, the solution operator should be a 'positive' operator. Since we are talking about general Banach spaces, we have to define what we mean by a nonnegative element of a Banach space. Though in all cases discussed here our Banach space is an $L_1(\Omega, \mu)$ space, where the nonnegativity of a function f is understood as $f(x) \geq 0$ μ-almost everywhere, it is more convenient to work in a more abstract setting.

2.1.1 Defining Order

In a given vector space X an order can be introduced either geometrically, by defining the so-called *positive cone* (in other words, what it means to be a *positive element* of X), or through the axiomatic definition:

Definition 1 Let X be an arbitrary set. A partial order (or simply, an order) on X is a binary relation, denoted here by '$\geq$', which is reflexive, transitive, and antisymmetric, that is,

(1) $x \geq x$ for each $x \in X$;
(2) $x \geq y$ and $y \geq x$ imply $x = y$ for any $x, y \in X$;
(3) $x \geq y$ and $y \geq z$ imply $x \geq z$ for any $x, y, z \in X$.

The *supremum* of a set is its least upper bound and the *infimum* is the greatest lower bound. The supremum and infimum of a set need not exist. For a two-point set $\{x, y\}$ we write $x \wedge y$ or $\inf\{x, y\}$ to denote its infimum and $x \vee y$ or $\sup\{x, y\}$ to denote supremum.

We say that X is a *lattice* if every pair of elements (and so every finite collection of them) has both supremum and infimum.

From now on, unless stated otherwise, any vector space X is real.

Definition 2 An ordered vector space is a vector space X equipped with partial order which is compatible with its vector structure in the sense that

(4) $x \geq y$ implies $x + z \geq y + z$ for all $x, y, z \in X$;
(5) $x \geq y$ implies $\alpha x \geq \alpha y$ for any $x, y \in X$ and $\alpha \geq 0$.

The set $X_+ = \{x \in X;\ x \geq 0\}$ is referred to as the positive cone of X.

If the ordered vector space X is also a lattice, then it is called a vector lattice or a Riesz space.

For an element x in a Riesz space X we can define its positive and negative part, and its absolute value, respectively, by

$$x_+ = \sup\{x, 0\}, \quad x_- = \sup\{-x, 0\}, \quad |x| = \sup\{x, -x\},$$

which are called lattice operations. We have

$$x = x_+ - x_-, \qquad |x| = x_+ + x_-. \tag{26}$$

The absolute value has a number of useful properties that are reminiscent of the properties of the scalar absolute value.

As the next step, we investigate the relation between the lattice structure and the norm, when X is both a normed and an ordered vector space.

Definition 3 A norm on a vector lattice X is called a lattice norm if

$$|x| \leq |y| \quad \text{implies} \quad \|x\| \leq \|y\|. \tag{27}$$

A Riesz space X complete under the lattice norm is called a Banach lattice.

Property (27) gives the important identity:

$$\|x\| = \||x|\|, \qquad x \in X. \tag{28}$$

2.1.2 *AM*- and *AL*-Spaces

Two classes of Banach lattices playing here a significant role are *AL*- and *AM*-spaces.

Definition 4 We say that a Banach lattice X is

(i) an AL-space if $\|x + y\| = \|x\| + \|y\|$ for all $x, y \in X_+$,
(ii) an AM-space if $\|x \vee y\| = \max\{\|x\|, \|y\|\}$ for all $x, y \in X_+$.

Example 3 Standard examples of AM-spaces are offered by the spaces $C(\overline{\Omega})$, where $\overline{\Omega}$ is either a bounded subset of $\mathbb{R}^n$, or in general, a compact topological space. Also the space $L_\infty(\Omega)$ is an AM-space. On the other hand, most known examples of AL-spaces are the spaces $L_1(\Omega)$. These examples exhaust all (up to a lattice isometry) cases of AM- and AL-spaces. However, particular representations of these spaces can be very different.

2.2 *Positive Operators*

Definition 5 A linear operator A from a Banach lattice X into a Banach lattice Y is called positive, denoted $A \geq 0$, if $Ax \geq 0$ for any $x \geq 0$.

An operator A is positive if and only if $|Ax| \leq A|x|$. This follows easily from $-|x| \leq x \leq |x|$ so, if A is positive, then $-A|x| \leq Ax \leq A|x|$. Conversely, taking $x \geq 0$, we obtain $0 \leq |Ax| \leq A|x| = Ax$.

A frequently used property of positive operators is given in

Theorem 1 *If A is an everywhere defined positive operator from a Banach lattice to a normed Riesz space, then A is bounded.*

Proof If A were not bounded, then we would have a sequence $(x_n)_{n\in\mathbb{N}}$ satisfying $\|x_n\| = 1$ and $\|Ax_n\| \geq n^3$, $n \in \mathbb{N}$. Because X is a Banach space, $x := \sum_{n=1}^{\infty} n^{-2}|x_n| \in X$. Because $0 \leq |x_n|/n^2 \leq x$, we have $\infty > \|Ax\| \geq \|A(|x_n|/n^2)\| \geq \|A(x_n/n^2)\| \geq n$ for all n, which is a contradiction. □

The norm of a positive operator can be evaluated by

$$\|A\| = \sup_{x\geq 0,\, \|x\|\leq 1} \|Ax\|. \tag{29}$$

Indeed, since $\|A\| = \sup_{\|x\|\leq 1} \|Ax\| \geq \sup_{x\geq 0, \|x\|\leq 1} \|Ax\|$, it is enough to prove the opposite inequality. For each x with $\|x\| \leq 1$ we have $|x| = x_+ + x_- \geq 0$ with $\|x\| = \||x|\| \leq 1$. On the other hand, $A|x| \geq |Ax|$, hence $\|A|x|\| \geq \||Ax|\| = \|Ax\|$. Thus $\sup_{\|x\|\leq 1} \|Ax\| \leq \sup_{x\geq 0, \|x\|\leq 1} \|Ax\|$ and the statement is proved.

As a consequence, we note that

$$0 \leq A \leq B \Rightarrow \|A\| \leq \|B\|. \tag{30}$$

Moreover, it is worthwhile to emphasize that if $A \geq 0$ and there exists K such that $\|Ax\| \leq K\|x\|$ for $x \geq 0$, then this inequality holds for any $x \in X$. Indeed, by (29), we have $\|A\| \leq K$ and using the definition of the operator norm, we obtain the desired statement.

2.3 Relation Between Order and Norm

There is a useful relation between the order, norm (absolute value) and the convergence of sequences if we are in $\mathbb{R}$—any monotonic sequence which is bounded (in absolute value), converges. One would like to have a similar result in Banach lattices. It turns out to be not so easy.

Existence of an order in some set X allows us to introduce in a natural way the notion of (order) convergence. Proper definitions of order convergence require nets of elements but we do not need to go to such details.

For a non-increasing sequence $(x_n)_{n\in\mathbb{N}}$ we write $x_n \downarrow x$ if $\inf\{x_n;\ n \in \mathbb{N}\} = x$. For a non-decreasing sequence $(x_n)_{n\in\mathbb{N}}$ the symbol $x_n \uparrow x$ have an analogous meaning. Then we say that an arbitrary sequence $(x_n)_{n\in\mathbb{N}}$ is order convergent to x if it can be sandwiched between two monotonic sequences converging to x. We write this as $x_n \overset{o}{\to} x$. One of the basic results is:

Proposition 1 *Let X be a normed lattice. Then:*

(1) The positive cone X_+ is closed.
(2) If $(x_n)_{n\in\mathbb{N}}$ is nondecreasing and $\lim_{n\to\infty} x_n = x$ in the norm of X, then

$$x = \sup\{x_n;\ n \in \mathbb{N}\}.$$

Analogous statement holds for nonincreasing sequences.

In general, the converse of Proposition 1(2) is false; that is, we may have $x_n \uparrow x$ but $(x_n)_{n\in\mathbb{N}}$ does not converge in norm. Indeed, consider $\mathbf{x}_n = (1, 1, 1 \ldots, 1, 0, 0, \ldots) \in l_\infty$, where 1 occupies only the n first positions. Clearly, $\sup_{n\in\mathbb{N}} \mathbf{x}_n = \mathbf{x} := (1, 1, \ldots, 1, \ldots)$ but $\|\mathbf{x}_n - \mathbf{x}\|_\infty = 1$.

However, such a converse holds in a special class of Banach lattices, called Banach lattices with order continuous norm. There we have, in particular, that $0 \le x_n \uparrow x$ and $x_n \le x$ for all $n \in \mathbb{N}$ if and only if $(x_n)_{n\in\mathbb{N}}$ is a Cauchy sequence, [1, Theorem 12.9].

All Banach lattices $L_p(\Omega)$ with $1 \le p < \infty$ have order continuous norms. On the other hand, neither $L_\infty(\Omega)$ nor $C(\bar{\Omega})$ (if Ω does not consist of isolated points) has order continuous norm.

The requirement that $(x_n)_{n\in\mathbb{N}}$ must be order dominated often is too restrictive. The spaces we are mostly concerned belong to a class which have a stronger property.

Definition 6 We say that a Banach lattice X is a *KB-space* (Kantorovič–Banach space) if every increasing norm bounded sequence of elements of X_+ converges in norm in X.

We observe that if $x_n \uparrow x$, then $\|x_n\| \le \|x\|$ for all $n \in \mathbb{N}$ and thus any *KB*-space has order continuous norm. Hence, spaces which do not have order continuous norm cannot be *KB*-spaces. This rules out the spaces of continuous functions, l_∞ and $L_\infty(\Omega)$, from being *KB*-spaces.

Any reflexive Banach space is a *KB*-space, [13, Theorem 2.82]. That *AL*-spaces (so, in particular, all L_1 spaces) are also *KB*-spaces follows from the following simple argument.

Theorem 2 *Any AL-space is a KB-space.*

Proof If $(x_n)_{n\in\mathbb{N}}$ is an increasing and norm bounded sequence, then for $0 \leq x_n \leq x_m$, we have

$$\|x_m\| = \|x_m - x_n\| + \|x_n\|$$

as $x_m - x_n \geq 0$ so that

$$\|x_m - x_n\| = \|x_m\| - \|x_n\| = |\|x_m\| - \|x_n\||\,.$$

Because, by assumption, $(\|x_n\|)_{n\in\mathbb{N}}$ is monotonic and bounded, and hence convergent, we see that $(x_n)_{n\in\mathbb{N}}$ is a Cauchy sequence and thus converges. □

2.3.1 Series of Positive Elements in Banach Lattices

We note the following two results which are series counterparts of the dominated and monotone convergence theorems in Banach lattices.

Theorem 3 ([14]) *Let $(x_n(t))_{n\in\mathbb{N}}$ be family of nonnegative sequences in a Banach lattice X, parameterized by a parameter $t \in T \subset \mathbb{R}$, and let $t_0 \in \overline{T}$.*

(i) If for each $n \in \mathbb{N}$ the function $t \to x_n(t)$ is non-decreasing and $\lim\limits_{t\nearrow t_0} x_n(t) = x_n$ in norm, then

$$\lim_{t\nearrow t_0} \sum_{n=0}^{\infty} x_n(t) = \sum_{n=0}^{\infty} x_n, \tag{31}$$

irrespective of whether the right hand side exists in X or $\|\sum\limits_{n=0}^{\infty} x_n\| := \sup\{\|\sum\limits_{n=0}^{N} x_n\|;\ N \in \mathbb{N}\} = \infty$. In the latter case the equality should be understood as the norms of both sides being infinite.

(ii) If $\lim\limits_{t\to t_0} x_n(t) = x_n$ in norm for each $n \in \mathbb{N}$ and there exists $(a_n)_{n\in\mathbb{N}}$ such that $x_n(t) \leq a_n$ for any $t \in T, n \in \mathbb{N}$ with $\sum\limits_{n=0}^{\infty} \|a_n\| < \infty$, then (31) holds as well.

Remark 2 Note that if X is a *KB*-space, then $\lim_{t\nearrow t_0}\sum_{n=0}^{\infty} x_n(t) \in X$ implies convergence of $\sum_{n=0}^{\infty} x_n$. In fact, since $x_n \geq 0$ (by closedness of the positive cone), $N \to \sum_{n=0}^{N} x_n$ is non-decreasing, and hence either $\sum_{n=0}^{\infty} x_n \in X$, or $\|\sum_{n=0}^{\infty} x_n\| = \infty$ and, in the latter case, $\left\|\lim_{t\nearrow t_0}\sum_{n=0}^{\infty} x_n(t)\right\| = \infty$.

2.4 Complexification

Due to the construction, solutions to all our models must be real. Thus, our problems should be posed in real Banach spaces. However, to take full advantage of the tools of functional analysis, such as the spectral theory, it is worthwhile to extend our spaces to include also complex valued functions, so that they become complex Banach spaces. While the algebraic and metric structure of Banach spaces can be easily extended to the complex setting, the extension of the order structure must be done with more care. This is done by the procedure called *complexification.*

Definition 7 Let X be a real vector lattice. The complexification X_C of X is the set of pairs $(x, y) \in X \times X$ where, following the scalar convention, we write $(x, y) = x + iy$. Vector operations are defined as in scalar case while the partial order is defined by

$$x_0 + iy_0 \leq x_1 + iy_1 \quad \text{if and only if} \quad x_0 \leq x_1 \text{ and } y_0 = y_1. \tag{32}$$

Remark 3 Note that from the definition it follows that $x \geq 0$ in X_C is equivalent to $x \in X$ and $x \geq 0$ in X. In particular, X_C with partial order (32) is not a lattice.

Example 4 Any positive linear operator A on X_C is a real operator; that is, $A : X \to X$. In fact, let $X \ni x = x_+ - x_-$. By definition, $Ax_+ \geq 0$ and $Ax_- \geq 0$ so $Ax_+, Ax_- \in X$ and thus $Ax = Ax_+ - Ax_- \in X$.

It is a more complicated task to introduce a norm on X_C because standard product norms, in general, fail to preserve the homogeneity of the norm, see [13, Example 2.88].

Since X_C is not a lattice, we cannot define the modulus of $z = x + iy \in X_C$ in a usual way. However, following an equivalent definition of the modulus in the scalar case, for $x + iy \in X_C$ we define

$$|x + iy| = \sup_{\theta\in[0,2\pi]} \{x \cos\theta + y \sin\theta\}.$$

It can be proved that this element exists.

Such a defined modulus has all standard properties of the scalar complex modulus. Thus, one can define a norm on the complexification X_C by

$$\|z\|_c = \|x + iy\|_c = \||x + iy|\|. \tag{33}$$

Properties (a)–(c) and $|x| \leq |z|$, $|y| \leq |z|$ imply that $\|\cdot\|_c$ is a norm on X_C, which is equivalent to the Euclidean norm on $X \times X$, denoted by $\|\cdot\|_C$. As the norm $\|\cdot\|$ is a lattice norm on X, we have $\|z_1\|_c \leq \|z_2\|_c$, whenever $|z_1| \leq |z_2|$, and $\|\cdot\|_c$ becomes a lattice norm on X_C.

Definition 8 A complex Banach lattice is an ordered complex Banach space X_C that arises as the complexification of a real Banach lattice X, according to Definition 7, equipped with the norm (33).

Remember: a complex Banach lattice is not a Banach lattice!

Any linear operator A on X can be extended to X_C according to

$$A_C(x + iy) = Ax + iAy.$$

We observe that if A is a positive operator between real Banach lattices X and Y then, for $z = x + iy \in X_C$, we have

$$(Ax)\cos\theta + (Ay)\sin\theta = A(x\cos\theta + y\sin\theta) \leq A|z|,$$

therefore $|A_C z| \leq A|z|$. Hence for positive operators

$$\|A_C\|_c = \|A\|. \tag{34}$$

There are, however, examples where $\|A\| < \|A_C\|_c$.

Note that the standard $L_p(\Omega)$ and $C(\Omega)$ norms are of the type (33). These spaces have a nice property of preserving the operator norm even for operators which are not necessarily positive, see [13, p. 63].

Remark 4 If for a linear operator A, we prove that it generates a semigroup of say, contractions, in X, then this semigroup will be also a semigroup of contractions on X_C, hence, in particular, A is a dissipative operator in the complex setting. Due to this observation we confine ourselves to real operators in real spaces.

2.5 First Semigroups

As mentioned before, we are concerned with methods of finding solutions of the Cauchy problem:

Definition 9 Given a complex or real Banach space and a linear operator $\mathcal{A}$ with domain $D(\mathcal{A})$ and range Ran$\mathcal{A}$ contained in X, and also given an element $\overset{\circ}{u} \subset X$, find a function $u(t) = u(t, \overset{\circ}{u})$ such that

1. $u(t)$ is continuous on $[0,\infty[$ and continuously differentiable on $]0,\infty[$ in the norm of X,
2. for each $t > 0$, $u(t) \in D(\mathcal{A})$ and

$$\partial_t u(t) = \mathcal{A}u(t), \quad t > 0, \tag{35}$$

3.

$$\lim_{t\to 0+} u(t) = u_0 \tag{36}$$

in the norm of X.

A function satisfying all conditions above is called the classical solution of (35), (36). If $u(t) \in D(\mathcal{A})$ (and thus $u \in C^1([0,\infty[, X))$, then such a function is called a strict solution to (35), (36).

To shorten notation, we denote by $C^k(I,X)$ a space of functions which, for each $t \in I \subset \mathbb{R}$ satisfy $u(t) \in X$ and are continuously differentiable k times in t with respect to the norm of X. Thus, e.g. a classical solution u satisfies $u \in C([0,\infty[, X) \cap C^1(]0,\infty[, X)$.

2.5.1 Definitions and Basic Properties

If the solution to (35), (36) is unique, then we can introduce a family of operators $(G(t))_{t\geq 0}$ such that $u(t,u_0) = G(t)u_0$. Ideally, $G(t)$ should be defined on the whole space for each $t > 0$, and the function $t \to G(t)u_0$ should be continuous for each $u_0 \in X$, leading to well-posedness of (35), (36). Moreover, uniqueness and linearity of $\mathcal{A}$ imply that $G(t)$ are linear operators. A fine-tuning of these requirements leads to the following definition.

Definition 10 A family $(G(t))_{t\geq 0}$ of bounded linear operators on X is called a C_0-semigroup, or a strongly continuous semigroup, if

(i) $G(0) = I$;
(ii) $G(t+s) = G(t)G(s)$ for all $t,s \geq 0$;
(iii) $\lim_{t\to 0^+} G(t)x = x$ for any $x \in X$.

A linear operator A is called the (infinitesimal) generator of $(G(t))_{t\geq 0}$ if

$$Ax = \lim_{h\to 0+} \frac{G(h)x - x}{h}, \tag{37}$$

with $D(A)$ defined as the set of all $x \in X$ for which this limit exists. Typically the semigroup generated by A is denoted by $(G_A(t))_{t\geq 0}$.

If $(G(t))_{t\geq 0}$ is a C_0-semigroup, then the local boundedness and (ii) lead to the existence of constants $M > 0$ and ω such that for all $t \geq 0$

$$\|G(t)\|_X \leq Me^{\omega t} \tag{38}$$

(see, e.g., [41, p. 4]). We say that $A \in \mathcal{G}(M,\omega)$ if it generates $(G(t))_{t\geq 0}$ satisfying (38). The *type* of $(G(t))_{t\geq 0}$ is defined as

$$\omega_0(G) = \inf\{\omega;\ \text{there is } M \text{ such that (38) holds}\}. \tag{39}$$

Let $(G_A(t))_{t\geq 0}$ be the semigroup generated by A. The following properties of $(G_A(t))_{t\geq 0}$ are frequently used, [41, Theorem 2.4].

(a) For $x \in X$

$$\lim_{h\to 0}\frac{1}{h}\int\limits_t^{t+h} G_A(s)xds = G_A(t)x. \tag{40}$$

(b) For $x \in X$, $\int_0^t G_A(s)xds \in D(A)$ and

$$A\int\limits_0^t G_A(s)xds = G_A(t)x - x. \tag{41}$$

(c) For $x \in D(A)$, $G_A(t)x \in D(A)$ and

$$\frac{d}{dt}G_A(t)x = AG_A(t)x = G_A(t)Ax. \tag{42}$$

(d) For $x \in D(A)$,

$$G_A(t)x - G_A(s)x = \int\limits_s^t G_A(\tau)Axd\tau = \int\limits_s^t AG_A(\tau)xd\tau. \tag{43}$$

From (42) and condition (iii) of Definition 10 we see that if A is the generator of $(G_A(t))_{t\geq 0}$, then for $\mathring{u}\in D(A)$ the function $t \to G_A(t)\mathring{u}$ is a classical solution of the following Cauchy problem,

$$\partial_t u(t) = A(u(t)), \quad t > 0, \tag{44}$$

$$\lim_{t\to 0^+} u(t) = \mathring{u}\,. \tag{45}$$

We note that ideally the generator A should coincide with $\mathcal{A}$ but in reality very often it is not so. However, for most of this chapter we are concerned with solvability of (44), (45); that is, with the case when $\mathcal{A}$ of (35) is the generator of a semigroup.

We noted above that for $\overset{\circ}{u} \in D(A)$ the function $u(t) = G_A(t)\overset{\circ}{u}$ is a classical solution to (44), (45). For $\overset{\circ}{u} \in X \setminus D(A)$, however, the function $u(t) = G_A(t)\overset{\circ}{u}$ is continuous but, in general, not differentiable, nor $D(A)$-valued, and, therefore, not a classical solution. Nevertheless, from (41), it follows that $v(t) = \int_0^t u(s)ds \in D(A)$ and therefore it is a strict solution of the integrated version of (44), (45):

$$\begin{aligned} \partial_t v &= Av + \overset{\circ}{u}, \quad t > 0 \\ v(0) &= 0, \end{aligned} \tag{46}$$

or equivalently,

$$u(t) = A \int_0^t u(s)ds + \overset{\circ}{u} . \tag{47}$$

We say that a function u satisfying (46) (or, equivalently, (47)) is a *mild solution* or *integral solution* of (44), (45). It can be proved that $t \to G(t)\overset{\circ}{u}$, $\overset{\circ}{u} \in D(A)$, is the only solution of (44), (45) taking values in $D(A)$. Similarly, for $\overset{\circ}{u} \in X$, the function $t \to G(t)\overset{\circ}{u}$ is the only mild solution to (44), (45).

Thus, if we have a semigroup, we can identify the Cauchy problem of which it is a solution. Usually, however, we are interested in the reverse question; that is, in finding the semigroup for a given equation. The answer is given by the Hille–Yoshida theorem (or, more properly, the Feller–Miyadera–Hille–Phillips–Yosida theorem). Before, however, we need to recall some terminology related to the spectrum of an operator.

2.5.2 Interlude: The Spectrum of an Operator

Let us recall that the resolvent set of A is defined by

$$\rho(A) = \{\lambda \in \mathbf{C};\ (\lambda I - A)^{-1} \in \mathcal{L}(X)\}$$

and, for $\lambda \in \rho(A)$, we define the resolvent of A by

$$R(\lambda, A) = (\lambda I - A)^{-1}.$$

The complement of $\rho(A)$ in $\mathbb{C}$ is called the *spectrum* of A and denoted by $\sigma(A)$. In general, it is possible that either $\rho(A)$ or $\sigma(A)$ is empty. The spectrum is usually subdivided into several subsets.

- *Point spectrum* $\sigma_p(A)$ is the set of $\lambda \in \sigma(A)$ for which the operator $\lambda I - A$ is not one-to-one. In other words, $\sigma_p(A)$ is the set of all eigenvalues of A.
- *Residual spectrum* $\sigma_r(A)$ is the set of $\lambda \in \sigma(A)$ for which $\lambda I - A$ is one-to-one and Ran $(\lambda I - A)$ is not dense in X.
- *Continuous spectrum* $\sigma_c(A)$ is the set of $\lambda \in \sigma(A)$ for which the operator $\lambda I - A$ is one-to-one and its range is dense in, but not equal to, X

The resolvent of any operator A satisfies the *resolvent identity*

$$R(\lambda, A) - R(\mu, A) = (\mu - \lambda)R(\lambda, A)R(\mu, A), \qquad \lambda, \mu \in \rho(A). \tag{48}$$

For any bounded operator the spectrum is a compact subset of $\mathbb{C}$ so that $\rho(A) \neq \emptyset$. If A is bounded, then the limit

$$r(A) = \lim_{n\to\infty} \sqrt[n]{\|A^n\|} \tag{49}$$

exists and is called *the spectral radius*. Clearly, $r(A) \leq \|A\|$. Equivalently,

$$r(A) = \sup_{\lambda\in\sigma(A)} |\lambda|. \tag{50}$$

To show that $\lambda \in \mathbb{C}$ belongs to the spectrum we often use the following result.

Theorem 4 *Let A be a closed operator. If $\lambda \in \rho(A)$, then dist$(\lambda, \sigma(A)) = 1/r(R(\lambda, A)) \geq 1/\|R(\lambda, A)\|$. In particular, if $\lambda_n \to \lambda$, $\lambda_n \in \rho(A)$, then $\lambda \in \sigma(A)$ if and only if $\{\|R(\lambda_n, A)\|\}_{n\in\mathbb{N}}$ is unbounded.*

For an unbounded operator A the role of the spectral radius often is played by the *spectral bound* $s(A)$ defined as

$$s(A) = \sup\{\Re\lambda;\ \lambda \in \sigma(A)\}. \tag{51}$$

2.5.3 Hille–Yosida Theorem

Theorem 5 *$A \in \mathcal{G}(M, \omega)$ if and only if*

(a) A is closed and densely defined,
(b) there exist $M > 0, \omega \in \mathbb{R}$ such that $(\omega, \infty) \subset \rho(A)$ and for all $n \geq 1, \lambda > \omega$,

$$\|(\lambda I - A)^{-n}\| \leq \frac{M}{(\lambda - \omega)^n}. \tag{52}$$

If A is the generator of $(G_A(t))_{t\geq 0}$, then

$$R(\lambda, A)x = \int_0^\infty e^{-\lambda t} G_A(t)xdt, \Re\lambda > \omega, \tag{53}$$

is valid for all $x \in X$.

A widely used approximation formula, which can also be used in the generation proof, is the operator version of the well-known scalar formula. Precisely, [41, Theorem 1.8.3], if A is the generator of a C_0-semigroup $(G_A(t))_{t\geq 0}$, then for any $x \in X$,

$$G_A(t)x = \lim_{n\to\infty}\left(I - \frac{t}{n}A\right)^{-n} x = \lim_{n\to\infty}\left(\frac{n}{t}R\left(\frac{n}{t}, A\right)\right)^n x \tag{54}$$

and the limit is uniform in t on bounded intervals.

2.5.4 Dissipative Operators and Contractive Semigroups

Let X be a Banach space (real or complex) and X^* be its dual. From the Hahn–Banach theorem, for every $u \in X$ there exists $u^* \in X^*$ satisfying $<u^*, u>= \|u\|^2 = \|u^*\|^2$. Therefore the *duality set*

$$\mathcal{J}(u) = \{u^* \in X^*;\ <u^*, u>= \|u\|^2 = \|u^*\|^2\} \tag{55}$$

is nonempty for every $u \in X$.

Definition 11 We say that an operator $(A, D(A))$ is *dissipative* if for every $u \in D(A)$ there is $u^* \in \mathcal{J}(u)$ such that

$$\Re <u^*, Ax> \leq 0. \tag{56}$$

If X is a real space, then the real part in the above definition can be dropped. An important equivalent characterisation of dissipative operators, [41, Theorem 1.4.2], is that A is dissipative if and only if for all $\lambda > 0$ and $u \in D(A)$,

$$\|(\lambda I - A)u\| \geq \lambda\|u\|. \tag{57}$$

Combination of the Hille–Yosida theorem with the above property gives a generation theorem for dissipative operators, known as the Lumer–Phillips theorem ([41, Theorem 1.4.3] or [27, Theorem II.3.15]).

Theorem 6 *For a densely defined dissipative operator $(A, D(A))$ on a Banach space X, the following statements are equivalent.*

(a) The closure $\overline{A}$ generates a semigroup of contractions.
(b) $\overline{\text{Ran}(\lambda I - A)} = X$ for some (and hence all) $\lambda > 0$.

If either condition is satisfied, then A satisfies (56) for any $u^* \in \mathcal{J}(u)$.

We observe that a densely defined dissipative operator is closable, [41, Theorem 1.4.5], so that the statement in item (a) makes sense. In other words, to prove that (the closure of) a dissipative operators generates a semigroup, we only need to show that the equation

$$\lambda u - Au = f \tag{58}$$

is solvable for f from a dense subset of X for some $\lambda > 0$. We do not need to prove that the solution to (58) defines a resolvent satisfying (52).

In particular, if we know that A is closed, then the density of $\text{Ran}(\lambda I - A)$ is sufficient for A to be a generator. On the other hand, if we do not know a priori that A is closed, then $\text{Ran}(\lambda I - A) = X$ yields A being closed and consequently that it is a generator.

Example 5 Let us have a look at the classical problem which often is incorrectly solved. Consider

$$Au = -\partial_x u, \quad x \in (0, 1),$$

on $D(A) = \{u \in W_1^1(I);\ u(0) = 0\}$, where $I\ =]0, 1[$. The state space is real $X = L_1(I)$. For a given $u \in X$, we have

$$\mathcal{J}(u) = \begin{cases} \|u\| \text{sign} u(x) & \text{if } u(x) \neq 0, \\ \alpha \in [-\|u\|, \|u\|] & \text{if } u(x) = 0. \end{cases}$$

Note that $\mathcal{J}$ is a multivalued function. Further, [23], any element of $W_1^1(I)$ can be represented by an absolutely continuous function on I.

Now, for $v \in \mathcal{J}(u)$ we have

$$\begin{aligned} <-\partial_x u, v> &= -\|u\| \int\limits_0^1 \partial_x u(x) \text{sign} u(x) dx \\ &= -\|u\| \left(\int\limits_{\{x \in I;\, u(x) > 0\}} \partial_x u(x) dx - \int\limits_{\{x \in I;\, u(x) < 0\}} \partial_x u(x) dx \right). \end{aligned}$$

Since u is continuous, both sets $I_+ := \{x \in I;\ u(x) > 0\}$ and $I_- := \{x \in I;\ u(x) < 0\}$ are open. Then, see [2, p. 42],

$$I_\pm = \bigcup_n]\alpha_n^\pm, \beta_n^\pm[$$

where $]\alpha_n^\pm, \beta_n^\pm[$ are non overlapping open intervals. Then

$$\int_{I_\pm} \partial_x u(x)dx = \sum_n (u(\beta_n^\pm) - u(\alpha_n^\pm)) = \begin{cases} u(1) & \text{if } 1 \in I_\pm, \\ 0 & \text{if } 1 \notin I_\pm, \end{cases}$$

as 1 only can be the right end of the component intervals and we used $u(0) = 0$. Now, if $1 \in I_+$, then $u(1) > 0$, if $1 \in I_-$, then $u(1) < 0$, and if $1 \notin I_+ \cup I_-$, then $u(1) = 0$. In any case,

$$<-\partial_x u, v> \leq 0$$

and the operator $(A, D(A))$ is dissipative. Clearly, the solution of

$$\lambda u + \partial_x u = f, \qquad u(0) = 0,$$

is given by $u(x) = e^{-\lambda x} \int\limits_0^x e^{\lambda s} f(s)ds$ and, for $\lambda > 0$,

$$\|u\| \leq \int_0^1 e^{-\lambda x} \left(\int_0^x e^{\lambda s} |f(s)| ds \right) dx \leq \|f\|,$$

which gives solvability of (58) in X. We note that, of course, with a more careful integration we would be able to obtain the Hille–Yosida estimate (52). This additional work is, however, not necessary for dissipative operators.

2.5.5 Analytic Semigroups

In the previous paragraph we noted that if a closed operator is dissipative, then we can prove that it generates a semigroup, provided (58) is solvable. It turns out that the solvability of (58) can be used to prove that A generates a semigroup without assuming that it is dissipative but then we must consider complex λ. Note that the considerations below are valid for an arbitrary Banach space.

Hence, let the inverse $(\lambda I - A)^{-1}$ exists in the sector

$$S_{\frac{\pi}{2}+\delta} := \{\lambda \in \mathbb{C};\ |\arg \lambda| < \frac{\pi}{2} + \delta\} \cup \{0\}, \tag{59}$$

for some $0 < \delta < \frac{\pi}{2}$, and let there exist C such that for every $0 \neq \lambda \in S_{\frac{\pi}{2}+\delta}$ the following estimate holds:

$$\|R(\lambda, A)\| \leq \frac{C}{|\lambda|}. \tag{60}$$

Then A is the generator of a uniformly bounded semigroup $(G_A(t))_{t\geq 0}$ (the constant M in (38) not necessarily equals C) and $(G_A(t))_{t\geq 0}$ is given by

$$G_A(t) = \frac{1}{2\pi i} \int\limits_{\Gamma} e^{\lambda t} R(\lambda, A) \mathrm{d}\lambda, \tag{61}$$

where Γ is an unbounded smooth curve in $S_{\frac{\pi}{2}+\delta}$. The reason why $(G_A(t))_{t\geq 0}$ is called analytic is that it extends to an analytic function on S_δ.

The estimate (60) is sometimes awkward to prove as it requires the knowledge of the resolvent in the whole sector. The result given in [27, Theorem II 4.6] allows to restrict the estimates to a positive half plane.

Theorem 7 *An operator $(A, D(A))$ on a Banach space X generates a bounded analytic semigroup $(G_A(z))_{z\in S_\delta}$ in a sector Σ_δ if and only if A generates a bounded strongly continuous semigroup $(G_A(t))_{t\geq 0}$ and there exists a constant $C > 0$ such that*

$$\|R(r + is, A)\| \leq \frac{C}{|s|} \tag{62}$$

for all $r > 0$ and $0 \neq s \in \mathbb{R}$.

This result can be generalized to arbitrary analytic semigroups: $(A, D(A))$ generates an analytic semigroup $(G_A(z))_{z\in S_\delta}$ if and only if A generates a strongly continuous semigroup $(G_A(t))_{t\geq 0}$ and there exist constants $C > 0, \omega > 0$ such that

$$\|R(r + is, A)\| \leq \frac{C}{|s|} \tag{63}$$

for all $r > \omega$ and $0 \neq s \in \mathbb{R}$.

If A is the generator of an analytic semigroup $(G_A(t))_{t\geq 0}$, then $t \to G_A(t)$ has derivatives of arbitrary order on $]0, \infty[$. This shows that $t \to G_A(t)\overset{\circ}{u}$ solves the Cauchy problem (36) for arbitrary $\overset{\circ}{u} \in X$. This is a significant improvement upon the case of C_0-semigroup, for which $\overset{\circ}{u} \in D(A)$ was required.

Fractional Powers of Generators and Interpolation Spaces

If A generates an analytic semigroup, then (61) can be regarded as the extension of the definition of e^{tA} via the so-called Dunford integral type functional calculus, [41]. In a similar way we can define $f(A)$ where f is any scalar function which is analytic in an open neighbourhood of the spectrum of A and such that the integral (61) is convergent.

One of the most important choices of f is

$$f(\lambda) = (-\lambda)^{-\alpha} = e^{i\pi\alpha}\lambda^{-\alpha},$$

where λ^α is real for $\lambda > 0$ and has a cut along the positive real axis. This gives rise to bounded operators $(-A)^{-\alpha}$ and, by inversion, to unbounded operators $(-A)^\alpha$.

We denote by $D((-A)^\alpha)$ the domain of $(-A)^\alpha$. It follows that

$$D(A) \subset D((-A)^\alpha) \subset X$$

if $0 < \alpha < 1$. For example, if $A = \Delta$ on the maximal domain in $L_2(\mathbb{R}^n)$, then $D(A) = W_2^2(\mathbb{R}^n)$ and $D((-A)^\alpha) = W_2^{2\alpha}(\mathbb{R}^n)$, see [37, 38].

We note an important property of fractional powers of generators and of the corresponding analytic semigroups, which will used in the sequel. If $(G_A(t))_{t\geq 0}$ is an analytic semigroup, then for every $t > 0$ and $\alpha \geq 0$ the operator $(-A)^\alpha G_A(t)(= G_A(t)(-A)^\alpha)$ is bounded and

$$\|t^\alpha(-A)^\alpha G_A(t)\| \leq M_\alpha \tag{64}$$

for some constant M_α.

Example 6 Let us consider briefly the classical example of the Dirichlet problem for the heat equation in $X = L_2(\mathbb{R}^n)$:

$$\begin{aligned} \partial_t u &= \Delta u, \\ u(0,x) &= \mathring{u} \in X. \end{aligned} \tag{65}$$

If we define $A = \Delta$ on the domain $D(A) = \{u \in W_2^1(\mathbb{R}^n);\ \Delta u \in X\}$ then, using the integration by parts for Sobolev space functions, we obtain

$$\int\limits_{\mathbb{R}^n} (\lambda u - \Delta u)\bar{u} dx = \lambda \int\limits_{\mathbb{R}^n} u\bar{u} dx + \int\limits_{\mathbb{R}^n} \nabla u \overline{\nabla u} dx =: a(u,u)$$

and it is easy to see that for $u \in W_2^1(\mathbb{R}^n)$

$$a(u,u) \geq \alpha \|u\|^2_{W_2^1(\mathbb{R}^n)},$$

for $\alpha = \min\{\lambda, 1\}$ so that A is coercive. This, on the one hand gives the solvability of

$$\lambda u - Au = f$$

by the Lax–Milgram lemma; that is, the existence of $(\lambda I - A)^{-1}$, and on the other hand, by dropping the term $\|u\|^2_{W_2^1(\mathbb{R}^n)}$, we obtain the Hille–Yosida estimate

$$\|(\lambda - A)^{-1} f\|_X \leq \lambda^{-1}\|f\|_X, \quad \lambda > 0.$$

Since A is closed and $D(A)$ is dense in X, A generates a semigroup of contractions $(G_A(t))_{t\geq 0}$ on $X = L_2(\mathbb{R}^n)$.

With some more work, see e.g. [42, Chapter 4], one can show that the resolvent satisfies also the estimate (60) yielding the analyticity of $(G_A(t))_{t\geq 0}$. Hence, the problem (65) is solvable for any initial value $\overset{\circ}{u}\in L_2(\mathbb{R}^n)$ and the solution $G_A(t)\overset{\circ}{u}$ is infinitely many times differentiable for $t > 0$ and such that for $t > 0$

$$G_A(t)\overset{\circ}{u}\in \bigcap_{n\geq 0} D(A^n) \subset \bigcap_{n\geq 0} W_2^{2n}(\Omega) \subset C^\infty(\Omega).$$

The last property expresses the smoothing property of the heat semigroup—for any initial value the solution becomes instantly infinitely smooth.

Furthermore, $D((-A)^\alpha) = W_2^{2\alpha}(\mathbb{R}^n)$ for $\alpha > 0$, see e.g. [37,38].

The spaces $D((-A)^\alpha)$ form an important class of intermediate spaces between $D(A)$ and X. However, in some situations they are not sufficient as they are not a priori independent of the form of A. To remove this drawback, first we observe that (64) can be written as

$$\|t^{1-\alpha} AG_A(t)x\| \in L_\infty(]0, 1[),$$

whenever $x \in D((-A)^\alpha)$. Taking this formula as the starting point, let $(A, D(A))$ be the generator of an analytic semigroup $(G_A(t))_{t\geq 0}$ on a Banach space X. Then we construct a family of intermediate spaces, $D_A(\alpha, r)$, $0 < \alpha < 1$, $1 \leq r \leq \infty$ in the following way:

$$D_A(\alpha, r) := \{x \in X : t \to v(t) := \|t^{1-\alpha-1/r} AG_A(t)x\|_X \in L_r(I)\}, \quad (66)$$

$$\|x\|_{D_A(\alpha,r)} := \|x\|_X + \|v(t)\|_{L_r(I)}, \quad (67)$$

where $I := (0, 1)$; see [38, p. 45]. It turns out that these spaces can be identified with real interpolation spaces between X and $D(A)$ and one can use a rich theory of the latter. In particular, by [38, Corollary 2.2.3], these spaces do not depend explicitly on A, but only on $D(A)$ and its graph norm. This is in contrast to the property of $D((-A)^\alpha)$ mentioned above, where we only have

$$D_A(\alpha, 1) \subset D((-A)^\alpha) \subset D_A(\alpha, \infty) \tag{68}$$

so in general $D((-A)^\alpha)$ may depend on the particular form of A.

Which makes the spaces $D_A(\alpha, r)$ as useful as $D((-A)^\alpha)$ in dealing with the semigroup generated by A is the fact that

$$\begin{aligned}
\|R(\lambda, A)x\|_{D_A(\alpha,r)} &= \|R(\lambda, A)x\|_{\mathcal{L}(X)} + \left(\int_0^1 \|s^{1-\alpha-1/r} AG_A(s)R(\lambda, A)x\|_X^r\right)^{1/r} \\
&= \|R(\lambda, A)x\|_{\mathcal{L}(X)} + \left(\int_0^1 \|R(\lambda, A)(s^{1-\alpha-1/r} AG_A(s)x)\|_X^r ds\right)^{1/r} \\
&\leq \|R(\lambda, A)\|_{\mathcal{L}(X)} \left(\|x\|_X + \left(\int_0^1 \|(s^{1-\alpha-1/r} AG_A(s)x)\|_X^r ds\right)^{1/r}\right) \\
&\leq \|R(\lambda, A)\|_{\mathcal{L}(X)} \|x\|_{D_A(\alpha,r)}.
\end{aligned}$$

This leads to the following observation.

Proposition 2 *Let $A_{\alpha,r}$ be the part of A in $D_A(\alpha, r)$. Then $\rho(A_{\alpha,r}) \supset \rho(A)$, $\|R(\lambda, A_{\alpha,r})\|_{\mathcal{L}(D_A(\alpha,r))} \leq \|R(\lambda, A)\|_{\mathcal{L}(X)}$ for $\lambda \in \rho(A)$. Consequently, $A_{\alpha,r}$ generates an analytic semigroup in $D_A(\alpha, r)$.*

2.5.6 Nonhomogeneous Problems

Let us consider the problem of finding the solution to the Cauchy problem:

$$\begin{aligned}
\partial_t u &= Au + f(t), \quad 0 < t < T, \\
u(0) &= \overset{\circ}{u},
\end{aligned} \tag{69}$$

where $0 < T \leq \infty$, A is the generator of a semigroup and $f : (0, T) \to X$ is a known function.

If we are interested in classical solutions, then clearly f must be continuous. However, this condition proves to be insufficient. Thus we generalise the concept of the *mild solution* introduced in (47). We observe that if u is a classical solution of (69), then it must be given by

$$u(t) = G(t)\overset{\circ}{u} + \int_0^t G(t-s)f(s)ds \tag{70}$$

(see, e.g., [41, Corollary 4.2.2]). The integral is well defined even if $f \in L_1([0,T],X)$ and $\overset{\circ}{u} \in X$. We call u defined by (70) the *mild solution* of (69). For an integrable f such a u is continuous but not necessarily differentiable, and therefore it may be not a solution to (69).

We have the following theorem giving sufficient conditions for a mild solution to be a classical solution (see, e.g., [41, Corollaries 4.2.5 and 4.2.6]).

Theorem 8 *Let A be the generator of a C_0-semigroup $(G_A(t))_{t\geq 0}$ and $x \in D(A)$. Then (70) is a classical solution of (69) if either*

(i) $f \in C^1([0,T],X)$, *or*
(ii) $f \in C([0,T],X) \cap L_1([0,T],D(A))$.

The assumptions of this theorem are often too restrictive for applications. On the other hand, it is not clear exactly what the mild solutions solve. A number of weak formulations of (69) have been proposed (see e.g. [29, pp. 88–89] or [9]), all of them having (70) as their solutions. We present here a result from [27, p. 451] which is particularly suitable for applications.

Proposition 3 *A function $u \in C(\mathbb{R}_+,X)$ is a mild solution to (69) with $f \in L_1(\mathbb{R}_+,X)$ in the sense of (70) if and only if $\int_0^t u(s)ds \in D(A)$ and*

$$u(t) = \overset{\circ}{u} + A\int_0^t u(s)ds + \int_0^t f(s)ds, \qquad t \geq 0. \tag{71}$$

If the semigroup $(G_A(t))_{t\geq 0}$ generated by A is analytic, then the requirements imposed on f can be substantially weakened. We have then the following counterpart of Theorem 8.

Theorem 9 *Let A be the generator of an analytic semigroup $(G_A(t))_{t\geq 0}$, $\overset{\circ}{u} \in X$ and $f \in L_1([0,T],X)$. Then (70) is the classical solution of (69) if either*

(i) f is locally Hölder continuous on $]0,T[$, or
(ii) there exists $\alpha > 0$ such that $f \in C(]0,T],X) \cap L_1([0,T],D((-A)^\alpha))$ and $t \to \|(-A)^\alpha f(t)\|_X$ is bounded over compact subsets of $]0,T]$.

Part (ii) of this theorem has been proved in [10].

An important refinement of this result, which becomes very useful in nonlinear problems is that, actually, the solution has a better regularity. In fact, under assumption (i), we additionally have

$$u \in C^{0,1-r}_{\text{loc}}(]0, T[, D((-A)^r)), \qquad 0 \leq r < 1. \tag{72}$$

By (68), the statement of the above theorem holds if the domains of the fractional powers are replaced by appropriate intermediate spaces $D(\alpha, 1)$.

2.5.7 Positive Semigroups

Definition 12 Let X be a Banach lattice. We say that the semigroup $(G(t))_{t\geq 0}$ on X is positive if for any $x \in X_+$ and $t \geq 0$,

$$G(t)x \geq 0.$$

We say that an operator $(A, D(A))$ is resolvent positive if there is ω such that $(\omega, \infty) \subset \rho(A)$ and $R(\lambda, A) \geq 0$ for all $\lambda > \omega$.

Remark 5 In this section, because we address several problems related to spectral theory, we need complex Banach lattices. Let us recall, Definitions 7 and 8, that a complex Banach lattice is always a complexification X_C of an underlying real Banach lattice X. In particular, $x \geq 0$ in X_C if and only if $x \in X$ and $x \geq 0$ in X.

It is easy to see that a strongly continuous semigroup is positive if and only if its generator is resolvent positive. In fact, the positivity of the resolvent for $\lambda > \omega$ follows from (53) and the closedness of the positive cone; see Proposition 1. Conversely, the latter, together with the exponential formula (54), shows that resolvent positive operators generate positive semigroups.

Positivity of a semigroup allows for strengthening of several results pertaining to the spectrum of its generator.

Theorem 10 ([43]) *Let $(G_A(t))_{t\geq 0}$ be a positive semigroup on a Banach lattice, with the generator A. Then*

$$R(\lambda, A)x = \int_0^\infty e^{-\lambda t} G_A(t)x dt \tag{73}$$

for all $\lambda \in \mathbb{C}$ with $\Re\lambda > s(A)$. Furthermore,

(i) Either $s(A) = -\infty$ or $s(A) \in \sigma(A)$;
(ii) For a given $\lambda \in \rho(A)$, we have $R(\lambda, A) \geq 0$ if and only if $\lambda > s(A)$;
(iii) For all $\Re\lambda > s(A)$ and $x \in X$, we have $|R(\lambda, A)x| \leq R(\Re\lambda, A)|x|$.

We conclude this section by briefly describing an approach of [3] which leads to several interesting results for resolvent positive operators. To fix attention, assume for the time being that $\omega < 0$ (thus, in particular, A is invertible and $-A^{-1} = R(0, A)$) and $\lambda > 0$. The resolvent identity

$$-A^{-1} = (\lambda - A)^{-1} + \lambda(\lambda - A)^{-1}(-A^{-1}),$$

can be extended by induction to

$$-A^{-1} = R(\lambda, A) + \lambda R(\lambda, A)^2 + \cdots + \lambda^n R(\lambda, A)^n(-A^{-1}). \tag{74}$$

Now, because all terms above are nonnegative, we obtain

$$\sup_{n\in\mathbb{N},\lambda>\omega} \{\lambda^n \|(\lambda - A)^{-n}(-A^{-1})\|_X\} = M < +\infty.$$

This is 'almost' the Hille–Yosida estimate and allows us to prove that the Cauchy problem (44), (45) has a mild Lipschitz continuous solution for $\mathring{u} \in D(A^2)$. If, in addition, A is densely defined, then this mild solution is differentiable, and thus it is a strict solution (see, e.g., [4] and [6, pp. 191–200]). These results are obtained by means of the integrated, or regularised, semigroups, which are beyond the scope of this lecture and thus we do not enter into details of this very rich field. We mention, however, an interesting consequence of (74) for the semigroup generation, which has already found several applications and which we use later.

Theorem 11 ([4, 22]) *Let A be a densely defined resolvent positive operator. If there exist $\lambda_0 > s(A), c > 0$ such that for all $x \geq 0$,*

$$\|R(\lambda_0, A)x\|_X \geq c\|x\|_X, \tag{75}$$

then A generates a positive semigroup $(G_A(t))_{t\geq 0}$ on X.

Proof Let us take $s(A) < \omega \leq \lambda_0$ and set $B = A - \omega I$ so that $s(B) < 0$. Because $R(0, B) = R(\omega, A) \geq R(\lambda_0, A)$, it follows from (75) and (30) that

$$\|R(0, B)x\|_X \geq \|R(\lambda_0, A)x\|_X \geq c\|x\|_X$$

for $x \geq 0$. Using (74) for B and $x = \lambda^n R(\lambda, B)^n g$, $g \geq 0$, we obtain, by (75),

$$\|\lambda^n R(\lambda, B)^n g\|_X \leq c^{-1}\|R(0, B)\lambda^n R(\lambda, B)^n g\| \leq M\|g\|_X,$$

for $\lambda > 0$. Again, by (30), we can extend the above estimate onto X proving the Hille–Yosida estimate. Because B is densely defined, it generates a bounded positive semigroup and thus $\|G_A(t)\| \leq Me^{\omega t}$. □

2.6 Perturbation Techniques

Verifying conditions of the Hille–Yosida, or even of the Lumer–Phillips, theorems for a concrete problem is quite often a formidable task. On the other hand, in many cases the operator appearing in the evolution equation at hand is built as a combination of much simpler operators that are relatively easy to analyse. The question now is to what extent the properties of these simpler operators are inherited by the full equation. More precisely, we are interested in the problem:

Problem P. *Let $(A, D(A))$ be a generator of a C_0-semigroup on a Banach space X and $(B, D(B))$ be another operator in X. Under what conditions does $A + B$, or an extension K of $A + B$, generates a C_0-semigroup on X?*

We note that the situation when $K = A + B$ is quite rare. Usually at best we can show that there is an extension of $A + B$ (another realization of $\mathcal{K} = \mathcal{A} + \mathcal{B}$) which is the generator. The reason for this is that, unless B is in some sense strictly subordinated to A, adding B to A may significantly alter some vital properties of A. The identification of K in such cases usually is a formidable task.

A Spectral Criterion

Usually the first step in establishing whether $A + B$, or some of its extensions, generates a semigroup is to find if $\lambda I - (A + B)$ (or its extension) is invertible for all sufficiently large λ.

In all cases discussed here we have the generator $(A, D(A))$ of a semigroup and a perturbing operator $(B, D(B))$ with $D(A) \subseteq D(B)$.

We note that B is A-bounded; that is, for some $a, b \geq 0$ we have

$$\|Bx\| \leq a\|Ax\| + b\|x\|, \qquad x \in D(A), \tag{76}$$

if and only if $BR(\lambda, A) \in \mathcal{L}(X)$ for $\lambda \in \rho(A)$.

In what follows we denote by K an extension of $A + B$. We now present an elegant result relating the invertibility properties of $\lambda I - K$ to the properties of 1 as an element of the spectrum of $BR(\lambda, A)$, first derived in [30].

Theorem 12 *Assume that $\Lambda = \rho(A) \cap \rho(K) \neq \emptyset$.*

(a) $1 \notin \sigma_p(BR(\lambda, A))$ for any $\lambda \in \Lambda$;
(b) $1 \in \rho(BR(\lambda, A))$ for some/all $\lambda \in \Lambda$ if and only if $D(K) = D(A)$ and $K = A + B$;
(c) $1 \in \sigma_c(BR(\lambda, A))$ for some/all $\lambda \in \Lambda$ if and only if $D(A) \subsetneq D(K)$ and $K = \overline{A + B}$;
(d) $1 \in \sigma_r(BR(\lambda, A))$ for some/all $\lambda \in \Lambda$ if and only if $K \supsetneq \overline{A + B}$.

Corollary 1 *Under the assumptions of Theorem 12, $K = A + B$ if one of the following criteria is satisfied: for some $\lambda \in \rho(\Lambda)$ either*

(i) $BR(\lambda, A)$ is compact (or, if $X = L_1(\Omega, d\mu)$, weakly compact), or
(ii) the spectral radius $r(BR(\lambda, A)) < 1$.

Proof If (ii) holds, then obviously $I - BR(\lambda, A)$ is invertible by the Neumann series theorem:

$$(I - BR(\lambda, A))^{-1} = \sum_{n=0}^{\infty} (BR(\lambda, A))^n, \tag{77}$$

giving the thesis by Theorem 12(b). Additionally, we obtain

$$R(\lambda, A + B) = R(\lambda, A)(I - BR(\lambda, A))^{-1} = R(\lambda, A) \sum_{n=0}^{\infty} (BR(\lambda, A))^n. \tag{78}$$

If (i) holds, then either $BR(\lambda, A)$ is compact or, in L_1 setting, $(BR(\lambda, A))^2$ is compact, [26, p. 510], and therefore, if $I - BR(\lambda, A)$ is not invertible, then 1 must be an eigenvalue, which is impossible by Theorem 12(a). □

If we write the resolvent equation

$$(\lambda I - (A + B))x = y, \quad y \in X, \tag{79}$$

in the (formally) equivalent form

$$x - R(\lambda, A)Bx = R(\lambda, A)y, \tag{80}$$

then we see that we can hope to recover x provided the Neumann series

$$R(\lambda)y := \sum_{n=0}^{\infty} (R(\lambda, A)B)^n R(\lambda, A)y = \sum_{n=0}^{\infty} R(\lambda, A)(BR(\lambda, A))^n y. \tag{81}$$

is convergent. Clearly, if (77) converges, then we can factor out $R(\lambda, A)$ from the series above getting again (78). However, $R(\lambda, A)$ inside acts as a regularising factor and (81) converges under weaker assumptions than (77) and this fact is frequently used to construct the resolvent of an extension of $A + B$ (see e.g. Theorem 16 and, in general, results of Sect. 2.6.1).

The most often used perturbation theorem is the Bounded Perturbation Theorem, see e.g. [27, Theorem III.1.3]

Theorem 13 *Let $(A, D(A)) \in \mathcal{G}(M, \omega)$ for some $\omega \in \mathbb{R}$, $M \geq 1$. If $B \in \mathcal{L}(X)$, then $(K, D(K)) = (A + B, D(A)) \in \mathcal{G}(M, \omega + M\|B\|)$.*

In many cases the Bounded Perturbation Theorem gives insufficient information. Then it can be combined with the Trotter product formula, [27, 41]. Assume K_0 is of type $(1, \omega_0)$, $\omega \in \mathbb{R}$, and K_1 is of type $(1, \omega_1)$. If $(K, D(K_0) \cap D(K_1)) := (K_0 + K_2, D(K_0) \cap D(K_1))$ generates a semigroup, then

$$G_K(t)x = \lim_{n\to\infty} (G_{K_0}(t/n)G_{K_1}(t/n))^n x, \quad x \in X, \tag{82}$$

uniformly in t on compact intervals and K is of type $(1, \omega)$ with $\omega = \omega_0 + \omega_1$. Moreover, if both semigroups $(G_{K_0}(t))_{t\geq 0}$ and $(G_{K_1}(t))_{t\geq 0}$ are positive, then $(G_K(t))_{t\geq 0}$ is positive.

The assumption of boundedness of B, however, is often too restrictive. Another frequently used result uses special structure of dissipative operators.

Theorem 14 *Let A and B be linear operators in X with $D(A) \subseteq D(B)$ and $A+tB$ is dissipative for all $0 \leq t \leq 1$. If*

$$\|Bx\| \leq a\|Ax\| + b\|x\|, \tag{83}$$

for all $x \in D(A)$ with $0 \leq a < 1$ and for some $t_0 \in [0, 1]$ the operator $(A + t_0B, D(A))$ generates a semigroup (of contractions), then $A + tB$ generates a semigroup of contractions for every $t \in [0, 1]$.

2.6.1 Positive Perturbations of Positive Semigroups

Perturbation results can be significantly strengthened in the framework of positive semigroups. This approach goes back to the work of Kato [33]. His results were extended in [13, 44] and recently, in a more abstract setting, in [8, 39]. The presented results are based on the exposition of [13] which is sufficient for our purposes.

We have seen in (77) that the condition $r(BR(\lambda, A)) < 1$ implies invertibility of $\lambda I - (A+B)$. It turns out that this condition is equivalent to invertibility for positive perturbations of resolvent positive operators.

Theorem 15 ([45]) *Assume that X is a Banach lattice. Let A be a resolvent positive operator in X and $\lambda > s(A)$. Let $B : D(A) \to X$ be a positive operator. Then the following are equivalent,*

(a) $r(B(\lambda I - A)^{-1}) < 1$;
(b) $\lambda \in \rho(A + B)$ *and* $(\lambda I - (A + B))^{-1} \geq 0$.

If either condition is satisfied, then

$$(\lambda I - A - B)^{-1} = (\lambda I - A)^{-1} \sum_{n=0}^{\infty} (B(\lambda I - A - B)^{-1})^n \geq (\lambda I - A)^{-1}. \tag{84}$$

Kato–Voigt Type Results

Here we consider only $X = L_1(\Omega, \mu)$. Let $(G(t))_{t\geq 0}$ be a strongly continuous semigroup on X. We say that $(G(t))_{t\geq 0}$ is a *substochastic semigroup* if for any $t \geq 0$ and $f \geq 0$, $G(t)f \geq 0$ and $\|G(t)f\| \leq \|f\|$, and a *stochastic semigroup* if additionally $\|G(t)f\| = \|f\|$ for $f \in X_+$.

Theorem 16 *Let $X = L_1(\Omega)$ and let the operators A and B satisfy*

1. *$(A, D(A))$ generates a substochastic semigroup $(G_A(t))_{t\geq 0}$;*
2. *$D(B) \supset D(A)$ and $Bu \geq 0$ for $u \in D(B)_+$;*
3. *for all $u \in D(A)_+$*

$$\int_{\Omega} (Au + Bu)d\mu \leq 0. \tag{85}$$

Then there is an extension $(K, D(K))$ of $(A+B, D(A))$ generating a C_0-semigroup of contractions, say, $(G_K(t))_{t\geq 0}$. The generator K satisfies

$$R(\lambda, K)u = \sum_{k=0}^{\infty} R(\lambda, A)(BR(\lambda, A))^k u, \quad \lambda > 0. \tag{86}$$

Proof First, assumption (85) gives us dissipativity on the positive cone. Next, let us take $u = R(\lambda, A)x = (\lambda I - A)^{-1}x$ for $x \in X_+$ so that $u \in D(A)_+$. Because $R(\lambda, A)$ is a surjection from X onto $D(A)$, by

$$(A + B)u = (A + B)R(\lambda, A)x = -x + BR(\lambda, A)x + \lambda R(\lambda, A)x,$$

we have

$$-\int_{\Omega} x\, d\mu + \int_{\Omega} BR(\lambda, A)x\, d\mu + \lambda \int_{\Omega} R(\lambda, A)x\, d\mu \leq 0. \tag{87}$$

Rewriting the above in terms of the norms, we obtain

$$\lambda\|R(\lambda, A)x\| + \|BR(\lambda, A)x\| - \|x\| \leq 0, \qquad x \in X_+, \tag{88}$$

from which $\|BR(\lambda, A)\| \leq 1$.

We define operators $K_r, 0 \leq r < 1$ by $K_r = A + rB$, $D(K_r) = D(A)$. We see that the spectral radius of $rBR(\lambda, A)$ does not exceed $r < 1$, the resolvent $(\lambda I - (A + rB))^{-1}$ exists and is given by

$$R(\lambda, K_r) := (\lambda I - (A + rB))^{-1} = R(\lambda, A)\sum_{n=0}^{\infty} r^n (BR(\lambda, A))^n, \tag{89}$$

where the series converges absolutely and each term is positive. Hence,

$$\|R(\lambda, K_r)y\| \leq \lambda^{-1}\|y\| \tag{90}$$

for all $y \in X$. Therefore, by the Lumer–Phillips theorem, for each $0 \leq r < 1$, $(K_r, D(A))$ generates a contraction semigroup which we denote $(G_r(t))_{t\geq 0}$. Since $(R(\lambda, K_r)x)_{0\leq r<1}$ is increasing as $r \uparrow 1$ for each $x \in X_+$, $\{\|R(\lambda, K_r)x\|\}_{0\leq r<1}$ is bounded and $X = L_1(\Omega)$ is a *KB*-space, there is an element $y_{\lambda,x} \in X_+$ such that

$$\lim_{r\to 1^-} R(\lambda, K_r)x = y_{\lambda,x}$$

in X. By the Banach–Steinhaus theorem we obtain the existence of a bounded positive operator on X, denoted by $R(\lambda)$, such that $R(\lambda)x = y_{\lambda,x}$. We use the Trotter–Kato theorem to obtain that $R(\lambda)$ is defined for all $\lambda > 0$ and it is the resolvent of a densely defined closed operator K which generates a semigroup of contractions $(G_K(t))_{t\geq 0}$. Moreover, for any $x \in X$,

$$\lim_{r\to 1^-} G_r(t)x = G_K(t)x, \tag{91}$$

and the limit is uniform in t on bounded intervals and, provided $x \geq 0$, monotone as $r \uparrow 1$. By the monotone convergence theorem, Theorem 3,

$$R(\lambda, K)x = \sum_{k=0}^{\infty} R(\lambda, A)(BR(\lambda, A))^k x, \qquad x \in X \tag{92}$$

and $R(\lambda, K)(\lambda I - (A + B))x = x$ which shows that $K \supseteq A + B$. □

The identification of K is a much more difficult task. We note that (85) can be written as

$$\int_{\Omega} (Au + Bu)\, d\mu = -c(u) \leq 0, \tag{93}$$

where c is a positive functional on $D(A)$. We assume that c has a monotone convergence property: if $c(u_n) \to c_u$ for $u_n \uparrow u$, then $c_u = c(u)$ (for instance, it is an integral functional). We note that a more general version of this assumption is considered in [8,39]. The following theorem is fundamental for characterizing the generator of the semigroup.

Theorem 17 *For any fixed $\lambda > 0$, there is $0 \leq \beta_\lambda \in X^*$ with $\|\beta_\lambda\| \leq 1$ such that for any $f \in X_+$,*

$$\lambda\|R(\lambda, K)f\| = \|f\| - \langle\beta_\lambda, f\rangle - c\,(R(\lambda, K)f) \tag{94}$$

and c extends to a nonnegative continuous linear functional on $D(K)$.

It turns out that

$$(BR(\lambda, A))^* \beta_\lambda = \beta_\lambda. \tag{95}$$

and hence, if $\beta_\lambda \neq 0$, then $\sigma_p(BR(\lambda, A))^* \neq \emptyset$. This implies that $\sigma_r(BR(\lambda, A)) \neq \emptyset$ and, by Theorem 12(d), $K \neq \overline{A + B}$.

Another result, though not as elegant, is often more useful. It is based on the observation that the following are equivalent:

(a) $K = \overline{T + B}$.
(b) $\int\limits_{\Omega} Ku\, d\mu \geq -c(u), \quad u \in D(K)_+$.

Though the implication (b) $\Rightarrow$ (a) seems to be useless as it requires the knowledge of K which is what we are looking for, we note that if we can prove it for an extension of K (for instance $K_{\max}$), then it will be valid for K. Hence

Theorem 18 ([7]) *If there exists an extension* $\mathcal{K}$ *such that* $\int_\Omega \mathcal{K}u\, d\mu \geq -c(u)$ *for all* $u \in D(\mathcal{K})_+$, *then* $K = \overline{A + B}$.

Arendt–Rhandi Theorem

Theorem 19 ([5]) *Assume that* X *is a Banach lattice,* $(A, D(A))$ *is a resolvent positive operator which generates an analytic semigroup and* $(B, D(A))$ *is a positive operator. If* $(\lambda_0 I - (A + B), D(A))$ *has a nonnegative inverse for some* λ_0 *larger than the spectral bound* $s(A)$ *of* A, *then* $(A + B, D(A))$ *generates a positive analytic semigroup.*

Proof The proof is an application of Theorem 15. Under assumptions of this theorem, we obtain that $r(BR(\lambda_0, A)) < 1$. In particular, the series $\sum\limits_{n=0}^{\infty} (BR(\lambda_0, A))^n$ converges in the uniform operator topology. Next, by Theorem 10, $R(\lambda, A) \geq 0$ if and only if $\rho(A) \ni \lambda > s(A)$. Thus, using the resolvent identity we have

$$R(\lambda, A) = R(\lambda_0, A) - (\lambda - \lambda_0) R(\lambda_0, A) R(\lambda, A) \leq R(\lambda_0, A)$$

whenever $\lambda \geq \lambda_0$. Since $BR(\lambda, A)$ is bounded in X, see Theorem 1, $B : D(A) \to X$ is bounded in the graph norm of $D(A)$. Let us now take $\lambda \in \mathbb{C}$ with $\Re\lambda \geq \lambda_0$, $\mathbb{R} \ni \mu > \lambda_0$ and $f \in D(A)$. Then $\mu R(\mu, A) R(\lambda, A) f \to R(\lambda, A) f$ as $\mu \to \infty$ in the graph norm of $D(A)$, see e.g., [41, Lemmas 1.3.2 and 1.3.3] and we have, for $f \in D(A)$,

$$\begin{aligned}|BR(\lambda, A) f| &= \lim_{\mu \to \infty} |\mu BR(\mu, A) R(\lambda, A)| \leq \lim_{\mu \to \infty} \mu BR(\mu, A) R(\Re\lambda, A)|f| \\ &= BR(\mu, A) R(\Re\lambda, A)|f|,\end{aligned}$$

where we used $|R(\lambda,A)f| \le R(\Re\lambda,A)|f|$ for $\Re\lambda > s(A)$, see Theorem 10. Thus, by density,

$$|BR(\lambda,A)f| \le BR(\Re\lambda,A)|f| \tag{96}$$

for all $f \in X$ and therefore

$$r(BR(\lambda,A)) \le r(BR(\lambda_0,A)) < 1$$

for any $\lambda \in \mathbb{C}$ with $\Re\lambda \ge \lambda_0$. In particular, $\sum_{n=0}^{\infty}(BR(\lambda,A))^n$ converges to a bounded linear operator with

$$\left\|\sum_{n=0}^{\infty}(BR(\lambda,A))^n f\right\| \le \left\|\sum_{n=0}^{\infty}(BR(\lambda_0,A))^n|f|\right\| \le M_{\lambda_0}\|f\|,$$

uniformly for $\lambda \in \mathbb{C}$ with $\Re\lambda > \lambda_0$. This, in particular, shows that $A+B$ generates a C_0-semigroup. Indeed, from (84) and the above estimate we see that

$$\|R(\lambda,A+B)f\| \le M_{\lambda_0}\|R(\lambda,A)\|\|f\|$$

for $\lambda > \lambda_0$, so the claim follows since A satisfies the Hille–Yosida estimates (52) there.

Next we consider the analyticity issue. Using Theorem 7, for the operator A, there are ω_A and M_A such that

$$\|R(r+is,A)\| \le \frac{M_A}{|s|}$$

for $r > \omega_A$. Taking now $\omega > \max\{\lambda_0,\omega_A\}$ we have, by (84),

$$\begin{aligned}\|R(r+is,A+B)f\| &= \left\|R(\lambda,A)\sum_{n=0}^{\infty}(BR(\lambda,A))^n f\right\| \le \frac{M_A}{|s|}\left\|\sum_{n=0}^{\infty}(BR(\lambda,A))^n f\right\| \\ &\le \frac{M_A M_{\lambda_0}}{|s|}\|f\|, \qquad f \in X,\end{aligned}$$

for all $r > \omega$. Therefore $(A+B, D(A))$ generates an analytic semigroup. □

2.7 Semi-Linear Problems

Let us introduce now the simplest nonlinearity and consider the semilinear abstract Cauchy problem

$$\begin{aligned}\partial_t u &= Au + f(t,u), \quad t > 0, \\ u(0) &= \mathring{u}, \end{aligned} \tag{97}$$

where A is a generator of a C_0-semigroup $(G_A(t))_{t\geq 0}$ and $f : [0,T]\times X \to X$ is a known function. Since *a priori* we know no properties of the solution u (which may even fail to exist), it is plausible to start from a weaker formulation of the problem, i.e. from the integral equation:

$$u(t) = G_A(t)\overset{\circ}{u} + \int_0^t G_A(t-s)f(s,u(s))ds. \tag{98}$$

This form is typical for fixed point techniques. Here, depending on the properties of $(G_A(t))_{t\geq 0}$ and f, we can use two main fixed point theorems: the Banach contraction principle and Schauder's theorem.

We shall focus on the Banach contraction principle which leads to Theorem 20 below. It requires a relatively strong regularity from f.

We say that $f : [0,T]\times X \to X$ is locally Lipschitz continuous in u, uniformly in t on bounded intervals, if for all $t' \in [0,T[$ and $c > 0$ there exists $L(c,t')$ such that for all $t \in [0,t']$ and $\|u\|, \|v\| \leq c$ we have

$$\|f(t,u) - f(t,v)\|_X \leq L(c,t')\|u - v\|_X.$$

Theorem 20 *Let $f : [0,\infty[\times X \to X$ be continuous in $t \in [0,\infty[$ and locally Lipschitz continuous in u, uniformly in t, on bounded intervals. If A is the generator of a C_0-semigroup $(G_A(t))_{t\geq 0}$ on X, then for any $\overset{\circ}{u}\in X$ there exists $t_{max} > 0$ such that the problem (98) has a unique mild solution u on $[0,t_{max}[$. Moreover, if $t_{max} < +\infty$, then $\lim\limits_{t\to\infty}\|u(t)\|_X = \infty$.*

The proof is done by Picard iterations, as in the scalar case. Also, similarly to the scalar case, a sufficient condition for the existence of a global mild solution is that f be uniformly Lipschitz continuous on X. Uniform Lipschitz continuity yields at most linear growth in $\|x\|$ of $\|f(t,x)\|$. In fact, even for $f(u) = u^2$ and $A = 0$, the blow-up occurs in finite time.

There are two standard sufficient conditions ensuring that the mild solution, described in Theorem 20, is a strict solution. Both follow from the corresponding results for nonhomogeneous problems. They are either that $f : [0,\infty[\times X \to X$ is continuously differentiable with respect to both variables, or that $f : [0,\infty[\times D(A) \to D(A)$ is continuous. Certainly, in both cases to ensure that the solution is strict we must assume that $\overset{\circ}{u}\in D(A)$.

As in the subsection on nonhomogeneous problems, a substantial relaxation of requirements can be achieved if A generates an analytic semigroup. A crucial role is played by the domains of fractional powers of generators. Let us denote $X_\alpha = D((-A)^\alpha)$ with the usual graph norm. We have

Theorem 21 *Let $U \subseteq \mathbb{R} \times X_\alpha$, $a \in \mathbb{R}$, be an open set and $f : U \to X$ is such that for any $(t,x) \in U$ there exist $(t,x) \in V \subset U$, $L > 0$ and $0 < \theta \leq 1$ such that*

$$\|f(t_1,x_1) - f(t_2,x_2)\| \leq L(|t_1 - t_2|^\theta + \|x_1 - x_2\|_{X_\alpha}), \quad (t_i,x_i) \in V, i = 1,2,$$

and let an invertible A, satisfying $0 \in \rho(A)$ be the generator of a bounded analytic semigroup. If $(0, \mathring{u}) \in U$, then there is $\bar{t} = \bar{t}(\mathring{u})$ such that (97) has a unique local classical solution $u \in C([0,\bar{t}[, X) \cap C^1(]0,\bar{t}[, X)$. Moreover, the solution continuously depends on the initial data and is not global in time if it either reaches the boundary of U or its X_α norm blows up in finite time.

In fact, we have a better regularity result. If the constants θ and L are uniform in U, then $u \in C^{1+\nu}(]0,\bar{t}[, X)$; that is, $\partial_t u$ is Hölder continuous on $]0,\bar{t}[$ with $\nu = \min\{\theta, \beta\}$ with $0 < \beta < 1 - \alpha$.

We formulated the above theorem in the form usually found in the literature on dynamical systems. However, as was pointed out earlier, using $X_\alpha = D((-A)^\alpha)$ often is inconvenient as it may depend on the particular form of A.

However, in [38, Chapter 7] we can find a parallel theory in which X_α can be any interpolation space discussed in the section on analytic semigroups. In particular, under the assumptions of Theorem 21, the solution u is a strict solution; that is $u \in C([0,t_{\max}[, D(A)) \cap C^1([0,t_{max}[, X)$, provided $\mathring{u} \in D(A)$ and $A\mathring{u} + f(0,\mathring{u})$. The last condition follows from the fact that $\partial_t u$, if it exists, is a mild solution of the equation:

$$\partial_t u(t) = G_A(t)A(\mathring{u} + f(0,\mathring{u})) + \int_0^t AG_A(t-s)\phi(s)ds \tag{99}$$

where $\phi(s) = f(s,u(s))$ is Hölder continuous by the regularization property mentioned above. Then continuity of $\partial_t u$ follows from Theorem 9.

3 Transport on Graphs

Let us recall that we consider the system of equations

$$\begin{cases} \partial_t u_j(x,t) & = c_j \partial_x u_j(x,t), \quad x \in (0,1), \quad t \geq 0, \\ u_j(x,0) & = f_j(x), \\ \phi^-_{ij} \xi_j c_j u_j(1,t) & = w_{ij} \sum\limits_{k=1}^{m} \phi^+_{ik}(\gamma_k c_k u_k(0,t)), \end{cases} \tag{100}$$

where $\Phi^- = (\phi^-_{ij})_{1\leq i \leq n, 1\leq j \leq m}$ and $\Phi^+ = (\phi^+_{ij})_{1\leq i \leq n, 1\leq j \leq m}$ are, respectively, the outgoing and incoming incidence matrices, while $\gamma_j > 0$ and $\xi_j > 0$ are the absorption/amplification coefficients at, respectively, the head and the tail of the edge e_j.

Theorem 22 ([18]) *The following conditions are equivalent:*

1. *$(A, D(A))$ generates a C_0-semigroup;*
2. *Each vertex of G has an outgoing edge.*

Proof 1. $\Rightarrow$ 2. Assume that there is a semigroup $(T_A(t))_{t\geq 0}$ generated by A and consider a classical solution $\mathbf{u}(t) = T_A(t)\mathbf{f}$ with $\mathbf{f} \in D(A)$. Suppose that a vertex, say, v_i has no outgoing edge. Then, by (11),

$$0 = \sum_{k=1}^{m} \phi_{ik}^{+} \gamma_k c_k u_k(0,t), \quad t > 0.$$

In particular, $u_k(x,t) = f_k(x + c_k t)$ for $0 \leq x + c_k t \leq 1$ so $u_k(0,t) = f(c_k t)$ for $0 \leq t \leq \frac{1}{c_k}$. Thus

$$0 = \sum_{k=1}^{m} \phi_{ik}^{+} \gamma_k c_k f_k(c_k t), \quad 0 \leq t \leq c^{-1} := \min\{c_k^{-1}\}.$$

Let $(\mathbf{f}^r)_{r\in\mathbb{N}}, \mathbf{f}^r \in D(A)$, approximate $\mathbf{1} = (1, 1, \ldots, 1)$ in X. Then

$$0 \leq \|\mathbf{1}|_{(0,c^{-1})} - (f_k^r(c_k\cdot))_{1\leq k\leq m}\|_X = \sum_{k=1}^{m} \frac{1}{c_k} \int_0^{c^{-1}c_k} |1 - f_k^r(z)|dz$$

$$\leq \sum_{k=1}^{m} \frac{1}{c_k} \int_0^1 |1 - f_k^r(z)|dz \to 0$$

as $r \to \infty$. Since X-convergence implies convergence almost everywhere of a subsequence, we have $0 = \sum_{k=1}^{m} \phi_{ik}^{+} \gamma_k c_k$ almost everywhere on $(0, c^{-1})$, and thus everywhere. Since, however, the graph is connected and we assumed that there is no outgoing edge at v_i, there must be an incoming edge and thus at least one term of the sum is positive while all other terms are nonnegative. Thus, if there is a vertex with no outgoing edge, then the set of initial conditions satisfying the boundary conditions is not dense in X and thus $(A, D(A))$ cannot generate a C_0-semigroup.

1. $\Rightarrow$ 2. It can be proved that, under assumption 2., the boundary conditions can be incorporated into the domain of the operator in the following compact form:

$$D(A) = \{\mathbf{u} \in (W_1^1([0,1]))^m;\ \mathbf{u}(1) = \mathbb{K}^{-1}\mathbb{C}^{-1}\mathbb{B}\mathbb{G}\mathbb{C}\mathbf{u}(0)\}, \tag{101}$$

where $\mathbb{B}$ is the adjacency matrix defined in (9). Clearly, $(C_0^\infty((0,1)))^m \subset D(A)$ and hence $D(A)$ is dense in X. Let us consider the resolvent equation for A. We have to solve

$$\lambda u_j - c_j \partial_x u_j = f_j, \quad j = 1, \ldots, m, \quad x \in (0,1),$$

with $\mathbf{u} \in D(A)$. Integrating, we find the general solution

$$c_j u_j(x) = c_j e^{\frac{\lambda}{c_j}x} v_j + \int\limits_x^1 e^{\frac{\lambda}{c_j}(x-s)} f_j(s)ds, \tag{102}$$

where $\mathbf{v} = (v_1, \ldots, v_m)$ is an arbitrary vector. Let $\mathbb{E}_\lambda(s) = \operatorname{diag}\left(e^{\frac{\lambda}{c_j}s}\right)_{1\leq j\leq m}$. Then (102) takes the form

$$\mathbb{C}\mathbf{u}(x) = \mathbb{C}\mathbb{E}_\lambda(x)\mathbf{v} + \int\limits_x^1 \mathbb{E}_\lambda(x-s)\mathbf{f}(s)ds.$$

To determine $\mathbf{v}$ so that $\mathbf{u} \in D(A)$, we use the boundary conditions. At $x = 1$ and at $x = 0$ we obtain, respectively

$$\mathbb{C}\mathbf{u}(1) = \mathbb{C}\mathbb{E}_\lambda(1)\mathbf{v}, \qquad \mathbb{C}\mathbf{u}(0) = \mathbb{C}\mathbf{v} + \int\limits_0^1 \mathbb{E}_\lambda(-s)\mathbf{f}(s)ds$$

so that

$$\mathbb{K}\mathbb{C}\mathbb{E}_\lambda(1)\mathbf{v} = \mathbb{K}\mathbb{C}\mathbf{u}(1) = \mathbb{B}\mathbb{G}\mathbb{C}\mathbf{u}(0) = \mathbb{B}\mathbb{G}\left(\mathbb{C}\mathbf{v} + \int\limits_0^1 \mathbb{E}_\lambda(-s)\mathbf{E}_\lambda(-s)\mathbf{f}(s)ds\right),$$

which can be written as

$$(\mathbb{I} - \mathbb{E}_\lambda(-1)\mathbb{C}^{-1}\mathbb{K}^{-1}\mathbb{B}\mathbb{G}\mathbb{C})\mathbf{v} = \mathbb{E}_\lambda(-1)\mathbb{C}^{-1}\mathbb{K}^{-1}\mathbb{B}\mathbb{G}\int\limits_0^1 \mathbb{E}_\lambda(-s)\mathbf{f}(s)ds.$$

Since the norm of $\mathbb{E}_\lambda(-1)$ can be made as small as one wishes by taking large λ, we see that $\mathbf{v}$ is uniquely defined by the Neumann series provided λ is sufficiently large and hence the resolvent of A exists. We need to find an estimate for it. First we observe that the Neumann series expansion ensures that A is a resolvent positive operator and hence the norm estimates can be obtained using only nonnegative entries. Next, we recall that $\mathbb{B}$ is column stochastic; that is, each column sums to 1. Adding together the rows in

$$\mathbb{K}\mathbb{C}\mathbb{E}_\lambda(1)\mathbf{v} = \mathbb{B}\mathbb{G}\mathbb{C}\mathbf{v} + \mathbb{B}\mathbb{G}\int\limits_0^1 E_\lambda(-s)\mathbf{f}(s)ds.$$

we obtain

$$\sum_{j=1}^{m}\xi_j c_j e^{\frac{\lambda}{c_j}} v_j = \sum_{j=1}^{m}\gamma_j c_j v_j + \sum_{j=1}^{m}\gamma_j \int_0^1 e^{-\frac{\lambda}{c_j}s} f_j(s)ds.$$

By (102), we can evaluate, for $j \in \{1,\ldots,m\}$,

$$\int_0^1 u_j(x)dx = v_j \int_0^1 e^{\frac{\lambda}{c_j}x}dx + \frac{1}{c_j}\int_0^1\int_x^1 e^{\frac{\lambda}{c_j}(x-s)} f_j(s)dsdx$$

$$= \frac{v_j c_j}{\lambda}\left(e^{\frac{\lambda}{c_j}} - 1\right) + \frac{1}{\lambda}\int_0^1 \left(1 - e^{-\frac{\lambda}{c_j}s}\right) f_j(s)ds$$

so that, renorming X with the norm $\|\mathbf{u}\|_{\Xi} = \sum_{j=1}^{m}\xi_j\|u_j\|_{L_1([0,1])}$, we have

$$\|\mathbf{u}\|_{\Xi} = \sum_{j=1}^{m}\xi_j \int_0^1 u_j(x)dx \tag{103}$$

$$= \frac{1}{\lambda}\sum_{j=1}^{m}\xi_j v_j c_j\left(e^{\frac{\lambda}{c_j}} - 1\right) + \frac{1}{\lambda}\sum_{j=1}^{m}\xi_j\int_0^1\left(1 - e^{-\frac{\lambda}{c_j}s}\right) f_j(s)ds$$

$$= \frac{1}{\lambda}\sum_{j=1}^{m} c_j v_j(\gamma_j - \xi_j) + \frac{1}{\lambda}\sum_{j=1}^{m}(\gamma_j - \xi_j)\int_0^1 e^{-\frac{\lambda}{c_j}s} f_j(s)ds + \frac{1}{\lambda}\sum_{j=1}^{m}\xi_j\int_0^1 f_j(s)ds.$$

We consider three cases (the first one being similar to [25, Proposition 3.3]).

(a) $\gamma_j \le \xi_j$ for $j = 1,\ldots,m$. Let us consider the iterates in the Neumann series for $\mathbf{v}$, $(\mathbb{E}_\lambda(-1)\mathbb{C}^{-1}\mathbb{K}^{-1}\mathbb{B}\mathbb{G}\mathbb{C})^n$. Using the fact that $\mathbb{C}$, $\mathbb{G}$ and $\mathbb{K}$ are diagonal so that they commute, we find

$$\mathbb{E}_\lambda(-1)\mathbb{C}^{-1}\mathbb{K}^{-1}\mathbb{B}\mathbb{G}\mathbb{C} \le (\mathbb{C}\mathbb{K})^{-1}\mathbb{E}_\lambda(-1)\mathbb{B}(\mathbb{C}\mathbb{K}).$$

Since $\mathbb{B}$ is (column) stochastic, $r(\mathbb{E}_\lambda(-1)\mathbb{C}^{-1}\mathbb{K}^{-1}\mathbb{B}\mathbb{G}\mathbb{C}) < 1$ for any $\lambda > 0$. Hence $R(\lambda, A)$ is defined and positive for any $\lambda > 0$. Under the assumption of this item, by dropping two first terms in the second line, (103) gives

$$\|\mathbf{u}\|_{\Xi} \le \frac{1}{\lambda}\sum_{j=1}^{m}\xi_j\int_0^1 f_j(s)ds = \frac{1}{\lambda}\|\mathbf{f}\|_{\Xi}, \quad \lambda > 0.$$

Since $D(A)$ is dense in X, $(A, D(A))$ generates a positive semigroup of contractions in $(X, \|\cdot\|_{\Xi})$.

(b) $\gamma_j \geq \xi_j$ for $j = 1, \ldots, m$. Then (103) implies that for some $\lambda > 0$ and $c = 1/\lambda$ we have

$$\|R(\lambda, A)\mathbf{f}\|_{\Xi} \geq c\|\mathbf{f}\|_{\Xi}$$

and, by density of $D(A)$, the application of the Arendt–Batty–Robinson theorem, Theorem 11, gives the existence of a positive semigroup generated by A in $(X, \|\cdot\|_{\Xi})$. Since, however, the norm $\|\cdot\|_{\Xi}$ and the standard norm $\|\cdot\|$ are equivalent, we see that A generates a positive semigroup in X.

(c) $\gamma_j < \xi_j$ for $j \in I_1$ and $\gamma_j \geq \xi_j$ for $j \in I_2$, where $I_1 \cap I_2 = \emptyset$ and $I_1 \cup I_2 = \{1, \ldots, m\}$. Let $\mathbb{L} = \mathrm{diag}(l_j)$ where $l_j = \xi_j$ for $j \in I_1$ and $l_j = \gamma_j$ for $j \in I_2$. Then

$$\mathbb{E}_\lambda(-1)\mathbb{C}^{-1}\mathbb{K}^{-1}\mathbb{B}\mathbb{G}\mathbb{C} \leq (\mathbb{C}\mathbb{K})^{-1}\mathbb{E}_\lambda(-1)\mathbb{B}(\mathbb{C}\mathbb{L}).$$

Thus, denoting by $A_{\mathbf{L}}$ the operator given by the expression A restricted to

$$D(A) = \{\mathbf{u} \in (W_1^1([0,1]))^m;\ \mathbf{u}(1) = \mathbb{K}^{-1}\mathbb{C}^{-1}\mathbb{B}\mathbb{L}\mathbb{C}\mathbf{u}(0)\},$$

we see that

$$0 \leq R(\lambda, A) \leq R(\lambda, A_{\mathbf{L}}) \tag{104}$$

for any λ for which $R(\lambda, A_{\mathbf{L}})$ exists. But, by item (b), $A_{\mathbf{L}}$ generates a positive semigroup and thus satisfies the Hille–Yosida estimates. Since clearly (104) yields $R^k(\lambda, A) \leq R^k(\lambda, A_{\mathbf{L}})$ for any $k \in \mathbb{N}$, for some $\omega > 0$ and $M \geq 1$ we have

$$\|R^k(\lambda, A)\| \leq \|R^k(\lambda, A_{\mathbf{L}})\| \leq M(\lambda - \omega)^{-k}, \qquad \lambda > \omega$$

and hence we obtain the generation of a semigroup by A.

□

4 Epidemiology

This chapter is based on [19, 40]. To simplify the exposition we replace the SIRS model given by (18) with the SIS model describing the evolution of epidemics which does not convey any immunity. Setting $\gamma = 0$ in the system (18) and thus discarding the 'recovered' class, we have

$$\begin{aligned}
\partial_t s(a,t) + \partial_a s(a,t) &= -\mu(a)s(a,t) - \Lambda(a, i(\cdot,t))s(a,t) + \delta(a)i(a,t),\\
\partial_t i(a,t) + \partial_a i(a,t) &= -\mu(a)i(a,t) + \Lambda(a, i(\cdot,t))s(a,t) - \delta(a)i(a,t),
\end{aligned}$$

$$\begin{aligned}
s(0,t) &= \int_0^\omega \beta(a)\{s(a,t) + (1-q)i(a,t)\}\,da, \\
i(0,t) &= q\int_0^\omega \beta(a)i(a,t)\,da, \\
s(a,0) &= s_0(a) = \phi^s(a), \\
i(a,0) &= i_0(a) = \phi^i(a),
\end{aligned} \tag{105}$$

for $0 \le t \le T \le +\infty, 0 \le a \le \omega \le +\infty$.

The force of infection is defined by (19); the concrete assumptions will be introduced when needed. In both cases we deal with a semilinear problem; that is, with a nonlinear (algebraic) perturbation of a linear problem. As in Sect. 2.7, the decisive role is played by the semigroup generated by the linear part of the problem.

Problems like (105) have been relatively well-researched, including the cases where μ and β are nonlinear functions depending on the total population, see [24, 46] and reference therein. Our model most resembles that discussed in [46], the main difference being that in *op. cit.* the maximum age ω is infinite which makes it plausible to assume that μ is bounded. However, a biologically realistic assumption is that $\omega < +\infty$ which, however, necessitates building into the model a mechanism ensuring that no individual can live beyond ω. It follows, e.g. [31], that the probability of survival of an individual till age a is given by

$$\Pi(a) = e^{-\int_0^a \mu(s)\,ds}.$$

Thus $\Pi(\omega) = 0$ which requires

$$\int_0^\omega \mu(s)\,ds = +\infty. \tag{106}$$

Hence, μ cannot be bounded as $a \to \omega^-$. This is in contrast with the case $\omega = +\infty$, where commonly it is assumed that μ is a bounded function on $\mathbb{R}_+$, and introduces another unbounded operator in the problem. We note that this difficulty was circumvented by Inaba in [32] by introducing the maximum reproduction age $a_\dagger < \omega$ and ignoring the evolution of the post-reproductive part of the population. Also, in papers such as [24], though $\omega < +\infty$, the assumption that the population is constant removes the death coefficient from the equation. The analysis of the model without any simplification in the scalar and linear case was done in [31] by reducing it to an integral equation along characteristics. It can be proved that the solution of such a problem is given by a strongly continuous semigroup. Here we shall prove this directly by refining the argument of [32].

4.1 Notation and Assumptions

We will work in the space $\mathbf{X} = L_1\left([0,\omega],\mathbb{R}^2\right)$ with norm $\|(p_1,p_2)\|_{\mathbf{X}} = \|p_1\| + \|p_2\|$, where the norm $\|\cdot\|$ refers to the norm in $L_1([0,\omega])$; the relevant norm in $\mathbb{R}^2$ will be denoted by $|\cdot|$. We also introduce necessary assumptions (cf. [31]) on the coefficients of (105) where, in what follows, for any measurable function ϕ on $[0,\omega]$ we shall use the notation

$$\bar{\phi} = \operatorname*{ess\,sup}_{a\in[0,\omega]} \phi(a), \quad \underline{\phi} = \operatorname*{ess\,inf}_{a\in[0,\omega]} \phi(a). \tag{107}$$

(H1) $0 \le \mu \in L_{\infty,loc}\left([0,\omega)\right)$, satisfying (106), with $\underline{\mu} > 0$;
(H2) $0 \le \beta \in L_\infty\left([0,\omega]\right)$;
(H3) $0 \le \delta \in L_\infty\left([0,\omega]\right)$;
(H4) $0 \le K \in L^\infty\left([0,\omega]^2\right)$.

Let $W_1^1\left([0,\omega],\mathbb{R}^2\right)$ be the Sobolev space of vector valued functions. Further, we define $\mathbf{S} = \operatorname{diag}\{-\partial_a,-\partial_a\}$ on $D(\mathbf{S}) = W_1^1\left([0,\omega],\mathbb{R}^2\right)$, $\mathbf{M}_\mu(a) = \operatorname{diag}\{-\mu(a),-\mu(a)\}$ on $D(\mathbf{M}_\mu) = \{\boldsymbol{\varphi} \in \mathbf{X} : \mu\boldsymbol{\varphi} \in \mathbf{X}\}$,

$$\mathbf{M}_\delta(a) = \begin{pmatrix} 0 & \delta(a) \\ 0 & -\delta(a) \end{pmatrix}; \tag{108}$$

$\mathbf{M}_\delta \in \mathcal{L}(\mathbf{X})$. Further, for a fixed $q \in [0,1]$,

$$\mathbf{B}(a) = \begin{pmatrix} \beta(a) & (1-q)\beta(a) \\ 0 & q\beta(a) \end{pmatrix} \tag{109}$$

with

$$\boldsymbol{\mathcal{B}}\boldsymbol{\varphi} = \int_0^\omega \mathbf{B}(a)\boldsymbol{\varphi}(a)\,da;$$

the operator $\boldsymbol{\mathcal{B}}$ is bounded. Moreover, we introduce the linear operator $\mathbf{A}_\mu$ defined on the domain

$$D\left(\mathbf{A}_\mu\right) = \left\{\boldsymbol{\varphi} \in D(\mathbf{S}) \cap D(\mathbf{M}_\mu);\ \boldsymbol{\varphi}(0) = \boldsymbol{\mathcal{B}}\boldsymbol{\varphi}\right\} \tag{110}$$

by

$$\mathbf{A}_\mu = \mathbf{S} + \mathbf{M}_\mu. \tag{111}$$

Let $\mathcal{Q}$ be the linear operator defined on the domain $D(\mathcal{Q}) = D(\mathbf{A}_\mu)$ by $\mathcal{Q} = \mathbf{A}_\mu + \mathbf{M}_\delta$. Using this notation, we see that (105) can be rewritten in the following compact form

$$\begin{aligned} \partial_t \mathbf{u} &= \mathbf{A}_\mu \mathbf{u} + \mathbf{M}_\delta \mathbf{u} + \mathfrak{F}(\mathbf{u}), \\ \mathbf{u}(0,t) &= \int_0^\omega \mathbf{B}(a)\mathbf{u}(a,t)\,da, \\ \mathbf{u}(a,0) &= \mathbf{u}_0(a) = \boldsymbol{\varphi}(a), \end{aligned} \tag{112}$$

where $\mathbf{u} = (s,i)^T$ and $\mathfrak{F}$ is a nonlinear function defined by

$$\mathfrak{F}(\mathbf{u}) = \begin{pmatrix} -\Lambda(\cdot,i)s \\ \Lambda(\cdot,i)s \end{pmatrix}$$

with Λ defined by (19).

4.2 *The Linear Part*

To prove that (105) is well-posed in $\mathbf{X}$, first we consider the linear operator $\mathcal{Q}$ on $D(\mathcal{Q}) = D(\mathbf{A}_\mu)$.

Theorem 23 *The linear operator $\mathcal{Q}$ generates a strongly continuous positive semigroup $(\mathcal{T}_\mathcal{Q}(t))_{t\geq 0}$ in $\mathbf{X}$.*

To carry out the proof of Theorem 23, it is sufficient to prove the generation result for $\mathbf{A}_\mu$ and use Theorem 13 (the Bounded Perturbation Theorem) to prove the generation for $\mathcal{Q}$; then we use some other tools to show the positivity of the combined semigroup. In this setting, first we have

Theorem 24 *The linear operator $\mathbf{A}_\mu$ generates a strongly continuous positive semigroup $\left(\mathcal{T}_{\mathbf{A}_\mu}(t)\right)_{t\geq 0}$ in $\mathbf{X}$ such that*

$$\|\mathcal{T}_{\mathbf{A}_\mu}(t)\|_{\mathcal{L}(\mathbf{X})} \leq e^{(\overline{\beta}-\underline{\mu})t}. \tag{113}$$

We prove this theorem in a sequence of lemmas in which we construct and estimate the resolvent of $\mathbf{A}_\mu$. We begin with introducing the survival rate matrix $\mathbf{L}(a)$, which represents the survival rate function in a multi-state population. $\mathbf{L}(a)$ is a solution of the matrix differential equation:

$$\frac{d\mathbf{L}}{da}(a) = \mathbf{M}_\mu(a)\mathbf{L}(a), \quad \mathbf{L}(0) = \mathbf{I}, \tag{114}$$

where $\mathbf{I}$ denotes the 2×2 identity matrix. The solution of (114) is a diagonal matrix given by

$$\mathbf{L}(a) = e^{-\int_0^a \mu(r)\,dr}\mathbf{I}. \tag{115}$$

From the above formula, we see that $\mathbf{L}(a)$ is invertible for all $a \in [0, \omega)$; its inverse is denoted $\mathbf{L}^{-1}(a)$. The inverse satisfies

$$\frac{d\mathbf{L}^{-1}}{da}(a) = -\mathbf{M}_\mu(a)\mathbf{L}^{-1}(a), \quad \mathbf{L}^{-1}(0) = \mathbf{I}. \tag{116}$$

Hence, we can define the fundamental matrix $\mathbf{L}(a, b)$ by

$$\mathbf{L}(a, b) = \mathbf{L}(a)\mathbf{L}^{-1}(b).$$

Lemma 1 *If $\lambda > \overline{\beta} - \underline{\mu}$, then $\left(\lambda\mathbf{I} - \mathbf{A}_\mu\right)^{-1}$ is given by*

$$\begin{aligned}\boldsymbol{\varphi} &= \left(\lambda\mathbf{I} - \mathbf{A}_\mu\right)^{-1}\boldsymbol{\psi}\\ &= e^{-\lambda a}\mathbf{L}(a)\left(\mathbf{I} - \int_0^\omega e^{-\lambda\sigma}\mathbf{B}(\sigma)\mathbf{L}(\sigma)\,d\sigma\right)^{-1}\int_0^\omega \mathbf{B}(a)\int_0^a e^{\lambda(\sigma-a)}\mathbf{L}(a,\sigma)\boldsymbol{\psi}(\sigma)d\sigma da\\ &\quad + e^{-\lambda a}\mathbf{L}(a)\int_0^a e^{\lambda\sigma}\mathbf{L}^{-1}(\sigma)\boldsymbol{\psi}(\sigma)\,d\sigma. \end{aligned} \tag{117}$$

Proof Let $\lambda > \overline{\beta} - \underline{\mu}$ and $\boldsymbol{\psi} \in \mathbf{X}$. A function $\boldsymbol{\varphi} \in D(\mathbf{A}_\mu)$ if and only if

$$\begin{aligned}&\lambda\boldsymbol{\varphi}(a) + \frac{d}{da}\boldsymbol{\varphi}(a) - \mathbf{M}_\mu(a)\boldsymbol{\varphi}(a) = \boldsymbol{\psi}(a),\\ &\boldsymbol{\varphi}(0) = \int_0^\omega \mathbf{B}(a)\boldsymbol{\varphi}(a)\,da, \qquad \mu\boldsymbol{\varphi} \in \mathbf{X}.\end{aligned} \tag{118}$$

By Duhamel's formula, the first equation of (118) gives

$$\begin{aligned}\boldsymbol{\varphi}(a) &= e^{-\lambda a}\mathbf{L}(a)\boldsymbol{\varphi}(0) + \int_0^a e^{-\lambda(a-s)}\mathbf{L}(a,s)\boldsymbol{\psi}(s)\,ds,\\ &= e^{-\lambda a - \int_0^a \mu(r)\,dr}\boldsymbol{\varphi}(0) + \int_0^a e^{-\lambda(a-s) - \int_s^a \mu(r)\,dr}\boldsymbol{\psi}(s)\,ds,\end{aligned} \tag{119}$$

for some unspecified as yet initial condition $\boldsymbol{\varphi}(0)$. For a fixed $\boldsymbol{\varphi}(0)$, we denote by $\mathcal{R}_{\boldsymbol{\varphi}(0)}(\lambda)\boldsymbol{\psi}$ the operator $\boldsymbol{\psi}(s) \to \boldsymbol{\varphi}$ defined above; it is easy to see that

$$(\lambda I - \mathbf{S} - \mathbf{M}_\mu)\mathcal{R}_{\boldsymbol{\varphi}(0)}(\lambda)\boldsymbol{\psi} = \boldsymbol{\psi}, \tag{120}$$

for a.a. $a \in [0,\omega)$. The unknown $\boldsymbol{\varphi}(0)$ can be determined from the boundary condition (118) by substituting (119); we get

$$\boldsymbol{\varphi}(0) = \int_0^\omega e^{-\lambda a - \int_0^a \mu(r)\,dr}\mathbf{B}(a)\boldsymbol{\varphi}(0)\,da + \int_0^\omega e^{-\lambda a - \int_0^a \mu(r)\,dr}\mathbf{B}(a)\left(\int_0^a e^{\lambda s + \int_0^s \mu(r)\,dr}\boldsymbol{\psi}(s)\,ds\right) da.$$

Since

$$\left|\int_0^\omega e^{-\lambda a - \int_0^a \mu(r)\,dr}\mathbf{B}(a)\,da\right| \leq \overline{\beta}\int_0^\omega e^{-(\lambda+\underline{\mu})a}\,da \leq \frac{\overline{\beta}}{\lambda + \underline{\mu}} < 1 \tag{121}$$

for $\lambda > \bar{\beta} - \underline{\mu}$, $\mathbf{I} - \int_0^\omega e^{-\lambda a - \int_0^a \mu(r)\,dr}\mathbf{B}(a)\,da$ is invertible with the inverse satisfying

$$\left|\left(\mathbf{I} - \int_0^\omega e^{-\lambda s - \int_0^s \mu(r)\,dr}\mathbf{B}(s)\,ds\right)^{-1}\right| \leq \frac{\lambda + \underline{\mu}}{\lambda - (\overline{\beta} - \underline{\mu})}. \tag{122}$$

Hence

$$\boldsymbol{\varphi}(0) = \left(\mathbf{I} - \int_0^\omega e^{-\lambda a - \int_0^a \mu(r)\,dr}\mathbf{B}(a)\,da\right)^{-1} \int_0^\omega e^{-\lambda a - \int_0^a \mu(r)\,dr}\mathbf{B}(a)\left(\int_0^a e^{\lambda s + \int_0^s \mu(r)\,dr}\boldsymbol{\psi}(s)\,ds\right) da$$

and we can substitute that $\boldsymbol{\varphi}(0)$ in the operator $\mathcal{R}_{\boldsymbol{\varphi}(0)}(\lambda)$ to define

$$\begin{aligned}\mathcal{R}(\lambda)\boldsymbol{\psi}(a) &= e^{-\lambda a - \int_0^a \mu(r)\,dr}\left(\mathbf{I} - \int_0^\omega e^{-\lambda s - \int_0^s \mu(r)\,dr}\mathbf{B}(s)\,ds\right)^{-1}\int_0^\omega e^{-\lambda a - \int_0^a \mu(r)\,dr}\mathbf{B}(a)\\ &\times \int_0^a e^{\lambda s + \int_0^s \mu(r)\,dr}\boldsymbol{\psi}(s)\,ds\,da + e^{-\lambda a - \int_0^a \mu(r)\,dr}\int_0^a e^{\lambda s + \int_0^s \mu(r)\,dr}\boldsymbol{\psi}(s)\,ds.\end{aligned}$$

The above calculations show that $\lambda - \mathbf{A}_\mu$ is one-to-one for $\lambda > \overline{\beta} - \underline{\mu}$. Routine calculations show that

$$\|\mathcal{R}(\lambda)\boldsymbol{\psi}\|_{\mathbf{X}} \leq \frac{1}{\lambda - (\overline{\beta} - \underline{\mu})}\|\boldsymbol{\psi}\|_{\mathbf{X}}. \tag{123}$$

Thus $\mathcal{R}(\lambda)$ is a bounded operator in $\mathbf{X}$.

Further, by lengthy calculations we also can show that

$$\int_0^\omega |\mu(a)\mathcal{R}(\lambda)\boldsymbol{\psi}(a)|\, da \leq \left(1 + \frac{\overline{\beta}}{\lambda - (\overline{\beta} - \underline{\mu})}\right) \|\boldsymbol{\psi}\|_{\mathbf{X}}.$$

Hence $\mathcal{R}(\lambda)\mathbf{X} \subset D(\mathbf{M}_\mu)$. Further, since for any $\boldsymbol{\psi} \in \mathbf{X}$, $\boldsymbol{\varphi} = \mathcal{R}(\lambda)\boldsymbol{\psi}$ satisfies

$$\lambda\mathcal{R}(\lambda)\boldsymbol{\psi} + \frac{d}{da}\mathcal{R}(\lambda)\boldsymbol{\psi} - \mathbf{M}_\mu\mathcal{R}(\lambda)\boldsymbol{\psi} = \boldsymbol{\psi}$$

almost everywhere, we have

$$\frac{d}{da}\mathcal{R}(\lambda)\boldsymbol{\psi} = \boldsymbol{\psi} - \lambda\mathcal{R}(\lambda)\boldsymbol{\psi} + \mathbf{M}_\mu\mathcal{R}(\lambda)\boldsymbol{\psi},$$

where, by the above estimates, all terms on the right hand side are in $\mathbf{X}$. Hence $\mathcal{R}(\lambda)\boldsymbol{\psi} \in \mathrm{W}_1^1([0,\omega],\mathbb{R}^2)$ and, consequently, $\mathcal{R}(\lambda)\boldsymbol{\psi} \in D(\mathbf{A}_\mu)$.

Since the boundary condition holds, using the results above we see that the operator $\mathcal{R}(\lambda)$, given by $(\lambda I - \mathbf{A}_\mu)\mathcal{R}(\lambda)\boldsymbol{\psi} = \boldsymbol{\psi}$, is such that maps $\mathcal{R}(\lambda) : \mathbf{X} \to D(\mathbf{A}_\mu)$. Then $\mathcal{R}(\lambda)$ is a right-inverse of the operator $\lambda I - \mathbf{A}_\mu$.

Summarizing, $\mathcal{R}(\lambda)$ is the right inverse of $(\lambda I - \mathbf{A}_\mu, D(\mathbf{A}_\mu))$. To prove that it is also a left inverse, we repeat the standard argument. Assume that for some $\boldsymbol{\varphi} \in D(\mathbf{A}_\mu)$ we have

$$\mathcal{R}(\lambda)(\lambda I - \mathbf{A}_\mu)\boldsymbol{\varphi} = \tilde{\boldsymbol{\varphi}} \neq \boldsymbol{\varphi}.$$

Since $\mathcal{R}(\lambda) : \mathbf{X} \to D(\mathbf{A}_\mu)$, we can write

$$(\lambda I - \mathbf{A}_\mu)\boldsymbol{\varphi} = (\lambda I - \mathbf{A}_\mu)\mathcal{R}(\lambda)(\lambda I - \mathbf{A}_\mu)\boldsymbol{\varphi} = (\lambda I - \mathbf{A}_\mu)\tilde{\boldsymbol{\varphi}}$$

since $\mathcal{R}(\lambda)$ is a right inverse of $\lambda I - \mathbf{A}_\mu$. But we proved that the linear operator $(\lambda I - \mathbf{A}_\mu)$ is one-to-one for $\lambda > \overline{\beta} - \underline{\mu}$, $\boldsymbol{\varphi} = \tilde{\boldsymbol{\varphi}}$ and hence $\mathcal{R}(\lambda) = (\lambda I - \mathbf{A}_\mu)^{-1}$. □

Lemma 2 $\overline{D(\mathbf{A})_+} = \mathbf{X}_+$.

Proof A proof of this result (with some gaps) is provided in [32, p. 60]. A more comprehensive proof can be found in [46]. We present a simpler proof which, moreover, allows for an approximation of $\boldsymbol{f} \in \mathbf{X}_+$ by elements of $D(\mathbf{A}_\mu)_+$.

Fix $\boldsymbol{f} \in \mathbf{X}_+$. First we note that for any given ϵ there is $0 \leq \boldsymbol{\phi} \in C_0^\infty((0,\omega),\mathbb{R}^2)$ such that $\|\boldsymbol{f} - \boldsymbol{\phi}\|_{\mathbf{X}} \leq \epsilon$. Clearly, $\boldsymbol{\phi} \in D(\mathbf{M}_\mu)$ but typically

$$\boldsymbol{\varphi}(0) \neq \int_0^\omega \mathbf{B}(a)\boldsymbol{\varphi}(a)\, da.$$

Take a function $0 \leq \eta \in C_0^\infty([0,\omega))$ with $\eta(0) = 1$ and let $\eta_\epsilon(a) = \eta(a/\epsilon)$. Further, let $\boldsymbol{\alpha} = (\alpha_1, \alpha_2)$ be a vector and consider

$$\boldsymbol{\psi} = \boldsymbol{\varphi} + \eta_\epsilon \boldsymbol{\alpha}.$$

Clearly, $\boldsymbol{\psi} \in W_1^1([0,\omega], \mathbb{R}^2) \cap D(\mathbf{M}_\mu)$. As far as the boundary condition is concerned we have, by the properties of the involved functions,

$$\boldsymbol{\alpha} = \int_0^\omega \mathbf{B}(a)\boldsymbol{\varphi}(a)\,da + \left(\int_0^{\epsilon\omega} \begin{pmatrix} \beta(a)\eta_\epsilon(a) & (1-q)\beta(a)\eta_\epsilon(a) \\ 0 & q\beta(a)\eta_\epsilon(a) \end{pmatrix} da \right) \boldsymbol{\alpha}. \tag{124}$$

Now, since

$$0 \leq \int_0^{\epsilon\omega} \beta(a)\eta_\epsilon(a)da = \epsilon \int_0^{\omega} \beta(\epsilon s)\eta(s)ds \leq \epsilon\bar{\beta},$$

the matrix l_1–norm satisfies

$$\left\| \left(\int_0^{\epsilon\omega} \begin{pmatrix} \beta(a)\eta_\epsilon(a) & (1-q)\beta(a)\eta_\epsilon(a) \\ 0 & q\beta(a)\eta_\epsilon(a) \end{pmatrix} da \right) \right\|_{\mathcal{L}(\mathbb{R}^2)} \leq \epsilon\bar{\beta}.$$

Thus, (124) is solvable for sufficiently small ϵ. Further, by the positivity of the above matrix and the properties of the Neumann series expansion, $\boldsymbol{\alpha}$ is nonnegative and

$$|\boldsymbol{\alpha}| \leq \left| \int_0^\omega \mathbf{B}(a)\boldsymbol{\varphi}(a)\,da \right| (1-\epsilon\bar{\beta})^{-1} \leq C$$

for some constant C, which is independent of ϵ for sufficiently small ϵ (the norm of $\|\boldsymbol{\varphi}\|$, which depends on ϵ, can be bounded by e.g. $\|\boldsymbol{f}\| + 1$ for $\epsilon < 1$). Hence we have

$$\|\boldsymbol{f} - \boldsymbol{\psi}\|_{\mathbf{X}} \leq \|\boldsymbol{f} - \boldsymbol{\varphi}\|_{\mathbf{X}} + |\boldsymbol{\alpha}|\|\eta_\epsilon\| \leq (1+C)\epsilon.$$

Proof of Theorem 24 Since the inverse of a bounded operator is closed, we see that $\lambda I - \mathbf{A}_\mu$, and hence $\mathbf{A}_\mu$, are closed. Thus the above lemmas with the estimate (123) show that $\mathbf{A}_\mu$ satisfies the assumptions of the Hille–Yosida theory. Hence, it generates a semigroup satisfying (113). Since the resolvent is positive, the semigroup is positive as well. □

Proof of Theorem 23 Since $\mathbf{M}_\delta \in \mathcal{L}(\mathbf{X})$, with $\|\mathbf{M}_\delta(a)\|_{\mathcal{L}(\mathbf{X})} \leq 2\bar{\delta}$, Theorem 13 (the Bounded Perturbation Theorem) is applicable and states that the linear operator $(\mathcal{Q}, D(\mathbf{A}))$ generates a strongly continuous semigroup denoted by $\left(\mathcal{T}_{\mathcal{Q}}(t)\right)_{t\geq 0}$. Using the estimate (113) we have:

$$\|\mathcal{T}_{\mathcal{Q}}(t)\|_{\mathcal{L}(\mathbf{X})} \leq e^{t(\overline{\beta}-\underline{\mu}+2\bar{\delta})}.$$

Using the structure of $\mathbf{M}_\delta$ we can improve this estimate and also show that the semigroup $(\mathcal{T}_{\mathcal{Q}}(t))_{t\geq 0}$ is positive. Since the variable a plays in $\mathbf{M}_\delta$ the role of a parameter, we find

$$\mathcal{T}_{\mathbf{M}_\delta}(t) = \begin{pmatrix} 1 & 1-e^{-t\delta(a)} \\ 0 & e^{-t\delta(a)} \end{pmatrix}$$

and so

$$\|\mathcal{T}_{\mathbf{M}_\delta}(t)\|_{\mathcal{L}(\mathbf{X})} = 1.$$

Also, $(\mathcal{T}_{\mathbf{M}_\delta}(t))_{t\geq 0}$ is positive. Hence, by (82), we obtain

$$\|\mathcal{T}_{\mathcal{Q}}(t)\|_{\mathcal{L}(\mathbf{X})} \leq e^{t(\overline{\beta}-\underline{\mu})} \tag{125}$$

and $(\mathcal{T}_{\mathcal{Q}}(t))_{t\geq 0}$ is positive. □

Remark 6 The estimates (113) and (125) are not optimal. In fact, for the scalar problem

$$\begin{aligned} \partial_t n(a,t) &= -\partial_a n(a,t) - \mu(a)n(a,t), \quad t>0, a\in(0,\omega) \\ n(0,t) &= \int_0^\omega \beta(a)n(a,t)da, \\ n(a,0) &= \overset{\circ}{n}(a), \end{aligned} \tag{126}$$

it can be proved, [31], that there is a unique dominant eigenvalue λ^* of the problem, which is the solution of the renewal equation

$$1 = \int_0^\omega \beta(a)e^{-\lambda a - \int_0^a \mu(s)ds} da, \tag{127}$$

such that

$$\|n(t)\| \leq Me^{t\lambda^*} \tag{128}$$

for some constant M. This eigenvalue is, respectively, positive, zero or negative if and only if the basic reproduction number

$$R = \int_0^{\omega} \beta(a) e^{-\int_0^a \mu(s) ds} da \tag{129}$$

is bigger, equal, or smaller, than 1.

Consider now an initial condition $(\overset{\circ}{s}, \overset{\circ}{i}) \in D(\mathbf{A}_\mu)$. Since the semigroup $(\mathcal{T}_{\mathcal{Q}}(t))_{t\geq 0}$ is positive, the strict solution (s, i) of the linear part of (105) is nonnegative and the total population $0 \leq s(a,t) + i(a,t) = n(a,t)$ satisfies (126). Using nonnegativity, we find $s(a,t) \leq n(a,t)$ and $i(a,t) \leq n(a,t)$ and consequently

$$\|\mathcal{T}_{\mathcal{Q}}(t)(\overset{\circ}{s}, \overset{\circ}{i})\|_{\mathbf{X}} \leq M e^{t\lambda^*} \|(\overset{\circ}{s}, \overset{\circ}{i})\|_{\mathbf{X}}$$

for $(\overset{\circ}{s}, \overset{\circ}{i}) \in D(\mathbf{A}_\mu)_+$. However, by Lemma 2, the above estimate can be extended to $\mathbf{X}_+$ and, by (29), to $\mathbf{X}$.

Note that the crucial role in the above argument is played by the fact that (s, i) satisfies the differential equation (105)—if it was only a mild solution, it would be difficult to directly prove that the sum $s + i$ is the mild solution to (126).

4.3 The Nonlinear Problem

In the case of intercohort transmission, discussed in these lectures, individuals of any age can infect individuals of any age, though with possibly different intensity. Then

$$\Lambda(a, i) = \int_0^{\omega} K(a, a') i(a') \, da', \tag{130}$$

where $K(a, a')$ is a nonnegative bounded function on $[0, \omega] \times [0, \omega]$ which accounts for the age-specific probability of becoming infected through contact with infectives of a particular age.

Since the nonlinear term $\mathfrak{F}$ is quadratic, standard calculations, see [41], give

Proposition 4 *$\mathfrak{F}$ is continuously Fréchet differentiable with respect to $\boldsymbol{\phi} \in \mathbf{X}$ and for any $\boldsymbol{\phi} = (\phi^s, \phi^i)$, $\boldsymbol{\psi} = (\psi^s, \psi^i) \in \mathbf{X}$ the Fréchet derivative at $\boldsymbol{\phi}$, $\mathfrak{F}_{\boldsymbol{\phi}}$, is defined by*

$$\left(\mathfrak{F}_{\boldsymbol{\phi}} \boldsymbol{\psi}\right)(a) := \begin{pmatrix} -\psi^s(a) \int_0^{\omega} K(a, a') \phi^i(a') \, da' - \phi^s(a) \int_0^{\omega} K(a, a') \psi^i(a') \, da' \\ \\ \psi^s(a) \int_0^{\omega} K(a, a') \phi^i(a') \, da' + \phi^s(a) \int_0^{\omega} K(a, a') \psi^i(a') \, da' \end{pmatrix}.$$

Hence we can apply Theorem 20 to claim that for each $\mathring{\mathbf{u}} = (\mathring{s}, \mathring{i}) \in \mathbf{X}$, there is a $t(\mathring{\mathbf{u}})$ such that the problem (112) has a unique mild solution on $[0, t(\mathring{\mathbf{u}})[\ni t \to \mathbf{u}(t)$; this solution is strict if $\mathring{\mathbf{u}} \in D(\mathbf{A}_\mu)$.

We recall that the proof consists in showing that the Picard iterates

$$\mathbf{u}_0 = \mathring{\mathbf{u}}$$

$$\mathbf{u}_n(t) = \mathcal{T}_{\mathcal{Q}}(t)\mathring{\mathbf{u}} + \int_0^t \mathcal{T}_{\mathcal{Q}}(t-s)\mathfrak{F}(\mathbf{u}_{n-1}(s))ds \tag{131}$$

converge in $C([0, t(\mathring{\mathbf{u}})[, B(\mathring{\mathbf{u}}, \rho))$ where $B(\mathring{\mathbf{u}}, \rho) = \{\mathbf{u} \in \mathbf{X};\ \|\mathbf{u} - \mathring{\mathbf{u}}\|_{\mathbf{X}} \leq \rho\}$ for some constant ρ. Since the nonlinearity is quadratic, it is not globally Lipschitz continuous and thus the question whether this solution can be extended to $[0, \infty[$ requires employing positivity techniques.

Since $\mathfrak{F}$ is not positive on $\mathbf{X}_+$, we cannot claim that the constructed local solution is nonnegative, as the iterates defined by (131) need not be positive even if we start with $\mathring{\mathbf{u}} \geq 0$. Hence, we re-write (112) in the equivalent form

$$\begin{cases} \dfrac{d\mathbf{u}}{dt} = (\mathcal{Q} - \kappa I)\mathbf{u} + (\kappa I + \mathfrak{F})(\mathbf{u}), \quad t > 0, \\ \mathbf{u}(0) = \mathring{\mathbf{u}}, \end{cases} \tag{132}$$

for some $\kappa \in \mathbb{R}_+$ to be determined. Denote $\mathcal{Q}_\kappa = \mathcal{Q} - \kappa I$; then $(\mathcal{T}_{\mathcal{Q},\kappa}(t))_{t\geq 0} = (e^{-\kappa t}\mathcal{T}_{\mathcal{Q}})_{t\geq 0}$ and hence $(\mathcal{T}_{\mathcal{Q},\kappa}(t))_{t\geq 0}$ is positive. It is also easy to see that the following result holds.

Lemma 3 *For any ρ there exists κ such that $(\kappa\mathbf{I} + \mathfrak{F})(\mathbf{X}_+ \cap B(\mathring{\mathbf{u}}, \rho)) \subset \mathbf{X}_+$.*

Then the Picard iterates corresponding to (132),

$$\mathbf{u}(t) = e^{-\kappa t}\mathcal{T}(t)\mathring{\mathbf{u}} + \int_0^t e^{-\kappa(t-s)}\mathcal{T}_Q(t-s)(\kappa I + \mathfrak{F})(\mathbf{u}(s))\,ds, \quad 0 \leq t < t(\mathring{\mathbf{u}}),$$

are nonnegative and we can repeat the standard estimates to arrive at

Theorem 25 *Assume that $\mathring{\mathbf{u}} \in \mathbf{X}_+$ and let $\mathbf{u} : [0, t(\mathring{\mathbf{u}})[\to \mathbf{X}$ be the unique mild solution of (112). Then this solution is nonnegative on the maximal interval of its existence. Moreover, the solutions continuously depend on the initial conditions on every compact subinterval of their joint interval of existence.*

4.4 Global Existence

Since quadratic nonlinearities do not satisfy the uniform Lipschitz condition, we cannot immediately claim that the solutions to (112) are global in time. In fact, it is well known that, even for ordinary differential equations, the solution with a quadratic nonlinearity can blow up in a finite time. Here, we use positivity to show that positive solutions exist globally in time. For this, we have to show that $t \to \|\mathbf{u}(t)\|_{\mathbf{X}}$ does not blow up in finite time. We state the following result:

Theorem 26 *For any* $\mathring{\mathbf{u}} \in D(\mathbf{A}_\mu) \cap \mathbf{X}_+$*, the problem (112) has a unique strict positive solution* $\mathbf{u}(t)$ *defined on the whole time interval* $[0, \infty[$.

Proof The proof uses the ideas of Remark 6. Under the adopted assumptions, we have a positive strict solution $\mathbf{u}(t) = \big(s(t), i(t)\big)$ to (105) in $L_1([0,\omega], \mathbb{R}^2)$ defined on its maximum interval of existence $[0, t_{max}[$. But then

$$\|\mathbf{u}(t)\|_{\mathbf{X}} = \int_0^\omega (s(a,t) + i(a,t))\, da = \int_0^\omega u(a,t)\, da, \quad t \in [0, t_{max}[,$$

where $u(a,t)$ is the solution to the McKendrick equation (126). But then, as long as $0 \le t < t(\mathring{\mathbf{u}})$,

$$\|\mathbf{u}(t)\|_{\mathbf{X}} \le e^{(\overline{\beta}-\underline{\mu})t} \| \mathring{\mathbf{u}} \|_{\mathbf{X}}.$$

Hence $\|\mathbf{u}(t)\|_{\mathbf{X}}$ does not blow up in finite time and the solution is global. □

Corollary 2 *For any* $\mathring{\mathbf{u}} \in \mathbf{X}_+$*, the problem (112) has a unique mild positive solution* $\mathbf{u}(t)$ *defined on the whole time interval* $[0, \infty[$.

The proof follows from the fact that $D(\mathbf{A}_\mu)_+$ is dense in $\mathbf{X}_+$ and the continuous dependence on initial conditions.

5 Coagulation–Fragmentation Equation

Recall that we deal with the equation

$$\partial_t u(x,t) = -a(x)u(x,t) + \int_x^\infty a(y)b(x|y)u(y,t)dy \tag{133}$$

$$-u(x,t)\int_0^\infty k(x,y)u(y,t)dy + \frac{1}{2}\int_0^x k(x-y,y)u(x-y,t)u(y,t)dy,$$

where $x \in \mathbb{R}_+ := (0, \infty)$ is the size of the particles/clusters. Here u is the density of particles of mass/size x, a is the fragmentation rate and b describes the distribution of masses x of particles spawned by a particle of mass y.

The fragmentation rate a is assumed to satisfy

$$0 \le a \in L_{\infty,loc}([0, \infty[), \tag{134}$$

that is, we allow a to be unbounded as $x \to \infty$. Further, $b \ge 0$ is a measurable function satisfying $b(x|y) = 0$ for $x > y$ and (24).

The expected number of particles resulting from a fragmentation of a size y parent, denoted by $n_0(y)$, is assumed to satisfy

$$n_0(y) < +\infty \tag{135}$$

for any fixed $y \in \mathbb{R}_+$. Further, we assume that there are $j \in (0, \infty)$, $l \in [0, \infty)$ and $a_0, b_0 \in \mathbb{R}_+$ such that for any $x \in \mathbb{R}_+$

$$a(x) \le a_0(1 + x^j), \qquad n_0(x) \le b_0(1 + x^l). \tag{136}$$

The coagulation kernel $k(x, y)$ represents the likelihood of a particle of size x attaching itself to a particle of size y. We assume that it is a measurable symmetric function such that for some $K > 0$ and $0 \le \beta \le \alpha < 1$

$$0 \le k(x, y) \le K((1 + a(x))^{\alpha}(1 + a(y))^{\beta} + (1 + a(x))^{\beta}(1 + a(y))^{\alpha}). \tag{137}$$

This will suffice to show local in time solvability of (133), whereas to show that the solutions are global in time we need to strengthen (137) to

$$0 \le k(x, y) \le K((1 + a(x))^{\alpha} + (1 + a(y))^{\alpha}) \tag{138}$$

for some $0 \le \alpha < 1$.

In fragmentation and coagulation problems, two spaces are most often used due to their physical relevance. In the space $L_1(\mathbb{R}_+, xdx)$ the norm of a nonnegative element u, given by $\int_0^\infty u(x)xdx$, represents the total mass of the system, whereas the norm of a nonnegative element u in the space $L_1(\mathbb{R}_+, dx)$, $\int_0^\infty u(x)dx$, gives the total number of particles in the system.

We use the scale of spaces with finite higher moments

$$X_m = L_1(\mathbb{R}_+, dx) \cap L_1(\mathbb{R}_+, x^m dx) = L_1(\mathbb{R}_+, (1 + x^m)dx), \tag{139}$$

where $m \in \mathbb{M} := [1, \infty)$. We extend this definition to $X_0 = L_1(\mathbb{R}_+)$. We note that, due to the continuous injection $X_m \hookrightarrow X_1$, $m \ge 1$, any solution in X_m is also a solution in the basic space X_1.

Thus, we denote by $\| \cdot \|_m$ the natural norm in X_m defined in (139). To shorten notation, we define

$$w_m(x) := 1 + x^m.$$

5.1 Analyticity of the Fragmentation Operator

To formulate the main results, we have to introduce more specific assumptions and notation. First we define

$$n_m(y) := \int_0^y b(x|y)x^m dx$$

for any $m \in \mathbb{M}_0 := \{0\} \cup \mathbb{M}$ and $y \in \mathbb{R}_+$. Further, let

$$N_0(y) := n_0(y) - 1 \quad \text{and} \quad N_m(y) := y^m - n_m(y), \quad m \geq 1.$$

It follows that

$$N_0(y) = n_0(y) - 1 \geq 0$$

and

$$N_m(y) = y^m - \int_0^y b(x|y)x^m dx \geq y^m - y^{m-1}\int_0^y b(x|y)xdx = 0 \tag{140}$$

for $m \geq 1$ with $N_1 = 0$.

Next, for $m \in \mathbb{M}$, let $(A_m u)(x) := a(x)u(x)$ on $D(A_m) = \{u \in X_m : au \in X_m\}$ and let B_m be the restriction to $D(A_m)$ of the integral expression

$$[\mathcal{B}u](x) = \int_x^\infty a(y)b(x|y)u(y)dy.$$

Theorem 27 ([12]) *Let a, b satisfy (24), (135) and (136), and let m be such that $m \geq j + l$ if $j + l > 1$ and $m > 1$ if $j + l \leq 1$.*

(a) The closure $(F_m, D(F_m)) = \overline{(-A_m + B_m, D(A_m))}$ generates a positive quasi-contractive semigroup, say $(S_{F_m}(t))_{t\geq 0}$, of the type at most $4a_0b_0$ on X_m. Furthermore, if $u \in D(F_m)_+$, then

$$N_m(x)a(x)u(x) \in X_0, \qquad m \in \mathbb{M}_0. \tag{141}$$

(b) If, moreover, for some m there is $c_m > 0$ such that

$$\liminf_{x\to\infty} \frac{N_m(x)}{x^m} = c_m, \tag{142}$$

then $F_m = -A_m + B_m$ and $(S_{F_m}(t))_{t\geq 0}$ is an analytic semigroup on X_m.

(c) If (142) holds for some m_0, then it holds for all $m \geq m_0$.

We note that (142) cannot hold for $m = 1$ as $N_1 = 0$.

Proof We shall fix m satisfying $m \geq j + l$ if $j + l > 1$ and $m > 1$ otherwise; see (136). First we show that $B_m := \mathcal{B}|_{D(A_m)}$ is well defined. Next, direct integration gives for $u \in D(A_m)$

$$\int_0^\infty (-A_m + B_m)u(x)w_m(x)dx = -\phi_m(u) := \int_0^\infty (N_0(x) - N_m(x))a(x)u(x)dx. \tag{143}$$

If the term $N_0(x) > 0$ were not present, then (143) would allow a direct application of the substochastic semigroup theory. In the present case we note that for $u \in D(A_m)_+$ we have, by (140),

$$-\phi_m(u) \leq \int_0^\infty N_0(y)a(y)u(y)dy \leq 4a_0b_0 \int_0^\infty u(x)w_m(x)dx =: \eta\|u\|_m,$$

Then we have $\tilde{\phi}_m(u) := \phi_m(u) + \eta \int_0^\infty u(x)w_m(x)dx \geq 0$ for $0 \leq u \in D(A_m)_+$ and the operator $(\tilde{A}_m, D(A_m)) := (A_m + \eta I, D(A_m))$ satisfies

$$\int_0^\infty (-\tilde{A}_m + B_m)u(x)w_m(x)dx = -\tilde{\phi}_m(u)$$

$$= -\eta \int_0^\infty u(x)w_m(x)dx + \int_0^\infty (N_0(x) - N_m(x))a(x)u(x)dx \leq 0.$$

Hence an extension $\tilde{F}_m$ of $-\tilde{A}_m + B_m$ generates a substochastic semigroup $(S_{\tilde{F}_m}(t))_{t\geq 0}$ and thus there is an extension F_m of $(-A_m + B_m, D(A_m))$, given by $(F_m, D(F_m)) = (\tilde{F}_m + \eta I, D(\tilde{F}_m))$, generating a positive semigroup $(S_{F_m}(t))_{t\geq 0} = (e^{\eta t} S_{\tilde{F}_m}(t))_{t\geq 0}$ on X_m.

Furthermore, $\tilde{\phi}_m$ extends to $D(F_m)$ by monotone limits of elements of $D(A_m)$. Thus, let $u \in D(F_m)_+$ with $D(A_m) \ni u_n \nearrow u$. Then, since

$$\int_0^\infty N_0(x)a(x)u(x)dx < \infty, \quad \int_0^\infty u(x)w_m(x)dx < \infty,$$

by (136), $m \geq j + l$ and $D(F_m) \subset X_m$, and the fact that $\tilde{\phi}_m(u_n)$ tends to a finite limit, we have

$$\lim_{n\to\infty} \int_0^\infty N_m(x)a(x)u_n(x)dx = \int_0^\infty N_m(x)a(x)u(x)dx < +\infty.$$

To prove part (b), we begin by observing that inequality (140) implies that $0 \leq N_m(x) \leq x^m$. This, together with (142), yields $c_m x^m/2 \leq N_m(x) \leq x^m$ for large x which, by (141), establishes that if $u \in D(F_m)$, then $au \in X_m$ or, in other words, that $D(F_m) \subset D(A_m)$. Since $(F_m, D(F_m))$ is an extension of $(-A_m + B_m, D(A_m))$, we see that $D(F_m) = D(A_m)$.

It is clear that the semigroup generated by $-A_m$ is bounded. Furthermore, if $\lambda = r + is$ then $|\lambda + a(x)|^2 \geq s^2$ and therefore, for all $r > 0$

$$\|R(r+is, -A_m)f\|_m = \int_0^\infty \left|\frac{1}{r+is+a(x)}\right| |f(x)|(1+x^m)dx \leq \frac{1}{|s|}\|f\|_m.$$

The analyticity of the fragmentation semigroup then follows from the Arendt–Rhandi theorem, Theorem 19. □

Example 7 One of the forms of $b(x|y)$ most often used in applications is

$$b(x|y) = \frac{1}{y}h\left(\frac{x}{y}\right) \tag{144}$$

which is referred to as the homogeneous fragmentation kernel. In this case the distribution of the daughter particles does not depend directly on their relative sizes but on their ratio. In this case

$$n_m(y) = \frac{1}{y}\int_0^y h\left(\frac{x}{y}\right)x^m dx = y^m \int_0^1 h(z)z^m dz =: h_m y^m.$$

Since

$$y = n_1(y) = \frac{1}{y}\int_0^y h\left(\frac{x}{y}\right)x dx = y\int_0^1 h(z)z dz = h_1 y,$$

we have $h_1 = 1$ so that $h_m < 1$ for any $m > 1$ and $N_m(y) = y^m(1-h_m)$. Hence, (142) holds.

On the other hand, fragmentation processes in which daughter particles accumulate close both to 0 and to the parent's size may not satisfy (142), [17].

5.2 *Existence of Solutions to the Fragmentation–Coagulation Problem*

Next, we introduce a nonlinear operator C_m in X_m defined for u from a suitable subset of X_m by the formula

$$(C_m u)(x) := -u(x)\int_0^\infty k(x,y)u(y)dy + \frac{1}{2}\int_0^x k(x-y,y)u(x-y)u(y)dy$$

so that the initial value problem for (133) can be written as an abstract semilinear Cauchy problem in X_m,

$$\partial_t u = -A_m u + B_m u + C_m u, \qquad u(0) = \mathring{u}\ . \tag{145}$$

To formulate the main theorems we have to introduce a new class of spaces which, as we shall see later, is related to intermediate spaces associated with the fragmentation operator F_m and its fractional powers. We set

$$X_m^{(\alpha)} := \left\{ u \in X_m;\ \int_0^\infty |u(x)|(\omega + a(x))^\alpha (1+x^m)\,dx < \infty \right\}, \tag{146}$$

where ω is a sufficiently large constant. Then we have

Theorem 28 ([17]) *Assume that a, b, k satisfy (24), (135), (136), (137)and (142) for some $m_0 > 1$, and let $m \geq \max\{j+l, m_0\}$ hold. Then, for each $\mathring{u} \in X_{m,+}^{(\alpha)}$, there is $\tau > 0$ such that the initial value problem (145) has a unique nonnegative classical solution $u \in C([0,\tau], X_m^{(\alpha)}) \cap C^1((0,\tau), X_m) \cap C((0,\tau), D(A_m))$. Furthermore, there is a measurable representation of u which is absolutely continuous in $t \in (0,\tau)$ for any $x \in \mathbb{R}_+$ and which satisfies (133) almost everywhere on $\mathbb{R}_+ \times (0,\tau)$.*

Finally, for the global in time solvability we need to restrict the growth rate of k. Namely, we have

Theorem 29 ([17]) *Let the assumptions of Theorem 28 hold with $\beta = 0$, that is, let k satisfy (138). Furthermore, let the constant j from assumption (136) be such that $\alpha j \leq 1$. Then any local solution of Theorem 28 is global in time.*

5.2.1 Interlude: Intermediate Spaces Associated with F_m

From now one, we shall assume that $b(x|y)$ is such that (142) is satisfied. Define, for a fixed constant $\omega > 4a_0 b_0$,

$$\begin{aligned} F_{m,\omega} &:= F_m - \omega I, \quad A_{m,\omega} := A_m + \omega I, \\ D(F_{m,\omega}) &= D(F_m) = D(A_m) = D(A_{m,\omega}), \end{aligned} \tag{147}$$

The operators $(F_{m,\omega}, D(A_m))$ and $(-A_{m,\omega}, D(A_m))$ generate analytic semigroups $(S_{F_{m,\omega}}(t))_{t\geq 0} = (e^{-\omega t}S_{F_m}(t))_{t\geq 0}$ and, respectively, $(S_{-A_{m,\omega}}(t))_{t\geq 0} = (e^{-\omega t}S_{-A_m}(t))_{t\geq 0}$ on X_m. Since each operator is invertible, the norms $\|u\|_{m,A} := \|A_{m,\omega}u\|_m$ and $\|u\|_{m,F} := \|F_{m,\omega}u\|_m$, $u \in D(A_m)$ are equivalent to each other and also to the corresponding graph norms on $D(A_m)$. Then we have (up to the equivalence of the respective norms)

$$D_{F_{m,\omega}}(\alpha, r) = D_{-A_{m,\omega}}(\alpha, r). \tag{148}$$

We find it most convenient to use $D_{-A_{m,\omega}}(\alpha, 1)$ which equals the real interpolation space $(X_m, D(A_{m,\omega}))_{\alpha,1}$. It follows that

$$(X_m, D(A_{m,\omega}))_{\alpha,1} = X_m^{(\alpha)} = \left\{u \in X_m;\ \int_0^\infty |u(x)|(\omega + a(x))^\alpha(1 + x^m)\,dx < \infty\right\},$$

which hereafter we equip with the norm

$$\|u\|_m^{(\alpha)} := \int_0^\infty |u(x)|(\omega + a(x))^\alpha(1 + x^m)\,dx. \tag{149}$$

In other words, there is a constant $c_1 \geq 1$ such that

$$c_1^{-1}\|u\|_m^{(\alpha)} \leq \|u\|_{D_{F_{m,\omega}}(\alpha,1)} \leq c_1\|u\|_m^{(\alpha)}, \qquad \forall\, u \in D_{F_{m,\omega}}(\alpha, 1). \tag{150}$$

Proof of Theorem 28 Here we assume that a and b satisfy the assumptions of Theorem 27(b) so that, in particular, (142) holds for some $m \geq j + l$ or $m > 1$ if $j + l \leq 1$. Furthermore, the coagulation kernel is such that (137) is satisfied. We fix $\omega > \max\{4a_0b_0, 1\}$ and denote $a_\omega^\alpha(x) := (\omega + a(x))^\alpha$. Similarly to (132), we consider the following modified version of (133)

$$\begin{aligned}\partial_t u(x,t) = {} & -(a_\omega(x) + \gamma a_\omega^\alpha(x))u(x,t) + \int_x^\infty a(y)b(x|y)u(y,t)dy \\ & +(\gamma a_\omega^\alpha(x) + \omega)u(x,t) - u(x,t)\int_0^\infty k(x,y)u(y,t)dy \\ & +\frac{1}{2}\int_0^x k(x-y,y)u(x-y,t)u(y,t)dy,\end{aligned} \tag{151}$$

where γ is a constant to be determined and α is the index appearing in (137). Then, see [41, Corollary 3.2.4], $(F_\gamma, D(F_\gamma)) := (F_{m,\omega} - \gamma A_m^\alpha, D(A_m))$ generates an analytic semigroup, say $(S_{F_\gamma}(t))_{t\geq 0}$, on X_m. Since $(S_{F_{m,\omega}}(t))_{t\geq 0}$ and $(S_{-\gamma A_m^\alpha}(t))_{t\geq 0}$

are positive and contractive, we can use the Trotter product formula to deduce that $(S_{F_\gamma}(t))_{t\geq 0}$ is also a positive contraction on X_m. Furthermore, since $S_{-\gamma A_m^\alpha}(t) \leq I$ for $t \geq 0$, using again the Trotter formula

$$S_{F_\gamma}(t)u \leq S_{F_{m,\omega}}(t)u, \quad u \in X_{m,+}. \tag{152}$$

and thus, for $u \in X_m^{(\alpha)}$

$$\|S_{F_\gamma}(t)u\|_m^{(\alpha)} \leq c_1^2 \|u\|_m^{(\alpha)}. \tag{153}$$

Next consider the set

$$\mathcal{U} = \{u \in X_{m,+}^{(\alpha)} : \ \|u\|_m^{(\alpha)} \leq 1 + b\}, \tag{154}$$

for some arbitrary fixed $b > 0$ and set

$$\gamma = 2K(b+1). \tag{155}$$

Then on $\mathcal{U}$ we obtain

$$\begin{aligned}(C_\gamma u)(x) &:= -u(x)\int_0^\infty k(x,y)u(y)dy + (\gamma(a_\omega^\alpha(x) + \omega)u(x)\\ &+\frac{1}{2}\int_0^x k(x-y,y)u(x-y)u(y)dy \geq 0.\end{aligned}$$

Similarly, on $\mathcal{U}$ we have $\|C_\gamma u\|_m \leq K_1(\mathcal{U})$, as well as

$$\|(\gamma A_m^\alpha + \omega I)u - (\gamma A_m^\alpha + \omega I)v\|_m \leq (\omega + \gamma)\|u - v\|_m^{(\alpha)}$$

and

$$\|C_\gamma u - C_\gamma v\|_m \leq K_2(\mathcal{U})\|u - v\|_m^{(\alpha)}, \tag{156}$$

for some constants $K_1(\mathcal{U})$, $K_2(\mathcal{U})$. Hence, for $\mathring{u} \in X_{m,+}^{(\alpha)}$ satisfying $\| \mathring{u} \|_m^{(\alpha)} \leq c_1^{-2}b$, for b of (154) and c_1 from (153), there is $\tau = \tau(\mathring{u})$ such that the mapping

$$(\mathcal{T}u)(t) = S_{F_\gamma}(t)\,\mathring{u} + \int_0^t S_{F_\gamma}(t-s)C_\gamma u(s)ds$$

is a contraction on $Y = C([0,\tau],\mathcal{U})$, with $\mathcal{U}$ defined by (154) and the metric induced by the norm $\|u(t)\|_Y := \sup_{0\le t\le\tau} \|u(t)\|_m^{(\alpha)}$. Therefore, for any $\overset{\circ}{u} \in X_{m,+}^{(\alpha)}$, there is a unique mild solution u to (145) in $X_{m,+}^{(\alpha)}$ which, moreover, satisfies

$$u \in C^1((0,\tau), X_m) \cap C((0,\tau), D(A_m)).$$

□

Proof of Theorem 29 The local solution, constructed in the previous section, can be extended in a usual way to the maximal forward interval of existence $I_{max} = [0, \tau_{max}(\overset{\circ}{u}))$. We also denote $I_{max}^0 = (0, \tau_{max}(\overset{\circ}{u}))$ Thus, to show that u is globally defined, we need to show that $\|u(t)\|_m^{(\alpha)}$ is *a priori* bounded uniformly in time.

Let us denote by M_r the r-th moment of u,

$$M_r(u) := \int_0^\infty x^r u(x)dx.$$

Then, for some constant L,

$$\|u\|_m^{(\alpha)} \le L \int_0^\infty |u(x)|(1 + x^{m+j\alpha})dx = L(M_0(u) + M_{m+j\alpha}(u)). \tag{157}$$

Though for a given m, Theorem 28 does not ensure the differentiability of $M_{m+\alpha j}$, it is valid in the scale of spaces X_r with $r \ge m$ provided, of course, $\overset{\circ}{u} \in X_r^{(\alpha)}$. Since the embedding $X_r^{(\alpha)} \subset X_m^{(\alpha)}$ is continuous for $r \ge m$, the solutions emanating from the same initial value $\overset{\circ}{u} \in X_r^{(\alpha)} \subset X_m^{(\alpha)}$ in each space, by construction, must coincide. Hence, let $\overset{\circ}{u} \in X_{m+j\alpha}^{(\alpha)} \subset X_{m+j\alpha} \subset X_m^{(\alpha)}$ so that

$$u \in C(I_{max}, X_{m+j\alpha}^{(\alpha)}) \cap C^1(I_{max}^0, X_{m+j\alpha}) \cap C(I_{max}^0, D(A_{m+j\alpha})),$$

with possibly different, but still nonzero length of the interval I_{max}. This, in particular, yields differentiability of $\|u(\cdot)\|_0 = M_0(u(\cdot))$ and, consequently, of $M_{m+j\alpha}(u(\cdot))$. To get the moment estimates we use the inequality, [16],

$$(x+y)^r - x^r - y^r \le (2^r - 1)(x^{r-1}y + y^{r-1}x) =: G_r(x^{r-1}y + y^{r-1}x), \tag{158}$$

for $r \ge 1$, $x, y \in \mathbb{R}_+$. Then

$$\int_0^\infty x^r (Cu)(x)dx = G_r K L_\alpha (M_{r+j\alpha-1}M_1 + M_{r-1}M_{1+j\alpha} + 2M_{r-1}M_1). \tag{159}$$

For the particular cases $r = 0$ and $r = 1$ we obtain

$$\int_0^\infty (Cu)(x)dx = -\frac{1}{2}\int_0^\infty k(x,y)u(x,t)u(y,t)dxdy \leq 0,$$

$$\int_0^\infty x(Cu)(x)dx = 0.$$

Hence, using the estimates for the linear part, we obtain on I^0_{max}

$$\begin{aligned} M_{0,t} &\leq 4a_0b_0(M_0 + M_m), \\ M_{1,t} &= 0, \\ M_{m+j\alpha,t} &\leq G_{m+j\alpha}KL_\alpha(M_{m+2j\alpha-1}M_1 + M_{m+j\alpha-1}(M_{1+j\alpha} + 2M_1)). \end{aligned} \tag{160}$$

We see that if $1 \leq r \leq r'$, then

$$M_r \leq M_1 + M_{r'} \tag{161}$$

as $x^r \leq x$ on $[0,1]$ and $x^r \leq x^{r'}$ on $[1,\infty)$. Thus, we see that in order for the moment system (160) to be closed, we must assume that $j\alpha \leq 1$. This allows us to re-write (160) as

$$\begin{aligned} M_{0,t} &\leq 4a_0b_0(2M_0 + M_{m+j\alpha}), \\ M_{m+j\alpha,t} &\leq G_{m+j\alpha}KL_\alpha((M_{m+j\alpha} + M_1)M_1 + (M_{m+j\alpha} + M_1)(M_2 + 3M_1)), \end{aligned} \tag{162}$$

where M_1 is constant and where we used $j\alpha \leq 1$. To find the behaviour of M_2, again we use (143) and (159), with an obvious simplification of (158), to get the estimate for M_2 as

$$M_{2,t} \leq 4KL_\alpha(M_{1+j\alpha}M_1 + M_1^2) \leq 4KL_\alpha(M_2M_1 + 2M_1^2).$$

Hence, M_2 is bounded on its interval of existence. Then, from the second inequality in (162), we see that $M_{m+j\alpha}$ satisfies a linear inequality with bounded coefficients and thus it also is bounded on I^0_{max}. This in turn yields the boundedness of M_0. Hence, $\|u(\cdot)\|_m^{(\alpha)}$ is bounded and thus u exists globally.

To ascertain global existence of solutions emanating from any initial datum $\overset{\circ}{u} \in X_m^{(\alpha)}$ we observe that since $X_{m+j\alpha}^{(\alpha)}$ is dense in $X_m^{(\alpha)}$, finite blow-up of such a solution would contradict the theorem on continuous dependence of solutions on the initial data. □

References

1. Ch.D. Aliprantis, O. Burkinshaw, *Positive Operators* (Academic, Orlando, 1985)
2. T. Apostol, *Mathematical Analysis* (Addison-Wesley, Reading, 1957)
3. W. Arendt, Resolvent positive operators. Proc. Lond. Math. Soc. **54**(3), 321–349 (1987)
4. W. Arendt, Vector-valued Laplace transforms and Cauchy problems. Israel J. Math. **59**(3), 327–352 (1987)
5. W. Arendt, A. Rhandi, Perturbation of positive semigroups. Archiv der Mathematik **56**(2), 107–119 (1991)
6. W. Arendt, Ch.J.K. Batty, M. Hieber, F. Neubrander, *Vector-Valued Laplace Transforms and Cauchy Problems* (Birkäuser, Basel, 2001)
7. L. Arlotti, J. Banasiak, Strictly substochastic semigroups with application to conservative and shattering solutions to fragmentation equations with mass loss. J. Math. Anal. Appl. **293**(2), 693–720 (2004)
8. L. Arlotti, B. Lods, M. Mokhtar-Kharroubi, On perturbed substochastic semigroups in abstract state spaces. Zeitschrift fur Analysis und ihre Anwendung **30**(4), 457–495 (2011)
9. J.M. Ball, Strongly continuous semigroups, weak solutions, and the variation of constants formula. Proc. Amer. Math. Soc. **63**(2), 370–373 (1977)
10. J. Banasiak, Remarks on solvability of inhomogeneous abstract Cauchy problem for linear and semilinear equations. Questiones Mathematicae **22**(1), 83–92 (1999)
11. J. Banasiak, Kinetic-type models with diffusion: conservative and nonconservative solutions. Transp. Theory Stat. Phys. **36**(1–3), 43–65 (2007)
12. J. Banasiak, Transport processes with coagulation and strong fragmentation. Discrete Continuous Dyn. Syst. Ser B **17**(2), 445–472 (2012)
13. J. Banasiak, L. Arlotti, *Positive Perturbations of Semigroups with Applications* (Springer, London, 2006)
14. J. Banasiak, M. Lachowicz, Around the Kato generation theorem for semigroups. Studia Mathematica **179**(3), 217–238 (2007)
15. J. Banasiak, W. Lamb, Coagulation, fragmentation and growth processes in a size structured population. Discrete Continuous Dyn. Syst. Ser B **11**(3), 563–585 (2009)
16. J. Banasiak, W. Lamb, Global strict solutions to continuous coagulation–fragmentation equations with strong fragmentation. Proc. Roy. Soc. Edinburgh Sect. A **141**, 465–480 (2011)
17. J. Banasiak, W. Lamb, Analytic fragmentation semigroups and continuous coagulation–fragmentation equations with unbounded rates. J. Math. Anal. Appl. **391**, 312–322 (2012)
18. J. Banasiak, P. Namayanja, Asymptotic behaviour of flows on reducible networks. Networks and Heterogeneous Media **9**(2), 197–216, (2014)
19. J. Banasiak, R.Y. M'pika Massoukou, A singularly perturbed age structured SIRS model with fast recovery. Discrete Continuous Dyn. Syst. Ser B **19**(8), 2383–2399 (2014)
20. M.K. Banda, Nonlinear hyperbolic systems of conservation laws and related applications, in *Evolutionary Equations with Applications to Natural Sciences*, ed. by J. Banasiak, M. Mokhtar-Kharroubi. Lecture Notes in Mathematics (Springer, Berlin, 2014)
21. J. Bang-Jensen, G. Gutin, *Digraphs: Theory, Algorithms and Applications*, 2nd edn. (Springer, London, 2009)
22. C.J.K. Batty, D.W. Robinson, Positive one-parameter semigroups on ordered Banach spaces. Acta Appl. Math. **2**(3–4), 221–296 (1984)
23. H. Brezis, *Functional Analysis, Sobolev Spaces and Partial Differential Equations* (Springer, New York, 2011)
24. S. Busenberg, M. Iannelli, H. Thieme, Global behavior of an age-structured epidemic. SIAM J. Math. Anal. **22**(4), 1065–1080 (1991)
25. B. Dorn, Semigroups for flows in infinite networks. Semigroup Forum **76**, 341–356 (2008)
26. N. Dunford, J.T. Schwartz, *Linear Operators, Part I: General Theory* (Wiley, New York, 1988)
27. K.-J. Engel, R. Nagel, *One-Parameter Semigroups for Linear Evolution Equations* (Springer, New York, 1999)

28. E. Estrada, Dynamical and evolutionary processes on complex networks, in *Evolutionary Equations with Applications to Natural Sciences* ed. by J. Banasiak, M. Mokhtar-Kharroubi. Lecture Notes in Mathematics (Springer, Berlin, 2014)
29. H.O. Fattorini, *The Cauchy Problem* (Addison-Wesley, Reading, 1983)
30. G. Frosali, C. van der Mee, F. Mugelli, A characterization theorem for the evolution semigroup generated by the sum of two unbounded operators. Math. Meth. Appl. Sci. **27**(6), 669–685 (2004)
31. M. Iannelli, *Mathematical Theory of Age-Structured Population Dynamics*, Applied Mathematics Monographs, Consiglio Nazionale delle Ricerche C.N.R., vol. 7 (Giardini, Pisa, 1995)
32. H. Inaba, A semigroup approach to the strong ergodic theorem of the multistate stable population process. Math. Popul. Stud. **1**(1), 49–77 (1988)
33. T. Kato, On the semigroups generated by Kolmogoroff's differential equations. J. Math. Soc. Jap. **6**, 1–15 (1954)
34. W. Lamb, Applying functional analytic techniques to evolution equations, in *Evolutionary Equations with Applications to Natural Sciences* ed. by J. Banasiak, M. Mokhtar-Kharroubi. Lecture Notes in Mathematics (Springer, Berlin, 2014)
35. P. Laurençot, On a class of continuous coagulation–fragmentation equations. J. Differ. Equ. **167**, 245–274 (2000)
36. P. Laurençot, Weak compactness techniques and coagulation equations, in *Evolutionary Equations with Applications to Natural Sciences* ed. by J. Banasiak, M. Mokhtar-Kharroubi. Lecture Notes in Mathematics (Springer, Berlin, 2014)
37. J.L. Lions, E. Magenes, *Nonhomogeneous Boundary Value Problems and Applications*, vol. 1 (Springer, New York, 1972)
38. A. Lunardi, *Analytic Semigoups and Optimal Regularity in Parabolic Problems* (Birkhäuser, Basel, 1995)
39. M. Mokhtar-Kharroubi, J. Voigt, On honesty of perturbed substochastic C_0-semigroups in L_1-spaces. J. Operat. Theor. **64**(1), 131–147 (2010)
40. R.Y. M'pika Massoukou, Age structured models of mathematical epidemiology, Ph.D. thesis, UKZN, 2014
41. A. Pazy *Semigroups of Linear Operators and Applications to Partial Differential Equations* (Springer, Berlin, 1983)
42. R. Showalter, *Hilbert Space Methods for Partial Differential Equations* (Longman, Harlow, 1977)
43. J. van Neerven, *The Asymptotic Behaviour of Semigroups of Linear Operators* (Birkhäuser, Basel, 1996)
44. J. Voigt, On substochastic semigroups C_0-semigroups and their generators. Transp. Theory Stat. Phys. **16**, 453–466 (1987)
45. J. Voigt, On resolvent positive operators and positive C_0-semigroups on AL-spaces. Semigroup Forum **38**(2), 263–266 (1989)
46. G.F. Webb, *Theory of Nonlinear Age Dependent Population Dynamics* (Marcel Dekker, New York, 1985)
47. K. Yosida, *Functional Analysis*, 5th edn. (Springer, Berlin, 1978)

Weak Compactness Techniques and Coagulation Equations

Philippe Laurençot

1 Introduction

Coagulation is one of the driving mechanisms for cluster growth, by which clusters (or particles) increase their sizes by successive mergers. Polymer and colloidal chemistry, aerosol science, raindrops and soot formation, astrophysics (formation of planets and galaxies), hematology, and animal herding are among the fields where coagulation phenomena play an important role, see [18, 33, 37, 75] for instance. This variety of applications has generated a long lasting interest in the modeling of coagulation processes. One of the first contributions in that direction is due to the Polish physicist Smoluchowski who derived a model for the evolution of a population of colloidal particles increasing their sizes by binary coagulation while moving according to independent Brownian motion [78, 79]. Neglecting spatial variations he came up with the discrete Smoluchowski coagulation equations

$$\frac{df_1}{dt} = -\sum_{j=1}^{\infty} K(1,j) f_1 f_j \,, \quad t > 0 \,, \tag{1}$$

$$\frac{df_i}{dt} = \frac{1}{2}\sum_{j=1}^{i-1} K(j,i-j) f_{i-j} f_j - \sum_{j=1}^{\infty} K(i,j) f_i f_j \,, \quad i \ge 2 \,,\ t > 0 \,. \tag{2}$$

Here the sizes of the particles are assumed to be multiples of a minimal size normalized to one and the coagulation kernel $K(i,j)$ accounts for the rate at which

Ph. Laurençot (✉)
Institut de Mathématiques de Toulouse, UMR 5219, Université de Toulouse, CNRS, 31062 Toulouse Cedex 9, France
e-mail: laurenco@math.univ-toulouse.fr; philippe.laurencot@math.univ-toulouse.fr

© Springer International Publishing Switzerland 2015

J. Banasiak, M. Mokhtar-Kharroubi (eds.), *Evolutionary Equations with Applications in Natural Sciences*, Lecture Notes in Mathematics 2126,
DOI 10.1007/978-3-319-11322-7_5

a particle of size i and a particle of size j encounter and merge into a single particle of size $i + j$. In the derivation performed in [78, 79] K is computed to be

$$K_{sm}(x, y) := \left(x^{1/3} + y^{1/3}\right) \left(\frac{1}{x^{1/3}} + \frac{1}{y^{1/3}}\right) \tag{3}$$

and then reduced to $K_0(x, y) := 2$ to allow for explicit computations of solutions to (1)–(2).

The function f_i, $i \geq 1$, denotes the size distribution function of particles of size $i \geq 1$ at time $t \geq 0$ and the meaning of the reaction terms in (1)–(2) is the following: the second term on the right-hand side of (2) describes the depletion of particles of size i by coalescence with other particles of arbitrary size while the first term on the right-hand side of (2) accounts for the gain of particles of size i due to the merging of a particle of size $j \in \{1, \cdots, i-1\}$ and a particle of size $i - j$. Note that the assumption of a minimal size entails that there is no formation of particles of size 1 by coagulation. Schematically, if P_x stands for a generic particle of size x, the coagulation events taken into account in the previous model are:

$$P_x + P_y \longrightarrow P_{x+y} \,. \tag{4}$$

The formation of particles of size i corresponds to the choice $(x, y) = (j, i - j)$ in (4) with $i \geq 2$ and $1 \leq j \leq i - 1$ and the disappearance of particles of size i to $(x, y) = (i, j)$ in (4) with $i \geq 1$ and $j \geq 1$. A salient feature of the elementary coagulation reaction (4) is that no matter is lost and we shall come back to this point later on.

Smoluchowski's coagulation equation was later on extended to a continuous size variable $x \in (0, \infty)$ and reads [64]

$$\begin{aligned}\partial_t f(t,x) = & \frac{1}{2}\int_0^x K(y, x-y)\, f(t,y)\, f(t,x-y)\, dy \\ & - \int_0^\infty K(x,y)\, f(t,x)\, f(t,y)\, dy\,, \quad (t,x) \in (0,\infty)\times(0,\infty)\,.\end{aligned} \tag{5}$$

In contrast to (1)–(2) which is a system of countably many ordinary differential equations, Eq. (5) is a nonlinear and nonlocal integral equation but the two terms of the right-hand side of (5) have the same physical meaning as in (1)–(2). The coagulation kernel $K(x, y)$ still describes the likelihood that a particle of mass $x > 0$ and a particle of mass $y > 0$ merge into a single particle of mass $x + y$ according to (4). Besides Smoluchowski's coagulation kernel (3) [78, 79] and the constant coagulation kernel $K_0(x, y) = 2$, other coagulation kernels have been derived in the literature such as $K(x, y) = (ax + b)(ay + b)$, $a > 0$, $b \geq 0$ [84], $K(x, y) = \left(x^{1/3} + y^{1/3}\right)^3$, $K(x, y) = \left(x^{1/3} + y^{1/3}\right)^2 \left|x^{1/3} - y^{1/3}\right|$, and $K(x, y) = x^\alpha y^\beta + x^\beta y^\alpha$, $\alpha \leq 1$, $\beta \leq 1$, the latter being rather a model case which includes the constant coagulation kernel K_0 ($\alpha = \beta = 0$), the additive one $K_1(x, y) := x + y$ ($\alpha = 0$, $\beta = 1$), and the multiplicative one $K_2(x, y) := xy$ ($\alpha = \beta = 1$). Observe that this short list of coagulation kernels already reveals a wide variety of behaviours for large values or small values of (x, y) (bounded or unbounded) and on the diagonal $x = y$ (positive or vanishing).

A central issue is which predictions on the coagulation dynamics can be made from the analysis of Smoluchowski's coagulation equation (5) and to what extent these predictions depend upon the properties of the coagulation kernel K. It was uncovered several years ago that the evolution of (5) could lead to two different dynamics according to the properties of K and is closely related to the conservation of matter already alluded to. More precisely, recall that there is neither loss nor gain of matter during the elementary coagulation reaction (4) and this is expected to be true as well during the time evolution of (5). In terms of f, the total mass of the particles distribution at time $t \geq 0$ is

$$M_1(f(t)) := \int_0^\infty x f(t,x)\, dx$$

and mass conservation reads

$$M_1(f(t)) = M_1(f(0)) , \qquad t \geq 0 , \tag{6}$$

provided $M_1(f(0))$is finite. To check whether (6) is true or not, we first argue formally and observe that, if ϑ is an arbitrary function, multiplying (5) by $\vartheta(x)$ and integrating with respect to $x \in (0, \infty)$ give, after exchanging the order of integration,

$$\begin{aligned} &\frac{d}{dt} \int_0^\infty \vartheta(x) f(t,x)\, dx \\ &\quad = \frac{1}{2} \int_0^\infty \int_0^\infty [\vartheta(x+y) - \vartheta(x) - \vartheta(y)]\, K(x,y) f(t,x) f(t,y)\, dydx. \end{aligned} \tag{7}$$

Clearly, the choice $\vartheta(x) = x$ leads to a vanishing right-hand side of (7) and provides the expected conservation of mass. However, one has to keep in mind that the previous computation is only formal as it uses Fubini's theorem without justification. That some care is indeed needed stems from [52] where it is shown that (6) breaks down in finite time for the multiplicative kernel $K_2(x, y) = xy$ for all non-trivial solutions. An immediate consequence of this result is that the mass-conserving solution constructed on a finite time interval in [57, 58] cannot be extended forever. Soon after the publication of [52] a particular solution to (1)–(2) was constructed for $K(i, j) = (ij)^\alpha$, $\alpha \in (1/2, 1)$ which fails to satisfy (6) for all times [50]. At the same time, it was established in [53, 90] that the condition $K(i, j) \leq \kappa(i + j)$ was sufficient for (1)–(2) to have global mass-conserving solutions, that is, solutions satisfying (6). Thanks to these results, a distinction was made between the so-called non-gelling kernels for which all solutions to (1)–(2) and (5) satisfy (6) and gelling kernels for which (6) is infringed in finite time for all non-trivial solutions. The conjecture stated in the beginning of the 1980s is that a coagulation kernel satisfying

$$K(x,y) \le \kappa(2+x+y)\ , \quad (x,y) \in (0,\infty)\times(0,\infty)\ , \tag{8}$$

is non-gelling while a coagulation kernel satisfying

$$K(x,y) \ge \kappa(xy)^{\alpha}\ , \quad \alpha > \frac{1}{2}\ , \quad (x,y) \in (0,\infty)\times(0,\infty)\ , \tag{9}$$

is gelling [24, 38, 53]. Following the contributions [53, 90] the conjecture for non-gelling coagulation kernels (8) is completely solved in [4] for the discrete coagulation equations (1)–(2), an alternative proof being given in [45]. It took longer for this conjecture to be solved for the continuous coagulation equation (5), starting from the pioneering works [1, 61] for bounded kernels and continuing with [20, 46, 81, 83]. Of particular importance is the contribution by Stewart [81] where weak L^1-compactness techniques were used for the first time and turned out to be a very efficient tool which was extensively used in subsequent works. As for the conjecture for gelling kernels (9), it was solved rather recently in [26, 39]. An intermediate step is the existence of solutions to (1)–(2) and (5) with non-increasing finite mass, that is, satisfying $M_1(f(t)) \le M_1(f(0))$ for $t \ge 0$, see [23, 44, 46, 52, 70, 80] and the references therein.

The purpose of these notes is twofold: on the one hand, we collect in Sect. 2 several results on the weak compactness in L^1-spaces which are scattered throughout the literature and which have proved useful in the analysis of (5). We recall in particular the celebrated Dunford–Pettis theorem (Sect. 2.2) which characterizes weakly compact sequences in L^1 with the help of the notion of uniform integrability (Sect. 2.3). Several equivalent forms of the latter are given, including a refined version of the de la Vallée Poussin theorem [14, 16, 49] (Sect. 2.4). We also point out consequences of the combination of almost everywhere convergence and weak convergence (Sect. 2.5). On the other hand, we show in Sect. 3 how the results stated in Sect. 2 apply to Smoluchowski's coagulation equation (5) and provide several existence results including that of mass-conserving solutions (Sect. 3.2). For the sake of completeness, we supplement the existence results with the occurrence of gelation in finite time for gelling kernels [26] (Sect. 3.3) and with uniqueness results (Sect. 3.4). For further information on coagulation equations and related problems we refer to the books [9, 19] and the survey articles [2, 48, 51, 89].

We conclude the introduction with a few words on related interesting issues: we focus in these notes on the deterministic approach to the modeling of coagulation and leave aside the stochastic approach which has been initiated in [54, 55, 78, 79] and further developed in [2, 8, 9, 15, 22, 23, 29, 39, 40, 70] and the references therein.

Another important line of research is the dynamics predicted by Smoluchowski's coagulation equation (5) for large times for homogeneous non-gelling kernels (8) and at the gelation time for homogeneous gelling kernels (9). In both cases the expected behaviour is of self-similar form (except for some particular kernels with homogeneity 1) but the time and mass scales are only well identified for non-gelling kernels and for the multiplicative kernel $K_2(x,y) = xy$, see the survey articles [51, 65, 86] and the references therein. Existence of mass-conserving self-similar solutions for a large class of non-gelling kernels have been constructed recently [28, 30] and their properties studied in [11, 25, 31, 60, 66]. Still for non-gelling kernels, the existence of other self-similar solutions (with a different scaling and possibly infinite mass) is uncovered in [8] for the additive kernel $K_1(x,y) = x + y$ and in [62] for the constant kernel $K_0(x,y) = 2$, both results relying on the use of the Laplace transform which maps (5) either to Burgers' equation or to an ordinary differential equation. Since the use of the Laplace transform has not proved useful for other coagulation kernels, a much more involved argument is needed to cope with a more general class of kernels [67, 68].

Finally, coagulation is often associated with the reverse process of fragmentation and there are several results available for coagulation-fragmentation equations, including existence, uniqueness, mass conservation, and gelation. Actually, the approach described below in Sect. 3 works equally well for coagulation-fragmentation equations under suitable assumptions on the fragmentation rates. Besides the survey articles [19, 48, 89], we refer for instance to [4, 12, 39, 40, 45, 80] for the discrete coagulation-fragmentation equations and to [1, 5, 6, 20, 23, 26, 27, 34–36, 43, 44, 46, 56, 61, 81, 82] for the continuous coagulation-fragmentation equations.

2 Weak Compactness in L^1

Let $(\Omega, \mathscr{B}, \mu)$ be a σ-finite measure space. For $p \in [1, \infty]$, $L^p(\Omega)$ is the usual Lebesgue space and we denote its norm by $\|\cdot\|_p$. If $p \in (1, \infty)$, the reflexivity of the space $L^p(\Omega)$ warrants that any bounded sequence in $L^p(\Omega)$ has a weakly convergent subsequence. In the same vein, any bounded sequence in $L^\infty(\Omega)$ has a weakly-$\star$ convergent subsequence by a consequence of the Banach–Alaoglu theorem [10, Corollary 3.30] since $L^\infty(\Omega)$ is the dual of the separable space $L^1(\Omega)$. A peculiarity of $L^1(\Omega)$ is that a similar property is not true as a consequence of the following result [91, Appendix to Chap. V, Sect. 4], the space $L^1(\Omega)$ being not reflexive.

Theorem 1 (Eberlein–Šmulian) *Let E be a Banach space such that every bounded sequence has a subsequence converging in the $\sigma(E, E')$-topology. Then E is reflexive.*

2.1 *Failure of Weak Compactness in L^1*

In a simpler way, a bounded sequence in $L^1(\Omega)$ need not be weakly sequentially compact in $L^1(\Omega)$ as the following examples show:

Concentration Consider $f \in C^\infty(\mathbb{R}^N)$ such that $f \geq 0$, $\operatorname{supp} f \subset B(0,1) := \{x \in \mathbb{R}^N \ : \ |x| < 1\}$, and $\|f\|_1 = 1$. For $n \geq 1$ and $x \in \mathbb{R}^N$, we define $f_n(x) = n^N f(nx)$ and note that

$$\|f_n\|_1 = \|f\|_1 = 1\,. \tag{10}$$

Thus $(f_n)_{n\geq 1}$ is bounded in $L^1(\mathbb{R}^N)$. Next, the function f being compactly supported, we have

$$\lim_{n\to\infty} f_n(x) = 0 \quad \text{for} \quad x \neq 0\,.$$

The only possible weak limit of $(f_n)_{n\geq 1}$ in $L^1(\mathbb{R}^N)$ would then be zero which contradicts (10). Consequently, $(f_n)_{n\geq 1}$ has no cluster point in the weak topology of $L^1(\mathbb{R}^N)$. In fact, $(f_n)_{n\geq 1}$ converges narrowly towards a bounded measure, the Dirac mass, that is,

$$\lim_{n\to\infty} \int_{\mathbb{R}^N} f_n(x)\, \psi(x)\, dx = \psi(0) \quad \text{for all} \quad \psi \in BC(\mathbb{R}^N)\,,$$

where $BC(\mathbb{R}^N)$ denotes the space of bounded and continuous functions on $\mathbb{R}^N$. The sequence $(f_n)_{n\geq 1}$ is not weakly sequentially compact in $L^1(\mathbb{R}^N)$ because it *concentrates* in the neighbourhood of $x = 0$. Indeed, for $r > 0$, we have

$$\int_{\{|x|\leq r\}} f_n(x)\, dx = \int_{\{|x|\leq nr\}} f(x)\, dx \underset{n\to\infty}{\longrightarrow} 1\,,$$

and

$$\int_{\{|x|\geq r\}} f_n(x)\, dx = \int_{\{|x|\geq nr\}} f(x)\, dx \underset{n\to\infty}{\longrightarrow} 0\,.$$

Vanishing For $n \geq 1$ and $x \in \mathbb{R}$, we set $f_n(x) = \exp(-|x-n|)$. Then

$$\|f_n\|_1 = 2\,, \tag{11}$$

and $(f_n)_{n\geq 1}$ is bounded in $L^1(\mathbb{R})$. Next,

$$\lim_{n\to\infty} f_n(x) = 0 \quad \text{for all} \quad x \in \mathbb{R}\,,$$

and we argue as in the previous example to conclude that $(f_n)_{n\geq 1}$ has no cluster point for the weak topology of $L^1(\mathbb{R})$. In that case, the sequence $(f_n)_{n\geq 1}$ "escapes at infinity" in the sense that, for every $r > 0$,

$$\lim_{n\to\infty} \int_r^\infty f_n(x)\, dx = 2 \quad \text{and} \quad \lim_{n\to\infty} \int_{-\infty}^r f_n(x)\, dx = 0\,.$$

The above two examples show that the mere boundedness of a sequence in $L^1(\Omega)$ does not guarantee at all its weak sequential compactness and additional information is thus required for the latter to be true. It actually turns out that, roughly speaking, the only phenomena that prevent a bounded sequence in $L^1(\Omega)$ from being weakly sequentially compact in $L^1(\Omega)$ are the concentration and vanishing phenomena described in the above examples. More precisely, a necessary and sufficient condition for the weak sequential compactness in $L^1(\Omega)$ of a bounded sequence in $L^1(\Omega)$ is given by the Dunford–Pettis theorem which is recalled below.

2.2 *The Dunford–Pettis Theorem*

We first introduce the modulus of uniform integrability of a bounded subset $\mathscr{F}$ of $L^1(\Omega)$ which somehow measures how elements of $\mathscr{F}$ concentrate on sets of small measures.

Definition 2 Let $\mathscr{F}$ be a bounded subset of $L^1(\Omega)$. For $\varepsilon > 0$, we set

$$\eta\{\mathscr{F},\varepsilon\} := \sup\left\{ \int_A |f|\, d\mu \ :\ f \in \mathscr{F}\,,\ A \in \mathscr{B}\,,\ \mu(A) \le \varepsilon \right\}, \tag{12}$$

and we define the modulus of uniform integrability $\eta\{\mathscr{F}\}$ of $\mathscr{F}$ by

$$\eta\{\mathscr{F}\} := \lim_{\varepsilon\to 0} \eta\{\mathscr{F},\varepsilon\} = \inf_{\varepsilon>0} \eta\{\mathscr{F},\varepsilon\}\,. \tag{13}$$

With this definition, we can state the Dunford–Pettis theorem, see [7, Part 2, Chap. VI, Sect. 2], [16, pp. 33–44], and [21, IV.8] for instance.

Theorem 3 *Let $\mathscr{F}$ be a subset of $L^1(\Omega)$. The following two statements are equivalent:*

(a) $\mathscr{F}$ is relatively weakly sequentially compact in $L^1(\Omega)$.
(b) $\mathscr{F}$ is a bounded subset of $L^1(\Omega)$ satisfying the following two properties:

$$\eta\{\mathscr{F}\} = 0\,, \tag{14}$$

and, for every $\varepsilon > 0$, there is $\Omega_\varepsilon \in \mathscr{B}$ such that $\mu(\Omega_\varepsilon) < \infty$ and

$$\sup_{f\in\mathscr{F}} \int_{\Omega\setminus\Omega_\varepsilon} |f|\, d\mu \le \varepsilon. \tag{15}$$

As already mentioned, the two conditions required in Theorem 3b to guarantee the weak sequential compactness of $\mathscr{F}$ in $L^1(\Omega)$ exclude the concentration and vanishing phenomena: indeed, the condition (14) implies that no concentration can take place while (15) prevents the escape to infinity which arises in the vanishing phenomenon.

Remark 4 The condition (15) is automatically fulfilled as soon as $\mu(\Omega) < \infty$ (with $\Omega_\varepsilon = \Omega$ for each $\varepsilon > 0$).

Thanks to the Dunford–Pettis theorem, the weak sequential compactness of a subset of $L^1(\Omega)$ can be checked by investigating the behaviour of its elements on measurable subsets of Ω. However, the characteristic functions of these sets being not differentiable, applying the Dunford–Pettis theorem in the field of partial differential equations might be not so easy. Indeed, as (partial) derivatives are involved, test functions are usually required to be at least weakly differentiable (or in a Sobolev space) which obviously excludes characteristic functions. Fortunately, an alternative formulation of the condition $\eta\{\mathscr{F}\} = 0$ is available and turns out to be more convenient to use in this field.

2.3 *Uniform Integrability in* L^1

Definition 5 A subset $\mathscr{F}$ of $L^1(\Omega)$ is said to be uniformly integrable if $\mathscr{F}$ is a bounded subset of $L^1(\Omega)$ such that

$$\lim_{c\to\infty} \sup_{f\in\mathscr{F}} \int_{\{|f|\ge c\}} |f|\, d\mu = 0 . \tag{16}$$

Before relating the uniform integrability property with the weak sequential compactness in $L^1(\Omega)$, let us give some simple examples of uniformly integrable subsets:

- If $\mathscr{F}$ is a bounded subset of $L^p(\Omega)$ for some $p \in (1,\infty)$, then $\mathscr{F}$ is uniformly integrable as

$$\sup_{f\in\mathscr{F}} \int_{\{|f|\ge c\}} |f|\, d\mu \le \frac{1}{c^{p-1}} \sup_{f\in\mathscr{F}} \int_{\{|f|\ge c\}} |f|^p\, d\mu \le \frac{1}{c^{p-1}} \sup_{f\in\mathscr{F}} \{\|f\|_p^p\} .$$

- If $f_0 \in L^1(\Omega)$, the set $\mathscr{F} := \left\{ f \in L^1(\Omega) \ :\ |f| \le |f_0| \ \ \mu - \text{a.e.} \right\}$ is uniformly integrable.
- If $\mathscr{F}$ is a uniformly integrable subset of $L^1(\Omega)$, then so is the set $\mathscr{F}^*$ defined by $\mathscr{F}^* := \{|f| \ :\ f \in \mathscr{F}\}$.
- If $\mathscr{F}$ and $\mathscr{G}$ are uniformly integrable subsets of $L^1(\Omega)$, then so is the set $\mathscr{F}+\mathscr{G}$ defined by $\mathscr{F}+\mathscr{G} := \{f+g \ :\ (f,g) \in \mathscr{F}\times\mathscr{G}\}$.

We next state the connection between the Dunford–Pettis theorem and the uniform integrability property.

Proposition 6 *Let $\mathscr{F}$ be a subset of $L^1(\Omega)$. The following two statements are equivalent:*

(i) *$\mathscr{F}$ is uniformly integrable.*
(ii) *$\mathscr{F}$ is a bounded subset of $L^1(\Omega)$ such that $\eta\{\mathscr{F}\} = 0$.*

In other words, the uniform integrability property prevents the concentration on sets of arbitrary small measure. Proposition 6 is a straightforward consequence of the following result [77].

Lemma 7 *Let $\mathscr{F}$ be a bounded subset of $L^1(\Omega)$. Then*

$$\eta\{\mathscr{F}\} = \lim_{c\to\infty} \sup_{f\in\mathscr{F}} \int_{\{|f|\ge c\}} |f|\, d\mu\,. \tag{17}$$

Proof We put

$$\eta_\star := \lim_{c\to\infty} \sup_{f\in\mathscr{F}} \int_{\{|f|\ge c\}} |f|\, d\mu = \inf_{c\ge 0} \sup_{f\in\mathscr{F}} \int_{\{|f|\ge c\}} |f|\, d\mu\,.$$

We first establish that $\eta\{\mathscr{F}\} \le \eta_\star$. To this end, consider $\varepsilon > 0$, $A \in \mathscr{B}$, $g \in \mathscr{F}$, and $c \in (0,\infty)$. If $\mu(A) \le \varepsilon$, we have

$$\begin{aligned}
\int_A |g|\, d\mu &= \int_{A\cap\{|g|<c\}} |g|\, d\mu + \int_{\{A\cap|g|\ge c\}} |g|\, d\mu \\
&\le c\, \mu(A) + \int_{\{|g|\ge c\}} |g|\, d\mu \\
&\le c\, \varepsilon + \sup_{f\in\mathscr{F}} \int_{\{|f|\ge c\}} |f|\, d\mu\,,
\end{aligned}$$

whence

$$\eta\{\mathscr{F},\varepsilon\} \le c\, \varepsilon + \sup_{f\in\mathscr{F}} \int_{\{|f|\ge c\}} |f|\, d\mu\,.$$

Passing to the limit as $\varepsilon \to 0$ leads us to

$$\eta\{\mathscr{F}\} \le \sup_{f\in\mathscr{F}} \int_{\{|f|\ge c\}} |f|\, d\mu$$

for all $c \in (0,\infty)$. Letting $c \to \infty$ readily gives the inequality $\eta\{\mathscr{F}\} \le \eta_\star$.

We now prove the converse inequality. For that purpose, we put

$$\Lambda := \sup_{f\in\mathscr{F}} \{\|f\|_1\} < \infty\,,$$

and observe that

$$\mu\left(\{x \in \Omega\ ,\ |f(x)| \geq c\}\right) \leq \frac{\Lambda}{c}$$

for all $f \in \mathscr{F}$ and $c > 0$. Consequently,

$$\sup_{f \in \mathscr{F}} \mu\left(\{x \in \Omega\ ,\ |f(x)| \geq c\}\right) \leq \frac{\Lambda}{c}\ ,$$

from which we deduce that

$$\eta_\star \leq \sup_{f \in \mathscr{F}} \int_{\{|f| \geq c\}} |f|\ d\mu \leq \eta\left\{\mathscr{F}, \frac{\Lambda}{c}\right\}\ .$$

Since the right-hand side of the previous inequality converges towards $\eta\{\mathscr{F}\}$ as $c \to \infty$, we conclude that $\eta_\star \leq \eta\{\mathscr{F}\}$ and complete the proof of the lemma. □

Owing to Proposition 6, the property $\eta\{\mathscr{F}\} = 0$ can now be checked by studying the sets where the elements of $\mathscr{F}$ reach large values. This turns out to be more suitable in the field of partial differential equations as one can use the functions $r \longmapsto (r - c)_+ := \max\{0, r - c\}$. For instance, if Ω is an open set of $\mathbb{R}^N$ and u is a function in the Sobolev space $W^{1,p}(\Omega)$ for some $p \in [1, \infty]$, then $(u - c)_+$ has the same regularity with $\nabla(u - c)_+ = \mathrm{sign}((u - c)_+)\nabla u$. This allows one in particular to use $(u - c)_+$ as a test function in the weak formulation of nonlinear second order elliptic and parabolic equations and thereby obtain useful estimates. A broader choice of functions is actually possible as we will see in the next theorem.

2.4 The de la Vallée Poussin Theorem

Theorem 8 *Let $\mathscr{F}$ be a subset of $L^1(\Omega)$. The following two statements are equivalent:*

(i) $\mathscr{F}$ is uniformly integrable.
(ii) $\mathscr{F}$ is a bounded subset of $L^1(\Omega)$ and there exists a convex function $\Phi \in C^\infty([0, \infty))$ such that $\Phi(0) = \Phi'(0) = 0$, Φ' is a concave function,

$$\Phi'(r) > 0 \quad \textit{if} \quad r > 0, \tag{18}$$

$$\lim_{r \to \infty} \frac{\Phi(r)}{r} = \lim_{r \to \infty} \Phi'(r) = \infty, \tag{19}$$

and

$$\sup_{f \in \mathscr{F}} \int_\Omega \Phi\left(|f|\right)\ d\mu < \infty. \tag{20}$$

When $\mathscr{F}$ is a sequence of integrable functions $(f_n)_{n\ge1}$, Theorem 8 is established by de la Vallée Poussin [14, pp. 451–452] (without the concavity of Φ' and the regularity of Φ) and is stated as follows: a sequence $(f_n)_{n\ge1}$ is uniformly integrable (in the sense that the subset $\{f_n\ :\ n\ge1\}$ of $L^1(\Omega)$ is uniformly integrable) if and only if there is a non-decreasing function $\varphi\ :\ [0,\infty)\to[0,\infty)$ such that $\varphi(r)\to\infty$ as $r\to\infty$ and

$$\sup_{n\ge1}\int_\Omega \varphi(|f_n|)\,|f_n|\,d\mu<\infty\,.$$

This result clearly implies Theorem 8. Indeed, if Φ denotes the primitive of φ satisfying $\Phi(0)=0$, the function Φ is clearly convex and the convexity inequality $\Phi(r)\le r\,\varphi(r)$ ensures that $(\Phi(|f_n|))_{n\ge1}$ is bounded in $L^1(\Omega)$. When $\mu(\Omega)<\infty$, a proof of Theorem 8 may also be found in [16] and [72, Theorem I.1.2] but the first derivative of Φ is not necessarily concave. As we shall see in the examples below, the possibility of choosing Φ' concave turns out to be helpful. The version of the de la Vallée Poussin theorem stated in Theorem 8 is actually established in [49]. The proof given below is slightly different from those given in the above mentioned references and relies on the following lemma:

Lemma 9 *Let $\Phi\in C^1([0,\infty))$ be a non-negative and convex function with $\Phi(0)=\Phi'(0)=0$ and consider a non-decreasing sequence of integers $(n_k)_{k\ge0}$ such that $n_0=1$, $n_1\ge2$, and $n_k\to\infty$ as $k\to\infty$. Given $f\in L^1(\Omega)$ and $k\ge1$, we have the following inequality:*

$$\int_{\{|f|<n_k\}}\Phi(|f|)\,d\mu\le\Phi'(1)\int_\Omega|f|\,d\mu$$
$$+\sum_{j=0}^{k-1}\left(\Phi'(n_{j+1})-\Phi'(n_j)\right)\int_{\{|f|\ge n_j\}}|f|\,d\mu. \qquad (21)$$

Proof As Φ is convex with $\Phi'(0)=0$, Φ' is non-negative and non-decreasing and

$$\Phi(r)\le r\,\Phi'(r),\quad r\in[0,\infty)\,.$$

Fix $k\ge1$. We infer from the properties of Φ that

$$\int_{\{|f|<n_k\}}\Phi(|f|)\,d\mu\le\int_{\{|f|<n_k\}}\Phi'(|f|)\,|f|\,d\mu$$
$$=\int_{\{0\le|f|<1\}}\Phi'(|f|)\,|f|\,d\mu+\sum_{j=0}^{k-1}\int_{\{n_j\le|f|<n_{j+1}\}}\Phi'(|f|)\,|f|\,d\mu$$

$$
\begin{aligned}
&\leq \Phi'(1) \int_{\{0\leq |f|<1\}} |f|\, d\mu + \sum_{j=0}^{k-1} \Phi'(n_{j+1}) \int_{\{n_j\leq |f|<n_{j+1}\}} |f|\, d\mu \\
&= \Phi'(1) \int_{\{0\leq |f|<1\}} |f|\, d\mu + \sum_{j=0}^{k-1} \Phi'(n_{j+1}) \int_{\{|f|\geq n_j\}} |f|\, d\mu \\
&\quad - \sum_{j=1}^{k} \Phi'(n_j) \int_{\{|f|\geq n_j\}} |f|\, d\mu \\
&\leq \Phi'(1) \int_{\Omega} |f|\, d\mu \\
&\quad + \sum_{j=0}^{k-1} \left(\Phi'(n_{j+1}) - \Phi'(n_j)\right) \int_{\{|f|\geq n_j\}} |f|\, d\mu\,,
\end{aligned}
$$

whence (21). □

The inequality (21) gives some clue towards the construction of a function Φ fulfilling the requirements of Theorem 8. Indeed, it clearly follows from (21) that, in order to estimate the norm of $\Phi(f)$ in $L^1(\Omega)$ uniformly with respect to $f \in \mathcal{F}$, it is sufficient to show that one can find a function Φ and a sequence $(n_k)_{k\geq 0}$ such that the sum in the right-hand side of (21) is bounded independently of $f \in \mathcal{F}$ and $k \geq 1$. Observing that this sum is bounded from above by the series

$$
\sum_{j=0}^{\infty} \left(\Phi'(n_{j+1}) - \Phi'(n_j)\right)\, X_j \tag{22}
$$

with

$$
X_j := \sup_{f\in\mathcal{F}} \int_{\{|f|\geq n_j\}} |f|\, d\mu\,, \qquad j \geq 0\,,
$$

and that $X_j \to 0$ as $j \to \infty$ by (16), the proof of Theorem 8 amounts to showing that one can find Φ and $(n_k)_{k\geq 0}$ such that the series (22) converges.

Proof of Theorem 8 (i) $\Longrightarrow$ (ii). Consider two sequences of positive real numbers $(\alpha_m)_{m\geq 0}$ and $(\beta_m)_{m\geq 0}$ satisfying

$$
\sum_{m=0}^{\infty} \alpha_m = \infty \quad \text{and} \quad \sum_{m=0}^{\infty} \alpha_m \beta_m < \infty. \tag{23}
$$

It then follows from (16) that there exists a non-decreasing sequence of integers $(N_m)_m \geq 0$ such that $N_0 = 1$, $N_1 \geq 2$ and

$$N_{m+1} \geq \left(1 + \frac{\alpha_m}{\alpha_{m-1}}\right) N_m, \quad m \geq 1,, \tag{24}$$

$$\sup_{f\in\mathcal{F}} \int_{\{|f|\geq N_m\}} |f|\, d\mu \leq \beta_m, \quad m \geq 1. \tag{25}$$

Let us now construct the function Φ and first look for a C^1-smooth function which is piecewise quadratic on each interval $[N_m, N_{m+1}]$. More precisely, we assume that, for each $m \geq 0$,

$$\Phi'(r) = A_m\, r + B_m, \quad r \in [N_m, N_{m+1}],$$

the real numbers A_m and B_m being yet to be determined. In order that such a function Φ fulfills the requirements of Theorem 8, A_m and B_m should enjoy the following properties:

(c1) $(A_m)_{m\geq 0}$ is a non-increasing sequence of positive real numbers, which implies the convexity of Φ and the concavity of Φ',
(c2) $A_{m+1}\, N_{m+1} + B_{m+1} = A_m\, N_{m+1} + B_m$, $m \geq 0$, which ensures the continuity of Φ',
(c3) $A_m\, N_m + B_m \to \infty$ as $m \to \infty$, so that (19) is satisfied,
(c4) the series $\sum A_m\ (N_{m+1} - N_m)\ \beta_m$ converges, which, together with (25), ensures that the right-hand side of (21) is bounded uniformly with respect to $f \in \mathcal{F}$.

Let us now prove that the previously constructed sequence $(N_m)_{m\geq 0}$ allows us to find (A_m, B_m) complying with the four constraints (c1)–(c4). According to (23) and (c4), a natural choice for A_m is

$$A_m := \frac{\alpha_m}{N_{m+1} - N_m}, \quad m \geq 0.$$

The positivity of (α_m) and (24) then ensure that the sequence $(A_m)_{m\geq 0}$ satisfies (c1). Next, (c2) also reads

$$A_{m+1}\, N_{m+1} + B_{m+1} = A_m\, N_m + B_m + \alpha_m,$$

from which we deduce that

$$A_m\, N_m + B_m = \sum_{i=0}^{m-1} \alpha_i + A_0\, N_0 + B_0.$$

The above identity allows us to determine the sequence $(B_m)_{m\geq 0}$ by

$$B_0 := 0 \quad \text{and} \quad B_m := \sum_{i=0}^{m-1} \alpha_i + A_0\, N_0 - A_m\, N_m\,, \quad m \geq 1\,,$$

and (c3) is a straightforward consequence of (23).

We are now in a position to complete the definition of Φ. We set

$$\Phi'(r) := \begin{cases} \dfrac{\alpha_0}{N_1 - N_0}\, r & \text{for } r \in [0, N_1)\,, \\ \dfrac{\alpha_m\; (r - N_m)}{N_{m+1} - N_m} + \displaystyle\sum_{i=0}^{m-1} \alpha_i + \dfrac{\alpha_0}{N_1 - N_0} & \text{for } r \in [N_m, N_{m+1})\,, \\ & m \geq 1, \end{cases}$$

and

$$\Phi(r) := \int_0^r \Phi'(s)\, ds, \quad r \in [0, \infty)\,.$$

Clearly $\Phi(0) = \Phi'(0) = 0$ and, for $m \geq 1$,

$$\lim_{r \to N_m -} \Phi'(r) = \Phi'(N_m) = \sum_{i=0}^{m-1} \alpha_i + \frac{\alpha_0}{N_1 - N_0}. \tag{26}$$

Consequently, $\Phi' \in C([0, \infty))$ and thus $\Phi \in C^1([0, \infty))$. Moreover, Φ' is differentiable in $(0, N_1)$ and in each open interval (N_m, N_{m+1}) with

$$\Phi''(r) = \begin{cases} \dfrac{\alpha_0}{N_1 - N_0} & \text{for } r \in (0, N_1)\,, \\ \dfrac{\alpha_m}{N_{m+1} - N_m} & \text{for } r \in (N_m, N_{m+1}), \quad m \geq 1\,, \end{cases}$$

and (24) ensures that Φ'' is non-negative and non-increasing, whence the convexity of Φ and the concavity of Φ'. We then deduce from the monotonicity of Φ', (23), and (26) that Φ' fulfills (18) and

$$\lim_{r \to \infty} \Phi'(r) = \infty\,.$$

The property (19) then follows by the L'Hospital rule.

We finally infer from (21), (25), and (26) that, for $f \in \mathscr{F}$ and $m \geq 1$, we have

$$\begin{aligned}\int_{\{|f|<N_m\}} \Phi(|f|)\, d\mu &\leq \Phi'(1) \int_\Omega |f|\, d\mu + \sum_{j=0}^{m-1} \alpha_j\, \beta_j \\ &\leq \Phi'(1) \sup_{g\in\mathscr{F}} \int_\Omega |g|\, d\mu + \sum_{j=0}^{\infty} \alpha_j\, \beta_j\,,\end{aligned}$$

and the right-hand side of the above inequality is finite by (23) and the boundedness of $\mathscr{F}$ in $L^1(\Omega)$. We let $m \to \infty$ in the above inequality and conclude that $\{\Phi(|f|)\ :\ f \in \mathscr{F}\}$ is bounded in $L^1(\Omega)$.

We next modify the function Φ constructed above in order to improve its regularity. To this end, we define $\Phi_1 \in C^1(\mathbb{R})$ by

$$\Phi_1(r) := \Phi(r) \quad \text{for} \quad r \geq 0 \quad \text{and} \quad \Phi_1(r) = \Phi''(0)\, \frac{r^2}{2} \quad \text{for} \quad r \leq 0\,.$$

As $\Phi_1'(r) \leq 0 = \Phi_1'(0)$ for $r \leq 0$, Φ_1' is non-decreasing so that Φ_1 is convex. Similarly, $\Phi_1''(r) = \Phi''(0)$ for $r \leq 0$, which guarantees that Φ_1'' is non-increasing and thus the concavity of Φ_1'.

Consider next $\vartheta \in C_0^\infty(\mathbb{R})$ such that

$$\vartheta \geq 0\,, \qquad \operatorname{supp} \vartheta = (-1,1)\,, \qquad \int_{\mathbb{R}} \vartheta(r)\, dr = 1\,.$$

We define a function Ψ by

$$\Psi(r) := (\vartheta * \Phi_1)(r) - (\vartheta * \Phi_1)(0) - (\vartheta * \Phi_1')(0)\, r\,, \quad r \in \mathbb{R}\,.$$

Clearly, $\Psi \in C^\infty(\mathbb{R})$ satisfies $\Psi(0) = \Psi'(0) = 0$. Next, thanks to the non-negativity of ϑ, the convexity of Φ_1 and the concavity of Φ_1' imply the convexity of Ψ and the concavity of Ψ'. Moreover, we have $\Psi'(r) > 0$ for $r > 0$. Indeed, assume for contradiction that $\Psi'(r_0) = 0$ for some $r_0 > 0$. Then

$$0 = \Psi'(r_0) = \int_{\mathbb{R}} \vartheta(s)\, \left(\Phi_1'(r_0 - s) - \Phi_1'(-s)\right)\, ds\,,$$

from which we infer that $\vartheta(s)\, \left(\Phi_1'(r_0 - s) - \Phi_1'(-s)\right) = 0$ for $s \in \mathbb{R}$ by the non-negativity of ϑ and the monotonicity of Φ_1'. Taking $s = 0$, we conclude that $0 = \Phi_1'(r_0) = \Phi'(r_0)$, and a contradiction. Consequently, Ψ fulfills (18).

We next check that Ψ is superlinear at infinity. To this end, we consider $r \geq 2$ and deduce from the monotonicity of Φ_1' and (19) that

$$\begin{aligned}\Psi'(r) &= \int_{-1}^{1} \vartheta(s)\, \Phi_1'(r-s)\, ds - (\vartheta * \Phi_1')(0) \\ &\geq \Phi_1'(r-1) \int_{-1}^{1} \vartheta(s)\, ds - (\vartheta * \Phi_1')(0) \\ &\geq \Phi'(r-1) - (\vartheta * \Phi_1')(0) \underset{r\to\infty}{\longrightarrow} \infty\,,\end{aligned}$$

and we use again the L'Hospital rule to conclude that Ψ fulfills (19).

Let us finally show that there is a constant $C > 0$ such that

$$\Psi(r) \leq C\, (r + \Phi(r))\,, \quad r \geq 0\,. \tag{27}$$

Indeed, either $r > 1$ and $r - s \geq 0$ for all $s \in (-1,1)$ or $r \in [0,1]$. In the former case, as Φ_1 is non-decreasing in $[0,\infty)$ and non-negative in $\mathbb{R}$, we have

$$\begin{aligned}\Psi(r) &\leq \int_{-1}^{1} \vartheta(s)\, \Phi_1(r+1)\, ds - r \int_{-1}^{1} \vartheta(s)\, \Phi_1'(-s)\, ds \\ &\leq \Phi(r+1) + \sup_{[-1,1]} \{|\Phi_1'|\}\, r\,.\end{aligned}$$

On the other hand, the concavity of Φ', the convexity of Φ, and the property $\Phi(0) = \Phi'(0) = 0$ entail that

$$\begin{aligned}\Phi(r) &= \int_0^r \Phi'\left(\frac{s}{r+1}(r+1)\right)\, ds \geq \int_0^r \frac{s}{r+1}\Phi'(r+1)\, ds \\ &= \frac{r^2}{2(r+1)^2}(r+1)\Phi'(r+1) \geq \frac{\Phi(r+1)}{4}\,.\end{aligned}$$

Combining the previous two estimates gives (27) for $r \geq 1$. When $r \in [0,1]$, the convexity of Φ_1 and the concavity of Φ_1' ensure that

$$\begin{aligned}\Psi(r) &\leq \int_{-1}^{1} \vartheta(s)\, (r-s)\, \Phi_1'(r-s)\, ds - r \int_{-1}^{1} \vartheta(s)\, \Phi_1'(-s)\, ds \\ &\leq r \int_{-1}^{1} \vartheta(s)\, \left(\Phi_1'(r-s) - \Phi_1'(-s)\right)\, ds \\ &\leq r \int_{-1}^{1} \vartheta(s)\, r\, \Phi_1''(-s)\, ds \leq r\, \Phi''(0)\,,\end{aligned}$$

whence (27).

Now, since $\mathscr{F}$ and $\{\Phi(|f|) : f \in \mathscr{F}\}$ are two bounded subsets of $L^1(\Omega)$, the boundedness of $\{\Psi(|f|) : f \in \mathscr{F}\}$ in $L^1(\Omega)$ readily follows from (27), which completes the proof of (i) $\Longrightarrow$ (ii) in Theorem 8.

(ii) $\Longrightarrow$ (i). Let $c \in (0,\infty)$. Owing to the convexity of Φ, the function $r \mapsto \Phi(r)/r$ is non-decreasing and

$$\begin{aligned}\sup_{f\in\mathscr{F}} \int_{\{|f|\geq c\}} |f| \, d\mu &= \sup_{f\in\mathscr{F}} \int_{\{|f|\geq c\}} \frac{|f|}{\Phi(|f|)} \, \Phi(|f|) \, d\mu \\ &\leq \frac{c}{\Phi(c)} \sup_{f\in\mathscr{F}} \int_\Omega \Phi(|f|) \, d\mu \, .\end{aligned}$$

It then follows from (19) that

$$\lim_{c\to\infty} \sup_{f\in\mathscr{F}} \int_{\{|f|\geq c\}} |f| \, d\mu = 0 \, ,$$

whence (16). □

Remark 10 Notice that the sequences (α_m) and (β_m) used in the proof of Theorem 8 can be a priori chosen arbitrarily provided they fulfill the condition (23). In particular, with the choice $\alpha_m = 1$, the above construction of the function Φ is similar to that performed in [49].

For further use, we introduce the following notation:

Definition 11 We define $\mathscr{C}_{VP}$ as the set of convex functions $\Phi \in C^\infty([0,\infty))$ with $\Phi(0) = \Phi'(0) = 0$ and such that Φ' is a concave function satisfying (18). The set $\mathscr{C}_{VP,\infty}$ denotes the subset of functions in $\mathscr{C}_{VP}$ satisfying the additional property (19).

A first consequence of Theorem 8 is that every function in $L^1(\Omega)$ enjoys an additional integrability property in the following sense.

Corollary 12 *Let $f \in L^1(\Omega)$. Then there is a function $\Phi \in \mathscr{C}_{VP,\infty}$ such that $\Phi(|f|) \in L^1(\Omega)$.*

Proof Clearly $\mathscr{F} = \{f\}$ fulfills the assertion (i) of Theorem 8. □

Remark 13 If $\mu(\Omega) < \infty$, we have

$$\bigcup_{p>1} L^p(\Omega) \subset L^1(\Omega) \, ,$$

but this inclusion cannot be improved to an equality in general. For instance, the function $f : x \longmapsto x^{-1} (\ln x)^{-2}$ belongs to $L^1(0,1/2)$ but $f \notin L^p(0,1/2)$ as soon as $p > 1$. A consequence of Corollary 12 is that $L^1(\Omega)$ is the union of the Orlicz spaces L_Φ, see [72] for instance.

Let us mention here that Corollary 12 is also established in [42, pp. 60–61], [63, Proposition A1] and [73], still without the requirement that Φ has a concave first derivative. However, the convex function Φ constructed in [63, Proposition A1] and [73] enjoys the properties (28) and (31) stated below, respectively. In fact, it follows clearly from the proof of Theorem 8 that there is some freedom in the construction of the function Φ and this fact has allowed some authors to endow it with additional properties according to their purpose. In particular, the concavity of Φ' is useful to establish the existence of weak solutions to reaction-diffusion systems [71], while the property (28) is used to study the spatially homogeneous Boltzmann equation [63] and the property (31) to show the existence of solutions to the spatially inhomogeneous BGK equation [73]. The possibility of choosing Φ' concave is also useful in the proof of the existence of solutions to the continuous coagulation-fragmentation equation as we shall see in Sect. 3. We now check that all these properties are actually a consequence of the concavity of Φ'.

Proposition 14 *Consider $\Phi \in \mathscr{C}_{VP}$. Then*

$$r \mapsto \frac{\Phi(r)}{r} \ \text{ is concave in } \ (0,\infty)\,, \tag{28}$$

$$\Phi(r) \le r\,\Phi'(r) \le 2\,\Phi(r)\,, \tag{29}$$

$$s\,\Phi'(r) \le \Phi(r) + \Phi(s)\,, \tag{30}$$

$$\Phi(\lambda r) \le \max\{1,\lambda^2\}\,\Phi(r)\,, \tag{31}$$

$$(r+s)\,\left(\Phi(r+s) - \Phi(r) - \Phi(s)\right) \le 2\,\left(r\,\Phi(s) + s\,\Phi(r)\right)\,, \tag{32}$$

for $r \ge 0$, $s \ge 0$, and $\lambda \ge 0$.

Proof The inequalities (29)–(32) being obviously true when $r = 0$ or $s = 0$, we consider $r > 0$, $s > 0$, and $t \in [0,1]$. Thanks to the concavity of Φ', we have

$$\begin{aligned}\frac{\Phi(tr+(1-t)s)}{tr+(1-t)s} &= \int_0^1 \Phi'(z(tr+(1-t)s))\,dz \\ &\ge \int_0^1 \left(t\,\Phi'(zr) + (1-t)\,\Phi'(zs)\right)\,dz \\ &\ge t\,\frac{\Phi(r)}{r} + (1-t)\,\frac{\Phi(s)}{s}\,,\end{aligned}$$

whence (28).

Next, the convexity of Φ ensures that

$$\Phi(0) - \Phi(r) \ge -r\,\Phi'(r)\,, \quad r \ge 0\,,$$

from which the first inequality in (29) follows. Similarly, we deduce from (28) that, for $r \geq 0$, we have

$$\begin{aligned} \Phi'(0) - \frac{\Phi(r)}{r} &\leq -r \left(\frac{\Phi'(r)}{r} - \frac{\Phi(r)}{r^2} \right) \\ -\frac{\Phi(r)}{r} &\leq -\Phi'(r) + \frac{\Phi(r)}{r} \\ r\,\Phi'(r) &\leq 2\,\Phi(r)\,, \end{aligned}$$

which completes the proof of (29).

Combining the convexity of Φ with (29) gives

$$s\Phi'(r) = (s-r)\Phi'(r) + r\Phi'(r) \leq \Phi(s) - \Phi(r) + 2\Phi(r)$$

for $r \geq 0$ and $s \geq 0$, hence (30).

Consider now $r \geq 0$ and $\lambda \in [0,1]$. We infer from the monotonicity (18) of Φ that

$$\Phi(\lambda r) \leq \Phi(r) \leq \max\{1, \lambda^2\}\,\Phi(r)\,.$$

Next, for $r \geq 0$, $s \in [0,r]$, and $\lambda > 1$, it follows from the concavity and non-negativity of Φ' that

$$\Phi'(s) = \Phi'\left(\frac{\lambda s}{\lambda} + \left(1 - \frac{1}{\lambda}\right) 0 \right) \geq \frac{\Phi'(\lambda s)}{\lambda}\,.$$

We integrate this inequality with respect to s over $(0,r)$ to obtain

$$\Phi(r) \geq \frac{\Phi(\lambda r)}{\lambda^2}\,,$$

and complete the proof of (31).

Finally, let $r \geq 0$, $s \geq 0$, $\rho \in [0,r]$, and $\sigma \in [0,s]$. We infer from the concavity of Φ' that

$$\Phi'(\rho+\sigma) - \Phi'(\rho) \geq \sigma\,\Phi''(\rho+\sigma) \quad \text{and} \quad \Phi'(\rho+\sigma) - \Phi'(\sigma) \geq \rho\,\Phi''(\rho+\sigma)\,,$$

whence

$$(\rho+\sigma)\,\Phi''(\rho+\sigma) + 2\,\Phi'(\rho+\sigma) \leq 4\,\Phi'(\rho+\sigma) - \Phi'(\rho) - \Phi'(\sigma)\,. \tag{33}$$

We use once more the concavity of Φ' to obtain

$$\Phi''(\tau) \geq \Phi''(\tau+\sigma)\,, \quad \tau \geq 0\,.$$

Integrating this inequality with respect to τ over $(0,\rho)$ we conclude that

$$\Phi'(\rho+\sigma) \le \Phi'(\rho) + \Phi'(\sigma)\,, \tag{34}$$

since $\Phi'(0)=0$. It next follows from (33) and (34) that

$$(\rho+\sigma)\,\Phi''(\rho+\sigma) + 2\,\Phi'(\rho+\sigma) \le 3\left(\Phi'(\rho)+\Phi'(\sigma)\right)\,.$$

As

$$\begin{aligned}&(r+s)\,\Phi(r+s) - r\,\Phi(r) - s\,\Phi(s)\\ &= \int_0^r\int_0^s \left\{(\rho+\sigma)\,\Phi''(\rho+\sigma) + 2\,\Phi'(\rho+\sigma)\right\}\,d\sigma d\rho\,,\end{aligned}$$

the previous inequality gives the upper bound

$$\begin{aligned}(r+s)\,\Phi(r+s) - r\,\Phi(r) - s\,\Phi(s) &\le 3\int_0^r\int_0^s \left(\Phi'(\rho)+\Phi'(\sigma)\right)\,d\sigma d\rho\\ &= 3\left(s\,\Phi(r) + r\,\Phi(s)\right)\,,\end{aligned}$$

which we combine with

$$\begin{aligned}(r+s)\left(\Phi(r+s)-\Phi(r)-\Phi(s)\right) &= (r+s)\,\Phi(r+s) - r\,\Phi(r) - s\,\Phi(s)\\ &\quad -s\,\Phi(r) - r\,\Phi(s)\,,\end{aligned}$$

to obtain (32). □

Remark 15 The property (31) implies that Φ enjoys the so-called Δ_2-condition, namely, there exists $\ell > 1$ such that $\Phi(2r) \le \ell\,\Phi(r)$ for $r \ge 0$. It also follows from (29) that Φ grows at most quadratically at infinity.

2.5 Weak Convergence in L^1 and a.e. Convergence

There are several connections between weak convergence in L^1 and almost everywhere convergence. The combination of both is actually equivalent to the strong convergence in $L^1(\Omega)$ according to Vitali's convergence theorem, see [21, Theorem III.3.6] for instance.

Theorem 16 (Vitali) *Consider a sequence $(f_n)_{n\ge1}$ in $L^1(\Omega)$ and a function $f \in L^1(\Omega)$ such that $(f_n)_{n\ge1}$ converges μ-a.e. towards f. The following two statements are equivalent:*

(i) $(f_n)_{n\ge1}$ converges (strongly) towards f in $L^1(\Omega)$.
(ii) The set $\{f_n : n\ge1\}$ is bounded in $L^1(\Omega)$ and fulfills the conditions (14) *and* (15).

In other words, the weak convergence in $L^1(\Omega)$ coupled with the μ-almost everywhere convergence imply the convergence in $L^1(\Omega)$.

Proof As the proof that (i) $\Longrightarrow$ (ii) is obvious, we turn to the proof of the converse and fix $\varepsilon > 0$. On the one hand, we deduce from (15) and the integrability of f that there exist $\Omega_\varepsilon \in \mathscr{B}$ with $\mu(\Omega_\varepsilon) < \infty$ such that

$$\sup_{n\geq 1} \int_{\Omega\setminus\Omega_\varepsilon} (|f_n| + |f|)\ d\mu \leq \varepsilon\,.$$

On the other hand, since $f \in L^1(\Omega)$ and $(f_n)_{n\geq 1}$ is bounded in $L^1(\Omega)$, we have

$$\int_{\{|f_n - f|\geq R\}} |f_n - f|\ d\mu \leq \frac{1}{R}\left(\|f\|_1 + \sup_{m\geq 1}\{\|f_m\|_1\}\right) \quad \text{for} \quad R > 0\,.$$

Then,

$$\begin{aligned}
\|f_n - f\|_1 &\leq \int_{\Omega\setminus\Omega_\varepsilon} (|f_n| + |f|)\ d\mu + \int_{\Omega_\varepsilon} |f_n - f|\ \mathbf{1}_{\{|f_n - f|\leq\varepsilon^{-1}\}}\ d\mu \\
&\quad + \int_{\Omega_\varepsilon} |f_n - f|\ \mathbf{1}_{\{|f_n - f|>\varepsilon^{-1}\}}\ d\mu \\
&\leq \varepsilon\left(1 + \|f\|_1 + \sup_{m\geq 1}\{\|f_m\|_1\}\right) + \int_{\Omega_\varepsilon} |f_n - f|\ \mathbf{1}_{\{|f_n - f|\leq\varepsilon^{-1}\}}\ d\mu\,.
\end{aligned}$$

Since Ω_ε has a finite measure, we now infer from the almost everywhere convergence of $(f_n)_{n\geq 1}$ and the Lebesgue dominated convergence theorem that the last term of the right-hand side of the above inequality converges to zero as $n \to \infty$. Consequently,

$$\limsup_{n\to\infty} \|f_n - f\|_1 \leq \varepsilon\left(1 + \|f\|_1 + \sup_{m\geq 1}\{\|f_m\|_1\}\right)\,.$$

Letting $\varepsilon \to 0$ completes the proof. □

Remark 17 The μ-a.e. convergence of $(f_n)_{n\geq 1}$ in Theorem 16 can be replaced by the convergence in measure.

Another useful consequence is the following result which is implicitly used in [17, 81], for instance, see also [47, Lemma A.2]. It allows one to identify the limit of the product of a weakly convergent sequence in L^1 with a bounded sequence which has an almost everywhere limit.

Proposition 18 *Let $(f_n)_{n\geq 1}$ be a sequence of measurable functions in $L^1(\Omega)$ and $(g_n)_{n\geq 1}$ be a sequence of measurable functions in $L^\infty(\Omega)$. Assume further that there are $f \in L^1(\Omega)$ and $g \in L^\infty(\Omega)$ such that*

$$f_n \rightharpoonup f \quad in \quad L^1(\Omega)\,, \tag{35}$$

$$|g_n(x)| \leq M \quad and \quad \lim_{n\to\infty} g_n(x) = g(x) \quad \mu - a.e. \tag{36}$$

Then

$$\lim_{n\to\infty} \int_\Omega |f_n|\,|g_n - g|\,d\mu = 0 \quad and \quad f_n g_n \rightharpoonup fg \quad in \quad L^1(\Omega)\,. \tag{37}$$

The proof of Proposition 18 combines the Dunford–Pettis theorem (Theorem 3) with Egorov's theorem which we recall now, see [74, p. 73] for instance.

Theorem 19 (Egorov) *Assume that $\mu(\Omega) < \infty$ and consider a sequence $(h_n)_{n\geq 1}$ of measurable functions in Ω such that $h_n \to h$ μ–a.e. for some measurable function h. Then, for any $\delta > 0$, there is a measurable subset $A_\delta \in \mathscr{B}$ such that*

$$\mu(A_\delta) \leq \delta \quad and \quad \lim_{n\to\infty} \sup_{x\in\Omega\setminus A_\delta} |h_n(x) - h(x)| = 0\,.$$

Proof of Proposition 18 Let $\varepsilon \in (0,1)$. On the one hand, the Dunford–Pettis theorem and (35) ensure that there exist $\delta > 0$ and $\Omega_\varepsilon \subset \Omega$ such that $\mu(\Omega_\varepsilon) < \infty$,

$$\sup_{n\geq 1} \int_{\Omega\setminus\Omega_\varepsilon} |f_n|\,d\mu \leq \frac{\varepsilon}{4M}\,, \quad \text{and} \quad \eta\{(f_n)_{n\geq 1}, \delta\} \leq \frac{\varepsilon}{4M}\,.$$

On the other hand, since $\mu(\Omega_\varepsilon) < \infty$, we deduce from Egorov's theorem and (36) that there is $\mathscr{O}_\varepsilon \subset \Omega_\varepsilon$ such that

$$\mu(\Omega_\varepsilon \setminus \mathscr{O}_\varepsilon) \leq \delta \quad \text{and} \quad \lim_{n\to\infty} \sup_{x\in\mathscr{O}_\varepsilon} |(g_n - g)(x)| = 0\,.$$

Then

$$\begin{aligned}
\int_\Omega |f_n|\,|g_n - g|\,d\mu &\leq 2M \int_{\Omega\setminus\Omega_\varepsilon} |f_n|\,d\mu + 2M \int_{\Omega_\varepsilon\setminus\mathscr{O}_\varepsilon} |f_n|\,d\mu \\
&\quad + \int_{\mathscr{O}_\varepsilon} |f_n|\,|g_n - g|\,d\mu \\
&\leq \varepsilon + \sup_{m\geq 1} \|f_m\|_1 \sup_{x\in\mathscr{O}_\varepsilon} |(g_n - g)(x)|\,.
\end{aligned}$$

Consequently,

$$\limsup_{n\to\infty} \int_\Omega |f_n|\,|g_n - g|\,d\mu \le \varepsilon\,,$$

and $(f_n(g_n-g))_{n\ge1}$ converges strongly towards zero in $L^1(\Omega)$. Since $g \in L^\infty(\Omega)$, the second statement in Proposition 18 readily follows from the first one and (35). □

Remark 20 Proposition 18 is somehow an extension of the following classical result: Let $p \in (1,\infty)$. If $f_n \rightharpoonup f$ in $L^p(\Omega)$ and $g_n \longrightarrow g$ in $L^{p/(p-1)}(\Omega)$, then $f_n g_n \rightharpoonup fg$ in $L^1(\Omega)$.

The final result of this section is a generalization of Proposition 18 and Remark 20.

Proposition 21 *Let $\psi \in C([0,\infty)$ be a non-negative convex function satisfying $\psi(0) = 0$ and $\psi(r) \ge C_0 r$ for $r \ge 1$ and some $C_0 > 0$, and denote its convex conjugate function by ψ^*. Assume that $\mu(\Omega) < \infty$ and consider two sequences $(f_n)_{n\ge1}$ and $(g_n)_{n\ge1}$ of real-valued integrable functions in Ω enjoying the following properties: there are f and g in $L^1(\Omega)$ such that*

1. *$f_n \rightharpoonup f$ in $L^1(\Omega)$ and*

$$C_1 := \sup_{n\ge1} \int_\Omega \psi(|f_n|)\,d\mu < \infty\,,$$

2. *$g_n \longrightarrow g$ μ–a.e. in Ω,*
3. *for each $\varepsilon \in (0,1]$, the family $\mathcal{G}_\varepsilon := \{\psi^*(|g_n|/\varepsilon) : n \ge 1\}$ is uniformly integrable in $L^1(\Omega)$.*

Then

$$f_n g_n \rightharpoonup fg \quad in \quad L^1(\Omega)\,.$$

Proof We first recall that, given $\varepsilon \in (0,1]$, the uniform integrability of $\mathcal{G}_\varepsilon$ in $L^1(\Omega)$ ensures that

$$C_2(\varepsilon) := \sup_{n\ge1} \int_\Omega \psi^*\left(\frac{|g_n|}{\varepsilon}\right)\,d\mu < \infty\,, \tag{38}$$

and

$$\lim_{\delta\to0} \eta\{\mathcal{G}_\varepsilon, \delta\} = 0\,, \tag{39}$$

the modulus of uniform integrability η being defined in Definition 2. We next observe that, thanks to Young's inequality

$$rs \le \psi(r) + \psi^*(s)\,, \qquad (r,s) \in [0,\infty)^2\,, \tag{40}$$

which ensures, together with (38) (with $\varepsilon = 1$), that

$$\int_\Omega |f_n g_n|\, d\mu \le \int_\Omega \big(\psi(|f_n|) + \psi^*(|g_n|)\big)\, d\mu \le C_1 + C_2(1)\,.$$

Consequently, $(f_n g_n)_{n\ge1}$ is a bounded sequence in $L^1(\Omega)$. Furthermore, the convexity of ψ, the weak convergence of $(f_n)_{n\ge1}$, and a weak lower semicontinuity argument entail that

$$\psi(|f|) \in L^1(\Omega) \quad \text{with} \quad \int_\Omega \psi(|f|)\, d\mu \le C_1\,, \tag{41}$$

while the μ-almost everywhere convergence of $(g_n)_{n\ge1}$ along with (38) and the Fatou lemma ensure that, for each $\varepsilon \in (0,1]$ and $\delta > 0$,

$$\int_\Omega \psi^*\left(\frac{|g|}{\varepsilon}\right) d\mu \le C_2(\varepsilon) \quad \text{and} \quad \eta\left\{\left\{\psi^*\left(\frac{|g|}{\varepsilon}\right)\right\}, \delta\right\} \le \eta\{\mathcal{G}_\varepsilon, \delta\}\,. \tag{42}$$

In particular, $fg \in L^1(\Omega)$ as a consequence of (40)–(42).

We now fix $\varepsilon \in (0,1]$ and $\delta \in (0,1)$. On the one hand, since $\mu(\Omega) < \infty$, we infer from Egorov's theorem that there is a measurable subset A_δ of Ω such that

$$\mu(A_\delta) \le \delta \quad \text{and} \quad \lim_{n\to\infty} \sup_{x\in\Omega\setminus A_\delta} |g_n(x) - g(x)| = 0\,. \tag{43}$$

On the other hand, since $g \in L^1(\Omega)$, there exists $k_\delta \ge 1$ such that

$$\mu\left(\{x \in \Omega\ :\ |g(x)| \ge k_\delta\}\right) \le \delta\,, \tag{44}$$

and we define $g_\delta := g\,\mathbf{1}_{(-k_\delta,k_\delta)}(g)$.

Now, for $\xi \in L^\infty(\Omega)$, we define

$$I(n) := \int_\Omega (f_n g_n - fg)\xi\, d\mu\,,$$

which we estimate as follows:

$$\begin{aligned} |I(n)| &\le \left|\int_\Omega (f_n - f) g\xi\, d\mu\right| + \left|\int_\Omega f_n(g_n - g)\xi\, d\mu\right| \\ &\le \left|\int_\Omega (f_n - f) g_\delta \xi\, d\mu\right| + \int_\Omega (|f_n| + |f|)|g - g_\delta||\xi|\, d\mu \\ &\quad + \int_{\Omega\setminus A_\delta} |f_n||g_n - g||\xi|\, d\mu + \int_{A_\delta} |f_n|(|g_n| + |g|)|\xi|\, d\mu\,. \end{aligned}$$

It first follows from Young's inequality (40) and the convexity of ψ that

$$\begin{aligned} I_1(n,\delta) &:= \int_\Omega (|f_n|+|f|)|g-g_\delta||\xi|\,d\mu \le \int_{\Omega\cap\{|g|\ge k_\delta\}} (|f_n|+|f|)|g||\xi|\,d\mu \\ &\le \|\xi\|_\infty \int_{\Omega\cap\{|g|\ge k_\delta\}} \left[\psi(\varepsilon|f_n|)+\psi(\varepsilon|f|)+2\psi^*\left(\frac{|g|}{\varepsilon}\right)\right]\,d\mu \\ &\le \|\xi\|_\infty \left[\varepsilon\int_\Omega (\psi(|f_n|)+\psi(|f|))\,d\mu + 2\int_{\Omega\cap\{|g|\ge k_\delta\}} \psi^*\left(\frac{|g|}{\varepsilon}\right)\,d\mu\right]. \end{aligned}$$

We then infer from (41), (42), and (44) that

$$I_1(n,\delta) \le 2\|\xi\|_\infty \left(\varepsilon C_1 + \eta\{\mathcal{G}_\varepsilon,\delta\}\right). \tag{45}$$

Next,

$$I_2(n,\delta) := \int_{\Omega\setminus A_\delta} |f_n||g_n-g||\xi|\,d\mu \le \|\xi\|_\infty \sup_{\Omega\setminus A_\delta}\{|g_n-g|\} \sup_{m\ge1}\{\|f_m\|_1\}. \tag{46}$$

We finally infer from (40), (42), (43), and the convexity of ψ that

$$\begin{aligned} I_3(n,\delta) &:= \int_{A_\delta} |f_n|(|g_n|+|g|)|\xi|\,d\mu \\ &\le \|\xi\|_\infty \int_{A_\delta} \left[2\psi(\varepsilon|f_n|)+\psi^*\left(\frac{|g_n|}{\varepsilon}\right)+\psi^*\left(\frac{|g|}{\varepsilon}\right)\right]\,d\mu \\ &\le 2\|\xi\|_\infty \left[\varepsilon\int_\Omega \psi(|f_n|)\,d\mu + \eta\{\mathcal{G}_\varepsilon,\delta\}\right] \\ &\le 2\|\xi\|_\infty \left(\varepsilon C_1 + \eta\{\mathcal{G}_\varepsilon,\delta\}\right). \end{aligned} \tag{47}$$

Combining (45)–(47) we end up with

$$\begin{aligned} |I(n)| \le & \left|\int_\Omega (f_n-f)g_\delta\xi\,d\mu\right| + \|\xi\|_\infty \sup_{\Omega\setminus A_\delta}\{|g_n-g|\} \sup_{m\ge1}\{\|f_m\|_1\} \\ & + 4\|\xi\|_\infty \left(\varepsilon C_1 + \eta\{\mathcal{G}_\varepsilon,\delta\}\right). \end{aligned} \tag{48}$$

Now, we first let $n\to\infty$ in the above inequality and use the weak convergence of $(f_n)_{n\ge1}$ in $L^1(\Omega)$, the boundedness of g_δ, and the uniform convergence (43) to obtain

$$\limsup_{n\to\infty} \left|\int_\Omega (f_n g_n - fg)\xi\,d\mu\right| \le 4\|\xi\|_\infty \left(\varepsilon C_1 + \eta\{\mathcal{G}_\varepsilon,\delta\}\right).$$

We next use the uniform integrability (39) to pass to the limit as $\delta \to 0$ in the above estimate and find

$$\limsup_{n\to\infty} \left| \int_\Omega (f_n g_n - fg)\xi \, d\mu \right| \le 4\varepsilon C_1 \|\xi\|_\infty .$$

We finally let $\varepsilon \to 0$ to complete the proof. □

3 Smoluchowski's Coagulation Equation

We now turn to Smoluchowski's coagulation equation

$$\begin{aligned} \partial_t f(t,x) &= \frac{1}{2} \int_0^x K(y, x-y)\, f(t,y)\, f(t,x-y)\, dy \\ &\quad - \int_0^\infty K(x,y)\, f(t,x)\, f(t,y)\, dy , \qquad (t,x) \in (0,\infty) \times (0,\infty) , \end{aligned} \tag{49}$$

$$f(0,x) = f^{in}(x) , \qquad x \in (0,\infty) , \tag{50}$$

and collect and derive several properties of its solutions in the next sections. As outlined in the introduction, some of these properties depend heavily on the growth of the coagulation kernel K which is a non-negative and symmetric function. For further use, we introduce the following notation: For $\mu \in \mathbb{R}$, the space of integrable functions with a finite moment of order μ is denoted by

$$L^1_\mu(0,\infty) := \left\{ g \in L^1(0,\infty) \ : \ \|g\|_{1,\mu} := \int_0^\infty (1+x^\mu)|g(x)|\, dx < \infty \right\} , \tag{51}$$

and we define

$$M_\mu(g) := \int_0^\infty x^\mu g(x)\, dx , \qquad g \in L^1_\mu(0,\infty) .$$

Note that $L^1_0(0,\infty) = L^1(0,\infty)$ and $\|\cdot\|_{1,0} = \|\cdot\|_1$. Next, for a measurable function g and $x > 0$, we set

$$\begin{aligned} Q_1(g)(x) &:= \frac{1}{2} \int_0^x K(y, x-y)\, g(y)\, g(x-y)\, dy , \\ L(g)(x) &:= \int_0^\infty K(x,y)\, g(y)\, dy , \qquad Q_2(g)(x) := g(x) L(g(x)) , \end{aligned}$$

whenever it makes sense.

3.1 Existence: Bounded Kernels

The first step towards the existence of solutions to (49)–(50) is to handle the case of bounded coagulation kernels.

Proposition 22 *If there is $\kappa_0 > 0$ such that*

$$0 \leq K(x,y) = K(y,x) \leq \kappa_0 , \quad (x,y) \in (0,\infty) \times (0,\infty) , \tag{52}$$

and

$$f^{in} \in L^1(0,\infty) , \quad f^{in} \geq 0 \ \text{a.e. in} \ (0,\infty) , \tag{53}$$

then there is a unique global solution $f \in C^1([0,\infty); L^1(0,\infty))$ to (49)–(50) such that

$$f(t,x) \geq 0 \ \text{ for a.e. } \ x \in (0,\infty) \ \text{ and } \ \|f(t)\|_1 \leq \|f^{in}\|_1 , \quad t \geq 0 . \tag{54}$$

Furthermore, if $f^{in} \in L^1_1(0,\infty)$, then

$$f(t) \in L^1_1(0,\infty) \ \text{ and } \ M_1(f(t)) = M_1(f^{in}) , \quad t \geq 0 . \tag{55}$$

Proof **Step 1** We first consider an initial condition f^{in} satisfying (53) and prove the first statement of Proposition 22. We note that Q_1 and Q_2 are locally Lipschitz continuous from $L^1(0,\infty)$ to $L^1(0,\infty)$ with

$$\|Q_i(f) - Q_i(g)\|_1 \leq \kappa_0 \left(\|f\|_1 + \|g\|_1\right) \|f - g\|_1$$

for $(f,g) \in L^1(0,\infty) \times L^1(0,\infty)$ and $i = 1,2$. Then, denoting the positive part of a real number r by $r_+ := \max\{r,0\}$, the map $f \mapsto Q_1(f)_+$ is also locally Lipschitz continuous from $L^1(0,\infty)$ to $L^1(0,\infty)$ and it follows from classical results on the well-posedness of differential equations in Banach spaces (see [3, Theorem 7.6] for instance) that there is a unique solution $f \in C^1([0,T_m); L^1(0,\infty))$ defined on the maximal time interval $[0,T_m)$ to the differential equation

$$\frac{df}{dt} = Q_1(f)_+ - Q_2(f) , \quad t \in (0,T_m) , \tag{56}$$

with initial condition $f(0) = f^{in}$. Since the positive part is a Lipschitz continuous function and $f \in C^1([0,T_m); L^1(0,\infty))$, the chain rule gives

$$\partial_t(-f)_+ = -\mathrm{sign}_+(-f)\, \partial_t f ,$$

where $\text{sign}_+(r) = 1$ for $r \geq 0$ and $\text{sign}_+(r) = 0$ for $r < 0$. We then infer from (56) that

$$\partial_t(-f)_+ = -\text{sign}_+(-f)\ Q_1(f)_+ + \text{sign}_+(-f)\ Q_2(f) \leq (-f)_+\ L(f)$$

and thus

$$\frac{d}{dt}\|(-f)_+\|_1 \leq \int_0^\infty (-f)_+\ L(f) \leq \kappa_0 \|f\|_1 \|(-f)_+\|_1 .$$

Since $(-f)_+(0) = (-f^{in})_+ = 0$, we readily deduce that $(-f)_+(t) = 0$ for all $t \in [0, T_m)$, that is, $f(t) \geq 0$ a.e. in $(0, \infty)$. Consequently, $Q_1(f)_+ = Q_1(f)$ and it follows from (56) that f is a solution to (49)–(50) defined for $t \in [0, T_m)$. To show that $T_m = \infty$, it suffices to notice that, thanks to the just established non-negativity of f, Fubini's theorem gives

$$\begin{aligned}\frac{d}{dt}\|f(t)\|_1 &= \int_0^\infty [Q_1(f)(t,x) - Q_2(f)(t,x)]\ dx \\ &= -\frac{1}{2}\int_0^\infty \int_0^\infty K(x,y) f(t,x) f(t,y)\ dydx \leq 0 ,\end{aligned}$$

for $t \in [0, T_m)$, which prevents the blowup in finite time of the L^1-norm of f and thereby guarantees that $T_m = \infty$.

Step 2 A straightforward consequence of Fubini's theorem is the following identity for any $\vartheta \in L^\infty(0, \infty)$:

$$\frac{d}{dt}\int_0^\infty \vartheta(x) f(t,x)\ dx = \frac{1}{2}\int_0^\infty \int_0^\infty \tilde{\vartheta}(x,y) K(x,y) f(t,x) f(t,y)\ dydx \tag{57}$$

where

$$\tilde{\vartheta}(x,y) := \vartheta(x+y) - \vartheta(x) - \vartheta(y) , \quad (x,y) \in (0,\infty) \times (0,\infty) . \tag{58}$$

As a consequence of (57) (with $\vartheta \equiv 1$), we recover the already observed monotonicity of $t \mapsto M_0(f(t))$ and complete the proof of (54).

Step 3 We now turn to an initial condition f^{in} having a finite first moment and aim at proving (55). Formally, (55) follows from (57) with the choice $\vartheta(x) = x$ since $\tilde{\vartheta} \equiv 0$ in that case. However, $\text{id} : x \mapsto x$ does not belong to $L^\infty(0, \infty)$ and an approximation argument is required to justify (55). More precisely, given $A > 0$, define $\vartheta_A(x) := \min\{x, A\}$ for $x > 0$. The corresponding function $\tilde{\vartheta}_A$

given by (58) satisfies

$$\tilde{\vartheta}_A(x,y) = \begin{cases} 0 & \text{if } 0 \le x+y \le A\,, \\ A-x-y & \text{if } 0 \le \max\{x,y\} \le A < x+y\,, \\ -\min\{x,y\} & \text{if } 0 \le \min\{x,y\} \le A < \max\{x,y\} < x+y\,, \\ -A & \text{if } A \le \min\{x,y\}\,. \end{cases} \tag{59}$$

In particular $\tilde{\vartheta}_A \le 0$ and it follows from (57) that

$$\int_0^\infty \vartheta_A(x) f(t,x)\, dx \le \int_0^\infty \vartheta_A(x) f^{in}(x)\, dx \le M_1(f^{in})\,, \quad t \ge 0\,.$$

Since $\vartheta_A \to \mathrm{id}$ as $A \to \infty$, the Fatou lemma entails that

$$M_1(f(t)) \le M_1(f^{in})\,, \quad t \ge 0\,, \tag{60}$$

and thus that $f(t) \in L^1_1(0,\infty)$ for all $t \ge 0$. To prove the conservation of matter, we use again (57) with $\vartheta = \vartheta_A$ and find

$$\begin{aligned} &\int_0^\infty \vartheta_A(x) f(t,x)\, dx - \int_0^\infty \vartheta_A(x) f^{in}(x)\, dx \\ &\qquad = -\int_0^t \left(I_1(s,A) + I_2(s,A) + I_3(s,A)\right)\, dx \end{aligned} \tag{61}$$

with

$$\begin{aligned} I_1(s,A) &:= \frac{1}{2}\int_0^A \int_{A-x}^A (x+y-A) K(x,y) f(s,x) f(s,y)\, dydx\,, \\ I_2(s,A) &:= \int_0^A \int_A^\infty x K(x,y) f(s,x) f(s,y)\, dydx\,, \\ I_3(s,A) &:= \frac{A}{2}\int_A^\infty \int_A^\infty K(x,y) f(s,x) f(s,y)\, dydx\,. \end{aligned}$$

On the one hand, it readily follows from (52) and (60) that

$$I_2(s,A) + I_3(s,A) \le \frac{\kappa_0}{A} M_1(f(s))^2 \le \frac{\kappa_0}{A} M_1(f^{in})^2\,. \tag{62}$$

On the other hand, by (52),

$$\begin{aligned} 0 \le I_1(s,A) &\le \frac{\kappa_0}{2}\int_0^A \int_{A-x}^A y f(s,x) f(s,y)\, dydx \\ &\le \frac{\kappa_0}{2}\int_0^\infty \int_0^\infty \mathbf{1}_{(0,A)}(x)\mathbf{1}_{(A,\infty)}(x+y) y f(s,x) f(s,y)\, dydx\,. \end{aligned}$$

Owing to (54) and (60), the Lebesgue dominated convergence theorem guarantees that

$$\lim_{A\to\infty} \int_0^t I_1(s, A)\, ds = 0\,. \tag{63}$$

Thanks to (62) and (63) we may pass to the limit as $A \to \infty$ in (61) and conclude that $M_1(f(t)) = M_1(f^{in})$ for $t \geq 0$. This completes the proof.

□

Remark 23 Another formal consequence of (57) is that, whenever it makes sense, $t \mapsto M_\mu(f(t))$ is non-increasing for $\mu \in (-\infty, 1)$ and non-decreasing for $\mu \in (1, \infty)$. Similarly,

$$t \mapsto \int_0^\infty (e^{\alpha x} - 1)\, f(t, x)\, dx \quad \text{is non-decreasing for} \quad \alpha > 0\,.$$

3.2 Existence: Unbounded Kernels

As already mentioned, most of the coagulation rates encountered in the literature are unbounded and grow without bound as $(x, y) \to \infty$ or as $(x, y) \to (0, 0)$. In that case, there does not seem to be a functional framework in which Q_1 and Q_2 are locally Lipschitz continuous and implementing a fixed point procedure does not seem to be straightforward. A different approach is then required and we turn to a compactness method which can be summarized as follows:

1. Build a sequence of approximations of the original problem which depends on a parameter $n \geq 1$, for which the existence of a solution is simple to show, and which converges in some sense to the original problem as $n \to \infty$.
2. Derive estimates which are independent of $n \geq 1$ and guarantee the compactness with respect to the size variable x and the time variable t of the sequence of solutions to the approximations.
3. Show convergence as $n \to \infty$.

To be more precise, let K be a non-negative and symmetric locally bounded function and consider an initial condition

$$f^{in} \in L^1_1(0, \infty)\,, \quad f^{in} \geq 0 \ \text{ a.e. in } \ (0, \infty)\,. \tag{64}$$

Given an integer $n \geq 1$, a natural approximation is to truncate the coagulation kernel K and define

$$K_n(x, y) := \min\{K(x, y), n\}\,, \quad (x, y) \in (0, \infty) \times (0, \infty)\,. \tag{65}$$

Clearly, K_n is a non-negative, bounded, and symmetric function and we infer from (64) and Proposition 22 that the initial-value problem (49)–(50) with K_n

instead of K has a unique non-negative solution $f_n \in C^1([0,\infty); L^1(0,\infty))$ which satisfies

$$M_0(f_n(t)) \le M_0(f^{in}) \quad \text{and} \quad M_1(f_n(t)) = M_1(f^{in}), \quad t \ge 0. \tag{66}$$

The compactness properties provided by the previous estimates are rather weak and the next step is to identify an appropriate topology for the compactness approach to work. A key observation in that direction is that, though being nonlinear, Eq. (49) is a nonlocal quadratic equation, in the sense that it does not involve nonlinearities of the form $f(t,x)^2$ but of the form $f(t,x)f(t,y)$ with $x \ne y$. While the former requires convergence in a strong topology to pass to the limit, the latter complies well with weak topologies. As first noticed in [81] the weak topology of L^1 turns out to be a particularly well-suited framework to prove the existence of solutions to (49)–(50) for several classes of unbounded coagulation kernels. In the remainder of this section, we will show how to use the tools described in Sect. 2 to achieve this goal.

3.2.1 Sublinear Kernels

We first consider the case of locally bounded coagulation kernels with a sublinear growth at infinity. More precisely, we assume that there is $\kappa > 0$ such that

$$0 \le K(x,y) = K(y,x) \le \kappa(1+x)(1+y), \quad (x,y) \in (0,\infty)\times(0,\infty), \tag{67}$$

$$\omega_R(y) := \sup_{x\in(0,R)} \frac{K(x,y)}{y} \underset{y\to\infty}{\longrightarrow} 0. \tag{68}$$

The following existence result is then available, see [46, 52, 70, 80].

Theorem 24 *Assume that the coagulation kernel K satisfies* (67)–(68) *and consider an initial condition f^{in} satisfying* (64). *There is a non-negative function*

$$f \in C([0,\infty); L^1(0,\infty)) \cap L^\infty(0,\infty; L^1_1(0,\infty))$$

such that

$$\int_0^\infty \vartheta(x)\left(f(t,x) - f^{in}(x)\right) dx = \frac{1}{2}\int_0^t \int_0^\infty \int_0^\infty \tilde{\vartheta}(x,y)K(x,y)f(s,x)f(s,y)\, dydxds \tag{69}$$

for all $t > 0$ and $\vartheta \in L^\infty(0,\infty)$ (with $\tilde{\vartheta}$ given by (58)*) and*

$$M_0(f(t)) \le M_0(f^{in}) \quad \textit{and} \quad M_1(f(t)) \le M_1(f^{in}), \quad t \ge 0. \tag{70}$$

On the one hand, Theorem 24 excludes the two important (and borderline) cases $K_1(x,y) = x+y$ and $K_2(x,y) = xy$ which will be handled in Sect. 3.2.2 and Sect. 3.2.3, respectively. On the other hand, owing to the possible occurrence of the gelation phenomenon already mentioned in the introduction, it is not possible to improve the second inequality in (70) to an equality in general.

We now turn to the proof of Theorem 24: for $n \ge 1$ define K_n by (65) and let $f_n \in C^1([0,\infty); L^1(0,\infty))$ be the non-negative solution to (49)–(50) with K_n instead of K which satisfies (66). To prove the weak compactness in $L^1(0,\infty)$ of $(f_n(t))_{n\ge1}$ for each $t \ge 0$, we aim at using the Dunford–Pettis theorem (Theorem 3). To this end, we shall study the behaviour of $(f_n(t))_{n\ge1}$ on sets with small measure and for large values of x. Owing to (68), we shall see below that the boundedness of $(M_1(f_n(t)))_{n\ge1}$ guaranteed by (66) is sufficient to control the behaviour for large x. We are left with the behaviour on sets with small measure which we analyze in the next lemma.

Lemma 25 *Let $\Psi \in \mathscr{C}_{VP}$ be such that $\Psi(f^{in}) \in L^1(0,\infty)$, the set $\mathscr{C}_{VP}$ being defined in Definition 11. For each $R > 0$, there is $C_1(R) > 0$ depending only on K, f^{in}, and R such that*

$$\int_0^R \Psi(f_n(t,x))\,dx \le \left(\int_0^R \Psi(f^{in}(x))\,dx\right) e^{C_1(R)t}\,, \quad t \ge 0\,, \quad n \ge 1\,. \tag{71}$$

Proof Fix $R > 0$. Since Ψ', K_n, and f_n are non-negative functions and $K_n \le K$, we infer from (49) and Fubini's theorem that

$$\begin{aligned} &\frac{d}{dt}\int_0^R \Psi(f_n(t,x))\,dx \\ &\quad \le \frac{1}{2}\int_0^R \int_0^x K(x-y,y) f_n(t,x-y) f_n(t,y)\,dy\,\Psi'(f_n(t,x))\,dx \\ &\quad \le \int_0^R \int_y^R K(x-y,y) f_n(t,x-y)\,\Psi'(f_n(t,x))\,dx\,f_n(t,y)\,dy\,. \end{aligned}$$

Since $\Psi \in \mathscr{C}_{VP}$ we deduce from (30), (66), and (67) that

$$\begin{aligned} \frac{d}{dt}\int_0^R \Psi(f_n(t,x))\,dx \le & \int_0^R \int_y^R K(x-y,y)\Psi(f_n(t,x-y))\,dx\,f_n(t,y)\,dy \\ & + \int_0^R \int_y^R K(x-y,y)\Psi(f_n(t,x))\,dx\,f_n(t,y)\,dy \\ \le & 2\kappa(1+R)^2\,M_0(f_n) \int_0^R \Psi(f_n(t,x))\,dx \\ \le & 2\kappa(1+R)^2\,M_0(f^{in}) \int_0^R \Psi(f_n(t,x))\,dx\,. \end{aligned}$$

Setting $C_1(R) := 2\kappa(1+R)^2\, M_0(f^{in})$ we obtain (71) after integration with respect to time. □

The next step towards the proof of Theorem 24 is the time equicontinuity of the sequence $(f_n)_{n\ge1}$ which we prove now.

Lemma 26 *There is $C_2 > 0$ depending only on K and f^{in} such that*

$$\|f_n(t) - f_n(s)\|_1 \le C_2(t-s)\,, \quad 0 \le s \le t\,, \quad n \ge 1\,. \tag{72}$$

Proof Fix $R > 0$. We infer from (49), Fubini's theorem, (66), and (67) that

$$\begin{aligned}
\|\partial_t f_n(t)\|_1 \le & \frac{1}{2}\int_0^\infty \int_y^\infty K(y, x-y) f_n(t,x) f_n(t,y)\; dxdy \\
& + \int_0^\infty \int_0^\infty K(x,y) f_n(t,x) f_n(t,y)\; dydx \\
\le & \frac{3\kappa}{2}\int_0^\infty \int_0^\infty (1+x)(1+y) f_n(t,x) f_n(t,y)\; dydx \\
\le & \frac{3\kappa}{2}\|f_n(t)\|_{1,1}^2 \le C_2 := \frac{3\kappa}{2}\|f^{in}\|_{1,1}^2\,,
\end{aligned}$$

from which (72) readily follows. □

We are now in a position to complete the proof of Theorem 24.

Proof of Theorem 24 We first recall that, owing to the de la Vallée Poussin theorem (in the form stated in Corollary 12), the integrability of f^{in} ensures that there is $\Phi \in \mathscr{C}_{VP,\infty}$ such that

$$\int_0^\infty \Phi(f^{in}(x))\; dx < \infty\,. \tag{73}$$

We then combine Lemma 25 (with $\Psi = \Phi$) and (73) to conclude that, for each $t \ge 0, n \ge 1$, and $R > 0$,

$$\int_0^R \Phi(f_n(t,x))\; dx \le \|\Phi(f^{in})\|_1 e^{C_1(R)t}\,, \tag{74}$$

where $C_1(R)$ only depends on K, f^{in}, and R.

Step 1: Weak Compactness. According to a variant of the Arzelà–Ascoli theorem (see [88, Theorem 1.3.2] for instance), the sequence $(f_n)_{n\ge1}$ is relatively sequentially compact in $C([0,T]; w-L^1(0,\infty))$ for every $T > 0$ if it enjoys the following two properties:

$$\text{The sequence } (f_n(t))_{n\ge1} \text{ is weakly compact in } L^1(0,\infty) \text{ for each } t \ge 0, \tag{75}$$

and:

$$\text{The sequence } (f_n)_{n\ge1} \text{ is weakly equicontinuous in } L^1(0,\infty) \text{ at every } t \ge 0, \tag{76}$$

see [88, Definition 1.3.1]. Recall that the space $C([0,\infty); w - L^1(0,\infty))$ is the space of functions h which are continuous in time with respect to the weak topology of $L^1(0,\infty)$, that is,

$$t \mapsto \int_0^\infty \vartheta(x)h(t,x)\,dx \in C([0,\infty)) \quad \text{for all} \quad \vartheta \in L^\infty(0,\infty)\,.$$

We first prove (75). To this end, we recall that we have already established (74) and note that (66) entails that

$$\int_R^\infty f_n(t,x)\,dx \le \frac{1}{R}\int_R^\infty x f_n(t,x)\,dx \le \frac{M_1(f^{in})}{R}\,. \tag{77}$$

Since $\Phi \in C_{VP,\infty}$, the properties (74) and (77) imply that the sequence $(f_n(t))_{n\ge1}$ is uniformly integrable in $L^1(0,\infty)$ for each $t \ge 0$ while (77) ensures that the condition (15) of the Dunford–Pettis theorem (Theorem 3) is satisfied. We are thus in a position to apply the Dunford–Pettis theorem to obtain (75).

Let us now turn to (76) and notice that Lemma 26 entails that $(f_n)_{n\ge1}$ is equicontinuous for the strong topology of $L^1(0,\infty)$ at every $t \ge 0$ and thus also weakly equicontinuous in $L^1(0,\infty)$ at every $t \ge 0$, which completes the proof of (76).

We have thereby established that the sequence $(f_n)_{n\ge1}$ is relatively sequentially compact in $C([0,T]; w - L^1(0,\infty))$ for every $T > 0$ and a diagonal process ensures that there are a subsequence of $(f_n)_{n\ge1}$ (not relabeled) and $f \in C([0,\infty); w - L^1(0,\infty))$ such that

$$f_n \longrightarrow f \quad \text{in} \quad C([0,T]; w - L^1(0,\infty)) \quad \text{for all} \quad T > 0\,. \tag{78}$$

Since f_n is non-negative and satisfies (66) for each $n \ge 1$, we readily deduce from the convergence (78) that $f(t)$ is non-negative and satisfies (70).

Step 2: Convergence. We now check that the function f constructed in the previous step solves (49)–(50) in an appropriate sense. To this end, let us first consider $t > 0$ and a function $\vartheta \in L^\infty(0,\infty)$ with compact support included in $(0, R_0)$ for some $R_0 > 0$. By (57),

$$\int_0^\infty \vartheta(x)\left(f_n(t,x) - f^{in}(x)\right)\,dx = \frac{1}{2}\left(I_{1,n}(t) + I_{2,n}(t) + I_{3,n}(t)\right)\,, \tag{79}$$

with

$$I_{1,n}(t) := \int_0^t \int_0^{R_0} \int_0^{R_0} \tilde{\vartheta}(x,y) K_n(x,y) f_n(s,x) f_n(s,y)\,dydxds\,,$$

$$I_{2,n}(t) := \int_0^t \int_0^{R_0} \int_{R_0}^\infty \tilde{\vartheta}(x,y) K_n(x,y) f_n(s,x) f_n(s,y)\, dydxds\,,$$

$$I_{3,n}(t) := \int_0^t \int_{R_0}^\infty \int_0^\infty \tilde{\vartheta}(x,y) K(x,y) f_n(s,x) f_n(s,y)\, dydxds\,.$$

Let us first identify the limit of $I_{1,n}(t)$. Since $(K_n)_{n\ge1}$ is a bounded sequence of $L^\infty((0,R_0)\times(0,R_0))$ by (67) and converges a.e. towards K in $(0,R_0)\times(0,R_0)$, we infer from Proposition 18 and the convergence (78) that

$$\lim_{n\to\infty} I_{1,n}(t) = \int_0^t \int_0^{R_0} \int_0^{R_0} \tilde{\vartheta}(x,y) K(x,y) f(s,x) f(s,y)\, dydxds\,. \tag{80}$$

Next, $\tilde{\vartheta}(x,y) = -\vartheta(x)$ for $(x,y) \in (0,R_0)\times(R_0,\infty)$ and

$$I_{2,n}(t) = -\int_0^t \int_0^{R_0} \int_{R_0}^\infty \vartheta(x) K_n(x,y) f_n(s,x) f_n(s,y)\, dydxds\,.$$

For $R > R_0$, we split $I_{2,n}(t)$ into two parts

$$I_{2,n}(t) = I_{21,n}(t,R) + I_{22,n}(t,R) \tag{81}$$

with

$$I_{21,n}(t,R) := -\int_0^t \int_0^{R_0} \int_{R_0}^R \vartheta(x) K_n(x,y) f_n(s,x) f_n(s,y)\, dydxds\,,$$

$$I_{22,n}(t,R) := -\int_0^t \int_0^{R_0} \int_R^\infty \vartheta(x) K_n(x,y) f_n(s,x) f_n(s,y)\, dydxds\,.$$

On the one hand we argue as for $I_{1,n}(t)$ to conclude that

$$\lim_{n\to\infty} I_{21,n}(t,R) = -\int_0^t \int_0^{R_0} \int_{R_0}^R \vartheta(x) K(x,y) f(s,x) f(s,y)\, dydxds\,. \tag{82}$$

On the other hand, using (66) and (68), we find

$$\begin{aligned}
|I_{22,n}(t,R)| \le& \|\vartheta\|_\infty \int_0^t \int_0^{R_0} \int_R^\infty \omega_{R_0}(y) y f_n(s,x) f_n(s,y)\, dydxds \\
\le& \|\vartheta\|_\infty \int_0^t M_0(f_n(s)) M_1(f_n(s))\, ds \sup_{y\in(R,\infty)} \{\omega_{R_0}(y)\} \\
\le& t \|\vartheta\|_\infty M_0(f^{in}) M_1(f^{in}) \sup_{y\in(R,\infty)} \{\omega_{R_0}(y)\} \underset{R\to\infty}{\longrightarrow} 0\,.
\end{aligned} \tag{83}$$

Similarly, owing to (68) and (70) (recall that (70) has been established at the end of Step 1),

$$\int_0^t \int_0^{R_0} \int_R^\infty \vartheta(x) K(x,y) f(s,x) f(s,y) \, dydxds$$
$$\leq t \|\vartheta\|_\infty M_0(f^{in}) M_1(f^{in}) \sup_{y\in(R,\infty)} \{\omega_{R_0}(y)\} \underset{R\to\infty}{\longrightarrow} 0 \,. \tag{84}$$

Combining (80)–(84) and letting first $n \to \infty$ and then $R \to \infty$, we end up with

$$\lim_{n\to\infty} I_{2,n}(t) = \int_0^t \int_0^{R_0} \int_{R_0}^\infty \tilde{\vartheta}(x,y) K(x,y) f(s,x) f(s,y) \, dydxds \,. \tag{85}$$

Finally, $\tilde{\vartheta}(x,y) = -\vartheta(y)$ for $(x,y) \in (R_0,\infty) \times (0,R_0)$ and $\tilde{\vartheta}(x,y) = 0$ if $(x,y) \in (R_0,\infty) \times (R_0,\infty)$ so that

$$I_{3,n}(t) = -\int_0^t \int_{R_0}^\infty \int_0^{R_0} \vartheta(y) K_n(x,y) f_n(s,x) f_n(s,y) \, dydxds \,,$$

and we argue as for $I_{2,n}(t)$ to obtain

$$\lim_{n\to\infty} I_{3,n}(t) = \int_0^t \int_{R_0}^\infty \int_0^\infty \tilde{\vartheta}(x,y) K(x,y) f(s,x) f(s,y) \, dydxds \,. \tag{86}$$

Using once more the convergence (78), we may use (80), (85), and (86) to pass to the limit as $n \to \infty$ in (79) and conclude that f satisfies (69) for all functions $\vartheta \in L^\infty(0,\infty)$ with compact support. Thanks to (67) and (70), a classical density argument allows us to extend the validity of (69) to arbitrary functions $\vartheta \in L^\infty(0,\infty)$.

Step 3: Strong Continuity. We now argue as in the proof of Lemma 26 to strengthen the time continuity of f. More precisely, let $t \geq 0$, $s \in [0,t]$, and $\vartheta \in L^\infty(0,\infty)$. We infer from (67), (69), and (70) that

$$\left| \int_0^\infty (f(t,x) - f(s,x)) \vartheta(x) \, dx \right|$$
$$\leq \frac{3\kappa}{2} \|\vartheta\|_\infty \int_s^t \int_0^\infty \int_0^\infty (1+x)(1+y) f(\sigma,x) f(\sigma,y) \, dydxd\sigma$$
$$\leq \frac{3\kappa}{2} \|\vartheta\|_\infty \int_s^t \|f(\sigma)\|_{1,1}^2 \, d\sigma$$
$$\leq \frac{3\kappa}{2} \|\vartheta\|_\infty \|f^{in}\|_{1,1}^2 \, |t-s| \,.$$

Therefore,

$$\|f(t) - f(s)\|_1 = \sup_{\vartheta \in L^\infty(0,\infty)} \left\{ \frac{1}{\|\vartheta\|_\infty} \left| \int_0^\infty (f(t,x) - f(s,x))\vartheta(x)\, dx \right| \right\}$$
$$\leq \frac{3\kappa}{2} \|f^{in}\|_{1,1}^2 \, |t-s| ,$$

which completes the proof of Theorem 24.

□

3.2.2 Linearly Growing Kernels

We next turn to coagulation kernels growing at most linearly at infinity, that is, we assume that there is $\kappa_1 > 0$ such that

$$0 \leq K(x,y) = K(y,x) \leq \kappa_1(2 + x + y), \quad (x,y) \in (0,\infty) \times (0,\infty). \tag{87}$$

Observe that coagulation kernels satisfying (87) also satisfy (67) but need not satisfy (68).

For this class of coagulation kernels, we establish the existence of mass-conserving solutions to (49)–(50) [4, 20, 46, 81, 83].

Theorem 27 *Assume that the coagulation kernel K satisfies* (87) *and consider an initial condition f^{in} satisfying* (64). *There is a non-negative function*

$$f \in C([0,\infty); L^1(0,\infty)) \cap L^\infty(0,\infty; L_1^1(0,\infty))$$

satisfying (69) *and*

$$M_0(f(t)) \leq M_0(f^{in}) \quad \textit{and} \quad M_1(f(t)) = M_1(f^{in}), \quad t \geq 0. \tag{88}$$

The main difference between the outcomes of Theorem 24 and Theorem 27 is the conservation of mass $M_1(f(t)) = M_1(f^{in})$ for all $t \geq 0$ for coagulation kernels satisfying (87).

Since the growth assumption (87) is more restrictive than (67), it is clear that the proof of Theorem 27 has some common features with that of Theorem 24. In particular, both Lemma 25 and 26 are valid in that case as well. The main difference lies actually in the control of the behaviour of $f_n(t,x)$ for large values of x which is provided by the boundedness of $M_1(f_n(t))$ in (66). This turns out to be not sufficient for coagulation kernels satisfying (87) and we first show that this assumption is particularly well-suited to control higher moments.

Lemma 28 *Let $\psi \in \mathscr{C}_{VP}$ be such that $x \mapsto \psi(1+x) f^{in}(x) \in L^1(0,\infty)$. There is a positive constant $C_3 > 0$ depending only on K and f^{in} such that, for $t \ge 0$ and $n \ge 1$,*

$$\int_0^\infty \psi(1+x) f_n(t,x)\,dx \le \left(\int_0^\infty \psi(1+x) f^{in}(x)\,dx \right) e^{C_3 t}\,, \tag{89}$$

the function f_n being still the solution to (49)–(50) *with K_n instead of K, the coagulation kernel K_n being defined by* (65).

Proof Fix $A > 0$ and define $\psi_A(x) := \min\{\psi(1+x), \psi(1+A)\} = \psi(1+\min\{x,A\})$ for $x > 0$. Then $\psi_A \in L^\infty(0,\infty)$ and it follows from (57) that

$$\frac{d}{dt} \int_0^\infty \psi_A(x) f_n(t,x)\,dx = \frac{1}{2} \int_0^\infty \int_0^\infty \tilde{\psi}_A(x,y) K_n(x,y) f_n(t,x) f_n(t,y)\,dydx\,.$$

Owing to the monotonicity and non-negativity of ψ, we note that

$$\begin{aligned}
\tilde{\psi}_A(x,y) &= \psi(1+x+y) - \psi(1+x) - \psi(1+y) \\
&\le \psi(2+x+y) - \psi(1+x) \\
&\qquad -\psi(1+y) \quad \text{if} \quad y \in (0, A-x) \quad \text{and} \quad x \in (0,A)\,, \\
\tilde{\psi}_A(x,y) &= \psi(1+A) - \psi(1+x) - \psi(1+y) \\
&\le \psi(2+x+y) - \psi(1+x) \\
&\qquad -\psi(1+y) \quad \text{if} \quad y \in (A-x, A) \quad \text{and} \quad x \in (0,A)\,, \\
\tilde{\psi}_A(x,y) &= -\psi(1+x) \le 0 \quad \text{if} \quad (x,y) \in (0,A) \times (A,\infty)\,, \\
\tilde{\psi}_A(x,y) &= -\psi(1+y) \le 0 \quad \text{if} \quad (x,y) \in (A,\infty) \times (0,A)\,, \\
\tilde{\psi}_A(x,y) &= -\psi(1+A) \le 0 \quad \text{if} \quad (x,y) \in (A,\infty) \times (A,\infty)\,.
\end{aligned}$$

Consequently,

$$\begin{aligned}
&\frac{d}{dt} \int_0^\infty \psi_A(x) f_n(t,x)\,dx \\
&\quad \le \frac{1}{2} \int_0^A \int_0^A (\psi(2+x+y) - \psi(1+x) \\
&\qquad\qquad -\psi(1+y))\, K_n(x,y) f_n(t,x) f_n(t,y)\,dydx\,,
\end{aligned}$$

and we infer from (32) and (87) that

$$\begin{aligned}
&(\psi(2+x+y) - \psi(1+x) - \psi(1+y))\, K_n(x,y) \\
&\quad \le \kappa_1 (\psi(2+x+y) - \psi(1+x) - \psi(1+y))\, (2+x+y) \\
&\quad \le 2\kappa_1 \left((1+x)\psi(1+y) + (1+y)\psi(1+x) \right)\,.
\end{aligned}$$

Combining the above two estimates leads us to

$$\frac{d}{dt}\int_0^\infty \psi_A(x) f_n(t,x)\, dx \le 2\kappa_1 \| f_n(t)\|_{1,1} \int_0^A \psi(1+x) f_n(t,x)\, dx\,.$$

Using (66), we end up with

$$\frac{d}{dt}\int_0^\infty \psi_A(x) f_n(t,x)\, dx \le 2\kappa_1 \| f^{in}\|_{1,1} \int_0^\infty \psi_A(x) f_n(t,x)\, dx\,,$$

and (89) follows by integration with $C_3 := 2\kappa_1 \| f^{in}\|_{1,1}$. □

Proof of Theorem 27 As in the proof of Theorem 24, the de la Vallée Poussin theorem (Theorem 8) ensures the existence of a function $\Phi \in \mathscr{C}_{VP,\infty}$ such that $\Phi(f^{in}) \in L^1(0,\infty)$. Moreover, observing that the property $f^{in} \in L^1(0,\infty;(1+x)\,dx)$ also reads $x \mapsto 1+x \in L^1(0,\infty; f^{in}(x)\,dx)$, we use once more the de la Vallée Poussin theorem (now with $\mu = f^{in}dx$) to obtain a function $\varphi \in \mathscr{C}_{VP,\infty}$ such that $x \mapsto \varphi(1+x)$ belongs to $L^1(0,\infty; f^{in}(x)\,dx)$. Summarizing we have established that there are two functions Φ and φ in $\mathscr{C}_{VP,\infty}$ such that

$$\int_0^\infty \left[\Phi(f^{in}(x)) + \varphi(1+x) f^{in}(x)\right]\, dx < \infty\,. \tag{90}$$

We now infer from (90), Lemma 25 (with $\Psi = \Phi$) and Lemma 28 (with $\psi = \varphi$) that, for each $t \ge 0$, $n \ge 1$, and $R > 0$,

$$\int_0^R \Phi(f_n(t,x))\, dx \le \|\Phi(f^{in})\|_1 e^{C_1(R)t}\,, \tag{91}$$

$$\int_0^\infty \varphi(1+x) f_n(t,x)\, dx \le \left(\int_0^\infty \varphi(1+x) f^{in}(x)\, dx\right) e^{C_3 t}\,, \tag{92}$$

where $C_1(R)$ only depends on K, f^{in}, and R and C_3 on K and f^{in}.

Step 1: Weak Compactness. We argue as in the first step of the proof of Theorem 24 with the help of (66) and (91) to conclude that there are a subsequence of $(f_n)_{n\ge1}$ (not relabeled) and a non-negative function $f \in C([0,\infty); w - L^1(0,\infty))$ such that

$$f_n \longrightarrow f \quad \text{in} \quad C([0,T]; w - L^1(0,\infty)) \quad \text{for all} \quad T > 0\,. \tag{93}$$

Step 2: Convergence. We keep the notation used in the proof of Theorem 24 and notice that (80) and (82) are still valid. A different treatment is required for

$I_{22,n}(t,R)$: thanks to (66), (87), (90), (92), and the monotonicity of $r \mapsto \varphi(r)/r$, we obtain

$$\begin{aligned} I_{22,n}(t,R) \le& \kappa_1 \|\vartheta\|_\infty \int_0^t \int_0^{R_0} \int_R^\infty (2+x+y) f_n(s,x) f_n(s,y)\, dydxds \\ \le& 2\kappa_1 \|\vartheta\|_\infty \int_0^t \int_0^{R_0} \int_R^\infty (1+x)(1+y) f_n(s,x) f_n(s,y)\, dydxds \\ \le& \frac{2\kappa_1 R}{\varphi(R)} \|\vartheta\|_\infty \int_0^t \|f_n(s)\|_{1,1} \int_R^\infty \varphi(1+y) f_n(s,y)\, dyds \\ \le& \frac{2\kappa_1 R}{\varphi(R)} \|\vartheta\|_\infty \|f^{in}\|_{1,1} \left(\int_0^\infty \varphi(1+y) f^{in}(y)\, dy \right) \frac{e^{C_3 t}}{C_3}, \end{aligned}$$

and the right-hand side of the above inequality converges to zero as $R \to \infty$ since $\varphi \in \mathscr{C}_{VP,\infty}$. Arguing in a similar way to handle $I_{3,n}(t)$, we complete the proof of (69) as in the proof of Theorem 24. Finally, the mass conservation $M_1(f(t)) = M_1(f^{in})$ for each $t > 0$ follows by passing to the limit as $n \to \infty$ in the equality $M_1(f_n(t)) = M_1(f^{in})$ from (66) with the help of (90), (92), (93), and the property $\varphi \in \mathscr{C}_{VP,\infty}$ to control the behaviour for large values of x.

□

3.2.3 Product Kernels

The last class of kernels we consider allows us to get rid of any growth condition on K provided it has a specific form. More precisely, we assume that there is a non-negative continuous function $r \in C([0,\infty))$ such that $r(x) > 0$ for $x > 0$ and

$$K(x,y) = r(x)r(y)\,, \quad (x,y) \in (0,\infty) \times (0,\infty)\,. \tag{94}$$

The celebrated multiplicative kernel $K_2(x,y) = xy$ fits into this framework with $r(x) = x$ for $x > 0$. Observe that no growth condition is required on r.

For this class of kernels, the existence result is similar to Theorem 24 and reads:

Theorem 29 *Assume that the coagulation kernel K satisfies* (94) *and consider an initial condition f^{in} satisfying* (64). *There is a non-negative function*

$$f \in C([0,\infty); L^1(0,\infty)) \cap L^\infty(0,\infty; L^1_1(0,\infty))$$

satisfying (69) *and*

$$M_0(f(t)) \le M_0(f^{in}) \quad \textit{and} \quad M_1(f(t)) \le M_1(f^{in})\,, \quad t \ge 0\,. \tag{95}$$

To prove Theorem 29, it turns out that it is easier to work with the following truncated version of K which differs from (65). Given an integer $n \geq 1$ and $x > 0$, we define $r_n(x) := \min\{r(x), n\}$ and

$$\tilde{K}_n(x,y) := r_n(x) r_n(y)\ , \quad (x,y) \in (0,\infty) \times (0,\infty)\ . \tag{96}$$

Clearly, $\tilde{K}_n$ is a non-negative, bounded, and symmetric function and we infer from (64) and Proposition 22 that the initial-value problem (49)–(50) with $\tilde{K}_n$ instead of K has a unique non-negative solution $g_n \in C^1([0,\infty); L^1(0,\infty))$ which satisfies

$$M_0(g_n(t)) \leq M_0(f^{in}) \quad \text{and} \quad M_1(g_n(t)) = M_1(f^{in})\ , \quad t \geq 0\ . \tag{97}$$

Without a control on the growth of K, the estimate (97) on $\|g_n(t)\|_{1,1}$ is not sufficient to control the behaviour of $g_n(t,x)$ for large values of x. As we shall see now, it is the specific form (94) of K which provides this control.

Lemma 30 *For $t > 0$, $A > 0$, and $n \geq 1$,*

$$\int_0^t \left(\int_0^\infty r_n(x) g_n(s,x)\, dx \right)^2 ds \leq 2M_0(f^{in})\ , \tag{98}$$

$$\int_0^t \left(\int_A^\infty r_n(x) g_n(s,x)\, dx \right)^2 ds \leq \frac{2M_1(f^{in})}{A}\ . \tag{99}$$

Proof On the one hand, the bound (98) readily follows from (57) (with $\vartheta \equiv 1$), (96), and the non-negativity of g_n. On the other hand, let $A > 0$ and define $\vartheta_A(x) := \min\{x, A\}$ for $x > 0$ as in the proof of Proposition 22. Owing to (59), we deduce from (57) and the non-negativity of g_n that

$$\begin{aligned} \int_0^t \int_A^\infty \int_A^\infty \tilde{K}_n(x,y) g_n(s,x) g_n(s,y)\, dydxds &\leq \frac{2}{A} \int_0^\infty \vartheta_A(x) f^{in}(x)\, dx \\ &\leq \frac{2M_1(f^{in})}{A}\ . \end{aligned}$$

Combining (96) and the above inequality gives (99). □

We next derive the counterpart of the equicontinuity property established in Lemma 26.

Lemma 31 *There is a modulus of continuity ω (that is, a function $\omega : (0,\infty) \to [0,\infty)$ satisfying $\omega(z) \to 0$ as $z \to 0$) such that*

$$\|g_n(t) - g_n(s)\|_1 \leq \omega(t-s)\ , \quad 0 \leq s \leq t\ , \quad n \geq 1\ . \tag{100}$$

Proof For $R > 0$ we introduce

$$m(R) := \sup_{x\in(0,R)} \left\{ \frac{r(x)}{1+x} \right\} ,$$

which is well-defined according to the continuity of r. We infer from (49) and Fubini's theorem that

$$\begin{aligned}
&\int_0^R |g_n(t,x) - g_n(s,x)| \, dx \\
&\quad \le \int_s^t \int_0^R \int_y^R \tilde{K}_n(y, x-y) g_n(\sigma, x-y) g_n(\sigma, y) \, dxdyd\sigma \\
&\qquad + \int_s^t \int_0^R \int_0^\infty \tilde{K}_n(x,y) g_n(\sigma,x) g_n(\sigma,y) \, dydxd\sigma \\
&\quad = \int_s^t \int_0^R \int_0^{R-y} r_n(x) r_n(y) g_n(\sigma,x) g_n(\sigma,y) \, dxdyd\sigma \\
&\qquad + \int_s^t \int_0^R \int_0^\infty r_n(x) r_n(y) g_n(\sigma,x) g_n(\sigma,y) \, dydxd\sigma \\
&\quad \le 2 \int_s^t \int_0^R \int_0^\infty r_n(x) r_n(y) g_n(\sigma,x) g_n(\sigma,y) \, dydxd\sigma \, .
\end{aligned}$$

We next use (97), Hölder's inequality, and (98) to obtain

$$\begin{aligned}
&\int_0^R |g_n(t,x) - g_n(s,x)| \, dx \\
&\quad \le 2 \int_s^t m(R) \|g_n(\sigma)\|_{1,1} \int_0^\infty r_n(y) g_n(\sigma,y) \, dyd\sigma \\
&\quad \le 2m(R) \|f^{in}\|_{1,1} \sqrt{t-s} \left[\int_s^t \left(\int_0^\infty r_n(y) g_n(\sigma,y) \, dy \right)^2 d\sigma \right]^{1/2} \\
&\quad \le \left(2\|f^{in}\|_{1,1} \right)^{3/2} m(R) \sqrt{t-s} \, .
\end{aligned}$$

Combining (97) and the above inequality leads us to

$$\begin{aligned}
\|g_n(t) - g_n(s)\|_1 &\le \int_0^R |g_n(t,x) - g_n(s,x)| \, dx + \int_R^\infty (g_n(t,x) + g_n(s,x)) \, dx \\
&\le \left(2\|f^{in}\|_{1,1} \right)^{3/2} m(R) \sqrt{t-s} + \frac{2}{R} M_1(f^{in}) \, ,
\end{aligned}$$

hence

$$\|g_n(t) - g_n(s)\|_1 \le C_4 \frac{1 + Rm(R)\sqrt{t-s}}{R} \tag{101}$$

for some positive constant C_4 depending only on f^{in}.

Now, $R \mapsto Rm(R)$ is a non-decreasing and continuous function such that $Rm(R) \to 0$ as $R \to 0$, $Rm(R) > 0$ for $R > 0$, and $Rm(R) \to \infty$ as $R \to \infty$. Introducing its generalized inverse

$$Q(z) := \inf\{R \ge 0 \ : \ Rm(R) \ge z\} \ , \quad z \ge 0 \ ,$$

the function Q is also non-decreasing with

$$Q(0) = 0 \ , \quad Q(z) > 0 \quad \text{for} \quad z > 0 \ , \quad \text{and} \quad Q(z) \to \infty \quad \text{as} \quad z \to \infty \ .$$

Setting

$$\frac{1}{\omega(z)} := \frac{3}{4} Q\left(\frac{1}{z}\right) \ , \quad z > 0 \ ,$$

and choosing $R = 1/\omega(\sqrt{t-s})$ in (101) we end up with

$$\|g_n(t) - g_n(s)\|_1 \le 2C_4 \omega(\sqrt{t-s}) \ .$$

The properties of Q ensure that $2C_4\omega(\sqrt{\cdot})$ is a modulus of continuity and the proof of Lemma 31 is complete. □

Proof of Theorem 29 The proof proceeds along the same lines as that of Theorem 24, the control on the behaviour for large x and the time equicontinuity being provided by Lemmas 30 and 31 instead of (66) and Lemma 26. Note that Lemma 25 is still valid owing to the local boundedness of r. □

Remark 32 As in [44], it is possible to extend Theorem 29 to perturbations of product kernels of the form $K(x,y) = r(x)r(y) + \tilde{K}(x,y)$ provided $0 \le \tilde{K}(x,y) \le \kappa_1 r(x) r(y)$ for $x > 0$, $y > 0$, and some $\kappa_1 > 0$.

Another peculiar extension of Theorem 29 is the possibility of constructing mass-conserving solutions for coagulation kernels of the form (94) satisfying $r(x)/\sqrt{x} \to \infty$ as $x \to \infty$.

Proposition 33 *Assume that the coagulation kernel K satisfies* (94) *and that $r \in C([0,\infty)) \cap C^1((0,\infty))$ is a concave and positive function such that*

$$\int_1^\infty \frac{dx}{r(x)^2} = \infty \ . \tag{102}$$

Consider an initial condition f^{in} satisfying (64) *together with $f^{in} \in L^1_2(0,\infty)$. Then there exists a solution f to* (49)–(50) *such that $M_1(f(t)) = M_1(f^{in})$ for each $t \geq 0$.*

The function $r(x) = \sqrt{2+x}\,(\ln(2+x))^\alpha$ satisfies the assumptions of Proposition 33 for $\alpha \in (0, 1/2]$.

Proof of Proposition 33 We keep the notations of the proof of Theorem 29, the existence part of Proposition 33 being a consequence of it. To show that the solution f constructed in the proof of Theorem 29 is mass-conserving, we check that (102) allows us to control the time evolution of the second moment $M_2(f(t))$ for all times $t \geq 0$. Indeed, for $n \geq 1$, we deduce from (57) that

$$\frac{d}{dt} M_2(g_n(t)) = \left(\int_0^\infty r_n(x)\, x\, g_n(t,x)\, dx \right)^2 .$$

Since r is concave, so is r_n and Jensen's inequality and (97) ensure that

$$\frac{d}{dt} M_2(g_n(t)) \leq M_1(f^{in})^2 \left[r\left(\frac{M_2(g_n(t))}{M_1(f^{in})} \right) \right]^2 .$$

Now, the assumption (102) guarantees that the ordinary differential equation

$$\frac{dY}{dt} = M_1(f^{in})^2 \left[r\left(\frac{Y}{M_1(f^{in})} \right) \right]^2 , \quad t \geq 0 ,$$

has a global solution $Y \in C^1([0,\infty))$ satisfying $Y(0) = M_2(f^{in})$ which is locally bounded. The comparison principle then implies that $M_2(g_n(t)) \leq Y(t)$ for all $t \geq 0$ and $n \geq 1$. We next use the convergence of $(g_n)_{n\geq 1}$ towards f to conclude that $M_2(f(t)) \leq Y(t)$ for all $t \geq 0$. We finally combine this information with (97) to show that $M_1(f(t)) = M_1(f^{in})$ for $t > 0$ and complete the proof. □

3.3 Gelation

In Sect. 3.2.2 we have shown the existence of mass-conserving solutions to (49)–(50) for coagulation kernels satisfying the growth condition (87). As already mentioned this property fails to be true in general for coagulation kernels which grows sufficiently fast for large x and y, a fact which has been known/conjectured since the early 1980s [24,38,53,92] but only proved recently in [26,39]. In fact, the occurrence of gelation was first shown for the multiplicative kernel $K_2(x,y) = xy$ by an elementary argument [52] and conjectured to take place for coagulation kernels K satisfying $K(x,y) \geq \kappa_m (xy)^{\lambda/2}$ for some $\lambda \in (1,2]$ and $\kappa_m > 0$ [24,38,53,92]. This conjecture was supported by a few explicit solutions constructed in [13,50,85]. A first breakthrough was made in [39] where a stochastic approach is used to show that, for a dense set of initial data, there exists at least one gelling solution to the discrete coagulation equations. A definitive and positive answer is provided in [26] where the occurrence of gelation in finite time is proved for all weak solutions starting from an arbitrary initial condition f^{in} satisfying (64) and $f^{in} \not\equiv 0$.

More precisely, let K be a non-negative and symmetric function such that

$$K(x, y) \geq r(x)r(y) , \quad (x, y) \in (0, \infty) \times (0, \infty) , \tag{103}$$

for some non-negative function r. Consider an initial condition f^{in}, $f^{in} \not\equiv 0$, satisfying (64) and let

$$f \in C([0, \infty); L^1(0, \infty)) \cap L^\infty(0, \infty; L^1_1(0, \infty)) , \quad f \geq 0 ,$$

be a solution to (49)–(50) satisfying (69) and (70) (such a solution exists for a large class of coagulation kernels, see Theorems 24 and 29).

Theorem 34 ([26]) *If there are $\lambda \in (1, 2]$ and $\kappa_m > 0$ such that $r(x) = \kappa_m x^{\lambda/2}$, $x > 0$, then gelation occurs in finite time, that is, there is $T_{gel} \in [0, \infty)$ such that*

$$M_1(f(t)) < M_1(f^{in}) \quad \textit{for} \quad t > T_{gel} .$$

The cornerstone of the proof of Theorem 34 is the following estimate.

Proposition 35 ([26]) *Let $\xi : [0, \infty) \longrightarrow [0, \infty)$ be a non-decreasing differentiable function satisfying $\xi(0) = 0$ and*

$$I_\xi := \int_0^\infty \xi'(A)\, A^{-1/2}\, dA < \infty . \tag{104}$$

Then, for $t > 0$,

$$\int_0^t \left(\int_0^\infty r(x)\xi(x)\, f(s, x)\, dx \right)^2 ds \leq 2I_\xi^2\, M_1(f^{in}) . \tag{105}$$

Let us mention at this point that, besides paving the way to a proof of the occurrence of gelation in finite time, other important consequences can be drawn from Proposition 35 due to the possibility of choosing different functions ξ. These consequences include temporal decay estimates for large times as well as more precise information on f across the gelation time [26].

Before proving Proposition 35, let us sketch how to use it to establish Theorem 34. Since $r(x) = \kappa_m x^{\lambda/2}$, a close look at (105) indicates that the choice $\xi(x) = x^{(2\ \lambda)/2}$ gives

$$\int_0^t M_1(f(s))^2\, ds = \int_0^t \left(\int_0^\infty x^{\lambda/2}\, \xi(x)\, f(s, x)\, dx \right)^2 ds \leq \frac{2I_\xi^2\, M_1(f^{in})}{\kappa_m^2}$$

for all $t > 0$. Consequently, $t \mapsto M_1(f(t))$ belongs to $L^2(0,\infty)$ and thus $M_1(f(t))$ cannot remain constant throughout time evolution. However this choice of ξ does not satisfy (104) as

$$\int_1^\infty \xi'(A)\, A^{-1/2}\, dA < \infty \quad \text{but} \quad \int_0^1 \xi'(A)\, A^{-1/2}\, dA = \infty\,.$$

We shall see below that a suitable choice is $\xi(x) = (x-1)_+^{(2-\lambda)/2}$.

Proof of Proposition 35 Fix $A > 0$ and take $\vartheta_A(x) = \min\{x, A\}$, $x > 0$, in (69). Recalling (59), we deduce from the non-negativity of f that

$$\begin{aligned}\int_0^t \int_A^\infty \int_A^\infty K(x,y) f(s,x) f(s,y)\, dydxds &\le \frac{2}{A} \int_0^\infty \vartheta_A(x) f^{in}(x)\, dx \\ &\le \frac{2M_1(f^{in})}{A}\,.\end{aligned}$$

Using (103) we end up with

$$\int_0^t \left(\int_A^\infty r(x) f(s,x)\, dx \right)^2 ds \le \frac{2M_1(f^{in})}{A}\,. \tag{106}$$

We then infer from (106), Fubini's theorem, and Cauchy–Schwarz' inequality that

$$\begin{aligned}&\int_0^t \left(\int_0^\infty r(x)\xi(x)\, f(s,x)\, dx \right)^2 ds \\ &= \int_0^t \left(\int_0^\infty \int_0^x r(x)\xi'(A)\, f(s,x)\, dAdx \right)^2 ds \\ &= \int_0^t \left(\int_0^\infty \xi'(A) \int_A^\infty r(x) f(s,x)\, dxdA \right)^2 ds \\ &\le I_\xi \int_0^t \int_0^\infty \xi'(A)\sqrt{A} \left(\int_A^\infty r(x) f(s,x)\, dx \right)^2 dAds \\ &\le 2M_1(f^{in}) I_\xi \int_0^\infty \frac{\xi'(A)}{\sqrt{A}}\, dA\,,\end{aligned}$$

hence (105). □

Proof of Theorem 34 It first follows from (69) with $\vartheta \equiv 1$, (103), the choice of r, and the non-negativity of f that

$$\int_0^t \left(\int_0^\infty x^{\lambda/2} f(s,x)\, dx \right)^2 ds \le \frac{2M_0(f^{in})}{\kappa_m^2}\,,$$

which implies, since $x \le 2^{(2-\lambda)/2} x^{\lambda/2}$ for $x \in (0,2)$,

$$\int_0^t \left(\int_0^2 x f(s,x)\, dx \right)^2 ds \le \frac{2^{3-\lambda} M_0(f^{in})}{\kappa_m^2} . \tag{107}$$

We next take $\xi(A) = (A-1)_+^{(2-\lambda)/2}$, $A > 0$, in Proposition 35 and note that

$$I_\xi = \frac{2-\lambda}{2} \int_1^\infty (A-1)^{-\lambda/2} A^{-1/2}\, dA < \infty$$

as $\lambda > 1$. Since $x - 1 \ge x/2$ for $x \ge 2$, we infer from (105) that

$$\begin{aligned} \frac{1}{2^{2-\lambda}} \int_0^t \left(\int_2^\infty x f(s,x)\, dx \right)^2 ds &\le \int_0^t \left(\int_0^\infty x^{\lambda/2} (x-1)_+^{(2-\lambda)/2} f(s,x)\, dx \right)^2 ds \\ &\le \frac{2 I_\xi^2\, M_1(f^{in})}{\kappa_m^2} . \end{aligned} \tag{108}$$

Combining (107) and (108) implies that $t \mapsto M_1(f(t))$ belongs to $L^2(0,\infty)$ and thus $M_1(f(t))$ cannot remain constant throughout time evolution. Since $M_1(f(t)) \le M_1(f^{in})$ for all $t > 0$ by (70), we conclude that $T_{gel} < \infty$. □

Using the same approach, we can actually extend Theorem 34 to a slightly wider setting encompassing the power functions.

Proposition 36 *Assume that $r \in C([0,\infty)) \cap C^1((0,\infty))$ is a concave and positive function which satisfies also*

$$\int_1^\infty \frac{dx}{x^{1/2}\, r(x)} < \infty \quad \text{and} \quad \lim_{x\to\infty} \frac{r(x)}{x^{1/2}} = \lim_{x\to\infty} \frac{x}{r(x)} = \infty , \tag{109}$$

as well as $r(x) \ge \delta x$ for $x \in (0,1)$ for some $\delta > 0$. Then $T_{gel} < \infty$.

A typical example of function r satisfying all the assumptions of Proposition 36 is a positive and concave function behaving as $\sqrt{x}\,(\ln x)^{1+\alpha}$ for large x for some $\alpha > 0$.

Proof The proof is similar to that of Theorem 34, the main difference being the choice of the function ξ in the use of Proposition 35. As r is concave and positive, the function $x \mapsto x/r(x)$ is non-decreasing and we set

$$\xi(x) := \left(\frac{x}{r(x)} - \frac{1}{r(1)} \right)_+ , \quad x > 0 .$$

Then ξ is a positive and non-decreasing differentiable function with $\xi(0) = 0$. Moreover, (109) guarantees that

$$\begin{aligned} I_\xi &= \left[\frac{A}{r(A)} A^{-1/2} \right]_{A=1}^{A=\infty} + \frac{1}{2} \int_1^\infty \frac{A}{r(A)} A^{-3/2}\, dA \\ &= -\frac{1}{r(1)} + \frac{1}{2} \int_1^\infty \frac{dA}{r(A)\, A^{1/2}} < \infty\,. \end{aligned}$$

We are therefore in a position to apply Proposition 35. Since there is $x_\star > 1$ such that $x/r(x) \ge 2/r(1)$ for $x \ge x_\star$ by (109), we deduce from (105) that

$$\int_0^t \left(\int_{x_\star}^\infty x\, f(s,x)\, dx \right)^2 ds \le 8 I_\xi^2 M_1(f^{in})\,. \tag{110}$$

We finally infer from (69) with $\vartheta \equiv 1$, (103), and the non-negativity of f that

$$\int_0^t \left(\int_0^\infty r(x)\, f(s,x)\, dx \right)^2 ds \le 2 M_0(f^{in})\,,$$

and the assumptions on r ensure that there is $\delta_\star \in (0,\delta)$ such that $\delta_\star x \le r(x)$ for $x \in (0, x_\star)$. Consequently,

$$\int_0^t \left(\int_0^{x_\star} x f(s,x)\, dx \right)^2 ds \le \frac{2 M_0(f^{in})}{\delta_\star^2}\,.$$

Combining this estimate with (110) allows us to conclude that $t \mapsto M_1(f(t)) \in L^2(0,\infty)$ and complete the proof. □

3.4 Uniqueness

The uniqueness issue has been investigated by several authors but the results obtained so far are restricted to mass-conserving solutions, an exception being the multiplicative kernel $K_2(x,y) = xy$. Actually two approaches have been developed to establish the uniqueness of solutions to (49)–(50): a direct one which consists in taking two solutions and estimating a weighted L^1-norm of their difference and another one based on a kind of Wasserstein distance. To be more specific, since the pioneering works [59, 61], uniqueness has been proved in [4, 20, 34, 40, 45, 70, 82] by the former approach and is summarized in the next result.

Proposition 37 *Assume that there is a non-negative subadditive function φ (that is, $\varphi(x+y) \le \varphi(x) + \varphi(y)$, $x > 0$, $y > 0$), such that*

$$K(x,y) \le \varphi(x)\varphi(y)\,, \quad x > 0\,,\ y > 0\,.$$

Let $T > 0$ and $f^{in} \in L^1(0,\infty;\varphi(x)\,dx)$, $f^{in} \geq 0$. There is at most one solution

$$f \in C([0,T]; L^1(0,\infty;\varphi(x)dx)) \cap L^1(0,T; L^1(0,\infty;\varphi(x)^2dx))$$

to (49)–(50).

Let us point out two immediate consequences of Proposition 37. First, if $\varphi(x) \leq x^{1/2}$, Proposition 37 implies the uniqueness of the solution constructed in Theorem 27 since it belongs to $L^\infty(0,\infty;L^1_1(0,\infty))$. Next, if $\varphi(x) = x$, it gives the uniqueness of solutions to (49)–(50) as long as $M_2(f) \in L^1(0,T)$. While this is only true up to a finite time T in the general framework considered in Theorem 24, it follows from Theorem 27 and Lemma 28 (with $\psi(x) = x^2$) that, if f^{in} belongs to $L^1_2(0,\infty)$ and satisfies (64), then the solution f to (49)–(50) constructed in Theorem 27 is such that $t \mapsto M_2(f(t)) \in L^\infty(0,T)$ for any $T > 0$. According to Proposition 37, this solution is unique.

Proof of Proposition 37 Let f_1 and f_2 be two solutions to (49)–(50) enjoying the properties listed in Proposition 37. We infer from (49) that

$$\begin{aligned}
&\frac{d}{dt}\int_0^\infty |(f_1 - f_2)(t,x)|\varphi(x)\,dx \\
&\quad = \frac{1}{2}\int_0^\infty \int_0^\infty K(x,y)(f_1+f_2)(t,y)(f_1-f_2)(t,x)\Theta(t,x,y)\,dydx
\end{aligned}$$

with $\sigma := \operatorname{sign}(f_1 - f_2)$ and

$$\Theta(t,x,y) := [(\varphi\sigma)(t,x+y) - (\varphi\sigma)(t,x) - (\varphi\sigma)(t,y)] \ .$$

Observing that

$$\begin{aligned}
(f_1 - f_2)(t,x)\Theta(t,x,y) &= |(f_1-f_2)(t,x)|\sigma(t,x)\Theta(t,x,y) \\
&= |(f_1-f_2)(t,x)|\left[(\varphi\sigma)(t,x+y)\sigma(t,x) - \varphi(x)\sigma(t,x)^2 - (\varphi\sigma)(t,y)\sigma(t,x)\right] \\
&\leq |(f_1-f_2)(t,x)|\,[\varphi(x+y) - \varphi(x) + \varphi(y)] \\
&\leq 2\varphi(y)|(f_1-f_2)(t,x)| \ ,
\end{aligned}$$

we further obtain

$$\begin{aligned}
&\frac{d}{dt}\int_0^\infty |(f_1 - f_2)(t,x)|\varphi(x)\,dx \\
&\quad \leq \int_0^\infty \int_0^\infty K(x,y)(f_1+f_2)(t,y)|(f_1-f_2)(t,x)|\varphi(y)\,dydx
\end{aligned}$$

$$\leq \int_0^\infty \int_0^\infty \varphi(x)\varphi(y)^2 (f_1+f_2)(t,y)|(f_1-f_2)(t,x)|\, dydx$$

$$= \int_0^\infty \varphi(y)^2 (f_1+f_2)(t,y)\, dy \int_0^\infty \varphi(x)|(f_1-f_2)(t,x)|\, dx\,,$$

and the conclusion follows by Gronwall's inequality. □

The second approach stems from the study of the well-posedness of (49)–(50) for the constant kernel $K_0(x,y)=2$, the additive kernel $K_1(x,y)=x+y$, and the multiplicative kernel $K_2(x,y)=xy$ performed in [62]. Roughly speaking, there is a uniqueness result for the kernel K_i in the weighted space $L^1(0,\infty;x^i dx)$, $i=0,1,2$. It is worth pointing out that the homogeneity of the weight matches that of the coagulation kernel. Though the main tool used in [62] is the Laplace transform, an alternative argument involving a weighted Wasserstein distance has been developed in [32] to extend this uniqueness result to a wider class of homogeneous kernels with arbitrary homogeneity. For simplicity, we restrict ourselves to the coagulation kernel

$$K(x,y) = x^\alpha y^\beta + x^\beta y^\alpha\,, \quad x>0\,,\ y>0\,, \tag{111}$$

and assume that its homogeneity $\lambda := \alpha+\beta$ lies in $(0,1]$. We refer to [32] for more general assumptions on K and f^{in} and homogeneities in $(-\infty,0)$ or in $(1,2)$.

Proposition 38 *Let $T>0$ and $f^{in}\in L^1_\lambda(0,\infty)$, $f^{in}\geq 0$. There is at most one solution $f\in C([0,T);w-L^1_\lambda(0,\infty))$ to* (49)–(50) *(recall that the space $L^1_\lambda(0,\infty)$ is defined in* (51)*).*

Proof Let f_1 and f_2 be two solutions to (49)–(50) enjoying the properties listed in Proposition 38. For $i=1,2$, we introduce the cumulative distribution function F_i of f_i given by

$$F_i(t,x) := \int_x^\infty f_i(t,y)\, dy\,, \quad t>0\,,\ x>0\,,$$

and set $E := F_1 - F_2$ and

$$R(t,x) := \int_0^x z^{\lambda-1}\, \mathrm{sign}(F_1-F_2)(t,z)\, dz\,, \quad t>0\,,\ x>0\,.$$

We infer from (49) after some computations (see [32, Proposition 3.3]) that

$$\frac{d}{dt}\int_0^\infty x^{\lambda-1}|E(t,x)|\, dx \leq \frac{1}{2}\left(A_1(t)+A_2(t)\right)\,, \tag{112}$$

where

$$A_1(t) := \int_0^\infty \int_0^\infty K(x,y)\left[(x+y)^{\lambda-1} - x^{\lambda-1}\right](f_1+f_2)(t,y)|E(t,x)|\, dydx$$

and

$$A_2(t) := \int_0^\infty \int_0^\infty \partial_x K(x,y)\tilde{R}(t,x,y)(f_1+f_2)(t,y)E(t,x)\,dydx\,,$$
$$\tilde{R}(t,x,y) := R(t,x+y) - R(t,x) - R(t,y)\,.$$

On the one hand, since $\lambda < 1$,

$$K(x,y)\left[(x+y)^{\lambda-1} - x^{\lambda-1}\right] \le 0\,, \quad x>0\,,\ y>0\,,$$

so that

$$A_1(t) \le 0\,. \tag{113}$$

On the other hand, it follows from the definition of R and the subadditivity of $x \mapsto x^\lambda$ that

$$\begin{aligned}
\left|\tilde{R}(t,x,y)\right| &= \left|\int_{\max\{x,y\}}^{x+y} z^{\lambda-1}\,\mathrm{sign}(E)(t,z)\,dz + \int_0^{\min\{x,y\}} z^{\lambda-1}\,\mathrm{sign}(E)(t,z)\,dz\right| \\
&\le \frac{1}{\lambda}\left[(x+y)^\lambda - \max\{x,y\}^\lambda + \min\{x,y\}^\lambda\right] \\
&\le \frac{2}{\lambda}\min\{x,y\}^\lambda\,.
\end{aligned}$$

We deduce from the previous inequality and (111) that there is $C_5 > 0$ depending only on α and β such that

$$|\partial_x K(x,y)|\left|\tilde{R}(t,x,y)\right| \le C_5\, x^{\lambda-1} y^\lambda\,.$$

Consequently,

$$A_2(t) \le C_5\, M_\lambda((f_1+f_2)(t)) \int_0^\infty x^{\lambda-1}|E(t,x)|\,dx\,. \tag{114}$$

Collecting (112)–(114) we end up with

$$\frac{d}{dt}\int_0^\infty x^{\lambda-1}|E(t,x)|\,dx \le \frac{C_5}{2}\, M_\lambda((f_1+f_2)(t)) \int_0^\infty x^{\lambda-1}|E(t,x)|\,dx\,,$$

and the conclusion follows by integration. □

Even though there is a version of Proposition 38 when K is given by (111) with $\lambda \in (1,2]$ (and is thus a gelling kernel), the requirement $f \in C([0,T); w-L^1_\lambda(0,\infty))$ is only true for $T < T_{gel}$ and thus provides no clue about uniqueness past the gelation time.

The only result we are aware of which deals with the uniqueness of solutions exhibiting a gelation transition is available for the multiplicative kernel $K_2(x,y)=xy$. In that particular case, using the Laplace transform, it is possible to characterize $M_1(f(t))$ for all times, prior and past the gelation time, and this information allows one to prove uniqueness [19, 41, 69, 76, 87].

Acknowledgements These notes grew out from lectures I gave at the Institut für Angewandte Mathematik, Leibniz Universität Hannover, in September 2008 and at the African Institute for Mathematical Sciences, Muizenberg, in July 2013. I thank Jacek Banasiak, Joachim Escher, Mustapha Mokhtar-Kharroubi, and Christoph Walker for their kind invitations as well as both institutions for their hospitality and support. This work was completed while visiting the Institut Mittag-Leffler, Stockholm.

References

1. M. Aizenman, T.A. Bak, Convergence to equilibrium in a system of reacting polymers. Commun. Math. Phys. **65**, 203–230 (1979)
2. D.J. Aldous, Deterministic and stochastic models for coalescence (aggregation, coagulation): a review of the mean-field theory for probabilists. Bernoulli **5**, 3–48 (1999)
3. H. Amann, *Ordinary Differential Equations* (Walter de Gruyter, Berlin/New York, 1990)
4. J.M. Ball, J. Carr, The discrete coagulation-fragmentation equations: existence, uniqueness, and density conservation. J. Stat. Phys. **61**, 203–234 (1990)
5. J. Banasiak, W. Lamb, Global strict solutions to continuous coagulation-fragmentation equations with strong fragmentation. Proc. R. Soc. Edinb. Sect. A **141**, 465–480 (2011)
6. J. Banasiak, W. Lamb, Analytic fragmentation semigroups and continuous coagulation-fragmentation equations with unbounded rates. J. Math. Anal. Appl. **391**, 312–322 (2012)
7. B. Beauzamy, *Introduction to Banach Spaces and their Geometry* (North-Holland, Amsterdam, 1985)
8. J. Bertoin, Eternal solutions to Smoluchowski's coagulation equation with additive kernel and their probabilistic interpretation. Ann. Appl. Probab. **12**, 547–564 (2002)
9. J. Bertoin, *Random Fragmentation and Coagulation Processes*. Cambridge Studies in Advanced Mathematics, vol. 102 (Cambridge University Press, Cambridge, 2006)
10. H. Brezis, *Functional Analysis, Sobolev Spaces and Partial Differential Equations*. Universitext (Springer, New York, 2011)
11. J.A. Cañizo, S. Mischler, Regularity, local behavior and partial uniqueness for self-similar profiles of Smoluchowski's coagulation equation. Rev. Mat. Iberoam. **27**, 803–839 (2011)
12. F.P. da Costa, Existence and uniqueness of density conserving solutions to the coagulation-fragmentation equations with strong fragmentation. J. Math. Anal. Appl. **192**, 892–914 (1995)
13. F.P. da Costa, A finite-dimensional dynamical model for gelation in coagulation processes. J. Nonlinear Sci. **8**, 619–653 (1998)
14. C.J. de la Vallée Poussin, Sur l'intégrale de Lebesgue. Trans. Am. Math. Soc. **16**, 435–501 (1915)
15. M. Deaconu, N. Fournier, E. Tanré, A pure jump Markov process associated with Smoluchowski's coagulation equation. Ann. Probab. **30**, 1763–1796 (2002)
16. C. Dellacherie, P.A. Meyer, *Probabilités et Potentiel*, Chapitres I à IV (Hermann, Paris, 1975)
17. R.J. DiPerna, P.-L. Lions, On the Cauchy problem for Boltzmann equations: global existence and weak stability. Ann. Math. **130**, 321–366 (1989)
18. R.L. Drake, A general mathematical survey of the coagulation equation, in *Topics in Current Aerosol Research (Part 2)*. International Reviews in Aerosol Physics and Chemistry (Pergamon Press, Oxford, 1972), pp. 203–376

19. P.B. Dubovskiĭ, *Mathematical Theory of Coagulation*. Lecture Notes Series, vol. 23 (Seoul National University, Seoul, 1994)
20. P.B. Dubovskiĭ, I.W. Stewart, Existence, uniqueness and mass conservation for the coagulation-fragmentation equation. Math. Methods Appl. Sci. **19**, 571–591 (1996)
21. N. Dunford, J.T. Schwartz, *Linear Operators. Part I: General Theory* (Interscience Publishers, New York, 1957)
22. R. Durrett, B.L. Granovsky, S. Gueron, The equilibrium behavior of reversible coagulation-fragmentation processes. J. Theor. Probab. **12**, 447–474 (1999)
23. A. Eibeck, W. Wagner, Stochastic particle approximations for Smoluchowski's coagulation equation. Ann. Appl. Probab. **11**, 1137–1165 (2001)
24. M.H. Ernst, R.M. Ziff, E.M. Hendriks, Coagulation processes with a phase transition. J. Colloid Interface Sci. **97**, 266–277 (1984)
25. M. Escobedo, S. Mischler, Dust and self-similarity for the Smoluchowski coagulation equation. Ann. Inst. H. Poincaré Anal. Non Linéaire **23**, 331–362 (2006)
26. M. Escobedo, S. Mischler, B. Perthame, Gelation in coagulation-fragmentation models. Commun. Math. Phys. **231**, 157–188 (2002)
27. M. Escobedo, Ph. Laurençot, S. Mischler, B. Perthame, Gelation and mass conservation in coagulation-fragmentation models. J. Differ. Equ. **195**, 143–174 (2003)
28. M. Escobedo, S. Mischler, M. Rodriguez Ricard, On self-similarity and stationary problem for fragmentation and coagulation models. Ann. Inst. H. Poincaré Anal. Non Linéaire **22**, 99–125 (2005)
29. N. Fournier, J.S. Giet, Convergence of the Marcus-Lushnikov process. Methodol. Comput. Appl. Probab. **6**, 219–231 (2004)
30. N. Fournier, Ph. Laurençot, Existence of self-similar solutions to Smoluchowski's coagulation equation. Commun. Math. Phys. **256**, 589–609 (2005)
31. N. Fournier, Ph. Laurençot, Local properties of self-similar solutions to Smoluchowski's coagulation equation with sum kernels. Proc. R. Soc. Edinb. Sect. A **136**, 485–508 (2006)
32. N. Fournier, Ph. Laurençot, Well-posedness of Smoluchowski's coagulation equation for a class of homogeneous kernels. J. Funct. Anal. **233**, 351–379 (2006)
33. S.K. Friedlander, *Smoke, Dust, and Haze: Fundamentals of Aerosol Dynamics*, 2nd edn. (Oxford University Press, New York, 2000)
34. A.K. Giri, On the uniqueness for coagulation and multiple fragmentation equation. Kinet. Relat. Models **6**, 589–599 (2013)
35. A.K. Giri, G. Warnecke, Uniqueness for the coagulation-fragmentation equation with strong fragmentation. Z. Angew. Math. Phys. **62**, 1047–1063 (2011)
36. A.K. Giri, Ph. Laurençot, G. Warnecke, Weak solutions to the continuous coagulation equation with multiple fragmentation. Nonlinear Anal. **75**, 2199–2208 (2012)
37. S. Gueron, S.A. Levin, The dynamics of group formation. Math. Biosci. **128**, 243–264 (1995)
38. E.M. Hendriks, M.H. Ernst, R.M. Ziff, Coagulation equations with gelation. J. Stat. Phys. **31** 519–563 (1983)
39. I. Jeon, Existence of gelling solutions for coagulation-fragmentation equations. Commun. Math. Phys. **194**, 541–567 (1998)
40. B.Jourdain, Nonlinear processes associated with the discrete Smoluchowski coagulation-fragmentation equation. Markov Process. Relat. Fields **9**, 103–130 (2003)
41. N.J. Kokholm, On Smoluchowski's coagulation equation. J. Phys. A **21**, 839–842 (1988)
42. M.A. Krasnosel'skiĭ, Ja.B. Rutickiĭ, *Convex Functions and Orlicz Spaces* (Noordhoff, Groningen, 1961)
43. W. Lamb, Existence and uniqueness results for the continuous coagulation and fragmentation equation. Math. Methods Appl. Sci. **27**, 703–721 (2004)
44. Ph. Laurençot, On a class of continuous coagulation-fragmentation equations. J. Differ. Equ. **167**, 245–274 (2000)
45. Ph. Laurençot, The discrete coagulation equation with multiple fragmentation. Proc. Edinb. Math. Soc. **45**(2), 67–82 (2002)

46. Ph. Laurençot, S. Mischler, From the discrete to the continuous coagulation-fragmentation equations. Proc. R. Soc. Edinb. Sect. A **132**, 1219–1248 (2002)
47. Ph. Laurençot, S. Mischler, The continuous coagulation-fragmentation equations with diffusion. Arch. Rational Mech. Anal. **162**, 45–99 (2002)
48. Ph. Laurençot, S. Mischler, On coalescence equations and related models, in *Modeling and Computational Methods for Kinetic Equations*, ed. by P. Degond, L. Pareschi, L. Russo (Birkhäuser, Boston, 2004), pp. 321–356
49. C.-H. Lê, Etude de la classe des opérateurs m-accrétifs de $L^1(\Omega)$ et accrétifs dans $L^\infty(\Omega)$. Thèse de 3ème cycle (Université de Paris VI, Paris, 1977)
50. F. Leyvraz, Existence and properties of post-gel solutions for the kinetic equations of coagulation. J. Phys. A **16**, 2861–2873 (1983)
51. F. Leyvraz, Scaling theory and exactly solved models in the kinetics of irreversible aggregation. Phys. Rep. **383**, 95–212 (2003)
52. F. Leyvraz, H.R. Tschudi, Singularities in the kinetics of coagulation processes. J. Phys. A **14**, 3389–3405 (1981)
53. F. Leyvraz, H.R. Tschudi, Critical kinetics near gelation. J. Phys. A **15**, 1951–1964 (1982)
54. A. Lushnikov, Coagulation in finite systems. J. Colloid Interface Sci. **65**, 276–285 (1978)
55. A.H. Marcus, Stochastic coalescence. Technometrics **10**, 133–143 (1968)
56. D.J. McLaughlin, W. Lamb, A.C. McBride, An existence and uniqueness result for a coagulation and multiple-fragmentation equation. SIAM J. Math. Anal. **28**, 1173–1190 (1997)
57. J.B. McLeod, On an infinite set of non-linear differential equations. Q. J. Math. Oxford Ser. **13**(2), 119–128 (1962)
58. J.B. McLeod, On an infinite set of non-linear differential equations (II). Q. J. Math. Oxford Ser. **13**(2), 193–205 (1962)
59. J.B. McLeod, On the scalar transport equation. Proc. Lond. Math. Soc. **14**(3), 445–458 (1964)
60. J.B. McLeod, B. Niethammer, J.J.L. Velázquez, Asymptotics of self-similar solutions to coagulation equations with product kernel. J. Stat. Phys. **144**, 76–100 (2011)
61. Z.A. Melzak, A scalar transport equation. Trans. Am. Math. Soc. **85**, 547–560 (1957)
62. G. Menon, R. Pego, Approach to self-similarity in Smoluchowski's coagulation equations. Commun. Pure Appl. Math. **LVII**, 1197–1232 (2004)
63. S. Mischler, B. Wennberg, On the spatially homogeneous Boltzmann equation. Ann. Inst. H. Poincaré Anal. Non Linéaire **16**, 467–501 (1999)
64. H. Müller, Zur allgemeinen Theorie der raschen Koagulation. Kolloidchemische Beihefte **27**, 223–250 (1928)
65. B. Niethammer, Self-similarity in Smoluchowski's coagulation equation. Jahresber. Deutsch. Math.-Verein. **116**, 43–65 (2014)
66. B. Niethammer, J.J.L. Velázquez, Optimal bounds for self-similar solutions to coagulation equations with product kernel. Commun. Partial Differ. Equ. **36**, 2049–2061 (2011)
67. B. Niethammer, J.J.L. Velázquez, Self-similar solutions with fat tails for Smoluchowski's coagulation equation with locally bounded kernels. Commun. Math. Phys. **318**, 505–532 (2013)
68. B. Niethammer, J.J.L. Velázquez, Erratum to: Self-similar solutions with fat tails for Smoluchowski's coagulation equation with locally bounded kernels. Commun. Math. Phys. **318**, 533–534 (2013)
69. R. Normand, L. Zambotti, Uniqueness of post-gelation solutions of a class of coagulation equations. Ann. Inst. H. Poincaré Anal. Non Linéaire **28**, 189–215 (2011)
70. J.R. Norris, Smoluchowski's coagulation equation: uniqueness, non-uniqueness and a hydrodynamic limit for the stochastic coalescent. Ann. Appl. Probab. **9**, 78–109 (1999)
71. M. Pierre, An L^1-method to prove global existence in some reaction-diffusion systems, in *Contributions to Nonlinear Partial Differential Equations*, vol. II (Paris, 1985). Pitman Research Notes in Mathematical Series, vol. 155 (Longman Science and Technology, Harlow, 1987), pp. 220–231
72. M.M. Rao, Z.D. Ren, *Theory of Orlicz Spaces*. Monographs and Textbooks in Pure Applied Mathematics, vol. 146 (Marcel Dekker, New York, 1991)

73. E. Ringeisen, Résultats d'existence pour le modèle de BGK de la théorie cinétique des gaz en domaine borné et non borné (Rapport interne MAB 96011, Bordeaux, 1996)
74. W. Rudin, *Real and Complex Analysis*, 3rd edn. (McGraw-Hill, New York 1987)
75. J.H. Seinfeld, S.N. Pandis, *Atmospheric Chemistry and Physics: From Air Pollution to Climate Change*, 2nd edn. (Wiley, New York, 2006)
76. M. Shirvani, H.J. van Roessel, Some results on the coagulation equation. Nonlinear Anal. **43**, 563–573 (2001)
77. M. Słaby, Strong convergence of vector-valued pramarts and subpramarts. Probab. Math. Stat. **5**, 187–196 (1985)
78. M. Smoluchowski, Drei Vorträge über Diffusion, Brownsche Molekularbewegung und Koagulation von Kolloidteilchen. Physik. Zeitschr. **17**, 557–599 (1916)
79. M. Smoluchowski, Versuch einer mathematischen Theorie der Koagulationskinetik kolloider Lösungen. Zeitschrift f. Physik. Chemie **92**, 129–168 (1917)
80. J.L. Spouge, An existence theorem for the discrete coagulation-fragmentation equations. Math. Proc. Camb. Philos. Soc. **96**, 351–357 (1984)
81. I.W. Stewart, A global existence theorem for the general coagulation-fragmentation equation with unbounded kernels. Math. Methods Appl. Sci. **11**, 627–648 (1989)
82. I.W. Stewart, A uniqueness theorem for the coagulation-fragmentation equation. Math. Proc. Camb. Philos. Soc. **107**, 573–578 (1990)
83. I.W. Stewart, Density conservation for a coagulation equation. Z. Angew. Math. Phys. **42**, 746–756 (1991)
84. W.H. Stockmayer, Theory of molecular size distribution and gel formation in branched-chain polymers. J. Chem. Phys. **11**, 45–55 (1943)
85. P.G.J. van Dongen , M.H. Ernst, Cluster size distribution in irreversible aggregation at large times. J. Phys. A **18**, 2779–2793 (1985)
86. P.G.J. van Dongen, M.H. Ernst, Scaling solutions of Smoluchowski's coagulation equation. J. Stat. Phys. **50**, 295–329 (1988)
87. H.J. van Roessel, M. Shirvani, A formula for the post-gelation mass of a coagulation equation with a separable bilinear kernel. Physica D **222**, 29–36 (2006)
88. I.I. Vrabie, *Compactness Methods for Nonlinear Evolutions*, 2nd edn. Pitman Monograph Surveys in Pure Applied Mathematics, vol. 75 (Longman, Harlow, 1995)
89. J.A.D. Wattis, An introduction to mathematical models of coagulation-fragmentation processes: a deterministic mean-field approach. Physica D **222**, 1–20 (2006)
90. W.H. White, A global existence theorem for Smoluchowski's coagulation equations. Proc. Am. Math. Soc. **80**, 273–276 (1980)
91. K. Yosida, *Functional Analysis* [Reprint of the sixth (1980) edn.]. Classics in Mathematics (Springer, Berlin, 1995)
92. R.M. Ziff, Kinetics of polymerization. J. Stat. Phys. **23**, 241–263 (1980)

Stochastic Operators and Semigroups and Their Applications in Physics and Biology

Ryszard Rudnicki

1 Introduction

Stochastic operators are positive linear operators defined on the space of integrable functions preserving the set of densities. They appear in ergodic theory of dynamical systems and iterated function systems. They also describe the evolution of Markov chains. The interested reader can find results concerning probabilistic and functional properties of stochastic operators in monographs [34, 74]. The book of Lasota and Mackey [55] is an excellent survey of many results concerning of their applications in ergodic theory. Applications of stochastic operators to statistical mechanics are presented in [65]. Stochastic semigroups are continuous semigroups of stochastic operators and they have been intensively studied because they play a special role in applications. They are generated by partial differential equations of different types and describe the behaviour of the distributions of Markov processes like diffusion processes, piecewise deterministic processes and hybrid stochastic processes. In this chapter we present many examples of stochastic operators and semigroups: the Frobenius–Perron operator, diffusion semigroups, flow semigroups with jumps and switching and semigroups related to hybrid systems. Then we present some results concerning their long-time behaviour: asymptotic stability, sweeping, completely mixing and convergence to self-similar solutions. The results concerning stochastic operators are applied to study ergodicity, mixing and exactness of dynamical systems and to integral operators appearing in the theory of cell cycle. The general results concerning stochastic semigroups are applied to diffusion processes, jump processes and biological models described by piecewise deterministic stochastic

R. Rudnicki (✉)
Institute of Mathematics, Polish Academy of Sciences, Bankowa 14, 40-007 Katowice, Poland
e-mail: rudnicki@us.edu.pl

© Springer International Publishing Switzerland 2015
J. Banasiak, M. Mokhtar-Kharroubi (eds.), *Evolutionary Equations with Applications in Natural Sciences*, Lecture Notes in Mathematics 2126,
DOI 10.1007/978-3-319-11322-7_6

processes: birth-death processes, the evolution of the genome, gene expression and physiologically structured models.

The organization of the chapter is as follows. Section 2 introduces stochastic operators and semigroups and results concerning generators of stochastic semigroups. We also present methods of constructing stochastic semigroups by perturbation theorems. All theoretical results are illustrated by examples of stochastic operators and semigroups. We show the utility of the Frobenius–Perron operator to study ergodic properties of dynamical systems. We present here stochastic operators related to iterated function systems and probability transition functions. We study in detail stochastic semigroups in the space l^1 corresponding to continuous time Markov chains. We introduce semigroups related to non-degenerate and degenerate diffusion processes and deterministic flows. As an illustration of perturbation theory we present semigroups related to piecewise deterministic Markov processes such as a pure jump process, flow with jumps and dynamics governed by a number of flows with switching. Another examples are stochastic hybrid systems: randomly flashing and multi-state diffusion. We also present some elementary nonlinear stochastic operators and semigroups.

In Sect. 3 we study the long-time behaviour of stochastic operators and semigroups. We start with the concept of asymptotic stability and illustrate it by showing that the tent and a logistic map are exact (have very strong ergodic property). Then we present the lower function theorem of Lasota–Yorke and its application to stochastic matrices. Our main results concern long-time behaviour of partially integral semigroups. We introduce the notion of sweeping and present some result which leads to the Foguel alternative, i.e. we find conditions when a partially integral semigroup is asymptotic stability or sweeping. We start with the definition of a partially integral semigroup. In particular we give sufficient and necessary conditions for asymptotic stability and sweeping of stochastic semigroups on the space l^1. We present also some auxiliary results which can be useful in studying of piecewise deterministic Markov processes. Then we introduce the notion of the Hasminskiĭ function. This notion is very useful in proofs of asymptotic stability of stochastic semigroups if we known that they fulfill the Foguel alternative. We also consider other long-time properties of stochastic semigroups as completely mixing property, sectorial limits, convergence after rescaling and self-similar solutions and illustrate them by applications to diffusion and jump processes.

Section 4 contains applications of general results to specific models. We present applications which come from population dynamics: cell cycle model, birth-death processes and structured-population models and from genetics: paralog families and gene expression. We also present some physical applications as the Ehrenfest model, diffusion and jump processes. The last part of this section is devoted to nonlinear stochastic operators and semigroups. Nonlinear stochastic operators and semigroups appear in models which contains binary operations. We present a stochastic operator which describes the relation between frequencies of genotypes in the parent and offspring generations. We also present some stochastic semigroups related to coagulation-fragmentation processes and to the Boltzmann kinetic theory of gases. It is interesting that similar semigroups can model aggregation processes in phytoplankton populations and phenotype-structured populations.

2 Stochastic Operators and Semigroups

In this section we give definitions and examples of stochastic operators and semigroups.

2.1 Definitions

Let the triple (X, Σ, m) be a σ-finite measure space. Denote by D the subset of the space $L^1 = L^1(X, \Sigma, m)$ which contains all densities

$$D = \{f \in L^1 : \ f \geq 0, \ \|f\| = 1\}.$$

A linear operator $P\colon L^1 \to L^1$ is called a *stochastic (or Markov) operator* if $P(D) \subset D$. If a linear operator $P\colon L^1 \to L^1$ is positive and if $\|Pf\| \leq \|f\|$ for $f \in L^1$ then P is called *substochastic*. In particular, each stochastic operator is substochastic. We can also consider *nonlinear stochastic operators* and define them as continuous operators $P\colon D \to D$.

One can define a stochastic operator by means of a *transition probability function*. We recall that $\mathcal{P}(x, A)$ is a transition probability function on (X, Σ) if $\mathcal{P}(x, \cdot)$ is a probability measure on (X, Σ) and $\mathcal{P}(\cdot, A)$ is a measurable function. Assume that $\mathcal{P}$ has the following property

$$m(A) = 0 \Longrightarrow \mathcal{P}(x, A) = 0 \text{ for } m\text{-a.e. } x \text{ and } A \in \Sigma. \tag{1}$$

Then for every $f \in D$ the measure

$$\mu(A) = \int f(x)\mathcal{P}(x, A)\, m(dx)$$

is absolutely continuous with respect to the measure m. This fact is a simple consequence of the Radon–Nikodym theorem, which says that the measure ν is absolutely continuous with respect to the measure m iff the following implication $m(A) = 0 \Rightarrow \nu(A) = 0$ holds for all sets $A \in \Sigma$. Now, the formula $Pf = d\mu/dm$ defines a stochastic operator $P : L^1 \to L^1$. Moreover, if $P^*\colon L^\infty \to L^\infty$ is the adjoint operator of P then $P^*g(x) = \int g(y)\, \mathcal{P}(x, dy)$. There are stochastic operators which are not given by transition probability functions [30]. But if X is a Polish space (i.e. a complete separable metric space), $\Sigma = \mathcal{B}(X)$ is the σ-algebra of

Borel subsets of X, and m is a probability Borel measure on X, then every stochastic operator on $L^1(X, \Sigma, m)$ is given by a transition probability function [40].

Before giving the definition of a stochastic semigroup we recall the definition of C_0-semigroups of operators in a Banach space and their elementary properties. Let $(E, \|\cdot\|)$ be a Banach space and let $\{P(t)\}_{t\geq 0}$ be a family of linear bounded operators on E. The family $\{P(t)\}_{t\geq 0}$ is called a *C_0-semigroup* or *strongly continuous semigroup* if it satisfies the following conditions:

(a) $P(0) = I$, i.e., $P(0)f = f$,
(b) $P(t+s) = P(t)P(s) \quad \text{for} \quad s, t \geq 0$,
(c) for each $f \in E$ the function $t \mapsto P(t)f$ is continuous.

Let $\mathfrak{D}(A)$ be the set of such $f \in E$, that there exists the limit

$$Af = \lim_{t \to 0+} \frac{P(t)f - f}{t}. \tag{2}$$

Then the set $\mathfrak{D}(A)$ is a linear subspace dense in E, and A is a linear operator from $\mathfrak{D}(A)$ to E. The operator A is called the *infinitesimal generator* (briefly the *generator*) of the semigroup $\{P(t)\}_{t\geq 0}$. We also say that the operator $A\colon \mathfrak{D}(A) \to E$ generates the semigroup $\{P(t)\}_{t\geq 0}$.

The notions of a C_0-semigroup and its generator is strictly connected with differential equations in the Banach space E. Let $\{P(t)\}_{t\geq 0}$ be C_0 semigroup on E and $A : \mathfrak{D}(A) \to E$ its generator. Then for every $f_0 \in \mathfrak{D}(A)$ the function $f\colon [0,\infty) \to E$ defined by the formula $f(t) = P(t)f_0$ is differentiable (in the sense of Fréchet) for $t \geq 0$ and satisfies the equation

$$f'(t) = Af(t), \quad \text{with initial condition } f(0) = f_0. \tag{3}$$

Equation of the form (3) is called an *evolution equation*. We will say that equation (3) *generates semigroup* $\{P(t)\}_{t\geq 0}$.

If $E = L^1(X, \Sigma, m)$ and $\{P(t)\}_{t\geq 0}$ is a *C_0-semigroup* of stochastic (sub-stochastic) operators on E then $\{P(t)\}_{t\geq 0}$ is called, respectively, a *stochastic (substochastic) semigroup*. We can also define a nonlinear stochastic semigroup. A family $\{P(t)\}_{t\geq 0}$ of nonlinear stochastic operators is called a *nonlinear stochastic semigroup* if it satisfies conditions (a), (b), and the map $(t, f) \mapsto P(t)f$ is continuous. We recall that in the definition of a nonlinear stochastic operator P we only require that it is defined on the set of densities.

Now we give some examples of stochastic operators and semigroups.

2.2 Frobenius–Perron Operator

Stochastic operators were introduced to study properties of dynamical systems. A dynamical system is a measurable transformation $S\colon X \to X$, where (X, Σ) is a measurable space. We are interested in the behaviour of trajectories

$\{x, S(x), S^2(x), \dots\}$ of points $x \in X$. If the system is chaotic, then it is very difficult to describe the behaviour of a single trajectory. Instead of this we can study stochastic properties of this system. That is, we choose a probability measure μ on (X, Σ) and observe the evolution of this measure under the action of the system. For example, if we start with the probability measure concentrated at point x, i.e. the Dirac measure δ_x, then under the action of the system we obtain the measure $\delta_{S(x)}$. In general, if a measure μ describes the distribution of points in the phase space X, then the measure ν given by the formula $\nu(A) = \mu(S^{-1}(A))$ describes the distribution of points after the action of the transformation S. In typical spaces we have some standard measure m, e.g., the Lebesgue on the space $\mathbb{R}^d$ or the counting measure on a countable space. Then we can only consider initial measures μ which are absolutely continuous with respect to m, i.e., there exists a density f such that $\mu(A) = \int_A f(x)\, m(dx)$, for $A \in \Sigma$. If the measure ν is also absolutely continuous with respect to m, and $g = d\nu/dm$, then we define an operator P_S by $P_S f = g$. This operator corresponds to the transition probability function $\mathscr{P}(x, A)$ on (X, Σ) given by

$$\mathscr{P}(x, A) = \begin{cases} 1, \text{ if } \mathrm{S}(x) \in A, \\ 0, \text{ if } \mathrm{S}(x) \notin A. \end{cases} \tag{4}$$

The operator P_S is correctly defined if the transition probability function $\mathscr{P}(x, A)$ satisfies condition (1). Now condition (1) takes the form

$$m(A) = 0 \Longrightarrow m(S^{-1}(A)) = 0 \text{ for } A \in \Sigma \tag{5}$$

and the transformation S which satisfies (5) is called *non-singular*. This operator can be extended to a bounded linear operator $P_S \colon L^1 \to L^1$, and P_S is a stochastic operator. The operator is called the *Frobenius–Perron operator* or the *transfer operator* or the *Ruelle operator*.

Now we give a formal definition of the Frobenius–Perron operator. Let (X, Σ, m) be a σ-finite measure space and let S be a measurable nonsingular transformation of X. An operator $P_S \colon L^1 \to L^1$ which satisfies the following condition

$$\int_A P_S f(x)\, m(dx) = \int_{S^{-1}(A)} f(x)\, m(dx) \text{ for } A \in \Sigma \text{ and } f \in L^1 \tag{6}$$

is called the *Frobenius–Perron operator* for the transformation S. The adjoint of the Frobenius–Perron operator $P^* \colon L^\infty \to L^\infty$ is given by $P^* g(x) = g(S(y))$ and is called the *Koopman operator* or the *composition operator*.

Observe that there are measurable transformations which do not satisfy (5). For example, if m is a Lebesgue measure on $X = \mathbb{R}^n$ and $S(x) = a$ for all $x \in X$, then $S^{-1}(A) = X$ if $A = \{a\}$ and (5) does not hold. It means that the Frobenius–Perron operator may not exist even for smooth transformations. It is not easy to check condition (5) and to find the Frobenius–Perron operator for an arbitrary transformation. We consider only piecewise smooth transformations of subsets of $\mathbb{R}^d$. Let X be a subset of $\mathbb{R}^d$ with nonempty interior and with the boundary of zero Lebesgue measure. Let $S\colon X \to X$ be a measurable transformation. We assume that there exists pairwise disjoint open subsets $U_1,\dots,U_n$ of X having the following properties:

(a) the sets $X_0 = X \setminus \bigcup_{i=1}^n U_i$ and $S(X_0)$ have zero Lebesgue measure,

(b) maps $S_i = S\Big|_{U_i}$ are diffeomorphisms from U_i onto $S(U_i)$, i.e., S_i are C^1 and invertible transformations and $\det S_i'(x) \neq 0$ at each point $x \in U_i$.

Then transformations $\varphi_i = S_i^{-1}$ are also diffeomorphisms from $S(U_i)$ onto U_i. Then the Frobenius–Perron operator P_S exists and is given by the formula

$$P_S f(x) = \sum_{i \in I_x} f(\varphi_i(x))|\det \varphi_i'(x)|, \tag{7}$$

where $I_x = \{i\colon \varphi_i(x) \in U_i\}$. Indeed,

$$\begin{aligned}\int_{S^{-1}(A)} f(x)\,dx &= \sum_{i=1}^n \int_{S^{-1}(A)\cap U_i} f(x)\,dx = \sum_{i=1}^n \int_{\varphi_i(A)} f(x)\,dx \\ &= \sum_{i=1}^n \int_{A\cap S(U_i)} f(\varphi_i(x))|\det \varphi_i'(x)|\,dx \\ &= \int_A \sum_{i\in I_x} f(\varphi_i(x))|\det \varphi_i'(x)|\,dx = \int_A P_S f(x)\,dx.\end{aligned}$$

Example 1 Let $S\colon [0,1] \to [0,1]$ be the transformation given by

$$S(x) = \begin{cases} 2x, & \text{for } x \in [0,1/2], \\ 2-2x, & \text{for } x \in (1/2,1]. \end{cases} \tag{8}$$

The transformation S is called the *tent map*. We have $U_1 = (0,1/2)$, $U_2 = (1/2,1)$ and the maps $\varphi_i\colon (0,1) \to (0,1)$ are given by $\varphi_1(x) = \frac{1}{2}x$ i $\varphi_2(x) = 1 - \frac{1}{2}x$. Thus the Frobenius–Perron operator P_S is of the form

$$P_S f(x) = \tfrac{1}{2} f(\tfrac{1}{2}) + \tfrac{1}{2} f(1 - \tfrac{1}{2}x). \tag{9}$$

Frobenius–Perron operators can be successfully used to study ergodic properties of transformations [55]. Before formulating results concerning this subject we recall some definitions. Let $S\colon X \to X$ be a measurable transformation of some measurable space (X,Σ). A measure μ defined on σ-algebra Σ is called *invariant* with respect to the transformation S if $\mu(S^{-1}(A)) = \mu(A)$ for every set $A \in \Sigma$. An invariant measure μ is called *ergodic* if $\mu(A) = 0$ or $\mu(X \setminus A) = 0$ for any measurable set A such that $A = S^{-1}(A)$. Ergodic transformations have many interesting properties. One of them is the following seminal theorem.

Theorem 1 (Birkhoff Ergodic Theorem) *Let $S: X \to X$ be a measurable transformation of (X, Σ, μ) and μ be a probability ergodic measure invariant with respect to S. Then for every $f \in L^1(X, \Sigma, \mu)$*

$$\lim_{T\to\infty} \frac{1}{T} \sum_{n=0}^{T-1} f(S^n(x)) = \int_X f(x)\,\mu(dx) \quad \text{for } \mu\text{- a.e. } x. \tag{10}$$

If $f = 1_A$ then the formula has an interesting interpretation. It takes the form

$$\lim_{T\to\infty} \frac{\#\{n \in \{0, \dots, T-1\}: S^n(x) \in A\}}{T} = \mu(A) \quad \text{for } \mu\text{- a.e. } x. \tag{11}$$

where $\#E$ is the number of elements of E. Formula (11) means that the average time spent in a measurable set A by the trajectory $\{x, S(x), S^2(x), \dots\}$ of almost every point x equals the measure of this set.

A stronger property than ergodicity is mixing. Let (X, Σ, μ, S) be a dynamical system with an invariant probability measure μ. This system is called *mixing* if

$$\lim_{n\to\infty} \mu(A \cap S^{-n}(B)) = \mu(A)\mu(B) \quad \text{for all } A, B \in \Sigma. \tag{12}$$

If $P = \mu$ and $P(B) > 0$ then condition (12) can be written in the following way

$$\lim_{n\to\infty} P(S^n(x) \in A | x \in B) = \mu(A) \quad \text{for all } A \in \Sigma,$$

which means that the trajectory of almost all points enters a set A with asymptotic probability $\mu(A)$. The stronger property than mixing is exactness. A system (X, Σ, μ, S) with a double measurable transformation S, i.e. $S(A) \in \Sigma$ and $S^{-1}(A) \in \Sigma$ for all $A \in \Sigma$, and an invariant probability measure μ is called *exact* if for every set $A \in \Sigma$ with $\mu(A) > 0$ we have $\lim_{n\to\infty} \mu(S^n(A)) = 1$. Observe that a mixing dynamical system (X, Σ, μ, S) with the double measurable and invertible transformation S is not exact because

$$\mu(S^n(A)) = \mu(S^{-n}(S^n(A))) = \mu(A).$$

For example, if $X = [0, 1]^2$, $\Sigma = \mathcal{B}(X)$, m is the Lebesgue measure, and $S: X \to X$ is the *baker transformation* given by

$$S(x, y) = \begin{cases} (2x, \frac{1}{2}y), & \text{for } x \in [0, \frac{1}{2}],\ y \in [0, 1], \\ (2x - 1, \frac{1}{2}y + \frac{1}{2}), & \text{for } x \in (\frac{1}{2}, 1],\ y \in [0, 1], \end{cases} \tag{13}$$

then the system is mixing but not exact.

Table 1 The relations between ergodic properties of the dynamical system (X, Σ, μ, S) and the Frobenius–Perron operator P_S

μ	f_*
Invariant	$P_S f_* = f_*$
Ergodic	f_* is a unique fixed point of P_S in D
Mixing	$\text{w-lim}_{n\to\infty} P_S^n f = f_*$ for every $f \in D$
Exact	$\lim_{n\to\infty} P_S^n f = f_*$ for every $f \in D$

Now let (X, Σ, m) be a σ-finite measure space, S be a measurable nonsingular transformation of X, and $P_S\colon L^1 \to L^1$ be the Frobenius–Perron operator. Let μ be a given probability measure absolutely continuous with respect to m and let f_* be the density of μ. Then the measure μ is invariant with respect to S iff $P_S f_* = f_*$; μ is ergodic iff f_* is a unique fixed point of P_S in the set of densities; μ is mixing iff for every $f \in D$ the density f_* is the weak limit of $P_S^n f$ as $n \to \infty$; μ is exact iff $\lim_{n\to\infty} P_S^n f = f_*$ for every $f \in D$. By the weak limit $\lim_{n\to\infty} P_S^n f$ we understand a function $h \in L^1$ such that for every $g \in L^\infty$ we have

$$\int_X P_S^n f(x)g(x)\, m(dx) = \int_X h(x)g(x)\, m(dx)$$

and we denote it by $\text{w-lim}_{n\to\infty} P_S^n f$. We collect the relations between ergodic properties of a dynamical system and the behavior of the Frobenius–Perron operator in Table 1.

2.3 *Iterated Function System*

In many applications, especially, in construction of fractals and in the methods of image compression, we use iterated function systems. An *iterated function system* on a complete metric space (X, ρ) is a sequence of maps $S_1, \dots, S_n$, where $S_i\colon X \to X$, for $i = 1, \dots, n$. Having this system we can define a transformation on some space $H(X)$ of subsets of X or on the set of probability measures on some σ-algebra of subsets of X [12, 42, 58]. In the first case we usually assume that maps S_i are contractions, $H(X)$ consists of all compact subsets of X and the map $F\colon H(X) \to H(X)$ is given by

$$F(A) = S_1(A) \cup \dots \cup S_n(A).$$

If we introduce the Hausdorff metric on $H(X)$ given by

$$h(A, B) = \max_{x\in A} \rho(x, B) + \max_{x\in B} \rho(x, A),$$

where $\rho(x, A) = \min\{\rho(x, y) : y \in A\}$, then $(H(X), h)$ is a complete metric space and the map F is a contraction. From the Banach contraction principle it follows that there exists a unique compact subset A_* of X such that $F(A_*) = A_*$ and we have $\lim_{k\to\infty} F^k(A) = A_*$ for each compact set A. In construction of fractals we usually assume that $X = \mathbb{R}^2$ and S_i are contractive similarity transformations, and from the formula $F(A_*) = A_*$ it follows that the limit set A_* is made up of the union of several smaller copies of itself, i.e., of the sets $S_i(A_*)$. It is why the set A_* has a self-similar fractal nature.

Since it is rather difficult, to iterate sets and go to the limit set, in practice, we construct fractals in a different way. Let $p_1(x), \dots, p_n(x)$ be non-negative continuous functions defined on X such that $p_1(x) + \cdots + p_n(x) = 1$ for all $x \in X$. We consider the following process. Take a point x_0. We choose a transformation S_i with probability $p_i(x_0)$ and we go to the point $x_1 = S_i(x_0)$. The point x_1 describes the new state of the iterated function system. We repeat the procedure but now with the initial point x_1, etc. In this way we obtain a random sequence of points (x_n). Precisely, for any initial point $x_0 \in X$ we can find a Markov process $\xi_n^{x_0}$, $n \in \mathbb{N}$, on X such that $\xi_0^{x_0} = x_0$ and with the transition probability function

$$\mathcal{P}(x, B) = \sum_{i \in I_B(x)} p_i(x), \quad \text{where } B \in \mathcal{B}(X) \text{ and } I_B(x) = \{i : S_i(x) \in B\}.$$

Under suitable assumptions on functions p_i, for every initial point $x_0 \in X$ the limit set of almost every sample path $\xi_n^{x_0}(\omega)$, $n \in \mathbb{N}$, is A_*. We recall that the limit set of a sequence is the set of all its accumulation points. Observe that if we start from the point a, i.e., the initial state is described by the Dirac measure δ_a, then the new state of the system is described by the probability measure $\sum_{i=1}^n p_i(a)\delta_{S_i(a)}$. If the initial state is random and described by a probability measure μ then the next state is given by the probability measure

$$P\mu(A) = \sum_{i=1}^n \int_{S_i^{-1}(A)} p_i(x)\,\mu(dx). \tag{14}$$

In this way we obtain a stochastic operator on probability measures. Under suitable assumptions on functions p_i (see e.g. [58]), the sequence of measures $P^k\mu$ is weakly convergent to a probability measure μ_* as $k \to \infty$ and the set A_* is the topological support of μ_*.

Though, in the theory of fractals it is more convenient to consider stochastic operators on measures, we can also study stochastic operators on densities corresponding to iterated function systems. Now, we assume that $S_1, \dots, S_n$ are non-singular transformations of the space (X, Σ, m). Let $P_1, \dots, P_n$ be the Frobenius–Perron operators corresponding to the transformations $S_1, \dots, S_n$. If the measure μ is absolutely continuous with respect to m and $f = d\mu/dm$, then from (14) it follows

$$P\mu(A) = \sum_{i=1}^n \int_{S_i^{-1}(A)} p_i(x) f(x)\, m(dx) = \sum_{i=1}^n \int_A P_i(p_i f)(x)\, m(dx),$$

which means that the measure $P\mu$ has the density $\sum_{i=1}^{n} P_i(p_i f)$. Thus, the evolution of densities is described by the stochastic operator

$$Pf = \sum_{i=1}^{n} P_i(p_i f).$$

2.4 Integral Stochastic Operators

Let (X, Σ, m) be a σ-finite measure space. If $k: X \times X \to [0, \infty)$ is a measurable function such that

$$\int_X k(x, y)\, m(dx) = 1$$

for almost all $y \in X$, then

$$Pf(x) = \int_X k(x, y) f(y)\, m(dy) \tag{15}$$

is a stochastic operator. The function k is called the *kernel* of the operator P and P is called an *integral* or *kernel* operator.

In finite or countable spaces all stochastic operators are integral. We consider, for example, the space $(\mathbb{N}, \Sigma, m)$, where $\mathbb{N} = \{0, 1, \ldots\}$, $\Sigma = 2^{\mathbb{N}}$ is the σ-algebra of all subsets of $\mathbb{N}$ and m is the counting measure, i.e., $m(A)$ is the number of elements of A. Any function $f: \mathbb{N} \to \mathbb{R}$ is represented as a sequence $x = (x_i)_{i \in \mathbb{N}}$. Thus, the integral in this space is given by

$$\int_{\mathbb{N}} x_i\, m(di) = \sum_{i=0}^{\infty} x_i$$

and a sequence $x = (x_i)_{i \in \mathbb{N}}$ is integrable iff

$$\sum_{i=0}^{\infty} |x_i| < \infty. \tag{16}$$

We use the notation $l^1 = L^1(\mathbb{N}, \Sigma, m)$. The elements of l^1 are real valued sequences $x = (x_i)_{i \in \mathbb{N}}$ satisfying (16) and the norm is given by $\|x\| = \sum_{i=0}^{\infty} |x_i|$. Let us observe that an arbitrary stochastic operator $P: l^1 \to l^1$ is an integral operator. Indeed, for each i the function $x \mapsto (Px)_i$ is a continuous linear functional from l^1 to $\mathbb{R}$. Thus, for each $i \in \mathbb{N}$ there is a sequence $(p_{ij})_{j \in \mathbb{N}} \in l^{\infty}$ such that

$$(Px)_i = \sum_{j=0}^{\infty} p_{ij} x_j = \int_{\mathbb{N}} p_{ij} x_j\, m(dj).$$

Moreover, the operator P is represented by a *stochastic matrix* $[p_{ij}]$, i.e., a matrix with nonnegative entries p_{ij} with each column summing to one: $\sum_{i=0}^{\infty} p_{ij} = 1$ for each $j \in \mathbb{N}$.

Many biological and physical processes can be modelled by means of stochastic integral operators. Some important examples are obtained by some random perturbation of dynamical systems. Let S be a (deterministic) dynamical system on a metric space X and assume that this system is perturbed by an external source of noise. The noise can be, for example, additive, i.e., $x_{n+1} = S(x_n) + \xi_n$ or multiplicative, i.e., $x_{n+1} = S(x_n)\xi_n$. Such systems are generally of the form

$$x_{n+1} = S(x_n, \xi_n), \tag{17}$$

where $(\xi_n)_{n=0}^{\infty}$ is a sequence of independent random variables (or elements) with the same distribution and the initial value of the system x_0 is independent of the sequence $(\xi_n)_{n=0}^{\infty}$. Studying systems of the form (17) we are often interested in the behaviour of the sequence of the measures (μ_k) defined by

$$\mu_n(A) = \text{Prob}\,(x_n \in A), \tag{18}$$

where $A \in \Sigma$, and Σ is the σ-algebra of Borel subset of X. The evolution of these measures can be described by a stochastic operator P on the space of probability Borel measures given by $\mu_{n+1} = P\mu_n$. Let m be a given Borel measure on the phase space X. Assume that for almost all y the distribution ν_y of the random variable $S(y, \xi_n)$ is absolutely continuous with respect to m. Let $k(x, y)$ be the density of ν_y and the operator P be given by (15). Then P is a stochastic integral operator. If the measure μ_0 is absolutely continuous with respect to m and has the density f, then the measures μ_n are also absolutely continuous with respect to m and has the density $P^n f$. It means that the operator P describes the evolution of the system (17).

2.5 Continuous Time Markov Chain

Now we give some examples of stochastic semigroups. The simplest example are stochastic semigroups on finite spaces. Such semigroups appears in the theory of continuous time Markov chains.

Now, we consider a measure space (X, Σ, m) with $X = \{1, 2, \ldots, n\}$, $\Sigma = 2^X$ and the measure m given by

$$m(A) = \sum_{i \in A} p_i, \tag{19}$$

where $(p_1, \ldots, p_n)$ is a sequence with positive terms p_i. In this space the function $f: X \to \mathbb{R}$ is represented as a sequence $y = (y_1, \ldots, y_n)$, the integral of y over X is given by $\sum_{i=1}^{n} y_i p_i$. Thus, the space $L^1(X)$ is isomorphic to the space $\mathbb{R}^n$ with the norm $\|y\| = |y_1|p_1 + \cdots + |y_n|p_n$. Now we construct a stochastic semigroup which corresponds to the system of linear differential equations

$$y_i'(t) = \sum_{j=1}^{n} a_{ij} y_j(t), \quad \text{dla} \quad i = 1, \ldots, n. \tag{20}$$

We assume that the coefficients a_{ij} have the following properties

(i) $a_{ij} \geq 0$ for $i \neq j$,
(ii) $\sum_{i=1}^{n} p_i a_{ij} = 0$ for $j = 1, \ldots, n$.

Let $y = (y_1, \ldots, y_n)$ and let $P(t)y = y(t)$ for $t \geq 0$, where $y(t) = (y_1(t), \ldots, y_n(t))$ is the solution of (20) with the initial condition $y(0) = y$. We show that $\{P(t)\}_{t\geq 0}$ is a stochastic semigroup. First, we check that if y is a sequence with nonnegative terms, then $y(t)$ is a sequence with nonnegative terms for $t > 0$. In order to do it we define $\lambda = -\min\{a_{ii} : i = 1, \ldots, n\}$, $b_{ij} = a_{ij}$ for $i \neq j$ and $b_{ii} = a_{ii} + \lambda$. Then $b_{ij} \geq 0$ for all i, j. Let $B = [b_{ij}]$. Then the function $y(t)$ satisfies equation

$$y'(t) = -\lambda y(t) + By(t). \tag{21}$$

The solution of this equation can be written in the form

$$y(t) = e^{-\lambda t} e^{Bt} y, \tag{22}$$

where

$$e^{Bt} = \sum_{k=0}^{\infty} \frac{t^k B^k}{k!}. \tag{23}$$

Since the matrix B has nonnegative entrances, also the matrix e^{Bt} has nonnegative entrances. From formula (22) it follows immediately that $y(t)$ is a nonnegative sequence for $t > 0$. We check that $\sum_{i=1}^{n} p_i y_i(t)$ does not depend on t:

$$\frac{d}{dt}\left(\sum_{i=1}^{n} p_i y_i(t)\right) = \sum_{i=1}^{n}\sum_{j=1}^{n} p_i a_{ij} y_j(t) = \sum_{j=1}^{n}\left(\sum_{i=1}^{n} p_i a_{ij}\right) y_j(t) = 0.$$

Thus, we have shown that if $y \in D$, then $y(t) \in D$, therefore, $P(t)(D) \subset D$ for $t > 0$. Since (20) is a homogeneous linear system of differential equations, $P(t)$ is a linear operator. Hence $P(t)$ is a stochastic operator for $t > 0$. Now we check conditions (a), (b), (c) of the definition of a stochastic semigroup. Condition (a) is obvious because $P(0)y = y(0) = y$. Condition (b) is due to the fact that (20) is an autonomous system (its right-hand side does not depend on t). Condition (c) is a simple consequence of the continuity of solutions. Thus, $\{P(t)\}_{t\ge 0}$ is a stochastic semigroup.

In a special case when m is a counting measure, i.e., $p_1 = \cdots = p_n = 1$, equation (20) generates a stochastic semigroup iff

(i) $a_{ij} \ge 0$ for $i \ne j$,
(ii) $\sum_{i=1}^{n} a_{ij} = 0$ for $j = 1, \ldots, n$.

2.6 Uniformly Continuous Stochastic Semigroups

The generator of the semigroup defined in 2.5 is a bounded operator A and it is of the form $A = -\lambda I + B$, where B is a positive and bounded operator. Generally, if the generator A of a C_0-semigroup $\{P(t)\}_{t\ge 0}$ in a Banach space E is bounded then the semigroup $\{P(t)\}_{t\ge 0}$ is *uniformly continuous*, i.e.,

$$\lim_{t\to t_0} \|P(t) - P(t_0)\| = 0 \text{ for } t_0 \ge 0. \tag{24}$$

This result is a simple consequence of the formula

$$P(t) = e^{At} = \sum_{k=0}^{\infty} \frac{t^k A^k}{k!}. \tag{25}$$

It is also a well known fact that the generator of a uniformly continuous semigroup is a bounded operator and that a bounded operator A is a generator of a positive semigroup on a Banach lattice iff $A + \|A\| I \ge 0$ (see e.g. Theorem 1.11, p. 255 [3]). In particular, if the operator A is a generator of a uniformly continuous stochastic semigroup $\{P(t)\}_{t\ge 0}$ then there exist a bounded and positive operator B and $\lambda \ge 0$ such that $A = -\lambda I + B$. On the other hand, since $P(t) = I + tA + o(t)$ and $P(t)$ preserves the integral we obtain

$$\int_X Af(x)\, m(dx) = 0 \quad \text{for } f \in L^1. \tag{26}$$

Thus $\int_X Bf(x)\, m(dx) = \lambda$ for $f \in D$. Assume that $\lambda \ne 0$ and let $P = B/\lambda$. Then $A = -\lambda I + \lambda P$, where P is a stochastic operator. Therefore, a generator of a uniformly bounded stochastic semigroup $\{P(t)\}_{t\ge 0}$ is of the form $A = -\lambda I + \lambda P$. Let $u_0 \in L^1$ and $u(t) = P(t)u_0$. Then $u(t)$ satisfies the evolution equation

$$u'(t) = -\lambda u(t) + \lambda P u(t). \tag{27}$$

For every $u_0 \in L^1$ equation (27) has the solution of the form

$$u(t) = \sum_{k=0}^{\infty} \frac{(\lambda t)^k e^{-\lambda t}}{k!} P^k u_0. \tag{28}$$

Since $u(t)$ is an element of L^1 we often use notation $u(t,x) = u(t)(x)$ and write equation (26) as the partial differential equation

$$\frac{\partial u(t,x)}{\partial t} = -\lambda u(t,x) + \lambda P u(t,x). \tag{29}$$

2.7 Generators of Substochastic and Stochastic Semigroups

As we have observed in 2.6, a bounded linear operator $A\colon L^1 \to L^1$ is a generator of a stochastic semigroup iff

$$A + \lambda I \geq 0 \text{ for some } \lambda > 0 \quad \text{and} \quad \int_X Af(x)\, m(dx) = 0 \text{ for all } f \in L^1. \tag{30}$$

Generally, if an operator $A\colon \mathscr{D}(A) \to L^1$, $\mathscr{D}(A) \subset L^1$, is a generator of a stochastic semigroup then $\int_X Af(x)\, m(dx) = 0$ for $f \in \mathscr{D}(A)$. If a linear operator $A\colon \mathscr{D}(A) \to L^1$ is unbounded and $\int_X Af(x)\, m(dx) = 0$ then condition $A + \lambda I \geq 0$ does not hold for any $\lambda > 0$ and the question whether this operator is a generator of a stochastic semigroup is highly non-trivial. We only recall here some needful results. The interested reader is referred to [6, 8, 82, 112].

Usually, in applications, we have an operator of the form $A + B$ and we know that A is a generator of a stochastic or substochastic semigroup and B is another linear operator and we want to check whether $A + B$ also generates a stochastic or substochastic semigroup. One of the answer on this questions the following Kato–Voigt–Banasiak theorem.

Theorem 2 *Assume that $(A, \mathscr{D}(A))$ is the generator of a substochastic semigroup on L^1 and $B\colon \mathscr{D}(A) \to L^1$ is a positive operator such that*

$$\int_E (Af(x) + Bf(x))\, m(dx) \leq 0 \quad \textit{for} \quad f \in \mathscr{D}(A),\ f \geq 0. \tag{31}$$

Then for each $r \in (0,1)$ the operator $(A + rB, \mathscr{D}(A))$ is the generator of a substochastic semigroup $\{P_r(t)\}_{t\geq 0}$ on L^1 and the family of operators $\{P(t)\}_{t\geq 0}$ defined by

$$P(t)f = \lim_{r\to 1^-} P_r(t)f, \quad f \in L^1,\ t > 0,$$

is a substochastic semigroup on L^1 with generator $(C, \mathscr{D}(C))$ being an extension of the operator $(A + B, \mathscr{D}(A))$:

$$\mathscr{D}(A) \subseteq \mathscr{D}(C) \quad \text{and} \quad Cf = Af + Bf \quad \text{for } f \in \mathscr{D}(A).$$

The semigroup $\{P(t)\}_{t\geq 0}$ is a minimal semigroup related to $A+B$, i.e., if $\{T(t)\}_{t\geq 0}$ is another positive semigroup generated by an extension of $(A + B, \mathscr{D}(A))$ then $T(t)f \geq P(t)f$ for all $f \in \mathscr{D}(A)$, $f \geq 0$.

The problem when the minimal substochastic semigroup defined in Theorem 2 is stochastic was studied in [7, 8, 108, 109]. In particular we have the following characterization.

Theorem 3 *Assume that $(A, \mathscr{D}(A))$ is the generator of a substochastic semigroup on L^1 and $B\colon \mathscr{D}(A) \to L^1$ is a positive operator such that*

$$\int_E (Af(x) + Bf(x))\, m(dx) = 0 \quad \text{for} \quad f \in \mathscr{D}(A),\ f \geq 0. \tag{32}$$

Let $\lambda > 0$ and $R(\lambda, A) = (\lambda I - A)^{-1}$. Then the following conditions are equivalent:

1. *The minimal semigroup $\{P(t)\}_{t\geq 0}$ related to $A + B$ is stochastic.*
2. *The generator of $\{P(t)\}_{t\geq 0}$ is the closure of $(A + B, \mathscr{D}(A))$.*
3. *If for some $f \in L^\infty$, $f \geq 0$, we have $(BR(\lambda, A))^* f = f$ then $f = 0$, where $(BR(\lambda, A))^*$ denotes the adjoint of $BR(\lambda, A)$.*

If A is a generator of a stochastic semigroup $\{S(t)\}_{t\geq 0}$ and $B\colon L^1 \to L^1$ is a bounded operator such that $\int_X Bf(x)\, dx = 0$ for $f \in L^1$, then from condition (2) of Theorem 3 it follows that $A+B$ is a generator of a stochastic semigroup $\{P(t)\}_{t\geq 0}$. The semigroup $\{P(t)\}_{t\geq 0}$ can be given by the Dyson–Phillips expansion

$$P(t)f = \sum_{n=0}^{\infty} S_n(t)f,$$

where

$$S_0(t)f = S(t)f, \quad S_{n+1}(t)f = \int_0^t S_n(t-s)BS(s)f\, ds, \quad n \geq 0.$$

A special role in applications is played by generators of the form $A+\lambda K-\lambda I$, where A is a generator of a stochastic semigroup $\{S(t)\}_{t\geq 0}$, K is a stochastic operator and $\lambda > 0$. The semigroup $\{P(t)\}_{t\geq 0}$ generated by $A + \lambda K - \lambda I$ is of the form

$$P(t)f = e^{-\lambda t} \sum_{n=0}^{\infty} \lambda^n S_n(t)f, \tag{33}$$

where

$$S_0(t)f = S(t)f, \quad S_{n+1}(t)f = \int_0^t S_n(t-s)KS(s)f\,ds, \quad n \ge 0. \tag{34}$$

2.8 Stochastic Semigroups on l^1

Now we study stochastic semigroups on the space $l^1 = L^1(\mathbb{N}, 2^{\mathbb{N}}, m)$, where m is the counting measure, and the elements of l^1 are real valued sequences $x = (x_i)_{i\in\mathbb{N}}$ such that $\sum_{i=0}^{\infty} |x_i| < \infty$. Any linear operator (bounded or unbounded) is given by an infinite dimensional matrix $Q = [q_{ij}]$, i.e., $(Qx)_i = \sum_{j=0}^{\infty} q_{ij}x_j$ for $j \in \mathbb{N}$. The question is what are necessary and sufficient conditions for the matrix Q to be a generator of a stochastic semigroup? One can expect that $Q = [q_{ij}]$ should be a *Kolmogorov matrix*, i.e., its entries have the following properties

(i) $q_{ij} \ge 0$ for $i \ne j$,
(ii) $\sum_{i=0}^{\infty} q_{ij} = 0$ for $j = 0, 1, 2, \dots$.

Indeed, if Q is a Kolmogorov matrix and the operator $Q: l^1 \to l^1$ is bounded, then Q is a generator of a stochastic semigroup $\{P(t)\}_{t\ge 0}$ given by $P(t) = e^{Qt}$. The matrix Q defines a bounded operator on l^1 if there exists a constant c such that for each j we have $\sum_{i=0}^{\infty} |q_{ij}| < c$. If Q is a Kolmogorov matrix then the boundedness of Q is equivalent to

$$\sup_{j\in\mathbb{N}} |q_{jj}| < \infty. \tag{35}$$

The problem when the unbounded operator given by a Kolmogorov matrix Q is is a generator of a stochastic semigroup is more complex and can be solved by using Theorems 2 and 3. In order to do it we need an auxiliary notion. A matrix $Q = [q_{ij}]$ is called a *sub-Kolmogorov matrix* if it satisfies condition (i) and the condition

(ii') $\sum_{i=0}^{\infty} a_{ij} \le 0$ for $j = 0, 1, 2, \dots$.

Corollary 1 *Let Q be a sub-Kolmogorov matrix. Then there is the minimal substochastic semigroup $\{P(t)\}_{t\ge 0}$ related to Q.*

Proof Let

$$\mathscr{D}_0(Q) = \{x \in l^1 : \sum_{j=0}^{\infty} |q_{jj}||x_j| < \infty\}. \tag{36}$$

The set $\mathscr{D}_0(Q)$ is dense in the space l^1 and the matrix Q defines a linear operator on $\mathscr{D}_0(Q)$ with values in l^1. Let A be the diagonal part of Q, i.e., $A = [a_{ij}]$, $a_{jj} = q_{jj}$ and $a_{ij} = 0$ for $i \ne j$, and let $B = [b_{ij}]$ be the off-diagonal part of Q, i.e.,

$B = Q - A$. The operator A with domain $\mathscr{D}(A) = \mathscr{D}_0(Q)$ is the generator of a substochastic semigroup $\{P_0(t)\}_{t\geq 0}$ on l^1 given by $(P_0(t)x)_i = x_i e^{-a_{ii}t}$ for $i \in \mathbb{N}$. The operator $B: \mathscr{D}(A) \to l^1$ is positive. By Theorem 2, there is the minimal substochastic semigroup $\{P(t)\}_{t\geq 0}$ related to Q. □

Now observe that in our case condition (3) of Theorem 3 can be written as $B^*x = (\lambda I - A)^*x$ or equivalently, $Q^*x = \lambda x$, where the symbol C^* denotes the transpose of the matrix C. If Q is a Kolmogorov matrix, condition (32) holds and from Theorem 3 we obtain the following Kato result [46].

Theorem 4 *Let Q be a Kolmogorov matrix and let $\lambda > 0$ be a positive constant. We denote $Q^* = (q^*_{i,j})_{i,j\geq 1}$, where $q^*_{i,j} = q_{j,i}$ for $i, j \geq 1$ The minimal semigroup related to Q is a stochastic semigroup on l^1 iff the equation $Q^*x = \lambda x$ has no nonzero solution $x \in l^\infty$ and $x \geq 0$.*

If the minimal semigroup $\{P(t)\}_{t\geq 0}$ related to a Kolmogorov matrix A is a stochastic semigroup then the matrix A is called *non-explosive*.

2.9 Continuity Equation

Now, we move on to semigroups related to partial differential equations. We start with the *continuity equation* called also the *transport equation* and the *Liouville equation*.

Consider a moving particle in an open set $G \subset \mathbb{R}^d$. We assume that if a particle is at point x then its velocity is $b(x)$. It means that if $x(t)$ is its position at time t then the function $x(t)$ satisfies the following equation

$$x'(t) = b(x(t)). \tag{37}$$

We also assume that the particle does not leave the set G. We choose the initial position x of the particle randomly with a density distribution function u_0. If $u(t,x)$ is the density of distribution of $x(t)$ then u satisfies the following equation

$$\frac{\partial u(t,x)}{\partial t} = -\operatorname{div}(b(x)u(t,x)), \tag{38}$$

where

$$\operatorname{div}(b(x)u(t,x)) = \sum_{i=1}^{d} \frac{\partial}{\partial x_i}(b_i(x)u(t,x)). \tag{39}$$

Indeed, given a domain $D \subset G$ with the smooth boundary S, consider the fluxes into the set D in the time interval of the length Δt:

$$I(\Delta t) = \int_D u(t + \Delta t, x)\,dx - \int_D u(t,x)\,dx. \tag{40}$$

Since the fluxes are through the surface S and since the speed at which particles cross the surface is $-n(x) \cdot b(x)$, where $n(x)$ is the outward-pointing unit normal vector to S, we have

$$I(\Delta t) = -\Delta t \int_S (n(x) \cdot b(x)u(t,x))\,d\sigma(x) + o(\Delta t). \tag{41}$$

According to the Gauss–Ostrogradski theorem we have

$$\int_S (n(x) \cdot b(x)u(t,x))\,d\sigma(x) = \int_D \operatorname{div}(b(x)u(t,x))\,dx. \tag{42}$$

Equations (40), (41) and (42) imply (38). Equation (38) generates a stochastic semigroup given by $P(t)u_0(x) = u(t,x)$.

The semigroup $\{P(t)\}_{t\ge 0}$ can be given explicitly. Namely, for each $\bar{x} \in G$ we denote by $\pi_t\bar{x}$ the solution $x(t)$ of (37) with the initial condition $x(0) = \bar{x}$. Let us fix $t > 0$ and define a transformation $S\colon G \to G$ by $S(\bar{x}) = \pi_t\bar{x}$. Then S is a nonsingular and invertible transformation and according to (7) the Frobenius–Perron operator corresponding to the map S is given by

$$P(t)f(x) = P_S f(x) = \begin{cases} f(\pi_{-t}x)\det\left[\dfrac{d}{dx}\pi_{-t}x\right], & \text{if } x \in \pi_t(G),\\ 0, \text{ if } x \notin \pi_t(G). \end{cases}$$

The adjoint semigroup $\{P^*(t)\}_{t\ge 0}$ of Koopman operators $P^*(t)\colon L^\infty \to L^\infty$ is given by $P^*(t)f(x) = f(\pi_t x)$. This semigroup $\{P^*(t)\}_{t\ge 0}$ is not strongly continuous but if we choose f sufficiently smooth then the function $u(t,x) = P^*(t)f(x)$ satisfies the following equation

$$\frac{\partial u(t,x)}{\partial t} = \sum_{i=1}^{d} b_i(x)\frac{\partial u(t,x)}{\partial x_i}. \tag{43}$$

2.10 *Diffusion Semigroup*

Semigroups generated by continuity equations are special subclass of stochastic semigroups related to diffusion processes. Consider the *Itô equation* of the form

$$dX_t = \sigma(X_t)\,dW_t + b(X_t)\,dt, \tag{44}$$

where W_t is a m-dimensional Brownian motion, $\sigma(x) = [\sigma_j^i(x)]$ is a $d \times m$ matrix and $b(x)$ is a vector in $\mathbb{R}^d$ for every $x \in \mathbb{R}^d$. We assume that for all $i = 1, \ldots, d$, $j = 1, \ldots, m$ the functions b_i, σ_j^i are sufficiently smooth and have bounded derivatives of all orders, and the function σ_j^i are also bounded. The process X_t, $t \geq 0$ is called *diffusion process*. Recall that the *Stratonovitch* equivalent *equation* is of the form

$$dX_t = \sigma(X_t) \circ dW_t + \sigma_0(X_t)\, dt, \tag{45}$$

where $\sigma_0^i = b_i - \frac{1}{2}\sum_{k=1}^m \sum_{j=1}^d \sigma_k^j \frac{\partial \sigma_k^i}{\partial x_j}$. Assume that X_t is a solution of (44) or (45) such that the distribution of X_0 is absolutely continuous and has the density $v(x)$. Then X_t has also the density $u(t,x)$ and u satisfies the *Fokker–Planck equation* (also called the *Kolmogorov forward equation*)

$$\frac{\partial u}{\partial t} = \sum_{i,j=1}^d \frac{\partial^2 (a_{ij}(x)u)}{\partial x_i \partial x_j} - \sum_{i=1}^d \frac{\partial (b_i(x)u)}{\partial x_i}, \tag{46}$$

where $a_{ij}(x) = \frac{1}{2}\sum_{k=1}^m \sigma_k^i(x)\sigma_k^j(x)$. Equation (46) can be written in another equivalent form

$$\frac{\partial u}{\partial t} = \sum_{i=1}^d \frac{\partial}{\partial x_i}\left(\sum_{j=1}^d a_{ij}(x)\frac{\partial u}{\partial x_j}\right) - \sum_{i=1}^d \frac{\partial(\sigma_0^i(x)u)}{\partial x_i}. \tag{47}$$

Note that the $d \times d$-matrix $a = [a_{ij}]$ is symmetric and nonnegative definite, i.e. $a_{ij} = a_{ji}$ and

$$\sum_{i,j=1}^d a_{ij}(x)\lambda_i\lambda_j \geq 0 \tag{48}$$

for every $\lambda \in \mathbb{R}^d$ and $x \in \mathbb{R}^d$, so we only assume weak ellipticity of the operator on the right hand side of equation (46). Let us consider the operator

$$Af = \sum_{i,j=1}^d \frac{\partial^2 (a_{ij}(x)u)}{\partial x_i \partial x_j} - \sum_{i=1}^d \frac{\partial (b_i(x)u)}{\partial x_i} \tag{49}$$

on the set $E = \{f \in L^1(\mathbb{R}^d) \cap C_b^2(\mathbb{R}^d) : Af \in L^1(\mathbb{R}^d)\}$, where $C_b^2(\mathbb{R}^d)$ denotes the set of all twice differentiable bounded functions whose derivatives of order ≤ 2 are continuous and bounded. If $v \in C_b^2(\mathbb{R}^d)$ then equation (46) has in any time interval $[0,T]$ a unique classical solution u which satisfies the initial condition $u(0,x) = v(x)$ and this solution and its spatial derivatives up to order 2 are uniformly bounded on $[0,T]\times\mathbb{R}^d$ (see [36, 104]). But if the initial function has a compact support, i.e. $v\in C_c^2(\mathbb{R}^d)$, then the solution $u(x,t)$ of (46) and its spatial derivatives converge exponentially to 0 as $\|x\|\to\infty$. From the Gauss–Ostrogradski theorem it follows that the integral $\int u(t,x)\,dx$ is constant. Let $P(t)v(x)=u(t,x)$ for $v\in C_c^2(\mathbb{R}^d)$ and $t\ge 0$. Since the operator $P(t)$ is a contraction on $C_c^2(\mathbb{R}^d)$ it can be extended to a contraction on $L^1(\mathbb{R}^d)$. Thus the operators $\{P(t)\}_{t\ge0}$ form a stochastic semigroup. We have $P(t)(C_c^2(\mathbb{R}^d))\subset C_b^2(\mathbb{R}^d)$ for $t\ge 0$. According to Proposition 1.3.3 of [33] the closure of the operator A generates the semigroup $\{P(t)\}_{t\ge0}$. The adjoint operators $\{P^*(t)\}_{t\ge0}$ form a semigroup on $L^\infty(\mathbb{R}^d)$ given by the formula

$$P^*(t)g(x)=\int_{\mathbb{R}^d} g(y)\mathcal{P}(t,x,dy)\quad \text{for } g\in L^\infty(\mathbb{R}^d),$$

where $\mathcal{P}(t,x,A)$ is the transition probability function for the diffusion process X_t, i.e. $\mathcal{P}(t,x,A)=\operatorname{Prob}(X_t\in A)$ and X_t is a solution of equation (45) with the initial condition $X_0=x$. If g is C^2-function, then the function $u(t,x)=P^*(t)f(x)$ satisfies the *Kolmogorov backward equation*

$$\frac{\partial u}{\partial t}=\sum_{i,j=1}^d a_{ij}(x)\frac{\partial^2 u}{\partial x_i\partial x_j}+\sum_{i=1}^d b_i(x)\frac{\partial u}{\partial x_i}. \tag{50}$$

If we assume that the functions a_{ij} satisfy the uniform elliptic condition

$$\sum_{i,j=1}^d a_{ij}(x)\lambda_i\lambda_j\ge\alpha|\lambda|^2 \tag{51}$$

for some $\alpha>0$ and every $\lambda\in\mathbb{R}^d$ and $x\in\mathbb{R}^d$ then the stochastic semigroup generated by the Fokker–Planck equation (46) is an integral semigroup. That is

$$P(t)f(x)=\int_{\mathbb{R}^d} q(t,x,y)f(y)\,dy,\quad t>0$$

and the kernel q is continuous and positive.

Now we consider degenerate diffusion processes, where instead of (51) we only assume (48). The fundamental theorem on the existence of smooth densities of the transition probability function for degenerate diffusion processes is due to Hörmander. In a series of papers [67, 68] Malliavin has developed techniques, called Malliavin calculus, to give probabilistic proof of this fact. Now we recall some results from this theory. Let $a(x)$ and $b(x)$ be two vector fields on $\mathbb{R}^d$. The *Lie bracket* $[a,b]$ is a vector field given by

$$[a,b]_j(x)=\sum_{k=1}^d\left(a_k\frac{\partial b_j}{\partial x_k}(x)-b_k\frac{\partial a_j}{\partial x_k}(x)\right).$$

We assume Hörmander's condition as in [73]

(H) For every $x \in \mathbb{R}^d$ vectors

$$\sigma_1(x), \ldots, \sigma_m(x),\ [\sigma_i, \sigma_j](x)_{0 \le i,j \le m},\ [\sigma_i, [\sigma_j, \sigma_k]](x)_{0 \le i,j,k \le m}, \ldots$$

span the space $\mathbb{R}^d$. Here $\sigma_1(x), \ldots, \sigma_m(x)$ are column vectors of the diffusion matrix $\sigma(x)$ and $\sigma_0(x)$ is the drift vector which occur in the Stratonovitch equation (45).

Note that the vector σ_0 appears only through brackets.

Theorem 5 (Hörmander) *Under hypothesis* (H) *the transition probability function* $\mathscr{P}(t, x, A)$ *has a density* $k(t, y, x)$ *and* $k \in C^\infty((0, \infty) \times \mathbb{R}^d \times \mathbb{R}^d)$.

There is one important difference between a non-degenerate diffusion and a degenerate diffusion which satisfies condition (H), namely, the kernel k is strictly positive if diffusion is non-degenerate, but in the degenerate case the kernel k can vanish on some subsets. We check where the kernel k is positive using a method based on support theorems [1, 16, 105]. Let $U(x_0, T)$ be the set of all points y for which we can find a $\phi \in L^2([0, T]; \mathbb{R}^m)$ such that there exists a solution of the equation

$$x_\phi(t) = x_0 + \int_0^t \big(\sigma(x_\phi(s))\phi(s) + \sigma_0(x_\phi(s))\big)\, ds \tag{52}$$

satisfying the condition $x_\phi(T) = y$. From the support theorem for diffusion processes it follows that the topological support of the measure $P(T, x_0, \cdot)$ coincides with closure in $\mathbb{R}^d$ of the set $U(x_0, T)$. Let $D_{x_0,\phi}$ be the Frechét derivative of the function $h \mapsto x_{\phi+h}(T)$ from $L^2([0, T]; \mathbb{R}^m)$ to $\mathbb{R}^d$. By $\tilde{U}(x_0, T)$ we denote all points y such that $x_\phi(T) = y$ and the derivative $D_{x_0,\phi}$ has rank d. Then

$$\tilde{U}(x_0, T) = \{y : k(T, y, x_0) > 0\} \quad \text{and} \quad \mathrm{cl}\, \tilde{U}(x_0, T) = \mathrm{cl}\, U(x_0, T),$$

where cl $=$ closure. The derivative $D_{x_0,\phi}$ can be found by means of the perturbation method for ordinary differential equations. Let

$$\Lambda(t) = \frac{d\sigma_0}{dx}(x_\phi(t)) + \sum_{i=1}^m \frac{d\sigma_i}{dx}(x_\phi(t))\phi_i(t) \tag{53}$$

and let $Q(t, t_0)$, for $T \ge t \ge t_0 \ge 0$, be a matrix function such that $Q(t_0, t_0) = I$ and $\dfrac{\partial Q(t, t_0)}{\partial t} = \Lambda(t) Q(t, t_0)$. Then

$$D_{x_0,\phi} h = \int_0^T Q(T, s)\sigma(x_\phi(s))h(s)\, ds. \tag{54}$$

Remark 1 In many applications diffusion processes are defined on some subset G of $\mathbb{R}^d$. In this case we should know the description of the process X_t when it hits (or approaches) the boundary ∂G of G. The simples situation is if the process X_t starting from any point G does not hit the boundary. Then such a boundary is called *a natural boundary* and then the theory of such a diffusion is the same as in the case $G = \mathbb{R}^d$, i.e., this diffusion is related to some stochastic semigroup and the both Kolmogorov equations can be considered without boundary conditions. If we consider a diffusion process which hits the boundary ∂G then, in order to still have a stochastic semigroup, we should assume that after hitting the boundary it returns to G. In this case we usually assume that the process is reflected at the boundary ∂G. Then the Kolmogorov backward equation should be considered with the boundary condition

$$\frac{\partial u(t,x)}{\partial n(x)} = 0 \quad \text{for } x \in \partial G,$$

where $n(x)$ is the normal vector to the boundary at $x \in \partial G$. It means that if A^* is the adjoint operator of A and if $g \in \mathscr{D}(A^*)$ then $\dfrac{\partial g}{\partial n} = 0$ on ∂G. Now from the formula

$$\int_G (Af)(x)g(x)\,dx = \int_G f(x)(A^*g)(x)\,dx$$

one can find the proper boundary condition for the Fokker–Planck equation.

Remark 2 In some applications, especially in population dynamics, the drift vector $b(x)$ does not have a globally bounded partial derivatives. The situation is similar to that in ordinary differential equations, namely, we are only able to prove the local existence of solutions of (44) in general case and, consequently, we cannot define a stochastic semigroup. But in some cases, for example in the stochastic version of the logistic equation

$$dX_t = (\alpha X_t - \beta X_t^2)\,dt + \gamma X_t\,dW_t, \quad \alpha, \beta, \gamma > 0, \tag{55}$$

the derivatives of $b(x) = \alpha x - \beta x^2$ are unbounded, but, using some comparison theorems, we are able to check the global existence of its solutions (precisely, if $X_0 \geq 0$). Hence there exists a stochastic semigroup defined on $L^1(0,\infty)$ related to (55).

2.11 Piecewise Deterministic Markov Process

Now, we consider stochastic semigroups related to piecewise deterministic Markov processes. According to a non formal definition by Davis [28], the class of *piecewise deterministic Markov processes* (PDMPs) is a general family of stochastic models covering virtually all non-diffusion applications. A more formal definition is the following: a continuous time Markov process $X(t)$, $t \geq 0$, is a PDMP if there is an increasing sequence of random times (t_n), called jumps, such that sample paths of $X(t)$ are defined in a deterministic way in each interval (t_n, t_{n+1}). We consider two types of behavior of the process at jump times: the process can jump to a new point or can change the dynamics which defines its trajectories.

We start with a continuous time version of the iterated function system. We have a set of flows π_t^i, $i \in I = \{1, \dots, k\}$. Each flow π_t^i is defined as the solution of a system of differential equations $x' = b^i(x)$ on $G \subset \mathbb{R}^d$ (see 2.9). The state of the system is a pair $(x, i) \in G \times I$. If the system is at state (x, i) then x can change according to the flow π_t^i and after time t reaches the state $(\pi_t^i(x), i)$ or jump to the state (x, j) with a bounded and continuous intensity $q_{ij}(x)$. The pair $(x(t), i(t))$ constitutes a Markov process $X(t)$ on $G \times I$. Let $\{S^i(t)\}_{t\geq 0}$ be the stochastic semigroup related to π_t^i, i.e., the semigroup of Frobenius-Perron operators and let the operator A_i be its generator. We assume that the random variable $X(0)$ has an absolutely continuous distribution, then so are $X(t)$ for $t > 0$. Define the functions u_i, $i \in I$, by the formula

$$\text{Prob}\,(x(t) \in E,\ \ i(t) = i) = \int_E u_i(t, y)\,dy.$$

and let $u = (u_1, \dots, u_k)$ be a vertical vector and $Au = (A_1 u_1, \dots, A_k u_k)$ is also a vertical vector. Let $q_{jj}(x) = -\sum_{i \neq j} q_{ij}(x)$ and denote by $Q(x)$ the matrix $[q_{ij}(x)]$. Then the vector u satisfies the following equation

$$\frac{\partial u}{\partial t} = Qu + Au, \tag{56}$$

Let $\mathcal{B}(G \times I)$ be the σ-algebra of Borel subsets of $G \times I$ and let m be the product measure on $\mathcal{B}(G \times I)$ given by $m(B \times \{i\}) = \mu(B)$ for each $B \in \mathcal{B}(G)$ and $1 \leq i \leq k$, where μ is the Lebesgue measure on G. The operator A generates a stochastic semigroup $\{S(t)\}_{t\geq 0}$ on the space $L^1(G \times I, \mathcal{B}(G \times I), m)$ given by the formula

$$S(t)f = (S^1(t) f_1, \dots, S^k(t) f_k),$$

where $f_i(x) = f(x, i)$ for $x \in G$, $1 \leq i \leq k$. Since

$$\int_{G\times I} Qf\,dm = \sum_{i=1}^k \int_G (Qf)_i(x)\,dx = \sum_{i=1}^k \int_G \Big(\sum_{j=1}^k q_{ij}(x) f_j(x)\Big)\,dx$$

$$= \sum_{j=1}^k \int_G \Big(\sum_{i=1}^k q_{ij}(x)\Big) f_j(x)\,dx = 0$$

the operator $A + Q$ is a generator of stochastic semigroup $\{P(t)\}_{t\geq 0}$. If we put $\lambda = \sup_{x\in G} \max\{-q_{11}(x), \ldots, -q_{kk}(x)\}$ and $K = \lambda^{-1}Q + I$. Then K is a stochastic operator and $Q = -\lambda I + \lambda K$. Thus the semigroup $\{P(t)\}_{t\geq 0}$ can be given by (33) and (34).

2.12 Flow with Jumps

Now we consider the second type of piecewise deterministic Markov processes a *flow with jumps*. A particle x moves on an open set $G \subset \mathbb{R}^d$ with velocity $b(x)$ (see 2.9) and with a measurable and bounded intensity $\lambda(x)$ it jumps to a point y. The location of y is described by a transition function $\mathcal{P}(x, A)$, i.e., $\mathcal{P}(x, A)$ is the probability that $y \in A$. We assume that $\mathcal{P}$ satisfies (1), therefore, there is a stochastic operator $P\colon L^1(G) \to L^1(G)$ such that

$$\int_G f(x)\mathcal{P}(x, A)\,dx = \int_A Pf(x)\,dx \quad \text{for } f \in L^1(G) \text{ and } A \in \mathcal{B}(G). \tag{57}$$

Let $x(t)$ be the location of a particle at time t. We also assume that the particle does not leave the set G. We choose the initial position x of the particle randomly with a density distribution u_0. If $u(t, x)$ is the density of distribution of $x(t)$ then u satisfies the following equation

$$\frac{\partial u(t,x)}{\partial t} = -\operatorname{div}(b(x)u(t,x)) - \lambda(x)u(t,x) + P(\lambda u)(t,x). \tag{58}$$

The proof of (58) is almost the same as the formula (38) we should only add to the right-hand side of (41) the term

$$\int_G \lambda(x)f(x)\mathcal{P}(x, D)\,dx\ \Delta t - \int_D \lambda(x)f(x)\,dx\ \Delta t + o(\Delta t)$$

and apply the formula (57). Since $Bf = -\lambda f + P(\lambda f)$ is a bounded operator on $L^1(G)$ and $\int_G Bf(x)\,dx = 0$ for $f \in L^1(G)$, equation (58) generates a stochastic semigroup on $L^1(G)$.

Equations of type (58) appear in such diverse areas as the theory of jump processes [86, 107], in astrophysics—where describes the fluctuations in the brightness of the Milky-Way [25] and in population dynamics. An example of application to population dynamics is presented in 4.8.

2.13 Stochastic Hybrid Systems

Stochastic hybrid systems (SHSs) is a general class of stochastic processes which includes continuous and discrete, deterministic and stochastic flows. Such systems have appeared as stochastic versions of deterministic hybrid systems—systems described by differential equations and jumps. Piecewise deterministic Markov processes and Markov process with diffusion belong to SHSs. One can find a definition of stochastic hybrid systems and many examples of their applications in [19, 24, 115]. We only present here two examples stochastic semigroups related to SHSs which can be obtained in the same way as those in 2.11 and in 2.12 but we need to replace deterministic flows by diffusion flows.

Example 2 The first example is a *randomly flashing diffusion*. Let us consider the stochastic equation

$$dX_t = (Y_t\sigma(X_t))\,dW_t + b(X_t)\,dt, \tag{59}$$

where W_t, $t \geq 0$ is a Wiener process, Y_t is a homogeneous Markov process with values 0, 1 independent of W_t and X_0. We assume that the process Y_t jumps from 0 to 1 with intensity q_0 and from 1 to 0 with intensity q_1. Equation (59) describes the process which randomly jumps between stochastic and deterministic states. Such processes appear in transport phenomena in sponge—type structures [5, 18, 64]. The pair (X_t, Y_t) is a Markov process on $\mathbb{R} \times \{0, 1\})$. Let $A_0 v(x) = -\big(b(x)v(x)\big)'$, $A_1 v(x) = \frac{1}{2}\big(\sigma^2(x)v(x)\big)' - \big(b(x)v(x)\big)'$ and

$$f = \begin{bmatrix} f_0 \\ f_1 \end{bmatrix}, \quad Af = \begin{bmatrix} A_0 f_0 \\ A_1 f_1 \end{bmatrix}, \quad Qf = \begin{bmatrix} -q_0 f_0 + q_1 f_1 \\ q_0 f_0 - q_1 f_1 \end{bmatrix},$$

where we identify the function $f\colon \mathbb{R} \times \{0, 1\} \to \mathbb{R}$ with a vertical vector (f_1, f_2) by $f_i(x) = f(x, i)$. Since A is a generator of a stochastic semigroup on $L^1(\mathbb{R}\times\{0, 1\})$, Q is a bounded operator on $L^1(\mathbb{R} \times \{0, 1\})$, and the integral of Qf is zero, $A + Q$ also generates a stochastic semigroup on $L^1(\mathbb{R} \times \{0, 1\})$.

Generally, if we replace in 2.11 the generators A_i of flows by the generators of the Fokker–Planck equations of the form (49), we introduce a stochastic semigroup [90] related to a *multi-state diffusion* on $\mathbb{R}^d$.

Example 3 We can also replace the generator A of a continuity equation by the generator of the Fokker–Planck equation in 2.12. In this way we introduce a stochastic semigroup on $\mathbb{R}^d$ related to *diffusion with jumps*.

2.14 Nonlinear Stochastic Operators

We recall that if $D \subset L^1(X, \Sigma, m)$ is the set of densities, then a nonlinear stochastic operator is any continuous operator $P: D \to D$. A large class of nonlinear stochastic operators can be obtained in the following way. Fix a positive integer r and consider a measurable function $q: X^{r+1} \to [0, \infty)$ such that

$$\int_X q(x, y_1, \dots, y_r)\, m(dx) = 1 \quad \text{for a.e. } y = (y_1, \dots, y_n). \tag{60}$$

Then we can define r-linear positive and continuous operator $Q: (L^1)^r \to L^1$ by

$$Q(f_1, \dots, f_r)(x) = \int_X \cdots \int_X f_1(y_1) \dots f_r(y_r) q(x, y_1, \dots, y_r)\, m(dy_1) \dots m(dy_r). \tag{61}$$

If $f_1, \dots, f_r \in D$ then $Q(f_1, \dots, f_r) \in D$. This implies that the operator

$$P(f) = Q(f, \dots, f), \quad f \in D, \tag{62}$$

is a nonlinear stochastic operator. The operator P is Lipschitz continuous:

$$\|P(f) - P(g)\| \le r\|f - g\| \quad \text{for } f, g \in D.$$

Indeed, let

$$P_i h = P(\underbrace{f, \dots, f}_{r-i}, h, \underbrace{g, \dots, g}_{i-1})(x)$$

for $i = 1, \dots, r$. Then P_i are (linear) stochastic operators on $L^1(X, \Sigma, m)$ and

$$\|Pf - Pg\| = \|\sum_{i=1}^{r} (P_{r-i} f - P_{r-i} g)\| \le \sum_{i=1}^{r} \|P_{r-i}(f - g)\| \le r\|f - g\|.$$

If $P_1, \dots, P_n$ are nonlinear stochastic operators on $D \subset L^1(X, \Sigma, m)$ and $c_1, \dots, c_n$ are nonnegative constants such that $c_1 + \dots + c_n = 1$ then the operator

$$P = c_1 P_1 + \dots + c_n P_n$$

is also a nonlinear stochastic operator.

A special role in applications play stochastic operators acting on l^1 and 1^1_d spaces and given by bilinear operators. Such operators are of the form

$$P(x)_k = \sum_{i \in X} \sum_{j \in X} a_{ij}^k x_i x_j,$$

for all $k \in X$, where $X = \mathbb{N}$ if P acts on $D \subset l^1$ and $X = \{1, \dots, d\}$ in the second case. We assume that the cubic matrix $\left[a_{ij}^k\right]$ has nonnegative entries a_{ij}^k and $\sum_{k \in X} a_{ij}^k = 1$ for all $i, j \in X$.

2.15 Nonlinear Stochastic Semigroups

We recall that a family $\{P(t)\}_{t \geq 0}$ of nonlinear stochastic operators is called a nonlinear stochastic semigroup if $P(0) = I$, $P(t+s) = P(t)P(s)$ for $s, t \geq 0$, and the map $(t, f) \mapsto P(t)f$ is continuous. Now, we present a general method of constructing nonlinear stochastic semigroups. Let $\{S(t)\}_{t \geq 0}$ be a (linear) stochastic semigroup on the space $L^1(X, \Sigma, m)$ and let A be its generator with domain $\mathfrak{D}(A)$. Let P be a nonlinear stochastic operator on the set of densities $D \subset L^1(X, \Sigma, m)$ and assume that P satisfies the global Lipschitz condition, i.e., $\|P(f) - P(g)\| \leq L\|f - g\|$ for all $f, g \in D$ and some $L > 0$. Let $\lambda > 0$. Consider the following evolution equation

$$u'(t) = Au(t) - \lambda u(t) + \lambda Pu(t), \quad u(0) = u_0, \tag{63}$$

where $u_0 \in D$. By a solution of (63) we understand a *mild solution*, i.e., a continuous function $u: [0, \infty) \to D$ such that

$$u(t) = T(t)u_0 + \int_0^t \lambda T(t-s)P(u(s))\, ds \quad \text{for } t \geq 0, \tag{64}$$

where the semigroup $\{T(t)\}_{t \geq 0}$ is given by $T(t) = e^{-\lambda t} S(t)$. The existence and uniqueness of solutions of equation (64) is a simple consequence of the method of variation of parameters (see e.g. [76]).

A special role in application plays equation (63) with $A = 0$. A linear change of time t leads to the case with $\lambda = 1$, thus, we can consider the equation

$$u'(t) + u(t) = Pu(t), \quad u(0) = u_0 \in D, \tag{65}$$

which has a strong solution $u: [0, \infty) \to D$ given by

$$u(t) = e^{-t} \sum_{n=0}^{\infty} (1 - e^{-t})^n u_n,$$

where the sequence (u_n), $n \in \mathbb{N}$ is defined by the recurrent formulae

$$u_n = \frac{1}{n} \sum_{k=0}^{n-1} Q(u_k, u_{n-1-k}) \quad \text{for } n \geq 1.$$

3 Asymptotic Properties of Stochastic Operators and Semigroups

Now we introduce some notions which characterize the behaviour of iterates of stochastic operators P^n, $n = 0, 1, 2, \ldots$, when $n \to \infty$ and stochastic semigroups $\{P(t)\}_{t\ge 0}$ when $t \to \infty$. Since the iterates of stochastic operators also form a (discrete time) semigroup we use notation $P(t) = P^t$ for their powers and we formulate most of definitions and results for both types of semigroups without distinguishing them.

3.1 *Asymptotic Stability*

Consider a stochastic semigroup $\{P(t)\}_{t\ge 0}$. A density f_* is called *invariant* if $P(t)f_* = f_*$ for each $t > 0$. The stochastic semigroup $\{P(t)\}_{t\ge 0}$ is called *asymptotically stable* if there is an invariant density f_* such that

$$\lim_{t\to\infty} \|P(t)f - f_*\| = 0 \quad \text{for} \quad f \in D.$$

If the semigroup $\{P(t)\}_{t\ge 0}$ is generated by some evolution equation $u'(t) = Au(t)$ then the asymptotic stability of $\{P(t)\}_{t\ge 0}$ means that the stationary solution $u(t) = f_*$ is asymptotically stable in the sense of Lapunov and this stability is global on the set D.

Remark 3 Let $\{P(t)\}_{t\ge 0}$ be a continuous time stochastic semigroup and $P = P(t_0)$ for some $t_0 > 0$. Then the semigroup $\{P(t)\}_{t\ge 0}$ is asymptotically stable iff the discrete semigroup $\{P^n\}_{n\in\mathbb{N}}$ is asymptotically stable. The proof of this fact goes as follows. Assume that the discrete semigroup $\{P^n\}_{n\in\mathbb{N}}$ is asymptotically stable. For $f \in D$ and $\varepsilon > 0$ we find $\delta > 0$ such that $\|P(t)f - P(s)f\| < \varepsilon$ if $|t-s| \le \delta$. Let $k \ge t_0/\delta$ be an integer and let $f_i = P(it_0/k)f$ for $i = 1, \ldots, k$. Then there exists $n_0 \in \mathbb{N}$ such that $\|P^n f_i - f_*\| < \varepsilon$ for $n \ge n_0$ and for $i = 1, \ldots, k$. Therefore, $\|P(t)f - f_*\| < 2\varepsilon$ for $t \ge n_0 t_0$.

Example 4 Let P_S be the Frobenius–Perron operator related to the tent map defined in Example 1, i.e., P_S is given by

$$P_S f(x) = \tfrac{1}{2} f(\tfrac{1}{2}x) + \tfrac{1}{2} f(1 - \tfrac{1}{2}x).$$

We check that the semigroup determined by P_S is asymptotically stable, which is equivalent to exactness of the tent map. The function $f_* = 1_{[0,1]}$ is an invariant density. Since any Frobenius–Perron operator is a contraction it is sufficient to check that $\lim_{n\to\infty} P^n f = f_*$ for f from a dense subset D_0 of D. We assume that D_0

is a set of densities which are Lipschitz continuous. Let $f \in D_0$ and let L be the Lipschitz constant for f. Then

$$|Pf(x) - Pf(y)| \le \tfrac{1}{2}|f(\tfrac{x}{2}) - f(\tfrac{y}{2})| + \tfrac{1}{2}|f(1 - \tfrac{1}{2}x) + f(1 - \tfrac{1}{2}y)| \le \tfrac{L}{2}|x - y|.$$

Thus $L/2$ is the Lipschitz constant for Pf and by induction we conclude that $L/2^t$ is the Lipschitz constant for $P^t f$. Hence, the sequence $P^t f$ converges uniformly to a constant function. Since $(P^t f)$ are densities, $(P^t f)$ converges to f_* uniformly, which implies the convergence in L^1.

Example 5 The second example is the logistic map $S(x) = 4x(1-x)$ on $[0,1]$. Denote by T the tent map and let $\Phi(x) = \frac{1}{2} - \frac{1}{2}\cos(\pi x)$. We check that $S \circ \Phi = \Phi \circ T$. Indeed

$$\begin{aligned} S(\Phi(x)) &= 4(\tfrac{1}{2} - \tfrac{1}{2}\cos(\pi x))(\tfrac{1}{2} + \tfrac{1}{2}\cos(\pi x)) = 1 - \cos^2(\pi x), \\ \Phi(T(x)) &= \tfrac{1}{2} - \tfrac{1}{2}\cos(\pi T(x)) = \tfrac{1}{2} - \tfrac{1}{2}\cos(2\pi x) = \tfrac{1}{2} - \tfrac{1}{2}(2\cos^2(\pi x) - 1) \\ &= 1 - \cos^2(\pi x). \end{aligned}$$

Since $S \circ \Phi = \Phi \circ T$ we have $P_S P_\Phi = P_\Phi P_T$. Which gives $P_S = P_\Phi P_T P_{\Phi^{-1}}$ and, by induction, $P_S^n = P_\Phi P_T^n P_{\Phi^{-1}}$. Let $f \in D$. Then $P_{\Phi^{-1}} f \in D$ and from the previous example $\lim_{n\to\infty} P_T^n P_{\Phi^{-1}} f = 1_{[0,1]}$. Therefore

$$\lim_{n\to\infty} P_S^n f = P_\Phi 1_{[0,1]} = 1_{[0,1]}(\Phi^{-1}(x))(\Phi^{-1}(x))' = (\Phi^{-1}(x))'.$$

Since $\Phi^{-1}(x) = \frac{1}{\pi}\arccos(1-2x)$ we have $(\Phi^{-1}(x))' = \dfrac{1}{\pi\sqrt{x(1-x)}}$. It means that the logistic map $S(x) = 4x(1-x)$ on $[0,1]$ is exact and has the invariant density

$$f_*(x) = \frac{1}{\pi\sqrt{x(1-x)}}.$$

3.2 Lower Function Theorem

Now we present a very useful result concerning asymptotic stability called the *lower function theorem* of Lasota and Yorke. A function $h \in L^1$, $h \ge 0$ and $h \ne 0$ is called a *lower function* for a stochastic semigroup $\{P(t)\}_{t\ge0}$ if

$$\lim_{t\to\infty} \|(P(t)f - h)^-\| = 0 \quad \text{for every } f \in D. \tag{66}$$

Here we use the notation $f^-(x) = 0$ if $f(x) \geq 0$ and $f^-(x) = -f(x)$ if $f(x) < 0$. The condition (66) can be written equivalently as: there are functions $\varepsilon(t) \in L^1$ such that $\lim_{t\to\infty} \|\varepsilon(t)\| = 0$ and $P(t)f \geq h - \varepsilon(t)$. Observe that if the semigroup is asymptotically stable then its invariant density f_* is a lower function for it. Lasota and Yorke [57] proved the following converse result.

Theorem 6 *Let $\{P(t)\}_{t\geq 0}$ be a stochastic semigroup. If there exists a lower function h for a stochastic semigroup $\{P(t)\}_{t\geq 0}$ then this semigroup is asymptotically stable.*

Now, we present an application of this result.

Example 6 Let $X = \{1, \dots, d\}$, $\Sigma = 2^X$, and m be the counting measure on X. Then we use the notation $l_d^1 = L^1(X, \Sigma, m)$. Let $P: l_d^1 \to l_d^1$ be a stochastic operator, i.e., the operator P is represented by a stochastic matrix $P: \mathbb{R}^d \to \mathbb{R}^d$ such that $p_{ij} \geq 0$ for each $i, j \in X$ and $\sum_{i=1}^d p_{ij} = 1$ for each $j \in X$. A sequence $(x_1, \dots, x_d)$ is a density if it has nonnegative elements and $x_1 + \cdots + x_d = 1$. We show that if for some $r \geq 1$ the matrix P^r has all positive entries, the operator P is asymptotically stable, i.e., there exists a density $\pi = (\pi_1, \dots, \pi_d)$ with positive elements π_i such that $\lim_{n\to\infty} P^n x = \pi$ for every density x. In order to prove this claim we observe that the entries of the matrix P^r are greater or equals some $c > 0$, which implies that $P^r x \geq [c, \dots, c]$ for any density x. Therefore, for $n \geq r$ and $x \in D$ we have

$$P^n x \geq P^r(P^{n-r}x) \geq [c, \dots, c]$$

because $P^{n-r}x$ is also a density. Thus, $[c, \dots, c]$ is a lower function and Theorem 6 completes the proof.

Remark 4 The result presented in Example 6 is a special case of the Perron–Frobenius theorem for non-negative matrices [71]. In particular, by spectral arguments, one can prove that the sequence $\{P^n x\}$ converges exponentially to π. We can also replace the assumption on positivity of the matrix P^r by a weaker one that P is irreducible, but in this case we can only prove asymptotic periodicity of the sequence $\{P^n x\}$.

The second application of the lower function theorem is given in 4.5.

3.3 *Partially Integral Semigroups*

Now and in 3.4 we present our main results concerning long-time behaviour of stochastic and substochastic semigroups. The proofs of these results are based on the theory of Harris operators [34, 44] and we do not give their here. We start with the definition of a partially integral semigroup.

A substochastic semigroup $\{P(t)\}_{t\geq 0}$ is called *partially integral* if there exists a measurable function $k\colon (0,\infty)\times X\times X\to [0,\infty)$, called a *kernel*, such that

$$P(t)f(x)\geq \int_X k(t,x,y)f(y)\,m(dy)$$

for every density f and

$$\int_X\int_X k(t,x,y)\,m(dy)\,m(dx)>0$$

for some $t>0$.

Theorem 7 ([90]) *Let $\{P(t)\}_{t\geq 0}$ be a partially integral stochastic semigroup. Assume that the semigroup $\{P(t)\}_{t\geq 0}$ has an invariant density f_*. Moreover, we assume that the semigroup $\{P(t)\}_{t\geq 0}$ has no other periodic points in the set of densities, i.e., if $P(t)f=f$ for some $f\in D$ and $t>0$, then $f=f_*$. If $f_*>0$ a.e. then the semigroup $\{P(t)\}_{t\geq 0}$ is asymptotically stable.*

For any $f\in L^1(X)$ the *support* of f is defined up to a set of measure zero by the formula

$$\operatorname{supp} f=\{x\in X: f(x)\neq 0\}.$$

We say that a stochastic semigroup $\{P(t)\}_{t\geq 0}$ *spreads supports* if for every set $A\in\Sigma$ and for every $f\in D$ we have

$$\lim_{t\to\infty} m(\operatorname{supp} P(t)f\cap A)=m(A)$$

and *overlaps supports,* if for every $f,g\in D$ there exists $t>0$ such that

$$m(\operatorname{supp} P(t)f\cap \operatorname{supp} P(t)g)>0.$$

Now we formulate corollaries which are often used in applications.

Corollary 2 ([90]) *A partially integral stochastic semigroup which spreads supports and has an invariant density is asymptotically stable.*

Corollary 3 ([90]) *A partially integral stochastic semigroup which overlaps supports and has an invariant density $f_*>0$ a.e. is asymptotically stable.*

Remark 5 The above corollaries generalize some earlier results [19, 66, 85, 89] for integral stochastic semigroups. Another proof of Corollary 3 is given in [11]. Corollary 2 remains true also for the Frobenius–Perron operators. Precisely, let S be a double-measurable transformation of a probability measure space (X,Σ,m). If S preserves the measure m and the iterates of the Frobenius–Perron operator P_S spread supports, then the semigroup of iterates of P_S is asymptotically stable [90]. It is interesting that if we assume only that a stochastic operator (or semigroup) P has an invariant density f_* and spreads supports, then P is weakly asymptotically stable (*mixing*). It means that for every $f \in D$ the sequence $P^n f$ converges weakly to f_*. One can expect that we can drop in Corollary 2 the assumption that the semigroup is partially integral, but its is no longer true. Indeed, in [92] we construct a stochastic operator $P\colon L^1[0,1] \to L^1[0,1]$ which spreads supports and $P\mathbf{1} = \mathbf{1}$ but it is not asymptotically stable.

If $\{P(t)\}_{t\geq 0}$ is a continuous time stochastic semigroup then we can strengthen considerably Theorem 7 and formulate the main result of this part.

Theorem 8 ([81]) *Let $\{P(t)\}_{t\geq 0}$ be a continuous time partially integral stochastic semigroup. Assume that the semigroup $\{P(t)\}_{t\geq 0}$ has a unique invariant density f_*. If $f_* > 0$ a.e., then the semigroup $\{P(t)\}_{t\geq 0}$ is asymptotically stable.*

Remark 6 The assumption that the invariant density is unique can be replaced by an equivalent one: that does not exist a set $E \in \Sigma$ such that $m(E) > 0$, $m(X \setminus E) > 0$ and $P(t)E = E$ for all $t > 0$. Here $P(t)$ is the operator acting on the σ-algebra Σ defined by: if $f \geq 0$, $\operatorname{supp} f = A$ and $\operatorname{supp} Pf = B$ then $PA = B$.

Remark 7 Theorem 8 is not longer true if we replace a continuous time semigroup by a discrete time semigroup (i.e., the iterates of a stochastic operator). Indeed, the stochastic operator P on the space l_2^1 given by $P(x_1, x_2) = (x_2, x_1)$ is an integral operator and it has a unique density $(\frac{1}{2}, \frac{1}{2})$ but it is not asymptotically stable.

3.4 Sweeping and Foguel Alternative

The second important notion which describes the long-time behaviour of stochastic semigroups is sweeping. The notion of sweeping was introduced by Komorowski and Tyrcha [49] and it is also known as *zero type property*. A stochastic semigroup $\{P(t)\}_{t\geq 0}$ is called *sweeping* with respect to a set $B \in \Sigma$ if for every $f \in D$

$$\lim_{t\to\infty} \int_B P(t)f(x)\, m(dx) = 0.$$

It is clear that if a stochastic semigroup is sweeping then it cannot be asymptotically stable. Our main aim in this part is to find such conditions on a semigroup to have alternative between asymptotic stability and sweeping.

The crucial role in results concerning sweeping plays the following condition:

(KT) There exists a measurable function f_* such that: $0 < f_* < \infty$ a.e., $P(t)f_* \leq f_*$ for $t \geq 0$, $f_* \notin L^1$ and $\int_A f_*\, dm < \infty$ for some set $A \in \Sigma$ with $m(A) > 0$.

Theorem 9 ([49]) *Let $\{P(t)\}_{t\geq 0}$ be an integral stochastic semigroup which has no invariant density. Assume that the semigroup $\{P(t)\}_{t\geq 0}$ and a set $A \in \Sigma$ satisfy condition (KT). Then the semigroup $\{P(t)\}_{t\geq 0}$ is sweeping with respect to A.*

In paper [90] it was shown that Theorem 9 holds for a wider class of operators than integral ones. In particular, the following result was proved (see [90] Corollary 4 and Remark 6).

Theorem 10 *Let $\{P(t)\}_{t\geq 0}$ be a stochastic semigroup which overlaps supports. Assume that the semigroup $\{P(t)\}_{t\geq 0}$ and a set $A \in \Sigma$ satisfy condition (KT). Then the semigroup $\{P(t)\}_{t\geq 0}$ is sweeping with respect to A.*

The main difficulty in applying Theorems 9 and 10 is to prove that a stochastic semigroup satisfies condition (*KT*). Now we formulate a criterion for sweeping which will be useful in applications.

Theorem 11 ([90]) *Let X be a metric space and $\Sigma = \mathcal{B}(X)$ be the σ—algebra of Borel subsets of X. We assume that a partially integral stochastic semigroup $\{P(t)\}_{t\geq 0}$ with the kernel k has the following properties:*

(a) for every $f \in D$ we have $\int_0^\infty P(t)f\,dt > 0$ a.e.,
(b) for every $y_0 \in X$ there exist $\varepsilon > 0$, $t > 0$, and a measurable function $\eta \geq 0$ such that $\int \eta\,dm > 0$ and

$$k(t, x, y) \geq \eta(x)$$

for $x \in X$ and $y \in B(y_0, \varepsilon)$, where $B(y_0, \varepsilon)$ is the open ball with center y_0 and radius ε. If the semigroup $\{P(t)\}_{t\geq 0}$ has no invariant density then it is sweeping with respect to compact sets.

Remark 8 If $\{P(t)\}_{t\geq 0}$ is a discrete time semigroup then condition (a) is of the form: for every $f \in D$ we have $\sum_{n=0}^{\infty} P^n f > 0$ a.e.

From Theorem 8 and Theorem 11 it follows

Corollary 4 *Let $\{P(t)\}_{t\geq 0}$ be a continuous time partially integral stochastic semigroup on $L^1(X, \mathcal{B}(X), m)$, where X is a metric space. Assume that conditions (a) and (b) of Theorem 11 hold. Then the semigroup $\{P(t)\}_{t\geq 0}$ is asymptotically stable if it has an invariant density and it is sweeping with respect to compact sets if it has no invariant density. In particular, if X is compact then the semigroup $\{P(t)\}_{t\geq 0}$ is asymptotically stable.*

The property that a stochastic semigroup $\{P(t)\}_{t\geq 0}$ is asymptotically stable or sweeping from a sufficiently large family of sets is called the *Foguel alternative* [55]. We use the notion of the Foguel alternative in a narrow sense, when the sweeping is from all compact sets.

From Theorems 7 and 11 it follows

Corollary 5 *The Foguel alternative holds for a partially integral stochastic semigroup* $\{P(t)\}_{t\ge 0}$ *on* $L^1(X, \mathcal{B}(X), m)$ *with a continuous and positive kernel* $k(t, x, y)$ *for* $t > 0$.

Now we present a result concerning long-time behaviour of stochastic semigroups given in [78] which generalizes Theorems 8 and 11.

Theorem 12 *Let* X *be a metric space and and* $\Sigma = \mathcal{B}(X)$. *Let* $\{P(t)\}_{t\ge 0}$ *be a continuous time partially integral substochastic semigroup on* $L^1(X)$ *with kernel* k *and which has a unique invariant density* f_*. *Let* $S = \operatorname{supp} f_*$. *We assume that for some* $t_0 > 0$

$$\int_S \int_S k(t_0, x, y)\, m(dx)\, m(dy) > 0.$$

Moreover, we assume that for some $t_1 > 0$

(a) there is no nonempty measurable set $B \subsetneq X \setminus S$ *such that* $P^*(t_1)\mathbf{1}_B \ge \mathbf{1}_B$,
(b) for every $y_0 \in X \setminus S$ *there exist* $\varepsilon > 0$ *and a measurable function* $\eta \ge 0$ *such that* $\int_{X\setminus S} \eta\, dm > 0$ *and*

$$k(t_1, x, y) \ge \eta(x) \tag{67}$$

for $x \in X$ *and* $y \in B(y_0, \varepsilon)$.

Then for every $f \in D$ *there exists a constant* $c(f)$ *such that*

$$\lim_{t\to\infty} \mathbf{1}_S P(t) f = c(f) f_*$$

and for every compact set $F \in \Sigma$ *and* $f \in D$ *we have*

$$\lim_{t\to\infty} \int_{F\cap X\setminus S} P(t) f(x)\, m(dx) = 0.$$

Remark 9 If we drop in Theorem 11 condition (b) then it is not longer true. Indeed, there is an integral stochastic operator with a strictly positive kernel which has no invariant density but it is not sweeping from compact sets (see [90], Remark 7). The notion of sweeping operators is similar to the notion of dissipative operators. A stochastic operator is called *dissipative* if $\sum_{n=0}^{\infty} P^n f(x) < \infty$ a.e. for a density with $f > 0$ a.e. This definition is independent of the choice of f. There are dissipative stochastic operators which are no sweeping (see [49] Example 1). It is interesting that a stochastic operator on $L^1(\mathbb{R})$ can be sweeping from compact sets but can be no sweeping from sets of finite Lebesgue measure (see [90], Remark 3).

From our general results concerning the Foguel alternative it is easy to prove the following well-known results on continuous time irreducible Markov chains (see [82] for the proof in the case l^1).

Theorem 13 *Let $\{P(t)\}_{t\ge 0}$ be a stochastic semigroup on l^1_d generated by the equation*

$$x'(t) = Qx(t).$$

Let us assume that the entries of the matrix Q satisfy the following condition

(T) *for all $1 \le i, j \le d$ there exists a sequence of integers $i_0, i_1, \ldots, i_r$ such that $i_0 = j$, $i_r = i$ and*

$$q_{i_r i_{r-1}} \cdots q_{i_2 i_1} q_{i_1 i_0} > 0. \tag{68}$$

Then the semigroup $\{P(t)\}_{t\ge 0}$ is asymptotically stable, i.e., there exists a density $\pi = (\pi_1, \ldots, \pi_d)$ with positive elements π_i such that $\lim_{t\to\infty}(P(t)x)_i = \pi_i$ for every density x and $1 \le i \le d$.

Proof We apply Corollary 4. Since $X = \{1, \ldots, d\}$ is a discrete space, the semigroup $\{P(t)\}_{t\ge 0}$ is integral with a continuous kernel k and the kernel k satisfies condition (b). From (T) it follows that condition (a) holds. Since the spaces X is compact, the semigroup $\{P(t)\}_{t\ge 0}$ cannot be sweeping from compact sets, and, consequently, it is asymptotically stable. □

Remark 10 By spectral arguments one can prove a stronger version of Theorem 13 with exponential convergence of $P(t)x$ to π as $t \to \infty$.

Theorem 14 *Let $\{P(t)\}_{t\ge 0}$ be a stochastic semigroup on l^1 generated by the equation*

$$x'(t) = Qx(t).$$

Let us assume that the entries of the matrix Q satisfy the following condition

(T) *for all $i, j \in \mathbb{N}$ there exists a sequence of nonnegative integers $i_0, i_1, \ldots, i_r$ such that $i_0 = j$, $i_r = i$ and*

$$q_{i_r i_{r-1}} \cdots q_{i_2 i_1} q_{i_1 i_0} > 0. \tag{69}$$

Then the semigroup $\{P(t)\}_{t\ge 0}$ satisfies the Foguel alternative:

(a) if the semigroup $\{P(t)\}_{t\ge 0}$ has an invariant density, then it is asymptotically stable,
(b) if the semigroup $\{P(t)\}_{t\ge 0}$ has no invariant density, then for every $x \in l^1$ and $i \in \mathbb{N}$ we have

$$\lim_{t\to\infty} (P(t)x)_i = 0. \tag{70}$$

Now we return to a piecewise deterministic Markov process considered in 2.11. The main problem with application of the Foguel alternative is to check that the stochastic semigroup related to this process is partially integral and its kernel satisfies condition (b). Now we discuss this subject. As in 2.11 we assume that the system is governed by k flows π_t^i and each flow π_t^i is defined as the solution of a system of differential equations $x' = b^i(x)$ on $G \subset \mathbb{R}^d$. We also assume that all transition intensities $q_{ij}(x)$ are continuous and positive functions. Denote by $\{P(t)\}_{t\ge 0}$ the semigroup corresponding to this system. Let $(i_1, \dots, i_{d+1})$ be a sequence of integers from the set $I = \{1, \dots, k\}$. For $x \in X$ and $t > 0$ we define the function $\psi_{x,t}$ on the set $\Delta_t = \{\tau = (\tau_1, \dots, \tau_d)\colon \tau_i > 0, \tau_1 + \dots + \tau_d \le t\}$ by

$$\psi_{x,t}(\tau_1, \dots, \tau_d) = \pi^{i_{d+1}}_{t-\tau_1-\tau_2-\dots-\tau_d} \circ \pi^{i_d}_{\tau_d} \circ \dots \circ \pi^{i_2}_{\tau_2} \circ \pi^{i_1}_{\tau_1}(x).$$

Assume that for some $y_0 \in X$, $t_0 > 0$ and $\tau^0 \in \Delta_{t_0}$ we have

$$\det\left[\frac{d\psi_{y_0,t_0}(\tau^0)}{d\tau}\right] \ne 0. \tag{71}$$

Then, according to [81], there exists a continuous function $k\colon G \times G \to [0,\infty)$ and a point $x_0 \in G$ such that $k(x_0, y_0) > 0$ and

$$P(t_0)f(x) \ge \int_G k(x,y)f(y)\,dy \quad \text{for } f \in D.$$

Remark 11 Condition (71) can be formulated using Lie brackets (see for their definition page 274). Assume that $q_{ij}(y_0) > 0$ for all $1 \le i, j \le k$. If vectors

$$b^2(y_0) - b^1(y_0), \dots, b^k(y_0) - b^1(y_0),\ [b^i, b^j](y_0)_{1\le i,j\le k},\ [b^i,[b^j,b^l]](y_0)_{1\le i,j,l\le k}, \dots$$

span the space $\mathbb{R}^d$ then (71) holds (see , e.g., [13] Theorem 4).

3.5 Hasminskiĭ Function

An advantage of the formulation of Corollary 4 in the form of an alternative is that in order to show asymptotic stability we do not need to prove the existence of an invariant density. It is enough to check that the semigroup is not sweeping with respect to compact sets then, automatically, the semigroup $\{P(t)\}_{t\ge 0}$ is asymptotically stable. We can eliminate the sweeping by means some method similar to that of Lyapunov function called Hasminskiĭ function.

Consider a continuous time stochastic semigroup $\{P(t)\}_{t\geq 0}$ and let A be its generator. Let $\mathscr{R} = (I - A)^{-1}$. A measurable function $V : X \to [0,\infty)$ is called a *Hasminskiĭ function* for the semigroup $\{P(t)\}_{t\geq 0}$ and a set $Z \in \Sigma$ if there exist $M > 0$ and $\varepsilon > 0$ such that

$$\int_X V(x)\mathscr{R}f(x)\,dm(x) \leq \int_X (V(x) - \varepsilon)f(x)\,dm(x) + \int_Z M\mathscr{R}f(x)\,dm(x). \tag{72}$$

Theorem 15 ([79]) *Let $\{P(t)\}$ be a stochastic semigroup generated by the equation*

$$\frac{\partial u}{\partial t} = Au.$$

Assume that there exists a Hasminskiĭ function for this semigroup and a set Z. Then the semigroup $\{P(t)\}$ is not sweeping with respect to the set Z.

In application we take V such that the function A^*V is "well defined" and it satisfies the following condition $A^*V(x) \leq -c < 0$ for $x \notin Z$. Then we check that V satisfies inequality (72). We called the function V the Hasminskiĭ function because he has showed [39] that the semigroup generated by a non-degenerate Fokker–Planck equation has an invariant density iff there exists a positive function V such that $A^*V(x) \leq -c < 0$ for $\|x\| \geq r$. We applied this method to multi-state diffusion processes [79] and diffusion with jumps [80], where inequality (72) was proved by using some generalization of the maximum principle. This method was also applied to flow with jumps (58) in [77] but the proof of inequality (72) is different and based on an approximation of V by a sequence of elements from the domain of the operator A^*.

Now, we present a result on asymptotic stability of stochastic semigroups on l^1 which is based on the idea of Hasminskiĭ function (see [82] for the proof).

Theorem 16 *Let $Q = [q_{ij}]$, $i, j = 0, 1, 2, \ldots$, be a non-explosive Kolmogorov matrix. We assume that there exist a sequence $v = (v_i)$ of nonnegative numbers and positive constants ε, m, and k such that*

$$\sum_{i=0}^{\infty} q_{ij}v_i \leq \begin{cases} m, & \text{for } j \leq k, \\ -\varepsilon, & \text{for } j > k. \end{cases} \tag{73}$$

Then the stochastic semigroup $\{P(t)\}_{t\geq 0}$ related to Q is not sweeping from the set $\{0, 1, \ldots, k\}$. In particular, if the matrix Q satisfies conditions (T) *and (73), then the semigroup $\{P(t)\}_{t\geq 0}$ is asymptotically stable.*

3.6 Completely Mixing

A stochastic semigroup $\{P(t)\}_{t\geq 0}$ is called *completely mixing* if

$$\lim_{t\to\infty} \|P(t)f - P(t)g\| = 0 \tag{74}$$

for any densities f and g. If a stochastic semigroup is completely mixing and has an invariant density f_* then it is asymptotically stable. However, the semigroup $\{P(t)\}_{t\geq 0}$ can be completely mixing, but it can have no invariant density. For example, the heat equation $\dfrac{\partial u}{\partial t} = \Delta u$ generates the semigroup which is completely mixing and has no invariant density. It is easy to check that if a stochastic semigroup $\{P(t)\}_{t\geq 0}$ is completely mixing then all fixed points of the semigroup $\{P^*(t)\}_{t\geq 0}$ are constant functions.

Remark 12 Let $\{P(t)\}_{t\geq 0}$ be a continuous time stochastic semigroup and $P = P(t_0)$ for some $t_0 > 0$. Then the semigroup $\{P(t)\}_{t\geq 0}$ is completely mixing iff the discrete semigroup $\{P^n\}_{n\in\mathbb{N}}$ is completely mixing.

Completely mixing property for non-degenerate Fokker–Planck equations was studied in the papers [23, 87]. The most general result in this direction was received in [14]:

Theorem 17 *Assume that all coefficients in the Fokker–Planck equation are bounded with their first and second partial derivatives, and the diffusion term satisfies uniform elliptic condition (51). Then the semigroup $\{P(t)\}_{t\geq 0}$ generated by this equation is completely mixing iff all fixed points of the semigroup $\{P^*(t)\}_{t\geq 0}$ are constant functions, i.e., if all bounded solutions of the elliptic equation*

$$\sum_{i,j=1}^{n} a_{ij}(x)\frac{\partial^2 u}{\partial x_i \partial x_j} + \sum_{i=1}^{n} b_i(x)\frac{\partial u}{\partial x_i} = 0$$

are constant.

It is worth pointing out that this theorem is no longer true if the boundedness of the drift coefficient $b(x)$ is replaced with its linear growth. A counter-example in one-dimensional case with constant diffusion is given in [87].

Completely mixing property is strictly connected with the notion of the relative entropy. For any continuous and convex function η and densities f, g the *η-entropy of f relative to g* is defined by

$$H_\eta(f \mid g) = \int_X g(x)\eta\left(\frac{f(x)}{g(x)}\right) m(dx).$$

Relative entropy is also called *statistical distance* and was introduced by Csiszár [27]. The most interesting examples of relative entropy are following:

1. if $\eta(u) = u \log u$, then $H_\eta(f \mid g) = \int (f \log f - f \log g)\, dm$ is called the *Kullback–Leibler entropy* or *information* of f relative to g,
2. if $\eta(u) = |1 - u|$, then $H_\eta(f|g) = \|f - g\|$,
3. if $\eta(u) = -u^a$, $a \in (0, 1)$, then $H_\eta(f|g) = -\int f^a g^{1-a}\, dm$.

The following result connects the notion of the relative entropy with stochastic operators and completely mixing property.

Theorem 18 *Let P be a stochastic operator and H_η be the relative entropy. Then*

$$H_\eta(Pf \mid Pg) \le H_\eta(f \mid g). \tag{75}$$

If $H_\eta(f_n \mid g_n) \to \eta(1)$, then $f_n - g_n \to 0$ in L^1.

A simple proof of this results can be found in [63].

Remark 13 Let P_S be the Frobenius–Perron operator for a measurable transformation S of a σ-finite measure space (X, Σ, m). Then P_S is completely mixing iff $\bigcap_{n=1}^{\infty} S^{-n}\Sigma = \{\emptyset, X\}$ (see [59]). If additionally the measure m is invariant then the transformation S is exact. In the paper [83] we give an example of a piecewise linear and expanding transformation of the interval $[0, 1]$, called the *one-dimensional Smale horseshoe*, which is completely mixing but for every density f the iterates $P_S^n f$ converge weakly to the standard Cantor measure. In particular, this transformation has no invariant density, and, therefore, it is not mixing. On the other hand the system given by the baker transformation (13) is mixing but not completely mixing.

Remark 14 Many abstract results concerning completely mixing property can be found in books [72, 74]. Completely mixing property of an integral stochastic operator appearing in a model of cell cycle was studied in [91].

3.7 Sectorial Limit

Now, we consider a stochastic semigroup $\{P(t)\}_{t\ge 0}$ corresponding to a diffusion process on $\mathbb{R}^d$. Let $S = \{x \in \mathbb{R}^d : \|x\| = 1\}$ and A be a measurable subset of S. Denote by $K(A)$ the cone spanned by A:

$$K(A) = \{x \in \mathbb{R}^d : x = \lambda y,\ y \in A,\ \lambda > 0\}.$$

Then the function

$$p_A(t) = \int_{K(A)} P(t)f(x)\,dx, \qquad f \in D,$$

describes the mass of particles which are in the cone $K(A)$. If the semigroup $\{P(t)\}_{t\geq 0}$ is completely mixing and sweeping from compact sets then the asymptotic behaviour of $p_A(t)$ as $t \to \infty$ does not depend on f. It is interesting when the following limit exists: $p_A = \lim_{t\to\infty} p_A(t)$. If this limit exists then p_A measures the *sectorial limit distribution* of particles.

The problem of finding the limit distribution p_A for arbitrary diffusion process in d—dimensional space is difficult. Some partial results can be obtained under additional assumption that all functions a_{ij} and b_i are periodic with the same periods (we recall that a function $f : \mathbb{R}^d \to \mathbb{R}$ is periodic if there exist independent vectors $v_1, \dots, v_d$ such that $f(x + v_i) = f(x)$ for each $x \in \mathbb{R}^d$ and $i = 1, \dots, d$).

In one-dimensional space we can consider the function $p_+(t) = \int_c^\infty u(x,t)\,dx$ which describes the mass of particles in the interval (c, ∞). The paper [88] provides a criterion for the existence of the limit $\lim_{t\to\infty} p_+(t)$ and the formula for its value. In particular, if the diffusion coefficient $\sigma \equiv 1$ and the finite limits

$$\lim_{x\to\infty} \int_0^x b(y)\,dy = r \text{ and } \lim_{x\to-\infty} \int_0^x b(y)\,dy = s$$

exist, then

$$\lim_{t\to\infty} p_+(t) = \frac{e^{2r}}{e^{2r} + e^{2s}}.$$

In the same paper an example is constructed such that the following condition holds

$$\limsup_{t\to\infty} \frac{1}{t} \int_0^t p_+(s)\,ds = 1 \qquad \text{and} \qquad \liminf_{t\to\infty} \frac{1}{t} \int_0^t p_+(s)\,ds = 0. \tag{76}$$

In this example $\sigma \equiv 1$ and $b(x) \to 0$ as $|x| \to \infty$. Condition (76) is rather surprising because even if diffusion is constant and the drift coefficient is small particles can synchronously oscillate between $+\infty$ and $-\infty$.

3.8 Convergence After Rescaling and Self-Similar Solutions

If a stochastic semigroup $\{P(t)\}_{t\geq 0}$ on the space $L^1(\mathbb{R}^d)$ has no invariant density one can investigate its convergence after rescaling. We say that a stochastic semigroup $\{P(t)\}_{t\geq 0}$ is *convergent after rescaling* if there exist a density g and functions $\alpha(t)$, $\beta(t)$ such that

$$\lim_{t\to\infty}\int_{\mathbb{R}^d}|P(t)f(x)-\alpha^d(t)g\big(\alpha(t)x+\beta(t)\big)|\,dx=0 \qquad \text{for every } f\in D. \tag{77}$$

If there exists a density f such that $P(t)f(x)=\alpha^d(t)g\big(\alpha(t)x+\beta(t)\big)$ then $u(t)=P(t)f$ is called a *self-similar solution* of the equation $u'(t)=Au(t)$, where A is the generator of the semigroup. For example, the stochastic semigroup $\{P(t)\}_{t\geq 0}$ generated by the heat equation $\dfrac{\partial u}{\partial t}=\Delta u$ satisfies (77) with $\alpha(t)=(1+t)^{-1/2}$ and $\beta(t)=0$. It has also a self-similar solution with $g(x)=(4\pi)^{-d/2}e^{-\|x\|^2/4}$.

Condition (77) implies completely mixing property. One of the weak versions of this condition is the central limit theorem. In papers [85, 86] it is shown that semigroups connected with processes with jumps satisfy condition (77), precisely, these processes are asymptotically log-normal.

3.9 Supplementary Remarks

In the previous subsections we have concentrated mainly on properties of partially integral stochastic semigroups and we omit a lot of important classical results concerning dynamical systems and Markov processes. Now we give some remarks concerning classical results on this subject. The interested reader in ergodic properties of dynamical systems is referred to the monographs [26, 51, 55, 84]. We begin with the well known Krylov–Bogoliubov theorem [22]:

Theorem 19 *Let $\{\pi_t\}_{t\geq 0}$ be a (discrete or continuous time) dynamical system on a compact metric space. Then there exists a probability Borel measure μ invariant and ergodic with respect to $\{\pi_t\}_{t\geq 0}$.*

Theorem 19 can be generalized to Markov processes which satisfy Feller property. A family of time homogeneous Markov processes on a metric space X with the transition probability function $\mathscr{P}(t,x,A)$ is called a *Feller family* if the operators $P^*(t)f(x)=\int_X f(y)\mathscr{P}(t,x,dy)$ form a C_0-semigroup on the space $C(X)$ of continuous and bounded functions $f\colon X\to\mathbb{R}$. Every Feller family on a compact metric space has an invariant probability measure μ, which also means that the Markov process with the initial distribution μ and with the transition probability function $\mathscr{P}(t,x,A)$ is stationary. This result can be generalized to the case when X is a Polish space (complete and separable) but then we need to assume additionally that there exists a point $x \in X$ such that the set of measures $\{\mathcal{P}(t,x,\cdot): t > 0\}$ is tight. We recall that a family of measures M on a Polish space X is *tight* if for any $\varepsilon > 0$ there is a compact subset K_ε of X such that, for all measures $m \in M$, $m(X \setminus K_\varepsilon) < \varepsilon$.

The Krylov–Bogoliubov theorem does not provide us information if the invariant measure μ has a density with respect to the standard measure m on X. One of the classical tool to prove the existence of invariant densities is the *abstract ergodic theorem* due to Kakutani and Yosida (see [55] Theorem 5.2.1):

Theorem 20 *Let P be a stochastic operator on the space $L^1(X,\Sigma,m)$. If for a given $f \in D$ the sequence*

$$A_n f = \frac{1}{n}\sum_{k=0}^{n-1} P^k f$$

is weakly precompact, then it converges strongly to some $f_* \in D$ and $Pf_* = f_*$.

In particular, if for some $f \in D$ there exists a $g \in L^1$ such that $A_n f \le g$ for all $n \in \mathbb{N}$, then the sequence $\{A_n f\}$ is weakly precompact and, consequently, the operator P has an invariant density.

A kind of uniform weak compactness of iterates of a stochastic operator P leads to another interesting asymptotic property. A stochastic operator $P: L^1 \to L^1$ is called *constrictive* if there exists a weakly compact subset F of L^1 such that

$$\lim_{n\to\infty} d(P^n f, F) = 0 \quad \text{for } f \in D,$$

where $d(P^n f, F)$ denotes the distance, in L^1 norm, between the element f and the set F. The importance of constrictiveness is a consequence of the following theorem of Komorník [48]:

Theorem 21 (Spectral Decomposition Theorem) *The iterates of a constrictive operator P can be written in the form*

$$P^n f = \sum_{i=1}^{r} \lambda_i(f) g_{\alpha^n(i)} + Q_n f \quad \textit{for } f \in L^1,$$

where:

- *$g_1,\dots,g_r$ are densities with disjoint supports;*
- *$\lambda_1,\dots\lambda_r$ are linear functionals on L^1;*
- *α is a permutation of numbers $1,\dots,r$ such that $Pg_i = g_{\alpha(i)}$ and α^n denotes the nth iterate of α; and*
- *Q_n is a sequence of operators such that $\lim_{n\to\infty} \|Q_n f\| = 0$ for $f \in L^1$.*

The operator P which fulfills the thesis of Theorem 21 is called *asymptotically periodic*. In particular, if there exists an *upper function* $f \in L^1$ for the operator P, i.e.,

$$\lim_{n\to\infty} \|(P^n f - h)^+\| = 0 \quad \text{for every } f \in D, \tag{78}$$

then the operator P is weakly constrictive and, consequently, P is asymptotically periodic.

4 Applications

In this section we give a number of applications of stochastic operators and semigroups to diffusion and jump processes, population dynamics, cell cycle models, gene expression and gene evolution. Our examples appear in an order related to Sect. 2.

4.1 Models of the Cell Cycle

We start with a simple model of the cell cycle which describes the relation between the size (mass, volume) x of a mother and a daughter cell. We assume that $g(x)$ is the *growth rate* of a cell with size x, i.e., $x(t)$ satisfies the differential equation

$$\frac{dx}{dt} = g(x(t)). \tag{79}$$

Denote by $\varphi(x)$ the *division rate* of a cell with size x, i.e., a cell with size x replicates during a small time interval of length Δt with probability $\varphi(x)\Delta t + o(\Delta t)$. Finally, we assume that a daughter cell has a half size of the mother cell. Given growth and division rate functions and the initial size x_0 of a cell it is not difficult to find the distribution of its life-span and the distribution of it size at the point of division. Let $\pi(t, x_0)$ be the size of a cell at age t if its initial size were x_0, i.e., $\pi(t, x_0) = x(t)$, where x is the solution of the equation $x' = g(x)$ with the initial condition $x(0) = x_0$. The *life-span* of a cell is a random variable T which depends on the initial size x_0 of a cell. Let $\Phi(t) = \text{Prob}\,(T > t)$ be the *survival function*, i.e., the probability that the life-span of a cell with initial size x_0 is greater than t. Then

$$\text{Prob}\,(t < T \le t + \Delta t \mid T > t) = \frac{\Phi(t) - \Phi(t + \Delta t)}{\Phi(t)} = \varphi(\pi(t, x_0))\Delta t + o(\Delta t).$$

From this equation we obtain

$$\Phi'(t) = -\Phi(t)\varphi(\pi(t, x_0))$$

and after simple calculations we get

$$\Phi(t) = \exp\left\{-\int_0^t \varphi(\pi(s, x_0))\, ds\right\}. \tag{80}$$

For $y \geq x_0$ we define $t(x_0, y)$ to be the time t such that $\pi(t, x_0) = y$. Since

$$\frac{\partial t}{\partial y} \cdot g(\pi(t, x_0)) = 1,$$

we see that

$$\frac{\partial t}{\partial y} = \frac{1}{g(y)}$$

and

$$\frac{\partial}{\partial y}\left(\int_0^{t(x_0,y)} \varphi(\pi(s, x_0))\, ds\right) = \frac{\varphi(y)}{g(y)}.$$

Let Y be the size of the cell at the moment of division. Then

$$\begin{aligned} \text{Prob}\,(Y > y) = \text{Prob}\,(\pi(t, x_0) > y) &= \exp\left\{-\int_0^{t(x_0,y)} \varphi(\pi(s, x_0))\, ds\right\} \\ &= \exp\left(-\int_{x_0}^{y} \frac{\varphi(r)}{g(r)}\, dr\right) = \exp(Q(x_0) - Q(y)), \end{aligned} \tag{81}$$

where $Q(x) = \int\limits_0^x \frac{\varphi(r)}{g(r)}\, dr$. Let ξ be a random variable with exponential distribution, i.e., $\text{Prob}\,(\xi > x) = e^{-x}$. Then

$$\begin{aligned} \text{Prob}\,(Y > y) &= \exp(Q(x_0) - Q(y)) = \text{Prob}\,\big(\xi > Q(y) - Q(x_0)\big) \\ &= \text{Prob}\,\big(Q^{-1}\big(Q(x_0) + \xi\big) > y\big), \end{aligned}$$

which means that the random variables Y and $Q^{-1}\big(Q(x_0) + \xi\big)$ have the same distribution. From this it follows that if the random variable x_0 and x_1 are initial sizes of a mother and a daughter cell, respectively, then

$$x_1 \overset{d}{=} \tfrac{1}{2} Q^{-1}\big(Q(x_0) + \xi\big), \tag{82}$$

where ξ is a random variable independent of x_0 and with exponential distribution, and the symbol $\stackrel{d}{=}$ means that both variable have the same distribution. In general, if random variable x_n is the initial size of the cell in the n-th generation then

$$x_{n+1} \stackrel{d}{=} \tfrac{1}{2}Q^{-1}\big(Q(x_n) + \xi_n\big), \tag{83}$$

where (ξ_n) is a sequence of independent random variables with exponential distribution and all variables (ξ_n) are also independent of x_0.

The model (83) is of the form (17) with $S(y,z) = \frac{1}{2}Q^{-1}\big(Q(y) + z\big)$. Now we will find the density of the random variable $S(y,\xi_n)$. We have

$$\begin{aligned}\text{Prob}\,\big(\tfrac{1}{2}Q^{-1}\big(Q(y)+\xi_n\big) \le x\big) &= \text{Prob}\,\big(Q(y)+\xi_n \le Q(2x)\big)\\ &= \text{Prob}\,(\xi_n \le Q(2x) - Q(y)) = 1 - e^{Q(y)-Q(2x)}\end{aligned}$$

for $x \ge y/2$. Thus the random variable $S(y,\xi_n)$ has the density

$$k(x,y) = -\tfrac{\partial}{\partial x}e^{Q(y)-Q(2x)} = 2Q'(2x)e^{Q(y)-Q(2x)}1_{\{y\le 2x\}}.$$

If f_n is the density of the distribution function of x_n, $n \in \mathbb{N}$, then $f_{n+1} = Pf_n$, where P is a stochastic operator P on $L^1[0,\infty)$ given by $Pf(x) = \int_0^\infty k(x,y)f(y)\,dy$.

Integral stochastic operators appear in a two phase model of cell cycle proposed by Tyrcha [110] which generalizes the model of Lasota–Mackey [54] and the tandem model of Tyson–Hannsgen [111]. The cell cycle is the series of events that take place in a cell leading to its replication. In Tyrcha model a cell cycle have two phases: A which has a random duration t_A and B with constant length t_B. A cell can move from phase A to B with rate $\varphi(x)$. Also Tyrcha model can be described by a randomly perturbed dynamical system (see e.g. [94]) and the relation between the size of cells in consecutive generations is given by a stochastic operator

$$Pf(x) = \int_0^{\lambda(x)} \lambda'(x)Q'(\lambda(x))e^{Q(y)-Q(\lambda(x))}f(y)\,dy, \tag{84}$$

where $\lambda(x) = \pi(-t_B, 2x)$. The operator P given by (84) can also be used in another one-phase model of the cell cycle in which a cell is characterized by its *maturity* x. In this model we have the same assumption concerning the growth of x and the division rate, but we assume that if the mother cell has maturity x at the moment of division then a new born daughter cell has maturity $\gamma(x)$. If λ is the inverse function to γ then the operator P describes the relation between the maturity of cells in consecutive generations.

The asymptotic properties of the operator P given by (84) depend on the function $\alpha(x) = Q(\lambda(x)) - Q(x)$. We have

(a) If $\alpha(x) > 1$ for sufficiently large x, then P is *asymptotically stable*, i.e., there exists a density f^* such that

$$\lim_{n\to\infty} \|P^n f - f^*\| = 0 \qquad \text{for } f \in D.$$

(b) If $\alpha(x) \le 1$ for sufficiently large x, then P is *sweeping* or *zero type*, i.e.,

$$\lim_{n\to\infty} \int_0^c P^n f(x)\,dx = 0 \qquad \text{for } f \in D \text{ and } c > 0.$$

(c) If $\inf \alpha(x) > -\infty$, then the operator P is *completely mixing*, i.e.,

$$\lim_{n\to\infty} \|P^n f - P^n g\| = 0 \qquad \text{for } f, g \in D.$$

These results were proved, respectively, (a) in [35], (b) in [62], and (c) in [91].

4.2 Ehrenfest Model

In 1907 Tatiana and Paul Ehrenfest proposed a simple model of diffusion to explain the second law of thermodynamics. We have two boxes containing d balls and they are labeled $1, \ldots, d$. The balls represent the molecules of gas in the process of diffusion or the number of balls corresponds to temperature in the description of the heat exchange. Balls independently change boxes at a rate λ. Let ξ_t, $t \ge 0$, be the number of balls in the first box and let $x_i(t) = \text{Prob}\,(\xi_t = i)$. Observe that

$$\begin{aligned}
&\text{Prob}\,(\xi_{t+\Delta t} = i \mid \xi_t = i) = 1 - \lambda d \Delta t + o(\Delta t) \quad \text{for } 0 \le i \le d,\\
&\text{Prob}\,(\xi_{t+\Delta t} = i \mid \xi_t = i+1) = \lambda(i+1)\Delta t + o(\Delta t) \quad \text{for } 0 \le i \le d-1,\\
&\text{Prob}\,(\xi_{t+\Delta t} = i \mid \xi_t = i-1) = \lambda(d+1-i)\Delta t + o(\Delta t) \quad \text{for } 1 \le i \le d,\\
&\text{Prob}\,(\xi_{t+\Delta t} = i \mid \xi_t = j) = o(\Delta t) \quad \text{for } |i-j| > 1.
\end{aligned}$$

From the law of total probability we obtain

$$x_i(t+\Delta t) = (1-\lambda d\Delta t)x_i(t) + \lambda(i+1)\Delta t x_{i+1}(t) + \lambda(d+1-i)\Delta t x_{i-1}(t) + o(\Delta t),$$

which leads to

$$\frac{x_i(t+\Delta t) - x_i(t)}{\Delta t} = -\lambda d x_i(t) + \lambda(i+1)x_{i+1}(t) + \lambda(d+1-i)x_{i-1}(t) + \frac{o(\Delta t)}{\Delta t},$$

and the limit passage $\Delta t \to 0$ gives

$$x_i'(t) = \lambda(d+1-i)x_{i-1}(t) - \lambda d x_i(t) + \lambda(i+1)x_{i+1}(t)$$

for $i = 0, 1, \ldots, d$. In this way we obtain a system $x' = Ax$ of $(d+1)$ equations, where the matrix A has entries $a_{i,i} = -\lambda d$, $a_{i,i-1} = \lambda(d+1-i)$, $a_{i,i+1} = \lambda(i+1)$ and $a_{i,j} = 0$ otherwise. This equation generates a stochastic semigroup on the space $X = \{0, 1, \ldots, d\}$ with the counting measure. Since condition (T) holds, according to Theorem 13 this semigroup is asymptotically stable. The invariant density $\pi = (\pi_0, \ldots, \pi_d)$ is given by $\pi_i = \binom{d}{i}2^{-d}$ for $0 \le i \le d$.

4.3 Pure Jump Process

We consider the following process which can serve as a model of a kangaroo movement [75]. A kangaroo is jumping on a plane. If it is at a position x then after jump its new location y is described by a transition function $\mathcal{P}(x, A)$, i.e., $\mathcal{P}(x, A)$ is the probability that $y \in A$. We assume that the kangaroo jumps with intensity λ, i.e., $\lambda \Delta t + o(\Delta t)$ is the probability that the kangaroo changes its position in the time interval $[t, t + \Delta t]$. We also assume that $\mathcal{P}$ satisfies (1), and, consequently, there exists a stochastic operator P corresponding to $\mathcal{P}$. Let $\{P(t)\}_{t\ge 0}$ be a uniformly continuous stochastic semigroup with the generator $A = -\lambda I + \lambda P$. If u_0 is a density of the distribution of the location of the kangaroo at time 0, then $P(t)u_0$ describes its location at time t. The domain of the generator A is the whole space L^1. Let $u_0 \in L^1$ and $u(t) = P(t)u_0$. Then $u(t)$ satisfies the evolution equation

$$u'(t) = -\lambda u(t) + \lambda P u(t). \tag{85}$$

and according to (28) the solution $u(t)$ is given by the formula

$$u(t) = \sum_{k=0}^{\infty} \frac{(\lambda t)^k e^{-\lambda t}}{k!} P^k u_0. \tag{86}$$

It is simple to check that if there exists a density f_* such that $\lim_{k\to\infty} P^k f = f_*$ for every density f, then $\lim_{t\to\infty} u(t) = f_*$ if $u_0 \in D$.

4.4 Birth-Death Process

Now, we give two examples of applications of stochastic semigroups acting on the space l^1. The first one is a birth-death process described by the following system of equations

$$x_i'(t) = -a_i x_i(t) + b_{i-1}x_{i-1}(t) + d_{i+1}x_{i+1}(t) \tag{87}$$

for $i \geq 0$, where $b_{-1} = d_0 = 0$, $b_i \geq 0$, $d_{i+1} \geq 0$ for $i \geq 0$, $a_0 = b_0$, $a_i = b_i + d_i$ for $i \geq 1$. There are many different interpretations of the system (87). For example it can describe the time evolution of the size of a given population. Let ξ_t, $t \geq 0$, be the number of individuals at time t. If the population size is i then b_i and d_i are, respectively, birth and death rates. It means that if $\xi_t = i$ then $\operatorname{Prob}(\Delta\xi_t = 1) = b_i \Delta t$ and $\operatorname{Prob}(\Delta\xi_t = -1) = d_i \Delta t$, where $\Delta\xi_t = \xi_{t+\Delta t} - \xi_t$.

Two special types of birth-death processes are *pure birth process* if $d_i = 0$ for all $i \geq 0$ and *pure death process* if $d_i = 0$ for all $i \geq 0$. A *Poisson process* $N(t)$ with intensity λ is a pure birth process with $b_i = \lambda$ for all $i \geq 0$. The process with $b_i = bi$ and $d_i = di$ is related to a population in which each individual dies or gives birth to a new individual with the rates d and b, respectively, and it is called a *simple birth-death processes*. Also the process considered in 4.2 is a finite state birth-death process.

The matrix Q corresponding to equation (87) is a Kolmogorov matrix. Assume that $b_i \leq \alpha i + \beta$ for all $i \geq 0$ and some α and β. It means that the birth sequence b_i does not grow too quickly. We check that if $x \in l^1$, $x \geq 0$, satisfies $Q^* x = \lambda x$ for some $\lambda > 0$ then $x = 0$. Indeed $Q^* x = \lambda x$ holds if

$$d_i x_{i-1} - (b_i + d_i) x_i + b_i x_{i+1} = \lambda x_i$$

for all $i \geq 0$. Then since $x \geq 0$, $b_i \geq 0$, and $d_i \geq 0$ we have

$$b_i x_{i+1} \geq (\lambda + b_i) x_i$$

for all $i \geq 0$. If $b_{i_0} = 0$ some $i_0 \geq 0$, then $x_0 = \cdots = x_{i_0} = 0$, and we can consider only this inequality for $i > i_0$. Thus, without loss of generality, we can assume that $b_i > 0$ for all $i \geq 0$. Then

$$x_{i+1} \geq \Big(1 + \frac{\lambda}{b_i}\Big) x_i,$$

and consequently

$$x_n \geq x_0 \prod_{i=0}^{n-1} \Big(1 + \frac{\lambda}{b_i}\Big) \quad \text{for } n \geq 1.$$

Since the product $\prod_{i=0}^{\infty} \Big(1 + \frac{\lambda}{b_i}\Big)$ diverges we have $x \notin l^\infty$ and according to Theorem 4 the matrix Q generates a stochastic semigroup.

Now, let us consider again a birth-death process with $b_i > 0$ and $d_{i+1} > 0$ for all $i \geq 0$. Let us assume that there exists $\varepsilon > 0$ such that $b_i \leq d_i - \varepsilon$ for $i \geq k$.

Then the system (87) generates a stochastic semigroup and condition (T) holds. Let $v_i = i$ for $i \geq 0$, then

$$\sum_{i=0}^{\infty} v_i q_{ij} = (j-1)d_j - j(b_j + d_j) + (j+1)b_j = b_j - d_j \leq -\varepsilon$$

for $j \geq k$, which implies condition (73). Hence, the stochastic semigroup generated by the system (87) is asymptotically stable.

4.5 Paralog Families

Now, we present a model describing the evolution of paralog families in a genome [97]. Two genes present in the same genome are said to be *paralogs* if they are genetically identical. It is not a precise definition of paralogs but it is sufficient for our purposes. We are interested in the size distribution of paralogous gene families in a genome. We divide genes into classes. The i-th class consists of all i-element paralog families. Let x_i be a number of families in the i-th class. Based on experimental data Słonimski et al. [103] suggested that

$$x_i \sim \frac{1}{2^i i}, \quad i = 2, 3, \dots.$$

On the other hand, Huynen and van Nimwegen [41] claimed that

$$x_i \sim i^{-\alpha}, \quad i = 1, 2, 3, \dots,$$

where $\alpha \in (2, 3)$ depends on the size of the genome and α decreases if the total number of genes increases. It is very difficult to decide which formula is correct if only experimental data are taken into account because one can compare only first few elements of both sequences. We present a simple model of the evolution of paralog families which can help to solve this problem.

The model is based on three fundamental evolutionary events: gene loss, duplication and accumulated change called for simplicity mutation. A single gene during the time interval of length Δt can be:

- *duplicated* with probability $d\Delta t + o(\Delta t)$ and duplication of it in a family of the i-th class moves this family to the $(i + 1)$-th class,
- *removed* from the genome with probability $r\Delta t + o(\Delta t)$. For $i > 1$, removal of a gene from a family of the i-th class moves this family to the $(i - 1)$-th class; removal of a gene from one-element family results in elimination of this family from the genome. A removed gene is eliminated permanently from the pool of all genes,
- *changed* with probability $m\Delta t + o(\Delta t)$ and the gene starts a new one-element family and it is removed from the family to which it belonged.

Moreover, we assume that all elementary events are independent of each other. Let $s_i(t)$ be the number of i-element families in our model at the time t. It follows from the description of our model that

$$s_1'(t) = -(d+r)s_1(t) + 2(2m+r)s_2(t) + m\sum_{k=3}^{\infty} ks_k(t), \tag{88}$$

$$s_i'(t) = d(i-1)s_{i-1}(t) - (d+r+m)is_i(t) + (r+m)(i+1)s_{i+1}(t) \tag{89}$$

for $i \geq 2$. Let $s(t) = \sum_{i=1}^{\infty} s_i(t)$ be the total number of families. Then the sequence $(p_i(t))$, where $p_i(t) = s_i(t)/s(t)$ is the size distribution of paralogous gene families in a genome at time t.

We construct a stochastic semigroup related to (88). First, we change variables. Let

$$y_i(t) = e^{(r-d)t} is_i(t).$$

Then

$$y_1' = -(2d+m)y_1 + (m+r)y_2 + \sum_{k=1}^{\infty} my_k, \tag{90}$$

$$y_i' = -(d+r+m+\tfrac{d-r}{i})iy_i + diy_{i-1} + (r+m)iy_{i+1} \tag{91}$$

for $i \geq 2$. We claim that the system (90)–(91) generates a stochastic semigroup on l^1. Indeed, the system (90)–(91) can be written in the following way $y'(t) = Qy(t)$ and Q is a Kolmogorov matrix. Let $\lambda > 0$ and $x \in l^{\infty}$, $x \geq 0$ satisfies $Q^*x = \lambda x$. Here

$$\begin{aligned}
(Q^*x)_1 &= -2dx_1 + 2dx_2,\\
(Q^*x)_2 &= (2m+r)x_1 - (r+2m+3d)x_2 + 3dx_3,\\
(Q^*x)_n &= mx_1 + (n-1)(r+m)x_{n-1} - (r(n-1) + d(n+1) + mn)x_n\\
&\quad + (n+1)dx_{n+1}
\end{aligned}$$

for $n \geq 3$. We consider the case of $d \neq 0$ (the case of $d = 0$ is trivial). The sequence $x = (x_i)_{i\geq 1}$ satisfies equation $Q^*x = \lambda x$ iff

$$\begin{aligned}
x_2 &= \Big(1 + \frac{\lambda}{2d}\Big)x_1,\\
x_3 &= \Big(1 + \frac{r+2m+\lambda}{3d}\Big)x_2 - \frac{r+2m}{3d}x_1,\\
x_{n+1} &= \Big(1 + \frac{(n-1)r + nm + \lambda}{(n+1)d}\Big)x_n - \frac{(n-1)(r+m)}{(n+1)d}x_{n-1} - \frac{m}{(n+1)d}x_1
\end{aligned}$$

for $n \geq 3$. The above system of equations can be replaced by one equation

$$x_{n+1} = \Big(1 + \frac{\lambda}{(n+1)d}\Big)x_n + \frac{(n-1)(r+m)}{(n+1)d}(x_n - x_{n-1}) + \frac{m}{(n+1)d}(x_n - x_1)$$

for $n \geq 1$. Hence, the sequence (x_n) is increasing. Thus

$$x_{n+1} \geq \Big(1 + \frac{\lambda}{(n+1)d}\Big)x_n,$$

and consequently

$$x_n \geq x_1 \prod_{i=2}^{n} \Big(1 + \frac{\lambda}{di}\Big) \quad \text{for } n \geq 2.$$

Since the product $\prod_{i=1}^{\infty}(1 + \lambda d^{-1} i^{-1})$ diverges we have $x \notin l^{\infty}$ and according to Theorem 4 the matrix Q generates a stochastic semigroup.

Now we prove the following result about asymptotic behaviour of the solution $(s_i(t))$ of (88) and (89).

Theorem 22 *Assume that $m > 0$. Let E be the space of sequences (x_i) which satisfy the condition $\sum_{i=1}^{\infty} i|x_i| < \infty$. There exists a sequence $(s_i^*) \in E$ such that for every solution $(s_i(t))$ of (88) and (89) with $(s_i(0)) \in E$ we have*

$$\lim_{t\to\infty} e^{(r-d)t} s_i(t) = Cs_i^* \tag{92}$$

for every $i = 1, 2, \ldots$ and C dependent only on the sequence $(s_i(0))$. Moreover if $d = r$ then

$$\lim_{t\to\infty} s_i(t) = C\frac{\alpha^i}{i}, \tag{93}$$

where $\alpha = \dfrac{r}{r+m}$.

In the case $d = r$ the total number of genes in a genome is constant. It means that the genome is in a stable state. In this case the distribution of paralog families is similar to that stated in Słonimski's conjecture, and both distributions are the same if $r = d = m$.

Proof First we check that the stochastic semigroup $\{P(t)\}_{t\geq 0}$ generated by the system (90)–(91) is asymptotically stable. From equation (90):

$$y_1' = -(2d + m)y_1 + (m + r)y_2 + \sum_{k=1}^{\infty} my_k$$

applied to densities $y(t)$, i.e., nonnegative sequences such that $\sum_{i=1}^{\infty} y_i(t) = 1$, we obtain

$$y_1'(t) \geq -(2d+m)y_1(t) + m.$$

This implies that

$$\liminf_{t\to\infty} y_1(t) \geq \frac{m}{2d+m}. \tag{94}$$

Let $h = (\frac{m}{2d+m}, 0, 0, \ldots)$. Then h is a lower function and the semigroup $\{P(t)\}_{t\geq 0}$ is asymptotically stable. Let $y^* = (y_i^*)$ be an invariant density for $\{P(t)\}_{t\geq 0}$. If we return to the original system we obtain (92) with $s_i^* = y_i^*/i$. If $r = d$, then the invariant density is of the form $y_i^* = \dfrac{m}{r}\Big(\dfrac{r}{r+m}\Big)^i$, which gives (93). □

Another proof of Theorem 22 can be done by Theorem 14. The Kolmogorov matrix Q related to the stochastic semigroup $\{P(t)\}_{t\geq 0}$ generated by the system (90)–(91) satisfies $q_{i-1,i} > 0$ for $i \geq 2$ and $q_{i+1,i} > 0$ for $i \geq 1$, therefore, condition (T) holds and, consequently, the semigroup satisfies the Foguel alternative. From inequality (94) we conclude that this semigroup is not sweeping from the set $\{1\}$. Hence, this semigroup is asymptotically stable.

4.6 *Examples of Diffusion Semigroups*

Though the semigroups generated by Fokker–Planck equations has originated from diffusion processes, now they are widely used to describe variety of phenomena with random noise. We present some applications of the theory developed in 3.3 and 3.4 to diffusion semigroups. We begin with one dimensional diffusion on some open interval $\Pi = (\alpha, \beta)$. The Fokker–Planck equation is the form

$$\frac{\partial u}{\partial t} = \frac{\partial^2}{\partial x^2}(a(x)u) - \frac{\partial}{\partial x}(b(x)u). \tag{95}$$

We assume that $a(x) > 0$ for all $x \in \Pi$. A stationary solution u of (95) is a solution of the ordinary differential equation

$$\big(a(x)u(x)\big)'' - \big(b(x)u(x)\big)' = 0. \tag{96}$$

Let $x_0 \in \Pi$ and

$$f_*(x) = \frac{1}{a(x)} \exp\Big(\int_{x_0}^{x} \frac{b(s)}{a(s)}\, ds\Big). \tag{97}$$

Then all solutions of (96) are of the form

$$u(x) = c_1 f_*(x) + c_2 f_*(x) \int_{x_0}^{x} \frac{dy}{f_*(y)}, \qquad c_1, c_2 \in \mathbb{R}.$$

Observe that if $\int_{-\infty}^{\infty} f_*(x)\,dx < \infty$ then the semigroup $\{P(t)\}_{t\geq 0}$ generated by (95) has an invariant density $f_*/\|f_*\|$ and consequently it is asymptotically stable. If $\int_{\alpha}^{\beta} f_*(x)\,dx = \infty$ then, after careful checking, we deduce that the semigroup $\{P(t)\}_{t\geq 0}$ has no invariant density, which means that it is sweeping from compact subsets of Π. In some cases we are able to show stronger results. For example, if f_* is not integrable but $\int_{c}^{\beta} f_*(x)\,dx < \infty$ then the semigroup $\{P(t)\}_{t\geq 0}$ is also sweeping from the interval $[c, \beta)$. Indeed, since $f_* > 0$, $Af_* = 0$ and $\int_{c}^{\beta} f_*(x)\,dx < \infty$ the semigroup $\{P(t)\}_{t\geq 0}$ and the set $[c, \beta)$ satisfy condition (KT). Thus Theorem 10 implies that the semigroup $\{P(t)\}_{t\geq 0}$ is sweeping from $[c, \beta)$.

Example 7 Consider the following equation

$$dX_t = \sigma X_t\,dW_t + bX_t\,dt. \tag{98}$$

Equation (98) has an universal character and can describe, e.g., the price of a stock or the size of the population. In the case of the population model, b is the growth rate which is perturbed by an external source of additive noise $\sigma\,dW_t$. The solution of (98) is given by

$$X_t = X_0 e^{(a-\sigma^2/2)t+\sigma W_t}. \tag{99}$$

This equations generates a stochastic semigroup on the space $L^1(0,\infty)$ which is related to the Fokker–Planck equation

$$\frac{\partial u}{\partial t} = \frac{\sigma^2}{2}\frac{\partial^2}{\partial x^2}(x^2 u) - \frac{\partial}{\partial x}(bxu). \tag{100}$$

We take $x_0 = 1$ and find the function f_* defined earlier

$$f_*(x) = \frac{2}{\sigma^2 x^2}\exp\left(\int_1^x \frac{2bs}{\sigma^2 s^2}\,ds\right) = \frac{2}{\sigma^2 x^2}\exp\left(\frac{2b}{\sigma^2}\log x\right) = \frac{2}{\sigma^2}x^{2b/\sigma^2-2}. \tag{101}$$

Since $\int_0^{\infty} f_*(x)\,dx = \infty$ the related stochastic semigroup is sweeping from compact subsets of $(0,\infty)$. Moreover, if $2b < \sigma^2$, then we have $\int_c^{\infty} f_*(x)\,dx < \infty$ for $c > 0$, thus this semigroup is also sweeping from all intervals $[c,\infty)$. If $2b > \sigma^2$, then we have $\int_0^{c} f_*(x)\,dx < \infty$ for $c > 0$, thus this semigroup is sweeping from all intervals $[0, c)$.

Example 8 Now we consider one dimensional *Langevin equation*

$$dV_t = -bV_t\,dt + \sigma\,dW_t, \tag{102}$$

which describes the velocity V_t of a moving particle. Here, $b > 0$ is a coefficient of friction, and σ is a diffusion coefficient. The solution of (102) is of the form

$$V_t = e^{-bt}V_0 + \sigma \int_0^t e^{-b(t-s)}\,dW_s. \tag{103}$$

The process V_t is called *Orstein–Uhlenbeck process* and if V_0 is a Gaussian random variable then so are $V_t, t > 0$. The Fokker–Planck equation is of the form

$$\frac{\partial u}{\partial t} = \frac{\sigma^2}{2}\frac{\partial^2 u}{\partial x^2} + \frac{\partial}{\partial x}(bxu). \tag{104}$$

We take $x_0 = 0$ and find

$$f_*(x) = \frac{2}{\sigma^2}\exp\Big(-\int_0^x \frac{2bs}{\sigma^2}\,ds\Big) = \frac{2}{\sigma^2}\exp\Big(-\frac{bx^2}{\sigma^2}\Big). \tag{105}$$

Since $\int_0^\infty f_*(x)\,dx < \infty$, the related stochastic semigroup has an invariant density $u_* = f_*/\|f_*\|$, therefore, it is asymptotically stable.

Example 9 In Remark 2 the following stochastic version of the logistic equation has appeared

$$dX_t = (\alpha X_t - \beta X_t^2)\,dt + \gamma X_t\,dW_t, \quad \alpha, \beta, \gamma > 0. \tag{106}$$

This equation generates a stochastic semigroup defined on $L^1(0,\infty)$. It is easy to check that the function

$$f_* = x^{2\alpha/\sigma^2 - 2}\exp\Big(-\frac{2\beta x}{\sigma^2}\Big)$$

is a stationary solution of the Fokker–Planck equation. If $2\alpha > \sigma^2$ then f_* is an integrable function. Hence, the semigroup is asymptotically stable in this case. If $2\alpha \le \sigma^2$ then the semigroup is sweeping from all sets $[c,\infty)$, $c > 0$.

Now we consider a non-degenerate diffusion in $\mathbb{R}^d$, i.e., with diffusion coefficients which satisfy (51). Then the stochastic semigroup generated by the Fokker–Planck equation (46) is an integral semigroup with the continuous and positive kernel. According to Corollary 5 this semigroup is asymptotically stable or is sweeping with respect to compact sets. Assume then there exist a non-negative C^2-function V, $\varepsilon > 0$ and $r \ge 0$ such that

$$A^*V(x) \le -\varepsilon \qquad \text{for} \quad \|x\| \ge r,$$

then V is a Hasminskiĭ function for our semigroup and, consequently, this semigroup is asymptotically stable. This theorem generalizes earlier results [32, 100].

Remark 15 The semigroup related to a non-degenerate diffusion in $\mathbb{R}^d$ is asymptotically stable or sweeping with respect to the family of sets with finite Lebesgue measures.

The problem of the long-time behaviour of stochastic semigroups related to a degenerate diffusion or a diffusion in subsets of $\mathbb{R}^d$ is much more difficult. Such diffusion processes appear in biological models and usually involve advanced method (see 2.10). Some techniques useful in studying semigroup related to degenerate diffusion are presented in the papers [93, 96] devoted to a prey-predator type stochastic system. The interested reader in applications of degenerate diffusion to biological models is also referred to the papers [45, 61, 102].

4.7 Gene Expression

Now we consider a simple piecewise deterministic Markov process modelling gene expression. Gene expression is a complex process which involves three processes: gene activation/inactivation, mRNA transcription/decay and protein translation/decay. Now we consider a very simple model when proteins production is regulated by a single gene and we omit the intermediate process of mRNA transcription. A gene can be in an active or an inactive state and it can be transformed into an active state or into an inactive state, with intensities q_0 and q_1, respectively. The rates q_0 and q_1 depend on the number of protein molecules $x(t)$. If the gene is active then proteins are produced with a constant speed p. In both states of the gene protein molecules undergo the process of degradation with rate μ. It means that the process $x(t)$, $t \geq 0$, satisfies the equation

$$x'(t) = pa(t) - \mu x(t), \tag{107}$$

where $a(t) = 1$ if the gene is active and $a(t) = 0$ in the opposite case. Since the right-hand side of equation (107) is negative for $x(t) > \frac{p}{\mu}$ we can restrict values of $x(t)$ to the invariant interval $\left[0, \frac{p}{\mu}\right]$. Thus, we have two one-dimensional flows on $\left[0, \frac{p}{\mu}\right]$ with $b^1(x) = -\mu x$ and $b^2(x) = p - \mu x$. The process $(x(t), a(t))$ has jump points when the gene changes its activity and it is an example of the process studied in 2.11. We check that if q_0 and q_1 are continuous and positive functions then the stochastic semigroup $\{P(t)\}_{t\geq 0}$ related to this process is asymptotically stable. The semigroup $\{P(t)\}_{t\geq 0}$ is defined on the space $L^1(X, \mathcal{B}(X), m)$, where $X = \left[0, \frac{p}{\mu}\right] \times \{0, 1\}$. Since the space X is compact, according to Corollary 4 it is sufficient to check that conditions (a) and (b) of Theorem 11 hold. In order to check condition (b) we apply the criterion from Remark 11. In our case $b_1(x) - b_0(x) = p \neq 0$ for all $x \in [0, 1]$, therefore, the vector $b_1(x) - b_0(x)$ span $\mathbb{R}$. Checking condition (a) is a little more technical but it follows from the fact that any two states (x,i), (y,j) can be joined by a path of the process $(x(t),a(t))$.

A more advance model which includes mRNA molecules and mRNA transcription was introduced by Lipniacki et al. [60] and studied in [20]. In this model $x_1(t)$ is the number of mRNA molecules and $x_2(t)$ the number of protein molecules. If the gene is active then mRNA transcript molecules are synthesized with a constant speed r. The protein translation proceeds with the rate $px_1(t)$, where p is a constant. The mRNA and protein degradation rates are μ_r and μ_p, respectively. Now, instead of (107) we have

$$\begin{cases} x_1'(t) = ra(t) - \mu_r x_1(t), \\ x_2'(t) = px_1(t) - \mu_p x_2(t) \end{cases} \tag{108}$$

and $a(t)$ the same function as in the previous model. The main result of the paper [20] is also the asymptotic stability of the semigroup $\{P(t)\}_{t\ge 0}$ related to this process. The proof is more advanced and its main idea is the following. First it is shown that the transition function of the related stochastic process has a kernel (integral) part. Then we find a set $\mathcal{E} \subset X$, which is a stochastic attractor and condition (a) and (b) hold on this set. Since $\mathcal{E}$ is a compact set, from Corollary 4 it follows that the semigroup is asymptotically stable.

4.8 Flows with Jumps in Population Dynamics

A large part of population dynamics are structured models. Models of this type describe the densities of distribution of some characteristics of individuals like their age, size or position in the space. Since living creatures reproduce from time to time, the evolution of these characteristics have a "jumping nature". Now, we present a very simple model, which leads directly to the process considered in 2.12. After that we provide some information about more advanced models.

Example 10 We consider a size structured model of a cellular population with conservation of the total number of cells. In this model a cell is characterized by its size $x(t)$ which grows according to the equation $x'(t) = b(x(t))$. We also assume that the death and division rates for a cell with size x are the same and given by $d(x)$. This assumption guarantees that the total number of cells is constant. We assume an equal division, i.e., a daughter cell has size a half of the size of the mother cell. By $u(t,x)$ we denote the density of distribution of x at time t. Then u satisfies equation (58) with $\lambda(x) = 2d(x)$ and $Pf(x) = 2f(2x)$ because

$$\mathcal{P}(x,B) = \begin{cases} 1, & \text{if } x/2 \in B, \\ 0, & \text{if } x/2 \notin B. \end{cases}$$

Very important role in population dynamics play two models. The first one is the Sharpe–Lotka–McKendrick *age-structured model* [69, 101] given by a partial differential equation with boundary and initial conditions:

$$\begin{aligned} &\frac{\partial u(t,a)}{\partial t} + \frac{\partial u(t,a)}{\partial a} = -\mu(a)u(t,a) \\ &u(t,0) = 2\int_0^\infty u(t,a)b(a)\,da, \quad u(0,a) = u_0, \end{aligned} \tag{109}$$

where $u(t,a)$ is a (non-probabilistic) density of age-distribution, μ and b are death and birth rates. The second model is a size-structure model of a cellular population introduced by Bell and Anderson [15]:

$$\frac{\partial}{\partial t}u(t,x) + \frac{\partial}{\partial x}(g(x)u(t,x)) = -(\mu(x)+b(x))u(t,x) + 4b(2x)u(t,2x) \tag{110}$$

with $u(0,x) = u_0(x)$, where u is the density of size distribution and g is the growth rate. Both models were intensively studied and generalized [29, 37, 70, 95, 114].

In [10] it is considered a model whose special cases are both models (109) and (110). The model is given by an evolution equation $u' = Au$. The main result of this paper is *asynchronous exponential growth* of the population. Precisely, under suitable assumptions there exist $\lambda \in \mathbb{R}$, an integrable function f_*, and a linear functional c on the set of initial conditions such that

$$e^{-\lambda t}u(t,x) \to c(u_0)f_*(x) \quad \text{in } L^1 \text{ as } t \to \infty. \tag{111}$$

The proof goes as follows. First we show that A is an infinitesimal generator of a continuous semigroup $\{T(t)\}_{t\ge0}$ of linear operators on the space $L^1(F)$ with the Lebesgue measure on some interval F. Then we prove that there exist $\lambda \in \mathbb{R}$, and continuous and positive functions v and w such that $Av = \lambda v$ and $A^*w = \lambda w$. From this it follows that the semigroup $\{P(t)\}_{t\ge0}$ given by $P(t) = e^{-\lambda t}T(t)$ is a stochastic semigroup on the space $L^1(F, \mathcal{B}(F), m)$, where m is a measure given by $m(B) = \int_B w(x)\,dx$. We can find $\alpha > 0$ such that the function $f_* = \alpha v$ is an invariant density with respect to $\{P(t)\}_{t\ge0}$. Then we check that the semigroup $\{P(t)\}_{t\ge0}$ is partially integral. Finally, from Theorem 8 we conclude that this semigroup is asymptotically stable. Since the Lebesgue measure and the measure m are equivalent we have (111).

4.9 Remarks on Applications of Nonlinear Stochastic Operators and Semigroups

Nonlinear stochastic operators and semigroups appear in models which contain binary operations, i.e., maps of the form $f: X \times X \to X$. This operations can be random. For example the genotype of a child is a result of some random operation—each chromosome of a child is chosen randomly from parents.

The simplest examples of binary operations appear in the classical Mendelian genetics [38]. Each gene can have some variant forms called alleles. We consider the simplest case of two alleles A and a. Assume also that each individual have a pair of alleles of a given gen: *AA* or *Aa* or *aa* called for simplicity the genotype and the genotype decides about some phenotypic trait. A child inherits randomly one allele from both his parents. For example if parents have genotype *AA* and *aa* then their children have only the pair *Aa*. Now we consider a population with random mating. If the frequencies of genotypes *AA*, *Aa*, and *aa* in the parent generation are x_1, y_1 and z_1, respectively, then in the offspring generation these frequencies are the following:

$$x_2 = x_1^2 + 2x_1\Big(\frac{y_1}{2}\Big) + \Big(\frac{y_1}{2}\Big)^2 = \Big(x_1 + \frac{y_1}{2}\Big)^2 = p^2,$$

$$y_2 = 2x_1z_1 + 2x_1\Big(\frac{y_1}{2}\Big) + 2\Big(\frac{y_1}{2}\Big)^2 + 2\Big(\frac{y_1}{2}\Big)z_1 = 2\Big(x_1 + \frac{y_1}{2}\Big)\Big(z_1 + \frac{y_1}{2}\Big) = 2pq,$$

$$z_2 = z_1^2 + 2z_1\Big(\frac{y_1}{2}\Big) + \Big(\frac{y_1}{2}\Big)^2 = \Big(z_1 + \frac{y_1}{2}\Big)^2 = q^2,$$

where $p = x_1 + \frac{y_1}{2}$ and $q = z_1 + \frac{y_1}{2}$ are, respectively, frequencies of alleles A and a in the population. The operator P given by $P(x_1, y_1, z_1) = (x_2, y_2, z_2)$ is a nonlinear stochastic operator and from the above formulas *Hardy–Weinberg principle* follows:

$$P^n(x_1, y_1, z_1) = P(x_1, y_1, z_1) = (p^2, 2pq, q^2), \quad \text{for } n \ge 1,$$

i.e., the frequencies of genotypes stabilize in the offspring generation and remains constant. Observe that the expected frequencies $(p^2, 2pq, q^2)$ depends on the initial frequencies (x_1, y_1, z_1), what makes the difference between nonlinear stochastic operator and linear one. In linear operators the limit densities were usually independent of the initial density. One can consider more advance models with the greater number of alleles and when a genotype depends on three or more alleles. Also in this case the frequencies of genotypes stabilize but not in one generation.

Examples of physical binary processes are coagulation and collision. Both phenomena leads to interesting mathematical models. The first model of coagulation processes is due to Smoluchowski [113] and it is given by the following equation

$$\frac{dx_i}{dt} = \frac{1}{2}\sum_{j=1}^{i-1} K_{j,i-j}x_jx_{i-j} - \sum_{j=1}^{\infty} K_{i,j}x_ix_j. \tag{112}$$

The matrix $K = [K_{ij}]$ describes the rate at which particles of size i coagulate with particles of size j and $x_i(t)$ is the frequency of particles of size i at time t. A continuous size version of this model is following:

$$\frac{\partial u(t,x)}{\partial t} = \frac{1}{2}\int_0^x K(x-y,y)u(t,x-y)u(t,y)\,dy - \int_0^\infty K(x,y)u(t,x)u(t,y)\,dy \tag{113}$$

Under suitable assumptions on the matrix or the function K, equations (112) and (113) generate nonlinear stochastic semigroups on l^1 and $L^1[0,\infty)$, respectively. Some modification of equation (112) is used to model aggregation processes in phytoplankton populations [43]. We cannot expect the existence of invariant densities for coagulation equations because the frequencies of particles with small size decrease. We can consider also models which describe both fragmentation and coagulation processes [2, 4, 9, 17, 52, 98]. In this case one can expect some results concerning size stabilization. In the coagulation-fragmentation model of size-structured phytoplankton aggregates [4] the four process are involved: growth, death, fragmentation, and coagulation of aggregates and the coagulation operator differs considerably from the operator K in (113) and it is of the form

$$Kf(x) = \frac{\int_0^x f(x-y)f(y)(x-y)yg(x-y)g(y)\,dy}{x\int_0^\infty zg(z)f(z)\,dz},$$

where $g(x)$ is the ability of an aggregate of size x to glue to another aggregate.

The second type of equations which generate nonlinear stochastic semigroups are related to the Boltzmann kinetic theory of gases, which describes a gas as a large number of moving particles which collide with each other. The fundamental object in this theory is the Boltzmann equation for the density distribution function of velocity of particles [15, 31]. One of versions of the Boltzmann equation is the Tjon–Wu equation (see [21, 50, 106]) on the density distribution function of the particles' energy. The Tjon–Wu equation is a special case of the equation (65)

$$u'(t) + u(t) = Pu(t), \quad u(0) = u_0 \in D, \tag{114}$$

considered in 2.15. The classical Tjon–Wu equation is defined on densities in the space $L^1[0,\infty)$ and the operator P is of the form

$$Pf(x) = \int_x^\infty \frac{1}{y}\int_0^y f(y-z)f(z)\,dz\,dy. \tag{115}$$

One can consider more general stochastic operator P of the form

$$Pf(x) = \int_0^\infty \int_0^\infty k(x,y,z)f(y)f(z)\,dy\,dz, \tag{116}$$

where $k\colon [0,\infty)^3 \to [0,\infty)$ is a measurable function such that

$$\int_0^\infty \int_0^\infty k(x,y,z)\,dx = 1 \quad \text{for } x, y \geq 0.$$

In the case of (115) we have

$$k(x,y,z) = \frac{1}{y+z} h\left(\frac{x}{y+z}\right) \tag{117}$$

with $h = 1_{[0,1]}$. The asymptotic stability of the classical Tjon–Wu equation was proven by Kiełek (see [47]) and when P is given by (116) and (117) with arbitrary h by Lasota and Traple (see [53,56]). It is interesting that equation (116) can be used in modelling of phenotype-structured populations [99]. In [99] one can find also a new result concerning asymptotic stability of equation (114) with a general kernel given by (116).

Acknowledgements This research was partially supported by the State Committee for Scientific Research (Poland) Grant No. N N201 608240. The author is a supervisor in the International Ph.D. Projects Programme of Foundation for Polish Science operated within the Innovative Economy Operational Programme 2007–2013 (Ph.D. Programme: Mathematical Methods in Natural Sciences).

References

1. S. Aida, S. Kusuoka, D. Strook, On the support of Wiener functionals, in *Asymptotic Problems in Probability Theory: Wiener Functionals and Asymptotic*, eds. by K.D. Elworthy, N. Ikeda, Pitman Research Notes in Math. Series, vol. 284 (Longman Scientific, Harlow, 1993) pp. 3–34
2. H. Amann, Coagulation-fragmentation processes. Arch. Rational Mech. Anal. **151**, 339–366 (2000)
3. W. Arendt et al., *One-Parameter Semigroups of Positive Operators*, ed. by R. Nagel, Lecture Notes Math., vol. 1184 (Springer, Berlin, 1986)
4. O. Arino, R. Rudnicki, Phytoplankton dynamics. C. R. Biol. **327**, 961–969 (2004)
5. V. Balakrishnan, C. Van den Broeck, P. Hanggi, First-passage times of non-Markovian processes: the case of a reflecting boundary. Phys. Rev. A **38**, 4213–4222 (1988)
6. J. Banasiak, On an extension of the Kato–Voigt perturbation theorem for substochastic semigroups and its application. Taiwanese J. Math. **5**, 169–191 (2001)
7. J. Banasiak, L. Arlotti, Strictly substochastic semigroups with application to conservative and shattering solutions to fragmentation equations with mass loss. J. Math. Anal. Appl. **293**, 693–720 (2004)
8. J. Banasiak, L. Arlotti, *Perturbations of Positive Semigroups with Applications*. Springer Monographs in Mathematics (Springer, London, 2006)
9. J. Banasiak, W. Lamb, Coagulation, fragmentation and growth processes in a size structured population. Discrete Contin. Dyn. Syst. Ser. B **11**, 563–585 (2009)
10. J. Banasiak, K. Pichór, R. Rudnicki, Asynchronous exponential growth of a general structured population model. Acta Appl. Math. **119**, 149–166 (2012)

11. W. Bartoszek, T. Brown, On Frobenius–Perron operators which overlap supports. Bull. Pol. Acad. **45**, 17–24 (1997)
12. M.F. Barnsley, *Fractals Everywhere* (Academic Press, New York, 1993)
13. Y. Bakhtin, T. Hurth, Invariant densities for dynamical system with random switching. Nonlinearity **25**, 2937–2952 (2012)
14. C.J.K. Batty, Z. Brzeźniak, D.A. Greenfield, A quantitative asymptotic theorem for contraction semigroups with countable unitary spectrum. Studia Math. **121**, 167–183 (1996)
15. G.I. Bell, E.C. Anderson, Cell growth and division I. A Mathematical model with applications to cell volume distributions in mammalian suspension cultures. Biophys. J. **7**, 329–351 (1967)
16. G. Ben Arous, R. Léandre, Décroissance exponentielle du noyau de la chaleur sur la diagonale (II). Probab. Theory Relat. Fields **90**, 377–402 (1991)
17. J. Bertoin, *Random Fragmentation and Coagulation Processes* (Cambridge University Press, Cambridge, 2006)
18. V. Bezak, A modification of the Wiener process due to a Poisson random train of diffusion-enhancing pulses. J. Phys. A **25**, 6027–6041 (1992)
19. H.A.P. Blom, J. Lygeros (eds.), *Stochastic Hybrid Systems: Theory and Safety Critical Applications*. Lecture Notes in Control and Information Sciences, vol. 337 (Springer, Berlin/Heidelberg, 2006)
20. A. Bobrowski, T. Lipniacki, K. Pichór, R. Rudnicki, Asymptotic behavior of distributions of mRNA and protein levels in a model of stochastic gene expression. J. Math. Anal. Appl. **333**, 753–769 (2007)
21. A.V. Bobylev, Exact solutions of the Boltzmann equation. Soviet Phys. Dokl. **20**, 822–824 (1976)
22. N.N. Bogoluboff, N.M. Kriloff, La théorie générale de la measure dans son application à l'étude des systèmes dynamiques de la méchanique non-linéare. Ann. Math. **38**, 65–113 (1937)
23. Z. Brzeźniak, B. Szafirski, Asymptotic behaviour of L^1 norm of solutions to parabolic equations. Bull. Pol. Acad. **39**, 1–10 (1991)
24. C.G. Cassandras, J. Lygeros (eds.), *Stochastic Hybrid Systems*. Control Engineering Series, vol. 24 (CRC, Boca Raton, 2007)
25. S. Chandrasekhar, G. Münch, The theory of fluctuations in brightness of the Milky-Way. Astrophys. J. **125**, 94–123 (1952)
26. I.P. Cornfeld, S.V. Fomin, Y.G. Sinai, *Ergodic Theory*. Grundlehren der Mathematischen Wissenschaften, vol. 245 (Springer, New York, 1982) x+486 pp.
27. I. Csiszár, Information–type measure of difference of probability distributions and indirect observations. Studia Sci. Math. Hungar. **2**, 299–318 (1967)
28. M.H.A. Davis, Piecewise-deterministic Markov processes: a general class of nondiffusion stochastic models. J. R. Stat. Soc. Ser. B **46**, 353–388 (1984)
29. O. Diekmann, H.J.A.M. Heijmans, H.R. Thieme, On the stability of the cell size distribution. J. Math. Biol. **19**, 227–248 (1984)
30. J. Dieudonne, Sur le théorème de Radon–Nikodym. Ann. Univ. Grenoble **23**, 25–53 (1948)
31. R.J. DiPerna, P.-L. Lions, On the Cauchy problem for Boltzmann equations: global existence and weak stability. Ann. Math. **130**, 321–366 (1989)
32. T. Dłotko, A. Lasota, Statistical stability and the lower bound function technique, in *Semigroups theory and applications*, vol. I, eds. by H. Brezis, M. Crandall, F. Kappel. Pitman Research Notes in Mathematics, vol. 141 (Longman Scientific & Technical, 1986)
33. S. Ethier, T. Kurtz, *Markov Processes: Characterization and Convergence* (Wiley, New York, 1986)
34. S.R. Foguel, *The Ergodic Theory of Markov Processes* (Van Nostrand Reinhold, New York, 1969)
35. H. Gacki, A. Lasota, Markov operators defined by Volterra type integrals with advanced argument. Ann. Polon. Math. **51** (1990), 155–166.
36. I.I. Gihman, A.V. Skorohod, *Stochastic Differential Equations* (Springer, New York, 1972)
37. M. Gyllenberg, H.J.A.M. Heijmans, An abstract delay-differential equation modelling size dependent cell growth and division. SIAM J. Math. Anal. **18**, 74–88 (1987)

38. M.B. Hamilton, *Population Genetics* (Wiley, Chichester, 2009)
39. R.Z. Hasminskiĭ, Ergodic properties of recurrent diffusion processes and stabilization of the solutions of the Cauchy problem for parabolic equations. Teor. Verojatn. Primenen. **5**, 196–214 (1960) (in Russian)
40. P. Hennequin, A. Tortrat, *Theorie des probabilities et quelques applications* (Masson et Cie, Paris, 1965)
41. M.A. Huynen, E. van Nimwegen, The frequency distribution of gene family size in complete genomes. Mol. Biol. Evol. **15**, 583–589 (1998)
42. J.E. Hutchinson, Fractals and self-similarity. Indian Univ. Math. J. **30**, 713–747 (1981)
43. G.A. Jackson, A model of the formation of marine algal flocs by physical coagulation processes. Deep Sea Res. **37**, 1197–1211 (1990)
44. B. Jamison, S. Orey, Markov chains recurrent in the sense of Harris. Z. Wahrsch. Verw. Gebiete **8**, 41–48 (1967)
45. C. Ji, D. Jiang, N. Shi, Analysis of a predator–prey model with modified Leslie–Gower and Holling-type II schemes with stochastic perturbation. J. Math. Anal. Appl. **359**, 482–498 (2009)
46. T. Kato, On the semi-groups generated by Kolmogoroff's differential equations. J. Math. Soc. Jpn. **6**, 1–15 (1954)
47. Z. Kiełek, Asymptotic behaviour of solutions of the Tjon–Wu equation. Ann. Polon. Math. **52** (1990), 109–118.
48. J. Komorník, Asymptotic periodicity of the iterates of Markov operators. Tôhoku Math. J. **38**, 15–27 (1986)
49. T. Komorowski, J. Tyrcha, Asymptotic properties of some Markov operators. Bull. Pol. Acad. **37**, 221–228 (1989)
50. M. Krook, T.T. Wu, Exact solutions of the Boltzmann equation. Phys. Fluids **20**, 1589–1595 (1977)
51. U. Krengel, *Ergodic theorems, de Gruyter Studies in Mathematics*, vol. 6 (Walter de Gruyter & Co., Berlin, 1985)
52. P. Laurençot, D. Wrzosek, The discrete coagulation equations with collisional breakage. J. Stat. Phys. **104**, 193–220 (2001)
53. A. Lasota, Asymptotic stability of some nonlinear Boltzmann-type equations. J. Math. Anal. Appl. **268**, 291–309 (2002)
54. A. Lasota, M.C. Mackey, Globally asymptotic properties of proliferating cell populations. J. Math. Biol. **19**, 43–62 (1984)
55. A. Lasota, M.C. Mackey, *Chaos, Fractals and Noise. Stochastic Aspects of Dynamics*. Springer Applied Mathematical Sciences, II edn, vol. 97 (Springer, New York, 1994)
56. A. Lasota, J. Traple, An application of the Kantorovich-Rubinstein maximum principle in the theory of the Tjon–Wu equation. J. Differ. Equ. **159**, 578–596 (1999)
57. A. Lasota, J.A. Yorke, Exact dynamical systems and the Frobenius–Perron operator. Trans. AMS **273**, 375–384 (1982)
58. A. Lasota, J.A. Yorke, Lower bound technique for Markov operators and iterated function systems. Random Computat. Dyn. **2**, 41–77 (1994)
59. M. Lin, Mixing for Markov operators. Z. Wahrsch. Verw. Gebiete **19**, 231–242 (1971)
60. T. Lipniacki, P. Paszek, A. Marciniak-Czochra, A.R. Brasier, M. Kimmel, Transcriptional stochasticity in gene expression. J. Theor. Biol. **238**, 348–367 (2006)
61. M. Liu, K. Wang, Q. Wu, Survival analysis of stochastic competitive models in a polluted environment and stochastic competitive exclusion principle. Bull. Math. Biol. **73**, 1969–2012 (2011)
62. K. Łoskot, R. Rudnicki, Sweeping of some integral operators. Bull. Pol. Acad. **37**, 229–235 (1989)
63. K. Łoskot, R. Rudnicki, Relative entropy and stability of stochastic semigroups. Ann. Pol. Math. **53**, 139–145 (1991)
64. J. Łuczka, R. Rudnicki, Randomly flashing diffusion: asymptotic properties. J. Stat. Phys. **83**, 1149–1164 (1996)

65. M.C. Mackey, *Time's Arrow: The Origins of Thermodynamic Behavior* (Springer, New York, 1992)
66. J. Malczak, An application of Markov operators in differential and integral equations. Rend. Sem. Mat. Univ. Padova **87**, 281–297 (1992)
67. P. Malliavin, Stochastic calculus of variations and hypoelliptic operators, in *Proc. Intern. Symp. Stoch. Diff. Equations of Kyoto 1976*, ed by K. Itô (Wiley, New York, 1978) pp. 195–263
68. P. Malliavin, C^k-hypoellipticity with degeneracy, in *Stochastic Analysis*, eds. by A. Friedman, M. Pinsky (Acadamic Press, New York, 1978) pp. 199–214
69. A.G. McKendrick, Application of mathematics to medical problems. Proc. Edinb. Math. Soc. **14** (1926), 98–130.
70. J. A. J. Metz, O. Diekmann (ed.), *The Dynamics of Physiologically Structured Populations*. Springer Lecture Notes in Biomathematics, vol. 68 (Springer, New York, 1986)
71. C. D. Meyer, *Matrix Analysis and Applied Linear Algebra* (SIAM, Philadelphia, 2000)
72. J. van Neerven, *The Asymptotic Behaviour of a Semigroup of Linear Operators* (Birkhäuser, Basel, 1996)
73. J. Norris, Simplified Malliavin calculus, in *Séminaire de probabilitiés XX*. Lecture Notes in Mathematics, vol. 1204 (Springer, New York, 1986) pp.101–130
74. E. Nummelin, *General Irreducible Markov Chains and Non-negative Operators*. Cambridge Tracts in Mathematics, vol. 83 (Cambridge University Press, Cambridge, 1984)
75. H.G. Othmer, S.R. Dunbar, W. Alt, Models of dispersal in biological systems. J. Math. Biol. **26**, 263–298 (1988)
76. A. Pazy, *Semigroups of Linear Operators and Applications to Partial Differential Equations*. Applied Mathematics Science, vol. 44 (Springer, New York, 1983)
77. K. Pichór, Asymptotic stability of a partial differential equation with an integral perturbation. Ann. Pol. Math. **68**, 83–96 (1998)
78. K. Pichór, Asymptotic stability and sweeping of substochastic semigroups. Ann. Polon. Math. **103**, 123–134 (2012)
79. K. Pichór, R. Rudnicki, Stability of Markov semigroups and applications to parabolic systems. J. Math. Anal. Appl. **215**, 56–74 (1997)
80. K. Pichór, R. Rudnicki, Asymptotic behaviour of Markov semigroups and applications to transport equations. Bull. Pol. Acad. **45**, 379–397 (1997)
81. K. Pichór, R. Rudnicki, Continuous Markov semigroups and stability of transport equations. J. Math. Anal. Appl. **249**, 668–685 (2000)
82. K. Pichór, R. Rudnicki, M. Tyran-Kamińska, Stochastic semigroups and their applications to biological models. Demonstratio Math. **45**, 463–495 (2012)
83. M. Pollicott, M. Yuri, *Dynamical systems and ergodic theory*. London Mathematical Society Student Texts, vol. 40 (Cambridge University Press, Cambridge, 1998)
84. R. Rudnicki, On a one-dimensional analogue of the Smale horseshoe. Ann. Pol. Math. **54**, 47–153 (1991)
85. R. Rudnicki, Asymptotic behaviour of an integro-parabolic equation. Bull. Pol. Acad. **40**, 111–128 (1992)
86. R. Rudnicki, Asymptotic behaviour of a transport equation. Ann. Pol. Math. **57**, 45–55 (1992)
87. R. Rudnicki, Asymptotical stability in L^1 of parabolic equations. J. Differ. Equ. **102**, 391–401 (1993)
88. R. Rudnicki, Strangely sweeping one-dimensional diffusion. Ann. Pol. Math. **58**, 37–45 (1993)
89. R. Rudnicki, Asymptotic properties of the Fokker–Planck equation, in *Chaos—the Interplay Between Stochastics and Deterministic Behaviour*, Karpacz'95 Proc., eds. by P. Garbaczewski, M. Wolf, A. Weron, Lecture Notes in Physics, vol. 457 (Springer, Berlin, 1995) pp. 517–521
90. R. Rudnicki, On asymptotic stability and sweeping for Markov operators. Bull. Pol. Acad. **43**, 245–262 (1995)
91. R. Rudnicki, Stability in L^1 of some integral operators. Integral Equ. Oper. Theory **24**, 320–327 (1996)
92. R. Rudnicki, Asymptotic stability of Markov operators: a counter-example. Bull. Pol. Acad. **45**, 1–5 (1997)

93. R. Rudnicki, Long-time behaviour of a stochastic prey-predator model. Stoch. Processes Appl. **108**, 93–107 (2003)
94. R. Rudnicki, Models of population dynamics and their applications in genetics, in *From Genetics to Mathematics*, eds. by M. Lachowicz, J. Miękisz, Series on Advances in Mathematics for Applied Sciences, vol. 79 (World Scientific, New Jersey, 2009), pp. 103–147.
95. R. Rudnicki, K. Pichór, Markov semigroups and stability of the cell maturation distribution. J. Biol. Syst. **8**, 69–94 (2000)
96. R. Rudnicki, K. Pichór, Influence of stochastic perturbation on prey-predator systems. Math. Biosci. **206**, 108–119 (2007)
97. R. Rudnicki, J. Tiuryn, D. Wójtowicz, A model for the evolution of paralog families in genomes. J. Math. Biol. **53**, 759–770 (2006)
98. R. Rudnicki, R. Wieczorek, Fragmentation—coagulation models of phytoplankton. Bull. Pol. Acad. Sci. Math. **54**, 175–191 (2006)
99. R. Rudnicki, P. Zwoleński, Model of phenotypic evolution in hermaphroditic populations. J. Math. Biol. in press. doi:10.1007/s00285-014-0798-3. http://arxiv.org/pdf/1309.3243v1.pdf
100. R. Sanders, L^1 stability of solutions to certain linear parabolic equations in divergence form. J. Math. Anal. Appl. **112**, 335–346 (1985)
101. F.R. Sharpe, A.J. Lotka, A problem in age-distributions. Philos. Mag. **21** (1911), 435–438.
102. U. Skwara, A stochastic symbiosis model with degenerate diffusion process. Ann. Polon. Math. **98**, 111–128 (2010)
103. P.P. Slonimski, M.O. Mosse, P. Golik, A. Henaût, Y. Diaz, J.L. Risler, J.P. Comet, J.C. Aude, A. Wozniak, E. Glemet, J.J. Codani, The first laws of genomics. Microb. Comp. Genomics **3**, 46 (1998)
104. D.W. Stroock, S.R.S. Varadhan, On degenerate elliptic-parabolic operators of second order and their associated diffusions. Commun. Pure Appl. Math. **24**, 651–713 (1972)
105. D.W. Stroock, S.R.S. Varadhan, On the support of diffusion processes with applications to the strong maximum principle, in *Proc. Sixth Berkeley Symposium on Mathematical Statistics and Probability*, vol. III (University of California Press, Berkeley, 1972) pp. 333–360
106. J.A. Tjon, T.T. Wu, Numerical aspects of the approach to a Maxwellian distribution. Phys. Rev. A. **19**, 883–888 (1979)
107. J. Traple, Markov semigroups generated by Poisson driven differential equations. Bull. Pol. Acad. **44**, 230–252 (1996)
108. M. Tyran-Kamińska, Substochastic semigroups and densities of piecewise deterministic Markov processes. J. Math. Anal. Appl. **357**, 385–402 (2009)
109. M. Tyran-Kamińska, Ergodic theorems and perturbations of contraction semigroups. Studia Math. **195**, 147–155 (2009)
110. J. Tyrcha, Asymptotic stability in a generalized probabilistic/deterministic model of the cell cycle. J. Math. Biol. **26**, 465–475 (1988)
111. J.J. Tyson, K.B. Hannsgen, Cell growth and division: a deterministic/probabilistic model of the cell cycle. J. Math. Biol. **23**, 231–246 (1986)
112. J. Voigt, On substochastic C_0-semigroups and their generators. Transp. Theory Stat. Phys. **16**, 453–466 (1987)
113. M. von Smoluchowski, Drei Vorträge über Diffusion, Brownsche Molekularbewegung und Koagulation von Kolloidteilchen. Phys. Z. **17**, 557–571, 585–599 (1916)
114. G.W. Webb, *Theory of Nonlinear Age-Dependent Population Dynamics* (Marcel Dekker, New York, 1985)
115. G.G. Yin, C. Zhu, *Hybrid Switching Diffusions: Properties and Applications*. Stochastic Modelling and Applied Probability, vol. 63 (Springer, New York, 2010)

Spectral Theory for Neutron Transport

Mustapha Mokhtar-Kharroubi

In memory of Seiji Ukaï

1 Introduction

These notes resume a lecture given in the Cimpa School "Evolutionary equations with applications in natural sciences" held in South Africa (Muizenberg, July 22–August 2, 2013). However, the oral style of the lecture has been changed and the bibliography augmented. This version benefited also from helpful remarks and suggestions of a referee whom I would like to thank. The notes deal with various functional analytic tools and results around spectral analysis of neutron transport-like operators. A first section gives a detailed introduction (mostly without proofs) to fundamental concepts and results on spectral theory of (non-selfadjoint) operators in Banach spaces; in particular, we provide an introduction to spectral analysis of semigroups in Banach spaces and its consequences on their time asymptotic behaviour as time goes to infinity. A special attention is paid to positive semigroups in ordered spaces (i.e. semigroups leaving invariant the cone of positive elements) because of their fundamental interest in neutron transport theory. We focus on the analysis of essential spectra and isolated eigenvalues with finite multiplicities. A second section deals with spectral analysis of weighted shift (or collisionless transport) semigroups. A third section is devoted to spectral analysis of perturbed semigroups in Banach spaces, in particular to stability of essential type for perturbed semigroups. A last section deals with a thorough analysis of compactness problems for general models of neutron transport; the results are very different depending on whether we work in L^p spaces ($1 < p < \infty$) or in (the physical) L^1 space; this issue is the very core of spectral analysis of neutron transport operators and allow the abstract theory to cover them.

M. Mokhtar-Kharroubi (✉)
Département de Mathématiques, CNRS UMR 6623, Université de Franche-Comté, 16 Route de Gray, 25030 Besançon, France
e-mail: mmokhtar@univ-fcomte.fr

© Springer International Publishing Switzerland 2015

J. Banasiak, M. Mokhtar-Kharroubi (eds.), *Evolutionary Equations with Applications in Natural Sciences*, Lecture Notes in Mathematics 2126,
DOI 10.1007/978-3-319-11322-7_7

Transport theory provides a *statistical* description of large populations of "particles" moving in a host medium (see e.g. [17]) and is of interest in various fields such as radiative transfer theory, nuclear reactor theory, gas dynamics, plasma physics, structured population models in mathematical biology etc. Among the most classical kinetic equations, we mention the one governing the transport of neutrons through the uranium fuel elements of a nuclear reactor. The aim of this lecture is to present various functional analytic tools and results motivated by this class of equations. In a nuclear reactor, the proportion of neutrons with respect to the atoms of the host medium, is infinitesimal (about 10^{-11}), so the possible collisions between neutrons are negligible in comparison with the collisions of neutrons with the atoms of the host material. Thus (in absence of feedback temperature) neutron transport equations as well as radiative transfer equations for photons are *genuinely linear*. The population of particles is described by a density function $f(t,x,v)$ of particles at time $t > 0$, at position x and with velocity v. In particular

$$\int\int f(t,x,v)dxdv$$

is the *expected* number of particles at time $t > 0$. One sees immediately that L^1 spaces are natural settings in transport theory! Various models are used in nuclear reactor theory:

(1) *Inelastic model for neutron transport*

$$\frac{\partial f}{\partial t} + v.\frac{\partial f}{\partial x} + \sigma(x,v)f(t,x,v) = \int_V k(x,v,v')f(t,x,v')dv'$$

where $(x,v) \in \Omega \times V$, $\Omega \subset \mathbb{R}^3$, $V = \{v \in \mathbb{R}^3;\ c_0 \leq |v| \leq c_1\}$ $(0 \leq c_0 < c_1 < \infty)$ and dv is Lebesgue measure, with initial condition $f(0,x,v) = f_0(x,v)$ and boundary condition

$$f(t,x,v)_{|\Gamma_-} = 0$$

where

$$\Gamma_- := \{(x,v) \in \partial\Omega \times V;\ v.n(x) < 0\}$$

and $n(x)$ is the unit exterior normal at $x \in \partial\Omega$. The collision frequency $\sigma(.,.)$ and the scattering kernel $k(.,.,.)$ are nonnegative.

(2) *Multiple scattering:* This physical model differs from the previous "reactor model" by the fact that $\Omega = \mathbb{R}^3$ (no boundary condition) but $\sigma(x,v)$ and $k(x,v,v')$ are compactly supported in space.

(3) *The presence of delayed neutrons*

Besides the prompt neutrons (appearing instantaneously in a fission process), some neutrons may appear after a time delay as a decay product of radioactive fission fragments and induce a suitable source term in the usual equation

$$\frac{\partial f}{\partial t} + v.\frac{\partial f}{\partial x} + \sigma(x,v)f(t,x,v) = \int_{\mathbb{R}^3} k(x,v,v')f(t,x,v')dv' + \sum_{i=1}^{m} \lambda_i g_i$$

which is thus coupled to m differential equations

$$\frac{dg_i}{dt} = -\lambda_i g_i + \int_{\mathbb{R}^3} k_i(x,v,v')f(t,x,v')dv' \quad (1 \le i \le m)$$

where $\lambda_i > 0\,(1 \le i \le m)$ are the radioactive decay constants; see [48, Chapter 4] and references therein.

(4) *Multigroup models* (motivated by numerical calculations)

$$\frac{\partial f_i}{\partial t} + v.\frac{\partial f_i}{\partial x} + \sigma_i(x,v)f_i(t,x,v) = \sum_{j=1}^{m} \int_{V_j} k_{i,j}(x,v,v')f_j(t,x,v')\mu_j(dv'),$$

$(1 \le i \le m)$ where the spheres

$$V_j := \left\{v \in \mathbb{R}^3,\ |v| = c_j\right\},\ \ 1 \le j \le m,\ \ (c_j > 0)$$

are endowed with surface Lebesgue measures μ_j and $f_i(t,x,v)$ is the density of neutrons (at time $t > 0$ located at $x \in \Omega$) with velocity $v \in V_i$.

(5) *Partly inelastic models*

$$\frac{\partial f}{\partial t} + v.\frac{\partial f}{\partial x} + \sigma(x,v)f(t,x,v) = K_e f + K_i f$$

in $L^p(\Omega \times V)$ where, e.g. $V = \left\{v \in \mathbb{R}^3;\ c_0 \le |v| \le c_1\right\}$. The inelastic scattering operator is just

$$K_i f = \int_V k(x,v,v')f(x,v')dv'$$

while the elastic scattering operator is given by

$$K_e f = \int_{S^2} k(x,\rho,\omega,\omega')f(x,\rho\omega')dS(\omega')$$

where $v = \rho\omega$. The presence of an elastic scattering operator acting only on the *angles* $\omega \in S^2$ of velocities changes strongly the spectral structure of neutron transport operators [35, 68].

(6) *Diffusive models*

$$\frac{\partial f}{\partial t} - \Delta_x f + \sigma(x, v) f = \int_0^{+\infty} k(x, v, v') f(t, x, v') dv'$$

(motivated also by numerical calculations) where the transport operator $\frac{\partial f}{\partial t} + v.\frac{\partial f}{\partial x}$ is replaced by the parabolic operator $\frac{\partial f}{\partial t} - \Delta_x f$ where Δ_x denotes the Laplacian in space variable $x \in \Omega$ with Dirichlet boundary condition, (here $v > 0$ denotes a " kinetic energy" instead of a velocity); see e.g. [14, p. 133]. In the same spirit, we mention that diffusion (i.e. heat) equations with Dirichlet boundary condition turn out to be *asymptotic approximations* (as $\varepsilon \to 0$) of usual neutron transport equations appropriately rescaled by means of a small parameter ε (typically the mean free path); see e.g. [6] and references therein. We find in [49] an approach of the diffusion approximation of neutron transport (on the torus) via spectral theory.

In this lecture, we ignore the presence of delayed neutrons but deal with an abstract velocity measure $\mu(dv)$ (with support V) covering a priori different models, e.g. Lebesgue measure on $\mathbb{R}^n$ or on spheres or even combinations of the two.

In absence of scattering event (i.e. $k(x, v, v') = 0$) the density of neutral particles (e.g. neutrons) is governed by

$$\frac{\partial f}{\partial t} + v.\frac{\partial f}{\partial x} + \sigma(x, v) f(t, x, v) = 0$$

with initial condition f_0 and is solved explicitly by the method of characteristics

$$f(t, x, v) = e^{-\int_0^t \sigma(x - \tau v, v) d\tau} f_0(x - tv, v) 1_{\{t \leq s(x,v)\}}$$

where

$$s(x, v) = \inf\{s > 0;\ x - sv \notin \Omega\}$$

is the first exit time function. This defines a positive C_0-semigroup $(U(t))_{t \geqslant 0}$ on $L^p(\Omega \times \mathbb{R}^3; dx \otimes d\mu)$

$$U(t) : g \to e^{-\int_0^t \sigma(x - \tau v, v) d\tau} g(x - tv, v) 1_{\{t \leq s(x,v)\}}$$

called the *advection semigroup*. Its generator T is given (at least for smooth domains Ω) by

$$Tg = -v.\frac{\partial g}{\partial x} - \sigma(x, v) g(x, v),\ g \in D(T)$$

$$D(T) = \left\{ g \in L^p(\Omega \times \mathbb{R}^3);\ v.\frac{\partial g}{\partial x} \in L^p,\ g_{|\Gamma_-} = 0 \right\};$$

(see e.g. [12, 13] for a trace theory in neutron transport theory). Then the treatment of the full equation follows naturally by perturbation theory. For instance, if the scattering operator

$$K : g \to \int_{\mathbb{R}^3} k(x, v, v')g(x, v')\mu(dv')$$

is bounded on $L^p(\Omega \times \mathbb{R}^3)$ then, by standard perturbation theory,

$$A := T + K \quad (D(A) = D(T))$$

generates a positive C_0-semigroup $(V(t))_{t \geqslant 0}$ which solves the full neutron transport equation.

There are two basic eigenvalue problems in nuclear reactor theory:

(1) *Criticality eigenvalue problem*

This problem consists in looking for (γ, g) where $\gamma > 0$ and g is a nontrivial nonnegative solution to

$$0 = -v.\frac{\partial g}{\partial x} - \sigma(x, v)g(x, v) + \int_V k_s(x, v, v')g(x, v')\mu(dv')$$
$$+\frac{1}{\gamma}\int_V k_f(x, v, v')g(x, v')\mu(dv'), \quad g_{|\Gamma_-} = 0;$$

here $k_s(x, v, v')$ and $k_f(x, v, v')$ are the scattering kernel and the fission kernel, see e.g. [41, 66].

(2) *The "time eigenelements"*

This problem consists in looking for (λ, g) with nontrivial g such that

$$-v.\frac{\partial g}{\partial x} - \sigma(x, v)g(x, v) + \int_V k(x, v, v')g(x, v')\mu(dv') = \lambda g(x, v), \quad g_{|\Gamma_-} = 0$$

and in relating them to time asymptotic behaviour ($t \to +\infty$) of the semigroup $(V(t))_{t \geqslant 0}$.

In this lecture, we focus on the second class of problems. There exists a considerable literature on the subject; we refer to [48] and references therein for the state of the art up to 1997. In these lecture, we present mostly new developments on this topic.

We note that this conventional neutron transport theory deals with the expected (or mean) behaviour of neutrons. In order to describe the fluctuations from the mean value of neutron populations, probabilistic formulations of neutron chain fissions were proposed very early, in particular in [7]. This leads to nonlinear problems governing *divergent* neutron chain fissions. Such problems are strongly related to spectral theory of usual (linear) neutron transport operators, see [30, 46, 57, 64].

We end this introduction by some historical notes. The beginning of spectral theory of neutron transport dates back to the beautiful and seminal paper by J. Lehner and M. Wing [36] devoted to a simplified model (constant cross sections) in slab geometry. The time asymptotic behaviour of neutron transport semigroups in bounded geometries is well-understood for a long time in the case when the velocities are *bounded away from zero*; this is a classical result by K. Jorgens:

Theorem 1 ([31]) *Let Ω be bounded and convex, let*

$$V = \left\{v \in \mathbb{R}^3;\ c_0 \leq |v| \leq c_1 < \infty\right\}$$

and let the scattering kernel $k(.,.,.)$ be bounded. If $c_0 > 0$ then $V(t)$ is compact on $L^2(\Omega \times V)$ for t large enough. In particular, for any $\alpha \in \mathbb{R}$

$$\sigma(A) \cap \{Re\lambda \geqslant \alpha\}$$

consists at most of finitely many eigenvalues with finite algebraic multiplicities $\{\lambda_1, \ldots .\lambda_m\}$ with spectral projections $\{P_1, \ldots .P_m\}$ and there exists $\beta < \alpha$ such that

$$V(t) = \sum_{j=1}^{m} e^{\lambda_j t} e^{tD_j} P_j + O(e^{\beta t})$$

where $D_j := (T - \lambda_j)P_j$.

The picture gets more complicated when arbitrarily small velocities must be taken into account. In this case, the (essential) spectrum of the generator T (of the advection semigroup $\{U(t); t \geqslant 0\}$) on $L^2(\Omega \times V)$ consists of a half-plane

$$\{\lambda \in \mathbb{C};\ Re\lambda \leq -\lambda^*\}$$

where "typically" $\lambda^* = \inf \sigma(x, v)$, see S. Albertoni and B. Montagnini [2]. Moreover, important compactness results were obtained very early, (see e.g. Demeru-Montagnini [16], Borysiewicz-Mika [8] and S. Ukai [74]) implying, for most physical scattering kernels, that the scattering operator K is T-compact on $L^2(\Omega \times V)$ i.e.

$$K : D(T) \to L^2(\Omega \times V)$$

is compact where $D(T)$ is endowed with the graph norm. It follows that the spectrum of $A = T + K$ consists of a left half-plane $\{\lambda \in \mathbb{C};\ Re\lambda \leq -\lambda^*\}$ and *at most* of isolated eigenvalues with finite algebraic multiplicities located in the right half-plane

$$\{\lambda \in \mathbb{C};\ Re\lambda > -\lambda^*\}.$$

(Note that it may happen that this set of isolated eigenvalues is empty for small bodies [2].) Then the time asymptotic behaviour of the solution is traditionally dealt with by means of inverse Laplace transform (Dunford calculus)

$$V(t)f = \lim_{\gamma \to +\infty} \frac{1}{2i\pi} \int_{\rho - i\gamma}^{\rho + i\gamma} e^{\lambda t}(\lambda - A)^{-1} f d\lambda$$

(with ρ large enough). If for some $\varepsilon > 0$

$$\sigma(T+K) \cap \{\lambda; Re\lambda > -\lambda^* + \varepsilon\} = \{\lambda_1, \ldots .\lambda_m\}$$

(with spectral projections $\{P_1, \ldots .P_m\}$) is finite and non-empty then, by shifting the path of integration and picking up the residues, we get an *asymptotic expansion*

$$V(t)f = \sum_{j=1}^{m} e^{\lambda_j t} e^{tD_j} P_j f + O_f(e^{\beta t}) \quad (\beta < -\lambda^* + \varepsilon);$$

for *smooth* initial data f; see, e.g. M. Borysiewicz and J. Mika [8] (see also M. Mokhtar-Kharroubi [45]). The drawbackof the approach is that we need very regular initial data (say $f \in D(A^2)$) to estimate the *transcient part* of the solution. To remedy this situation, a more relevant approach, initiated by I. Vidav [76], consists in studying the spectrum of the semigroup $(V(t))_{t \geqslant 0}$ itself instead of the spectrum of its generator because of the lack (in general) of a spectral mapping theorem relating spectra of semigroups and spectra of their generators. The perturbed semigroup $(V(t))_{t \geqslant 0}$ is expanded into a Dyson–Phillips series

$$V(t) = \sum_{n=0}^{\infty} U_n(t)$$

where $U_0(t) = U(t)$ is the advection semigroup and

$$U_{n+1}(t) = \int_0^t U(t-s) K U_n(s) ds \quad (n \geqslant 0).$$

Theorem 2 ([76]) *If some remainder term* $R_n(t) := \sum_{j=n}^{\infty} U_j(t)$ *is compact for large* t *then* $\sigma(V(t)) \cap \left\{\mu; |\mu| > e^{-\lambda^* t}\right\}$ *consists at most of isolated eigenvalues with finite multiplicities. In particular,* $\forall \varepsilon > 0$,

$$\sigma(T+K) \cap \{\lambda; Re\lambda \geqslant -\lambda^* + \varepsilon\} = \{\lambda_1, \ldots .\lambda_m\}$$

is finite and

$$V(t) = \sum_{j=1}^{m} e^{\lambda_j t} e^{tD_j} P_j + O(e^{\beta t})$$

in operator norm where $\beta < -\lambda^* + \varepsilon$.

Vidav's result had relevant applications to realistic models of kinetic theory much later; see Y. Shizuta [71], G. Greiner [25], J. Voigt [77, 79], P. Takak [72], M. Mokhtar-Kharroubi [42, 43] and L. Weiss [82]. The role of positivity in peripheral spectral theory of neutron transport was emphasized by I. Vidav [75], T. Hiraoka-S. Ukaï [29], Angelescu-Protopopescu [4] and more recently, in others directions, e.g. by G. Greiner [26], J. Voigt [78] and M. Mokhtar-Kharroubi [43–45, 47].

2 Fundamentals of Spectral Theory

This section is a crash course (mostly without proofs) on the fundamental concepts and results on spectral theory of closed linear operators on complex Banach spaces with a special emphasis on generators of strongly continuous semigroups. Because of their importance in transport theory, the basic spectral properties of positive operators (i.e. leaving invariant the positive cone of a Banach lattice) are also given. Finally, we show the role of peripheral spectral theory of positive semigroups in their time asymptotic behaviour as $t \to +\infty$. Apart from Subsection 2.10, the material of this section is widely covered by the general references [15, 20–22, 32, 61, 73] and will be used in the sequel without explicit mention. Subsection 2.10 presents a class of positive semigroups whose real spectra can be described completely; this class covers weighted shift (i.e. advection) semigroups we deal with in Sect. 3.

2.1 Basic Definitions and Results

We start with some basic definitions and results. Let X be a complex Banach space and let

$$T : D(T) \subset X \to X$$

be a linear operator defined on a subspace $D(T)$. We say that T is a closed operator if its graph

$$\{(x, Tx);\ x \in D(T)\}$$

is closed in $X \times X$. We define the resolvent set of T by

$$\rho(T) := \{\lambda \in \mathbb{C};\ \lambda - T : D(T) \to X \text{ is bijective}\},$$

the spectrum of T by

$$\sigma(T) := \{\lambda \in \mathbb{C};\ \lambda \notin \rho(T)\}$$

and the resolvent operator by

$$(\lambda - T)^{-1} : X \to X \ \ (\lambda \in \rho(T)).$$

In particular, if there exists $x \in D(T) - \{0\}$ and $\lambda \in \mathbb{C}$ such that $Tx = \lambda x$ then $\lambda \in \sigma(T)$. In this case, λ is an eigenvalue of T and

$$\ker(T) := \{x \in D(T);\ (T - \lambda)x = 0\}$$

is the corresponding eigenspace. In contrast to finite dimensional spaces, in general, $\sigma(T)$ is *not* reduced to eigenvalues! For instance, one can show that the spectrum of the multiplication operator on $C([0,1])$ (endowed with the sup norm)

$$T : f \in C([0,1]) \to Tf \in C([0,1])$$

where $Tf(x) = xf(x)$ is equal to $[0,1]$ and that T has no eigenvalue. For *unbounded* operators, the spectrum may be empty or equal to $\mathbb{C}$! For example, let $X = C([0,1];\ \mathbb{C})$ endowed with the sup-norm and

$$Tf = \frac{df}{dx},\quad D(T) = C^1([0,1]).$$

Then $\forall\lambda \in \mathbb{C}$, $x \in [0,1] \to e^{\lambda x} \in \mathbb{C}$ is an eigenfunction of T so $\sigma(T) = \mathbb{C}$. If we replace $(T, D(T))$ by

$$\hat{T}f = \frac{df}{dx},\quad D(\hat{T}) = \left\{f \in C^1([0,1]);\ f(0) = 0\right\}$$

then $\forall\lambda \in \mathbb{C}$ and $\forall g \in X$, the equation

$$\lambda f - \frac{df}{dx} = g,\ f(0) = 0$$

is uniquely solvable; thus $\rho(\hat{T}) = \mathbb{C}$ and $\sigma(\hat{T}) = \varnothing$.

It is useful to decompose the spectrum of T as follows: The point spectrum

$$\sigma_p(T) = \{\lambda \in \mathbb{C};\ \lambda - T : D(T) \to X \text{ is not injective}\}.$$

The approximate point spectrum

$$\sigma_{ap}(T) = \{\lambda \in \mathbb{C};\ \lambda - T : D(T) \to X \text{ not injective or } (\lambda - T)X \text{ not closed}\};$$

this terminology is motivated by the fact that $\lambda \in \sigma_{ap}(T)$ if and only if there exists a sequence $(x_n)_n \subset D(T)$ such that

$$\|x_n\| = 1, \ \|Tx_n - \lambda x_n\| \to 0.$$

The residual spectrum

$$\sigma_{res}(T) = \{\lambda \in \mathbb{C}; (\lambda - T)X \text{ is not dense}\}.$$

We note that

$$\sigma(T) = \sigma_{res}(T) \cup \sigma_{ap}(T)$$

is a non-disjoint union. Among the first results, we note:

- $(\lambda - T)^{-1} : X \to X$ is a bounded operator for $\lambda \in \rho(T)$, i.e. $(\lambda - T)^{-1} \in \mathcal{L}(X)$, (by the closed graph theorem).
- $\rho(T)$ is an open subset of $\mathbb{C}$ (so $\sigma(T)$ is closed) and

$$\lambda \in \rho(T) \to (\lambda - T)^{-1} \in \mathcal{L}(X)$$

is holomorphic.

More precisely, if $\mu \in \rho(T)$ then $\lambda \in \rho(T)$ if $|\lambda - \mu| < \left\|(\mu - T)^{-1}\right\|^{-1}$ and then

$$(\lambda - T)^{-1} = \sum_0^{+\infty} (\mu - \lambda)^n \left[(\mu - T)^{-1}\right]^{n+1}.$$

It follows that $|\lambda - \mu| \geqslant \left\|(\mu - T)^{-1}\right\|^{-1}$ for any $\lambda \in \sigma(T)$ and then

$$dist(\mu, \sigma(T)) \geqslant \left\|(\mu - T)^{-1}\right\|^{-1}.$$

In particular $\left\|(\mu - T)^{-1}\right\| \to \infty$ as $dist(\mu, \sigma(T)) \to 0$.

Bounded operators $T \in \mathcal{L}(X)$ enjoy specific properties:

- $\sigma(T)$ is bounded and non-empty.
- The spectral radius of $T \in \mathcal{L}(X)$, defined by

$$r_\sigma(T) := \sup\{|\lambda|;\ \lambda \in \sigma(T)\},$$

is equal to $\lim_{n\to\infty} \|T^n\|^{\frac{1}{n}} = \inf_n \|T^n\|^{\frac{1}{n}}$.

- In particular $r_\sigma(T) \leq \|T\|$ and $(\lambda - T)^{-1}$ is given by a Laurent's series

$$(\lambda - T)^{-1} = \sum_1^\infty \lambda^{-n} T^{n-1} \quad (|\lambda| > r_\sigma(T))$$

with $T^m = \frac{1}{2i\pi} \int_C \lambda^m (\lambda - T)^{-1} d\lambda$ where C is any circle (positively oriented) centered at the origin with radius $> r_\sigma(T)$.

If $T : D(T) \subset X \to X$ is densely defined linear operator, we can define its dual operator

$$T' : D(T') \subset X' \to X'$$

by

$$\langle Tx, y' \rangle_{X,X'} = \langle x, T'y' \rangle_{X,X'}$$

with domain

$$D(T') = \left\{ y' \in X'; \exists c \geqslant 0, \ \left| \langle Tx, y' \rangle \right| \leq c \|x\| \ \forall x \in D(T) \right\}.$$

We note that T' is closed but not necessarily densely defined. But if X is reflexive then T' is densely defined, $(T')' = T$, $\sigma(T') = \sigma(T)$ and $(\lambda - T')^{-1} = \left((\lambda - T)^{-1}\right)'$. In particular if $T \in \mathcal{L}(X)$ then $r_\sigma(T') = r_\sigma(T)$.

We end this section with a *spectral mapping theorem* for bounded operators. Let $T \in \mathcal{L}(X)$ and let $\Omega \ni \lambda \to f(\lambda) \in \mathbb{C}$ be holomorphic on some open neighborhood Ω of $\sigma(T)$. Then there exists an open set ω such that $\sigma(T) \subset \omega \subset \overline{\omega} \subset \Omega$ and $\partial\omega$ consists of finitely many simple closed curves that do not intersect. One defines a Dunford integral

$$f(T) = \frac{1}{2i\pi} \int_{\partial\omega} f(\lambda)(\lambda - T)^{-1} d\lambda \in \mathcal{L}(X)$$

where $\partial\omega$ is properly oriented (the definition does not depend on the choice of ω). In particular if $f(\lambda)$ is a polynomial then $f(T)$ coincides with the usual meaning of $f(T)$. Then we have a spectral mapping theorem

$$\sigma(f(T)) = f(\sigma(T)).$$

2.2 Spectral Decomposition and Riesz Projection

Let X be a complex Banach space such that

$$X = X_1 \oplus X_2$$

(direct sum) where X_i $(i = 1, 2)$ are *closed* subspaces. Let $P : x \in X \to Px$ be the (continuous) projection on X_1 along X_2. Let

$$T : D(T) \subset X \to X$$

be a closed linear operator such that $P(D(T)) \subset D(T)$ and X_i $(i = 1, 2)$ are invariant under T. The parts T_i $(i = 1, 2)$ of T on X_i $(i = 1, 2)$ are defined by

$$D(T_i) = D(T) \cap X_i, \quad T_i x = Tx \quad (x \in D(T_i)).$$

We say that T is *reduced* by X_i $(i = 1, 2)$. Then

$$\sigma(T) = \sigma(T_1) \cup \sigma(T_2)$$

(not necessarily a disjoint union),

$$\sigma_p(T) = \sigma_p(T_1) \cup \sigma_p(T_2) \text{ and } \sigma_{ap}(T) = \sigma_{ap}(T_1) \cup \sigma_{ap}(T_2).$$

Similar results hold for any finite direct sum: $X = X_1 \oplus \ldots \oplus X_n$; see e.g. [73, Theorem 5.4, p. 289].

Let now $T : D(T) \subset X \to X$ be a closed linear operator such that $\sigma(T)$ is a *disjoint* union of two non-empty closed subsets σ_1 and σ_2 and let σ_1 be compact. Then there exists Γ, a finite number of rectifiable simple closed curves properly oriented enclosing an open set O which contains σ_1 and such that σ_2 is included in the exterior of O. Then

$$P := \int_{\Gamma} (\lambda - T)^{-1} d\lambda; \quad P^2 = P$$

and $X = X_1 \oplus X_2$ ($X_1 = PX$ and $X_2 = (I - P)X = Ker P$) reduces T (i.e. X_i are T invariant), $\sigma(T_i) = \sigma_i$ where $T_i := T_{|X_i}$ and T_1 is bounded. P is the *spectral projection* associated with σ_1. If σ_1 consists of finitely many points $(\lambda_1, \ldots, \lambda_n)$ then

$$P = P_1 + \ldots + P_n, \quad P_j P_k = \delta_{jk} P_j$$

$$P_j := \int_{\Gamma_j} (\lambda - T)^{-1} d\lambda$$

(where Γ_j is e.g. a small circle enclosing λ_j). P_j is the spectral projection associated with λ_j. We study now the structure of the resolvent around an isolated singularity. Let $\mu \in \sigma(T)$ be an *isolated* point of $\sigma(T)$. There exists a Laurent's series around μ

$$(\lambda - T)^{-1} = \sum_{n=-\infty}^{+\infty} (\lambda - \mu)^n U_n$$

where

$$U_n = \frac{1}{2i\pi} \int_C \frac{(\lambda - T)^{-1}}{(\lambda - \mu)^{n+1}} d\lambda \quad (n \in \mathbb{Z})$$

where C is a small circle positively oriented centered at μ. In particular, the residues

$$U_{-1} = \frac{1}{2i\pi} \int_C (\lambda - T)^{-1} d\lambda$$

is the spectral projection P. In addition

$$U_{-(n+1)} = (-1)^n (\mu - T)^n P \quad (n \geqslant 0).$$

We have

$$U_{-(n+1)} U_{-(m+1)} = U_{-(n+m+1)}$$

so μ is a *pole* of the resolvent (i.e. there exists $k > 0$ such that $U_{-k} \neq 0$ and $U_{-n} = 0 \ \forall n > k$) if and only if there exists $k > 0$ such that $U_{-k} \neq 0$ and $U_{-(k+1)} = 0$. Then k is the *order* of the pole. In this case, μ is an eigenvalue of T and $PX = Ker(\mu - T)^k$. The *algebraic multiplicity* $m_a \leq +\infty$ of μ is the dimension of PX. Conversely, if $m_a < +\infty$, i.e. P is of finite rank, then $(\mu - T)^{m_a} P = 0$ and then μ is a pole of the resolvent of order $\leq m_a$. Actually, the order k of the pole is the smallest $j \in \mathbb{N}$ such that $(\mu - T)^j P = 0$. The subspace $Ker(\mu - T)^k$ contains the *generalized* eigenvectors; it coincides with the eigenspace if and only if $PX = Ker(\mu - T)$, i.e. $k = 1$ (simple pole); μ is said to be a semi-simple eigenvalue. We say that μ is *algebraically simple* if $m_a = 1$.

2.3 Application to Riesz-Schauder Theory

As a first illustration of the interest of Riesz projections, we show why the non zero eigenvalues of compact operators have finite algebraic multiplicities. Let $T : X \to X$ be a *compact* operator (i.e. maps bounded sets into relatively compact ones). Then $\sigma(T)/\{0\}$ consists at most of isolated eigenvalues. Let $\alpha \in \sigma(T)$ with $\alpha \neq 0$. Define T_λ (in the neighborhood of α) by

$$(\lambda - T)^{-1} = \lambda^{-1} + T_\lambda.$$

Then $(\lambda - T)(\lambda^{-1} + T_\lambda) = I$ implies that $T_\lambda = T(\lambda^{-1}T_\lambda + \lambda^{-2}I)$ is compact. So (C being a small circle around α positively oriented) the spectral projection

$$\begin{aligned} U_{-1} &= \frac{1}{2i\pi}\int_C (\lambda - T)^{-1} d\lambda = \frac{1}{2i\pi}\int_C \lambda^{-1} d\lambda + \frac{1}{2i\pi}\int_C T_\lambda d\lambda \\ &= \frac{1}{2i\pi}\int_C T_\lambda d\lambda \end{aligned}$$

is compact too. Since U_{-1} has a closed range then the open mapping theorem and Riesz theorem imply that U_{-1} has finite-dimensional range. Hence α has a finite algebraic multiplicity.

This result extends to *power compact* operators. Indeed, let $T \in \mathcal{L}(X)$ and $n \in \mathbb{N}$ ($n \geqslant 2$) such that T^n is compact. The spectral mapping theorem

$$\sigma(T^n) = (\sigma(T))^n$$

implies that $\sigma(T)/\{0\}$ consists at most of isolated points. Let $\alpha \in \sigma(T)$ with $\alpha \neq 0$. Then, for λ close to α, $(\lambda^n - T^n) = (\lambda^{n-1}I + \lambda^{n-2}T + \ldots + T^{n-1})(\lambda - T)$ implies

$$\begin{aligned} (\lambda - T)^{-1} &= (\lambda^n - T^n)^{-1}(\lambda^{n-1}I + \lambda^{n-2}T + \ldots + T^{n-1}) \\ &= [\lambda^{-n} + C_\lambda]\,(\lambda^{n-1}I + \lambda^{n-2}T + \ldots + T^{n-1}) \\ &= \lambda^{-n}(\lambda^{n-1}I + \lambda^{n-2}T + \ldots + T^{n-1}) \\ &\quad + C_\lambda(\lambda^{n-1}I + \lambda^{n-2}T + \ldots + T^{n-1}) \end{aligned}$$

(where C_λ is compact) so the spectral projection

$$U_{-1} = \frac{1}{2i\pi}\int_C (\lambda - T)^{-1} d\lambda = \frac{1}{2i\pi}\int_C C_\lambda(\lambda^{n-1}I + \lambda^{n-2}T + \ldots + T^{n-1}) d\lambda$$

is compact and we argue as previously.

2.4 Spectral Mapping Theorem for a Resolvent

Let $T : D(T) \subset X \to X$ be closed linear operator and $\lambda_0 \in \rho(T)$. The spectral links between T and its resolvent $(\lambda_0 - T)^{-1}$ are completely described by:

- $\sigma\left[(\lambda_0 - T)^{-1}\right]\backslash\{0\} = (\lambda_0 - \sigma(T))^{-1}$ (so $r_\sigma\left[(\lambda_0 - T)^{-1}\right] = [dist(\lambda_0, \sigma(T))]^{-1}$)
- $\sigma_p\left[(\lambda_0 - T)^{-1}\right]\backslash\{0\} = (\lambda_0 - \sigma_p(T))^{-1}$
- $\sigma_{ap}\left[(\lambda_0 - T)^{-1}\right]\backslash\{0\} = (\lambda_0 - \sigma_{ap}(T))^{-1}$
- $\sigma_{res}\left[(\lambda_0 - T)^{-1}\right]\backslash\{0\} = (\lambda_0 - \sigma_{res}(T))^{-1}$

- μ is an isolated point of $\sigma(T)$ if and only if $(\lambda_0 - \mu)^{-1}$ is an isolated point of $\sigma\left[(\lambda_0 - T)^{-1}\right]$. In this case, the residues and the orders of the pole of $(\lambda - T)^{-1}$ at μ and of $\left[\lambda - (\lambda_0 - T)^{-1}\right]^{-1}$ at $(\lambda_0 - \mu)^{-1}$ coincide.

See [20, Chapter IV]. These properties are of interest e.g. when we deal with Riesz-Schauder theory of operators with compact resolvent.

2.5 Fredholm Operators

A closed operator $T : D(T) \subset X \to X$ is said to be a Fredholm operator if $\dim Ker(T) < \infty$ and the range $R(T)$ of T is closed with finite codimension (i.e. $\dim \frac{X}{R(T)} < \infty$). Let $T : D(T) \subset X \to X$ be closed linear operator; its Fredholm domain is defined by

$$\rho_F(T) := \{\lambda \in \mathbb{C};\ \lambda - T : D(T) \to X \text{ is Fredholm}\}.$$

Then $\rho_F(T)$ is open and $\rho(T) \subset \rho_F(T)$. If λ_0 is an isolated eigenvalue of T with finite algebraic multiplicity then $\lambda_0 \in \rho_F(T)$, (see [32, Chapter IV]).

We recall that $T \in \mathcal{L}(X)$ is Fredholm if and only if there exists $S \in \mathcal{L}(X)$ such that $I - ST$ and $I - TS$ are finite rank operators (see [21] p. 190). The essential spectrum of T is defined by

$$\sigma_{ess}(T) := \mathbb{C} \backslash \rho_F(T).$$

Let $\mathcal{K}(X) \subset \mathcal{L}(X)$ be the closed ideal of compact operators. The Calkin algebra

$$\mathcal{C}(X) := \frac{\mathcal{L}(X)}{\mathcal{K}(X)}$$

is endowed with the quotient norm (for $\hat{T} := T + \mathcal{K}(X)$)

$$\left\| \hat{T} \right\|_{\mathcal{C}(X)} = \inf_{K \in \mathcal{K}(X)} \|T + K\| = dist(T, \mathcal{K}(X)).$$

Then

$$\rho_F(T) = \rho(\hat{T}) \text{ and } \sigma_{ess}(T) = \sigma(\hat{T}).$$

The essential norm of $T \in \mathcal{L}(X)$ is defined by

$$\|T\|_{ess} := \left\| \hat{T} \right\|_{\mathcal{C}(X)}.$$

In particular, $\|T\|_{ess} \leq \|T\|$ and the essential norm $\|.\|_{ess}$ is submultiplicative, i.e.

$$\|T_1T_2\|_{ess} \leq \|T_1\|_{ess}\, \|T_2\|_{ess} \quad (T_i \in \mathcal{L}(X),\ i = 1,2).$$

The *essential* radius of $T \in \mathcal{L}(X)$ is defined by

$$r_{ess}(T) := r_\sigma(\hat{T}).$$

Then

$$r_{ess}(T) = \sup\left\{|\lambda|\,;\ \lambda \in \sigma(\hat{T})\right\} = \sup\left\{|\lambda|\,;\ \lambda \in \sigma_{ess}(T)\right\}.$$

In addition

$$r_{ess}(T) = \lim_{n\to\infty} \left\| \left(\hat{T}\right)^n \right\|_{\mathcal{C}(X)}^{\frac{1}{n}} = \lim_{n\to\infty} \left\| \hat{T^n} \right\|_{\mathcal{C}(X)}^{\frac{1}{n}} = \lim_{n\to\infty} \|T^n\|_{ess}^{\frac{1}{n}}.$$

The unbounded component of $\rho_F(T)$ consists of resolvent set and at most of isolated eigenvalues with finite algebraic multiplicities, (see [21, p. 204]). Then the essential radius of $T \in \mathcal{L}(X)$ is also given by

$$\inf\{r > 0;\ \lambda \in \sigma(T),\ |\lambda| > r \Rightarrow \lambda \in \sigma_{discr}(T)\}$$

where $\sigma_{discr}(T)$ refers to the isolated eigenvalues of T with finite algebraic multiplicities. Note that for any $\varepsilon > 0$, $\sigma(T) \cap \{|\lambda| \geqslant r_{ess}(T) + \varepsilon\}$ consists at most of finitely many eigenvalues with finite algebraic multiplicities. We point out that there exist several non equivalent definitions of essential spectrum for bounded operators but the corresponding essential radius is the same for all them, see [19, Corollary 4.11, p. 44].

2.6 Semigroups and Generators

Let X be a complex Banach space. By a C_0-semigroup on X we mean a family $(S(t))_{t\geqslant 0}$ of bounded linear operators on X indexed by $t \geqslant 0$ such that $S(0) = I$, $S(t)S(s) = S(t+s)$ and such that the *strong* continuity condition holds:

$$[0,+\infty[\ \ni t \to S(t)x \in X$$

is continuous for all $x \in X$. By the uniform boundedness theorem, $(S(t))_{t\geqslant 0}$ is locally bounded in $\mathcal{L}(X)$. The infinitesimal generator of $(S(t))_{t\geqslant 0}$ is the unbounded linear operator defined by

$$T : x \in D(T) \subset X \to \lim_{t\to 0} \frac{S(t)x - x}{t} \in X$$

with domain

$$D(T) = \left\{ x;\ \lim_{t \to 0} \frac{S(t)x - x}{t} \text{ exists in } X \right\}.$$

Then T is closed and densely defined. In addition, $D(T)$ is invariant under $S(t)$ and $S(t)Tx = TS(t)x\ \ \forall x \in D(T)$. Finally, $\forall x \in D(T)$,

$$f : t \geqslant 0 \to S(t)x \in X \text{ is } C^1$$

and

$$f'(t) = Tf(t),\ f(0) = x;$$

see e.g. [15].

If

$$p : \mathbb{R}_+ \to [-\infty, +\infty[$$

is subadditive (i.e. $p(t+s) \le p(t) + p(s)$) and locally bounded from above then

$$\lim_{t \to +\infty} \frac{p(t)}{t} = \inf_{t>0} \frac{p(t)}{t}.$$

see e.g. [15]. Since

$$t \geqq 0 \to p(t) := \ln(\|S(t)\|) \in [-\infty, +\infty[$$

is subadditive and locally bounded from above then

$$\omega := \inf_{t>0} \frac{\ln(\|S(t)\|)}{t} = \lim_{t \to +\infty} \frac{\ln(\|S(t)\|)}{t} \in [-\infty, +\infty[.$$

In particular $(S(t))_{t \geqslant 0}$ is exponentially bounded, i.e.

$$\forall \alpha > \omega\ \exists M_\alpha \geqslant 1;\ \|S(t)\| \le M_\alpha e^{\alpha t}\ \forall t \geqq 0;$$

ω is called the *type* or growth bound of $(S(t))_{t \geqslant 0}$. In addition, for any $t > 0$

$$\begin{aligned} r_\sigma(S(t)) &= \lim_{n \to +\infty} \|S(t)^n\|^{\frac{1}{n}} = \lim_{n \to +\infty} \|S(nt)\|^{\frac{1}{n}} \\ &= \lim_{n \to +\infty} \exp \frac{1}{n} \ln \|S(nt)\| = \lim_{n \to +\infty} \exp t \frac{1}{nt} \ln \|S(nt)\| = e^{\omega t}. \end{aligned}$$

We recall that $\{Re\lambda > \omega\} \subset \rho(T)$ and

$$(\lambda - T)^{-1} = \int_0^{+\infty} e^{-\lambda t} S(t)dt \quad (Re\lambda > \omega)$$

where the integral converges in operator norm. Thus $\sigma(T) \subset \{Re\lambda \leq \omega\}$ and the *spectral bound* of T

$$s(T) := \sup\{Re\lambda;\ \lambda \in \sigma(T)\} \leq \omega.$$

We end this section with the famous Hille–Yosida–Phillips–Miyadera–Feller theorem (commonly called Hille–Yosida theorem) which provides a general framework for a huge amount of linear evolution equations of mathematical physics and probability theory [22].

Theorem 3 *Let $T : D(T) \subset X \to X$ be a closed densely defined linear operator. Then T is the generator of a C_0-semigroup $(S(t))_{t\geqslant 0}$ satisfying the estimate $\|S(t)\| \leq Me^{\alpha t}$ $\forall t \geqq 0$ if and only if $\sigma(T) \subset \{Re\lambda \leq \alpha\}$ and*

$$\left\|\left[(\lambda - T)^{-1}\right]^n\right\| \leq \frac{M}{(Re\lambda - \alpha)^n} \quad (Re\lambda > \alpha)\ \forall n \in \mathbb{N}.$$

We note that if X is a reflexive complex Banach space and if $(S(t))_{t\geqslant 0}$ is a C_0-semigroup with generator T then the dual semigroup $(S'(t))_{t\geqslant 0}$ is *strongly continuous* and its generator is given by T'. In particular $(S(t))_{t\geqslant 0}$ and $(S'(t))_{t\geqslant 0}$ have the same type while T and T' have the same spectral bound.

2.7 Partial Spectral Mapping Theorems for Semigroups

In general, there exist *partial* spectral links between a C_0-semigroup and its generator, see [20, Chapter IV].

Theorem 4 *Let X be a complex Banach space and $(S(t))_{t\geqslant 0}$ be a C_0-semigroup on X with generator T.Then:*

(i) $e^{t\sigma_{ap}(T)} \subset \sigma_{ap}(S(t))\backslash\{0\}$.
(ii) $e^{t\sigma_p(T)} = \sigma_p(S(t))\backslash\{0\}$.
(iii) $e^{t\sigma_{res}(T)} = \sigma_{res}(S(t))\backslash\{0\}$.
(iv) $m_g(\lambda, T) \leq m_g(e^{\lambda t}, S(t))$
(v) $m_a(\lambda, T) \leq m_a(e^{\lambda t}, S(t))$
(vi) $k(\lambda, T) \leq k(e^{\lambda t}, S(t))$.

Here m_g (resp. m_a, resp. k) refers to geometric multiplicity (resp. algebraic multiplicity, resp. multiplicity of a pole). We note that the possible *failure* of the spectral mapping theorem stems from the approximate point spectrum. The link between the eigenvalues of $(S(t))_{t\geqslant 0}$ and those of its generator T is clarified further by:

Theorem 5 *Let X be a complex Banach space and $(S(t))_{t\geqslant 0}$ be a C_0-semigroup on X with generator T. Then:*

(i) $Ker(\mu - T) = \cap_{t\geqslant 0} Ker(e^{\mu t} - S(t))$.
(ii) $Ker(e^{\mu t} - S(t)) = \overline{lin}_{n\in\mathbb{Z}} Ker(\mu + \frac{2i\pi n}{t} - T)\ \forall t > 0$.

Theorem 6 ([24] Proposition 1.10 or [20] p. 283) *Let X be a complex Banach space and $(S(t))_{t\geqslant 0}$ be a C_0-semigroup on X with generator T and let $t > 0$ be fixed. Let $e^{\mu t}$ be a pole of $S(t)$ of order k and let Q be the corresponding residue. Then*

(i) For every $n \in \mathbb{Z}$, $\mu + \frac{2i\pi n}{t}$ is (at most) a pole of $(\lambda - T)^{-1}$ of order at most k and residue P_n.
(ii) $QX = \overline{lin}_{n\in\mathbb{Z}} P_n X$.

Corollary 1 *Let X be a complex Banach space and $(S(t))_{t\geqslant 0}$ be a C_0-semigroup on X with generator T and let $t > 0$ be fixed. Let $\alpha \neq 0$ be an isolated eigenvalue of $S(t)$ with finite algebraic multiplicity and with residue Q. Then $Q = \sum_{j=1}^{n} P_j$ where the P_j are the residues of $(\lambda - T)^{-1}$ at $\{\lambda_1, \ldots, \lambda_n\}$, the (finite and nonempty) set of eigenvalues of T such that $e^{\lambda_i t} = \alpha$.*

2.8 Essentially Compact Semigroups

The fact that $\|.\|_{ess}$ is submultiplicative implies that

$$t \geqq 0 \to p_{ess}(t) := \ln(\|S(t)\|_{ess}) \in [-\infty, +\infty[$$

is subadditive. It is also locally bounded from above so

$$\omega_{ess} := \inf_{t>0} \frac{\ln(\|S(t)\|_{ess})}{t} = \lim_{t\to+\infty} \frac{\ln(\|S(t)\|_{ess})}{t} \in [-\infty, \omega].$$

In particular $\forall \alpha > \omega_{ess}\ \exists M_\alpha \geqslant 1$ such that

$$\|S(t)\|_{ess} \leq M_\alpha e^{\alpha t}\ \forall t \geqq 0;$$

ω_{ess} is called the *essential type* (or essential growth bound) of $(S(t))_{t\geqslant 0}$. For any $t > 0$

$$r_{ess}(S(t)) = \lim_{n\to+\infty} \|S(t)^n\|_{ess}^{\frac{1}{n}} = \lim_{n\to+\infty} \|S(nt)\|_{ess}^{\frac{1}{n}}$$
$$= \lim_{n\to+\infty} \exp\frac{1}{n}\ln\left(\|S(nt)\|_{ess}\right)$$
$$= \lim_{n\to+\infty} \exp t\frac{1}{nt}\ln\left(\|S(nt)\|_{ess}\right) = e^{\omega_{ess}t}.$$

A C_0-semigroup $(S(t))_{t\geq 0}$ on a complex Banach space X is said to be *essentially compact* if its essential type is less than its type (i.e. $\omega_{ess} < \omega$). Such semigroups have a nice finite-dimensional asymptotic structure.

Theorem 7 *Let X be a complex Banach space and $(S(t))_{t\geqslant 0}$ be an essentially compact C_0-semigroup on X with generator T. Then:*

(i) $\sigma(T) \cap \{Re\lambda > \omega_{ess}\}$ consists of a nonempty set of isolated eigenvalues with finite algebraic multiplicities.
(ii) For any ω' such that $\omega_{ess} < \omega' < \omega$, $\sigma(T) \cap \{Re\lambda \geqslant \omega'\}$ consists of a finite set (depending on ω') $\{\lambda_1, \dots, \lambda_m\}$ of eigenvalues of T.
(iii) Let P_j be the residues of $(\lambda - T)^{-1}$ at λ_j and let $P := \sum_{j=1}^m P_j$. Then the projector P reduces $(S(t))_{t\geqslant 0}$ and

$$S(t) = \sum_{j=1}^{m} e^{\lambda_j t} e^{tD_j} P_j + O(e^{(\omega'-\varepsilon)t})$$

(for some $\varepsilon > 0$) where $D_j := (T - \lambda_j)P_j$ are nilpotent bounded operators ($D_j^{k_j} = 0$ where k_j is the order of the pole λ_j).

Proof Let ω' be such that $\omega_{ess} < \omega' < \omega$. Let $t > 0$ be fixed. Then $e^{\omega_{ess}t} < e^{\omega' t} < e^{\omega t}$ and

$$\sigma(S(t)) \cap \left\{\mu;\ |\mu| \geqslant e^{\omega' t}\right\}$$

consists of a finite (and nonempty) set of eigenvalues with finite algebraic multiplicities $\{\mu_1, \dots, \mu_n\}$ while

$$\sigma(S(t)) \cap \left\{\mu;\ |\mu| < e^{\omega' t}\right\} \subset \left\{\mu;\ |\mu| < e^{(\omega'-\varepsilon)t}\right\}$$

for some $\varepsilon > 0$. For each j $(1 \leq j \leq n)$ let $\left\{\lambda_j^1, \dots, \lambda_j^{l_j}\right\}$ be the (finite and nonempty) set of eigenvalues λ of T such that $e^{\lambda t} = \mu_j$. Then the residue of the pole μ_j of the resolvent of $S(t)$ is given by

$$Q_j = \sum_{k=1}^{l_j} P_j^k$$

where P_j^k is the residue of the λ_j^k of the resolvent of T. Let $Q = \sum_{j=1}^{n} Q_j$ be the spectral projection corresponding to the eigenvalues $\{\mu_1, \ldots, \mu_n\}$ of $S(t)$ in $\left\{\mu;\ |\mu| \geqslant e^{\omega' t}\right\}$. One sees that $Q = \sum_{j=1}^{n} \sum_{k=1}^{l_j} P_j^k$ is nothing but the spectral projection corresponding to the eigenvalues of T in $\{Re\lambda \geqslant \omega'\}$. We decompose $S(t)$ as $S(t)Q + S(t)(I - Q)$. We know that $\sigma(S(t)_{|ImQ}) = \{\mu_1, \ldots, \mu_n\}$ while $\sigma(S(t)_{|KerQ}) \subset \left\{\mu;\ |\mu| < e^{(\omega' - \varepsilon)t}\right\}$ so the type of $S(t)_{|KerQ}$ is $\leq \omega' - \varepsilon$. Finally, $S(t)_{|ImQ}$ is generated by the bounded operator

$$T(\sum_{j=1}^{m} P_j) = \sum_{j=1}^{m} TP_j = \sum_{j=1}^{m} \left[\lambda_j P_j + (T - \lambda_j)P_j\right] = \sum_{j=1}^{m} \left[\lambda_j P_j + D_j\right]$$

so $S(t)_{|ImQ} = \sum_{j=1}^{m} e^{\lambda_j t} e^{tD_j} P_j$.

2.9 *Peripheral Spectral Theory and Applications*

In ordered Banach spaces, *positive* semigroups (i.e. leaving invariant the positive cone) enjoy nice spectral properties. For the sake of simplicity, we restrict ourselves to Lebesgue spaces

$$X = L^p(\Omega, \mathcal{A}, \mu) \quad (1 \leq p \leq +\infty)$$

where $(\Omega, \mathcal{A}, \mu)$ is a measure space (i.e. Ω is a set, $\mathcal{A}$ is a σ-algebra of subsets of Ω and μ is a σ-finite measure on $\mathcal{A}$) although most of the results hold in general Banach lattices. For short, we will write $L^p(\mu)$ (or just L^p) instead of $L^p(\Omega, \mathcal{A}, \mu)$. Let $L_+^p(\mu)$ be the cone of nonnegative a.e. functions. Then

$$L^p(\mu) = L_+^p(\mu) - L_+^p(\mu).$$

More precisely

$$f = f_+ - f_-, \quad \forall f \in L^p(\mu)$$

where

$$f_+ = \sup\{f, 0\}, \quad f_- = \sup\{-f, 0\}.$$

In particular

$$|f| = f_+ + f_-, \quad \|f\| = \||f|\|$$

where $|f|(x) := |f(x)|$. An operator $G \in L(X)$ is said to be *positive* if $Gf \in L_+^p(\mu)$ $\forall f \in L_+^p(\mu)$. We write $G \geqslant 0$. In this case

$$|Gf| = |Gf_+ - Gf_-| \leq Gf_+ + Gf_- = G(|f|)$$

and consequently

$$\|G\| = \sup_{\|f\|\leq 1,\ f\in L_+^p} \|Gf\|.$$

It follows that if $0 \leq G_1 \leq G_2$ with $G_i \in \mathcal{L}(L^p)(i = 1, 2)$ then $\|G_1\| \leq \|G_2\|$. This last property applied to the iterates shows that $r_\sigma(G_1) \leq r_\sigma(G_2)$. It is easy to see that $G \in \mathcal{L}(L^p)$ is positive if and only if its dual operator $G' \in \mathcal{L}(L^{p'})$ is positive.

A C_0-semigroup $(S(t))_{t\geqslant 0}$ on X is said to be positive if $\forall t > 0$, $S(t)$ is a positive operator. A C_0-semigroup $(S(t))_{t\geqslant 0}$ with type ω and generator T is positive if and only if the resolvent $(\lambda - T)^{-1}$ is positive for $\lambda > \omega$; this follows from

$$(\lambda - T)^{-1} f = \int_0^{+\infty} e^{-\lambda t} S(t) f dt \quad (\lambda > \omega)$$

and the exponential formula

$$S(t)f = \lim_{n\to+\infty} (I - \frac{t}{n} T)^{-n} f.$$

A fundamental result for positive C_0-semigroups $(S(t))_{t\geqslant 0}$ on *Lebesgue* spaces $L^p(\mu)$ is that *the type of* $(S(t))_{t\geqslant 0}$ *coincides with the spectral bound* $s(T)$ *of its generator* T (see e.g. *[20]*). Another fundamental spectral property of positive operators $G \in \mathcal{L}(X)$ (in general Banach lattices) is that the spectral radius belongs to the spectrum

$$r_\sigma(G) \in \sigma(G).$$

Let us show an analogous property for a generator T of a positive semigroup $(S(t))_{t\geqslant 0}$:

$$s(T) > -\infty \Rightarrow s(T) \in \sigma(T).$$

Indeed, note first that

$$\left|(\lambda - T)^{-1} f\right| \leq \int_0^{+\infty} e^{-Re\lambda t} S(t) |f| dt \quad (\forall Re\lambda > s(T))$$

so $\left\|(\lambda - T)^{-1}\right\| \leq \left\|(Re\lambda - T)^{-1}\right\|$ $(\forall Re\lambda > s(T))$. By assumption there exists a sequence $(\beta_n)_n \subset \sigma(T)$ such that $Re\beta_n \to s(T)$. We build a sequence $(\lambda_n)_n$

with $Re\lambda_n > s(T)$ (so $(\lambda_n)_n \subset \rho(T)$), $Im\lambda_n = Im\beta_n$ and $Re\lambda_n \to s(T)$. Then $|\lambda_n - \beta_n| \to 0$ and $\left\|(\lambda_n - T)^{-1}\right\| \to +\infty$. Thus $\left\|(Re\lambda_n - T)^{-1}\right\| \to +\infty$ and consequently $s(T) \in \sigma(T)$.

Let $G \in \mathcal{L}(L^p)$ be positive. We say that G is *irreducible* if $\forall f \in L^p_+(\mu),\ f \neq 0$ and $\forall g \in L^{p'}_+(\mu),\ g \neq 0$ there exists $n \in \mathbb{N}$ (depending a priori on f and g) such that

$$\langle G^n f, g\rangle_{L^p, L^{p'}} > 0.$$

For $p < +\infty$, this is equivalent to saying that there is no closed subspace $L^p(\Omega', \mu)$ (with $\mu(\Omega') > 0$ and $\mu(\Omega/\Omega') > 0$) invariant by G. For instance, if $Gf > 0$ a.e. $\forall f \in L^p_+(\Omega),\ f \neq 0$ (we say that G is *positivity-improving*) then G is irreducible. A positive C_0-semigroup $(S(t))_{t\geqslant 0}$ is said to be irreducible if $\forall f \in L^p_+(\mu),\ f \neq 0$ and $\forall g \in L^{p'}_+(\mu),\ g \neq 0$ there exists $t > 0$ (depending a priori on f and g) such that

$$\langle S(t) f, g\rangle_{L^p, L^{p'}} > 0.$$

For $p < +\infty$, this is equivalent to saying that there is no closed subspace $L^p(\Omega', \mu)$ (with $\mu(\Omega') > 0$ and $\mu(\Omega/\Omega') > 0$) invariant by *all* $S(t)$. A positive C_0-semigroup $(S(t))_{t\geqslant 0}$ with generator T is irreducible if and only if $(\lambda - T)^{-1}$ is positivity-improving for some $\lambda > s(T)$. This follows easily from

$$\langle (\lambda - T)^{-1} f, g\rangle = \int_0^{+\infty} e^{-\lambda t} \langle S(t) f, g\rangle dt.$$

We recall a useful result combining compactness and irreducibility:

Theorem 8 ([63]) *If $G \in \mathcal{L}(X)$ is compact and irreducible then $r_\sigma(G) > 0$.*

The fact that $r_\sigma(G^n) = r_\sigma(G)^n$ implies easily:

Corollary 2 *If some power of $G \in L(X)$ is compact and positivity-improving then $r_\sigma(G) > 0$.*

The following result can be found in [61, Chapter CIII].

Theorem 9 *Let $(S(t))_{t\geqslant 0}$ be a positive C_0-semigroup on $L^p(\mu)$ with generator T. If $s(T)$ is a pole of the $(\lambda - T)^{-1}$ then the boundary spectrum*

$$\sigma_b(T) := \sigma(T) \cap (s(T) + i\mathbb{R})$$

consists of poles of the resolvent and is cyclic in the sense that there exists $\alpha \geqslant 0$ such that

$$\sigma_b(T) := s(T) + i\alpha\mathbb{Z}.$$

Corollary 3 *Let $(S(t))_{t\geqslant 0}$ be a positive C_0-semigroup on $L^p(\Omega, \mathcal{A}, \mu)$ with generator T. We assume that $(S(t))_{t\geqslant 0}$ is essentially compact (i.e. $\omega_{ess} < \omega$). Then*

$$\sigma_b(T) = \{s(T)\}$$

i.e. $s(T)$ is the leading eigenvalue and is strictly dominant (i.e. $\exists\varepsilon > 0$; $Re\lambda \leq s(T) - \varepsilon\ \forall\lambda \in \sigma(T)$, $\lambda \neq s(T)$).

Proof According to the theorem above, $\sigma_b(T)$ is either unbounded or reduces to $\{s(T)\}$. The fact that $\omega_{ess} < \omega$ implies that $\sigma_b(T)$ is finite. ■

By combining essential compactness and positivity arguments we get a fundamental functional analytic result:

Theorem 10 ([61] Prop 3.5, p. 310) *Let $(S(t))_{t\geqslant 0}$ be an irreducible C_0-semigroup on $L^p(\Omega, \mathcal{A}, \mu)$ with generator T. We assume that $(S(t))_{t\geqslant 0}$ is essentially compact (i.e. $\omega_{ess} < \omega$). Then $s(T)$ is the leading eigenvalue, is strictly dominant and is algebraically simple. In particular there exists $\varepsilon > 0$ such that*

$$S(t)f = e^{s(T)t}\left(\int f(x)v(x)\mu(dx)\right)u + O(e^{(s(T)-\varepsilon)t})$$

where u is the (strictly positive almost everywhere) eigenfunction of T associated to $s(T)$ and v is the (strictly positive almost everywhere) eigenfunction of T' associated to $s(T') = s(T)$ with the normalization $\int u(x)v(x)\mu(dx) = 1$.

2.10 Semigroups with Dense Local Quasinilpotence Subspace

This subsection deals with a class of positive semigroups whose real spectra can be described completely. This class is well-suited to weighted shift semigroups we consider in the next section. We resume here some abstract results from [56]. For the sake of simplicity, we restrict ourselves to complex Lebesgue spaces $X = L^p(\mu)$ $(1 \leq p \leq \infty)$. Let $(S(t))_{t\geqslant 0}$ be a positive semigroup on $L^p(\mu)$. We define its local quasinilpotence subset by

$$Y = \left\{f \in L^p(\mu);\ \lim_{t\to+\infty} \|S(t)\,|f|\|^{\frac{1}{t}} = 0\right\}$$

where $|f|$ is the absolute value of $f \in L^p(\mu)$.

Lemma 1 *Y is a subspace of $L^p(\mu)$ invariant under $(S(t))_{t\geqslant 0}$.*

Proof

(i) Linearity: Clearly $\lambda f \in Y$ if $f \in Y$. Let $\varepsilon > 0,\ f, g \in Y$ be given. There exists $\overline{t} > 0$ depending on them such that

$$\|S(t)\,|f|\| \leq \varepsilon^t \text{ and } \|S(t)\,|g|\| \leq \varepsilon^t \quad \forall t \geqslant \overline{t}.$$

So $\|S(t)\,|f+g|\| \leq \|S(t)\,(|f|+|g|)\| \leq 2\varepsilon^t \quad \forall t \geqslant \overline{t}$ and

$$\|S(t)\,|f+g|\|^{\frac{1}{t}} \leq 2^{\frac{1}{t}}\varepsilon \leq 2\varepsilon \quad \forall t \geqslant \max(\overline{t}, 1).$$

(ii) Invariance: Let $\tau > 0,\ f \in Y$.

$$\begin{aligned}\|S(t)\,|S(\tau)f|\|^{\frac{1}{t}} &\leq \|S(t)\,(S(\tau)\,|f|)\|^{\frac{1}{t}} \\ &= \|S(t+\tau)\,(|f|)\|^{\frac{1}{t}} = \left(\|S(t+\tau)\,(|f|)\|^{\frac{1}{t+\tau}}\right)^{\frac{t+\tau}{t}} \to 0\end{aligned}$$

as $t \to +\infty$; i.e. $S(\tau)f \in Y$.

Theorem 11 *Let $(S(t))_{t\geqslant 0}$ be a positive semigroup on $L^p(\mu)$ with type ω. If its local quasinilpotence subspace is dense in $L^p(\mu)$ then $[0, e^{\omega t}] \subset \sigma_{ap}(S(t))$.*

Proof Let $t > 0$ be fixed. Let $0 < \mu < e^{\omega t}$ and $y \in Y$. The equation

$$\mu x - S(t)x = y; \quad (y \in Y,\ \|y\| = 1)$$

can be solved by

$$x = \frac{1}{\mu}\sum_{k=0}^{\infty}\frac{1}{\mu^k}S(t)^k y = \frac{1}{\mu}\sum_{k=0}^{\infty}\frac{1}{\mu^k}S(kt)y$$

provided that this series converges. This is the case since

$$\left\|\frac{1}{\mu^k}S(kt)y\right\|^{\frac{1}{k}} = \frac{1}{\mu}\left(\|S(kt)y\|^{\frac{1}{kt}}\right)^t \to 0 \text{ as } k \to +\infty.$$

In particular $x \geqslant 0$ for $y \geqslant 0$ and

$$\|x\| \geqslant \frac{1}{\mu^{k+1}}\left\|S(t)^k y\right\| \quad \forall k \in \mathbb{N}.$$

There exists $z_k \subset L^p_+(\mu)$ such that $\|z_k\| = 1$ and

$$\left\|S(t)^k z_k\right\| \geqslant \frac{1}{2}\left\|S(t)^k\right\|.$$

By the denseness of Y, $\exists\ y_k \in Y$ such that $\|y_k\| = 1$

$$\left\| S(t)^k y_k \right\| \geqslant \frac{1}{3} \left\| S(t)^k \right\|.$$

We may assume that $y_k \geqslant 0$ since $\left\| S(t)^k |y_k| \right\| \geqslant \left\| S(t)^k y_k \right\|$ and $|y_k| \in Y$. The solution $\hat{x}_k$ of

$$\mu \hat{x}_k - S(t)\hat{x}_k = y_k$$

satisfies

$$\|\hat{x}_k\| \geqslant \frac{1}{\mu^{k+1}} \left\| S(t)^k y_k \right\| \geqslant \frac{1}{3} \frac{1}{\mu^{k+1}} \left\| S(t)^k \right\|.$$

So

$$\lim_{k\to+\infty} \inf \|\hat{x}_k\|^{\frac{1}{k}} \geqslant \frac{1}{\mu} \lim_{k\to+\infty} \left\| S(t)^k \right\|^{\frac{1}{k}} = \frac{e^{\omega t}}{\mu} > 1$$

and then $\lim_{k\to+\infty} \|\hat{x}_k\| = \infty$. Finally $x_k := \frac{\hat{x}_k}{\|\hat{x}_k\|}$ is such that

$$\|x_k\| = 1 \text{ and } \|\mu x_k - S(t)x_k\| \to 0$$

i.e. $\mu \in \sigma_{ap}(S(t))$. The closedness of $\sigma_{ap}(S(t))$ ends the proof. ■

Lemma 2 *Let $(S(t))_{t\geqslant 0}$ be a positive semigroup on $L^p(\mu)$ with generator T. Let Y be the local quasinilpotence subspace of $(S(t))_{t\geqslant 0}$. Then, for any $\lambda > \omega$,*

$$\lim_{k\to\infty} \left\| (\lambda - T)^{-k} y \right\|^{\frac{1}{k}} = 0 \ \ \forall y \in Y.$$

Proof For any $y \in Y$ and any $\varepsilon > 0$ there exists $t_{y,\varepsilon} > 0$ such that

$$\|S(t)y\| \leq \varepsilon^t \ \ \forall t \geqslant t_{y,\varepsilon}$$

i.e. (write $\varepsilon = e^{-A}$)

$$\|S(t)y\| \leq e^{-At} \ \ \forall t \geqslant t_{y,\varepsilon}$$

so $\exists M_{y,A} \geqslant 0$ such that

$$\|S(t)y\| \leq M_{y,A} e^{-At} \ \ \forall t \geqslant 0.$$

Hence

$$\begin{aligned}\left\|(\lambda - T)^{-k} y\right\| &= \left\| \int_0^{+\infty} dt_1 \ldots \int_0^{+\infty} dt_k e^{-\lambda(t_1+\ldots+t_k)} S(t_1 + \ldots + t_k) y \right\| \\ &\leq \int_0^{+\infty} dt_1 \ldots \int_0^{+\infty} dt_k e^{-\lambda(t_1+\ldots+t_k)} \left\| S(t_1 + \ldots + t_k) y \right\| \\ &\leq M_{y,A} \int_0^{+\infty} dt_1 \ldots \int_0^{+\infty} dt_k e^{-\lambda(t_1+\ldots+t_k)} e^{-A(t_1+\ldots+t_k)} \\ &= \frac{M_{y,A}}{(\lambda + A)^k}\end{aligned}$$

and

$$\lim \sup_{k \to +\infty} \left\|(\lambda - T)^{-k} y\right\|^{\frac{1}{k}} \leq \frac{1}{\lambda + A}$$

which ends the proof since $A > 0$ is arbitrary. ■

Theorem 12 *Let $(S(t))_{t \geqslant 0}$ be a positive semigroup on $L^p(\mu)$ with generator T. Let $s(T)$ be the spectral bound of T. If the local quasinilpotence subspace of $(S(t))_{t \geqslant 0}$ is dense in $L^p(\mu)$ then*

$$(-\infty, s(T)] \subset \sigma_{ap}(T).$$

Proof Let $\lambda < s(T) < \mu$ be fixed. Consider

$$\frac{1}{\mu - \lambda} x - (\mu - T)^{-1} x = y \in Y.$$

Arguing as for the semigroup, we show the existence of $(x_k)_k$ with $\|x_k\| = 1$ and

$$\left\| \frac{1}{\mu - \lambda} x_k - (\mu - T)^{-1} x_k \right\| \to 0$$

i.e. $\frac{1}{\mu - \lambda} \in \sigma_{ap}((\mu - T)^{-1})$ or equivalently $\lambda \in \sigma_{ap}(T)$. The closedness of $\sigma_{ap}(T)$ ends the proof. ■

Corollary 4 *Let $(S(t))_{t \geqslant 0}$ be a positive semigroup on $L^p(\mu)$ with type ω and generator T. We assume that the local quasinilpotence subspace of $(S(t))_{t \geqslant 0}$ is dense in $L^p(\mu)$.*

(i) If $\sigma(T)$ is invariant under translations along the imaginary axis then

$$\sigma(T) = \{\lambda \in \mathbb{C};\ Re\lambda \leq \omega\}.$$

(ii) If $\sigma(S(t))$ is invariant under rotations then

$$\sigma(S(t)) = \{\mu \in \mathbb{C};\ |\mu| \leq e^{\omega t}\}.$$

3 Spectral Analysis of Advection Semigroups

Neutron transport theory is mainly a perturbation theory (by scattering operators) of suitable weighted shift semigroups called advection semigroups

$$U(t) : g \to e^{-\int_0^t \sigma(x-\tau v,v)d\tau} g(x-tv,v)1_{\{t\leq s(x,v)\}}$$

where

$$s(x,v) = \inf\{s > 0;\ x - sv \notin \Omega\}$$

is the (first) exit time function from the spatial domain Ω. We describe here the spectra of such semigroups. This section resumes essentially [56]; (an alternative approach is given in [78]).

3.1 On Advection Semigroups

Let $\Omega \subset \mathbb{R}^n$ be an open subset and let μ be a positive Borel measure on $\mathbb{R}^n$ with support V. Let

$$\sigma : \Omega \times V \to \mathbb{R}_+$$

be measurable and such that

$$\lim_{t\to 0} \int_0^t \sigma(x-\tau v,v)d\tau = 0 \ \text{ a.e.}$$

Let

$$s(x,v) = \inf\{s > 0;\ x - sv \notin \Omega\}$$

be the so-called exit time function. Then

$$S(t) : g \to e^{-\int_0^t \sigma(x-\tau v,v)d\tau} g(x-tv,v)1_{\{t\leq s(x,v)\}}$$

defines a positive semigroup on $L^p(\Omega \times V; dx \otimes \mu(dv))$ (for any $1 \leq p \leq +\infty$), strongly continuous when $p < +\infty$, see e.g. [78]. The dual streaming semigroups in $L^{p'}(\Omega \times V)$ are given by

$$S'(t) : f \to e^{-\int_0^t \sigma(x+\tau v,v)d\tau} f(x+tv,v)1_{\{t \leq s(x,-v)\}}.$$

3.2 Invariance Property of Transport Operators

Let $\mu\{0\} = 0$ and

$$\alpha : (x,v) \in \Omega \times V \to \frac{x.v}{|v|^2}.$$

For any $\eta > 0$

$$M_\eta : f \in L^p(\Omega \times V) \to e^{-i\eta\alpha(x,v)} f \in L^p(\Omega \times V)$$

is an isometric isomorphism.

Theorem 13 ([78]) $M_\eta^{-1} S(t) M_\eta = e^{i\eta t} S(t)$. *In particular* $\sigma(S(t))$ *is invariant by rotations.*

Proof We have

$$\begin{aligned} S(t)M_\eta f &= e^{-\int_0^t \sigma(x-\tau v,v)d\tau} M_\eta f(x-tv,v)1_{\{t\leq s(x,v)\}} \\ &= e^{-\int_0^t \sigma(x-\tau v,v)d\tau} e^{-i\eta\frac{(x-tv).v}{|v|^2}} f(x-tv,v)1_{\{t\leq s(x,v)\}} \end{aligned}$$

so

$$M_\eta^{-1} S(t) M_\eta f = e^{i\eta\frac{x.v}{|v|^2}} e^{-i\eta\frac{(x-tv).v}{|v|^2}} S(t)f = e^{i\eta t} S(t) f$$

so $M_\eta^{-1} S(t) M_\eta = e^{i\eta t} S(t)$. Hence $\sigma(M_\eta^{-1} S(t) M_\eta) = \sigma(e^{i\eta t} S(t))$. On the other hand, by similarity,

$$\sigma(M_\eta^{-1} S(t) M_\eta) = \sigma(S(t))$$

and

$$\sigma(e^{i\eta t} S(t)) = e^{i\eta t}\sigma(S(t))$$

so we are done. ∎

As previously we have:

Theorem 14 *Let T be the generator of a streaming semigroup $(S(t))_{t\geqslant 0}$. Then $M_\eta^{-1}TM_\eta = T + i\eta I$. In particular $\sigma(T)$ is invariant by translation along the imaginary axis.*

Proof Let $f \in D(T)$. Then

$$\begin{aligned}
\frac{S(t)M_\eta f - M_\eta f}{t} &= M_\eta \frac{M_\eta^{-1}S(t)M_\eta f - f}{t} \\
&= M_\eta \frac{e^{i\eta t}S(t)f - f}{t} \\
&= M_\eta \frac{e^{i\eta t}S(t)f - e^{i\eta t}f}{t} + M_\eta \frac{e^{i\eta t}f - f}{t} \\
&= e^{i\eta t}M_\eta \frac{S(t)f - f}{t} + \frac{e^{i\eta t} - 1}{t} M_\eta f \\
&\to M_\eta Tf + i\eta M_\eta f
\end{aligned}$$

so $M_\eta f \in D(T)$ and $TM_\eta f = M_\eta Tf + i\eta M_\eta f$ or $M_\eta^{-1}TM_\eta = T + i\eta I$. By similarity, $\sigma(T) = \sigma(M_\eta^{-1}TM_\eta) = \sigma(T) + i\eta \forall \eta \in \mathbb{R}$. ■

3.3 Decomposition of the Phase Space

We consider the partition of the phase space $\Omega \times V$ according to

$$E_1 = \{(x,v) \in \Omega \times V;\ s(x,-v) < +\infty\},$$

$$E_2 = \{(x,v) \in \Omega \times V;\ s(x,-v) = +\infty,\ s(x,v) < +\infty\},$$

$$E_3 = \{(x,v) \in \Omega \times V;\ s(x,-v) = +\infty,\ s(x,v) = +\infty\}.$$

This induces a direct sum

$$L^p(\Omega \times V; dx \otimes \mu(dv)) = L^p(E_1) \oplus L^p(E_2) \oplus L^p(E_3)$$

where, we identify $L^p(E_i)$ to the closed subspace of functions $f \in L^p(\Omega \times V)$ vanishing almost everywhere on $\Omega \times V \setminus E_i$. If some set E_i has zero measure then we drop out $L^p(E_i)$ from the direct sum above.

Theorem 15 *The subspaces $L^p(E_i)$ $(i = 1, 2, 3)$ are invariant under $(S(t))_{t\geqslant 0}$.. For each $i = 1, 2, 3$, we denote by $(S_i(t))_{t\geqslant 0}$ the part of $(S(t))_{t\geqslant 0}$. on $L^p(E_i)$ and* *by T_i its generator. Then*

$$\sigma(S(t)) = \sigma(S_1(t)) \cup \sigma(S_2(t)) \cup \sigma(S_3(t))$$

$$\sigma(T) = \sigma(T_1) \cup \sigma(T_2) \cup \sigma(T_3).$$

We have also similar results where $\sigma(.)$ is replaced by $\sigma_p(.)$ or $\sigma_{ap}(.)$. In addition, if $\sigma(.,.)$ is bounded then $(S_3(t))_{t\geqslant 0}$ extends to a positive group.

Proof We check that the direct sum $L^p(\Omega \times V) = L^p(E_1) \oplus L^p(E_2) \oplus L^p(E_3)$ reduces $(S(t))_{t\geqslant 0.}$ We restrict ourselves to $L^p(E_1)$. Let $f \in L^p(E_1)$, i.e. f vanishes almost everywhere on $E_2 \cup E_3$. We have to show that $S(t)f \in L^p(E_1)$ i.e. $S(t)f$ vanishes almost everywhere on $E_2 \cup E_3$. Since

$$S(t)f(x,v) = e^{-\int_0^t \sigma(x-\tau v,v)d\tau} f(x-tv,v)1_{\{t\leq s(x,v)\}}$$

is zero for $t > s(x,v)$, we assume from the start that $t \leq s(x,v)$. One notes that $(x,v) \in E_2 \cup E_3 \Leftrightarrow s(x,-v) = +\infty$ and

$$s(x-tv,-v) = t + s(x,-v)$$

so that $(x-tv,-v) \in E_2 \cup E_3$ and $f(x-tv,v) = 0$. Since the projection P_i on $L^p(E_i)$ along $L^p(\Omega \times V \setminus E_i)$ commutes with $(S(t))_{t\geqslant 0.}$ then the direct sum above reduces also the generator T. Finally, on E_3 (if $\sigma(.,.)$ is bounded) $(S_3(t))_{t\geqslant 0}$ extends to a positive group where

$$S_3(t)^{-1}g = e^{\int_0^t \sigma(x+\tau v,v)d\tau} f(x+tv,v) \quad (t > 0).$$

3.4 Spectra of the First Reduced Advection Semigroup

Lemma 3 *Let $t > 0$ be fixed. For any $f \in L^p(E_1)$*

$$\|S_1(t)f\|^p = \int_{\{t<s(y,-v)\}\cap\{s(y,-v)<\infty\}} e^{-p\int_0^t \sigma(y+\tau v,v)d\tau} |f(y,v)|^p \, dx\mu(dv).$$

Proof We have to compute the norm of $S_1(t)f$ on the set

$$\{t \leq s(x,v)\} \cap \{s(x,-v) < +\infty\},$$

so $\|S_1(t)f\|^p$ is equal to

$$\int_{\{t\leq s(x,v)\}\cap\{s(x,-v)<+\infty\}} e^{-p\int_0^t \sigma(x-\tau v,v)d\tau} |f(x-tv,v)|^p \, dx\mu(dv).$$

Since $s(x - tv, -v) = t + s(x, -v)$ is finite if and only if $s(x, -v)$ is finite then the change of variable

$$y := x - tv \in \Omega$$

gives $s(y, -v) > t$ and

$$\|S_1(t)f\|^p = \int_{\{t<s(y,-v)\}\cap\{s(y,-v)<\infty\}} e^{-p\int_0^t \sigma(y+\tau v,v)d\tau} |f(y,v)|^p \, dy\mu(dv).$$

The type of $(S_1(t))_{t\geqslant 0}$ is equal to $-\lambda_1^*$ where

$$\lambda_1^* = \lim_{t\to+\infty} \inf_{\{t<s(y,-v)\}\cap\{s(y,-v)<\infty\}} \frac{1}{t}\int_0^t \sigma(y + \tau v, v)d\tau.$$

because

$$\begin{aligned}\|S_1(t)\| &= \sup_{\{t<s(y,-v)\}\cap\{s(y,-v)<\infty\}} e^{-\int_0^t \sigma(y+\tau v,v)d\tau} \\ &= e^{-\inf_{\{t<s(y,-v)\}\cap\{s(y,-v)<\infty\}} \int_0^t \sigma(y+\tau v,v)d\tau}\end{aligned}$$

so

$$\frac{\ln \|S_1(t)\|}{t} = -\inf_{\{t<s(y,-v)\}\cap\{s(y,-v)<\infty\}} \frac{1}{t}\int_0^t \sigma(y + \tau v, v)d\tau$$

and

$$\omega_1 = -\lim_{t\to+\infty} \inf_{\{t<s(y,-v)\}\cap\{s(y,-v)<\infty\}} \frac{1}{t}\int_0^t \sigma(y + \tau v, v)d\tau.$$

We have

$$\sigma(S_1(t)) = \left\{\mu \in \mathbb{C}; |\mu| \leq e^{-\lambda_1^* t}\right\}, \; \sigma(T_i) = \{\lambda \in \mathbb{C}; Re\lambda \leq -\lambda_1^*\}.$$

Indeed, it suffices to show that the local quasinilpotence subspace of $(S_1(t))_{t\geqslant 0}$ is *dense* in $L^p(E_1)$. Let

$$O_m := \{x, v) \in \Omega \times V;\; s(x, -v) \leq m\}.$$

We note that $\cup_m L^p(O_m)$ is dense in $L^p(E_1)$ because of

$$\cup_m O_m = \{x, v) \in \Omega \times V;\; s(x, -v) < +\infty\}.$$

Finally

$$\|S_1(t)f\|^p = \int_{\{t<s(y,-v)\}\cap\{s(y,-v)<\infty\}} e^{-p\int_0^t \sigma(y+\tau v,v)d\tau} |f(y,v)|^p \, dy\mu(dv)$$

shows that, for $f \in L^p(O_m)$, $\|S_1(t)f\| = 0$ for $t > m$ so $\cup_m L^p(O_m)$ is included in the local quasinilpotence subspace of $(S_1(t))_{t\geq 0}$. ■

3.5 Spectra of the Second Reduced Advection Semigroup

We deal now with $(S_2(t))_{t\geq 0}$ on $L^p(E_2)$ where

$$E_2 = \{(x,v) \in \Omega \times V;\ s(x,-v) = +\infty,\ s(x,v) < +\infty\}.$$

We consider first the case $1 < p < +\infty$. Indeed, by duality $\sigma(S_2(t)) = \sigma(S_2'(t))$ where

$$S_2'(t)f = e^{-\int_0^t \sigma(x+\tau v,v)d\tau} f(x+tv,v).$$

Thus $\|S_2(t)f\|^{p'}$ is equal to

$$\int_{\{s(x,-v)=\infty,\ s(x,v)<\infty\}} e^{-p'\int_0^t \sigma(y+\tau v,v)d\tau} |f(x+tv,v)|^p \, dx\mu(dv)$$

$$= \int_{\{s(y,-v)=\infty,\ t\leq s(y,v)<\infty\}} e^{-p'\int_0^t \sigma(y+\tau v,v)d\tau} |f(y,v)|^p \, dy\mu(dv).$$

Introducing the sets

$$O_m' := \{x,v) \in E_2;\ s(y,v) \leq m\}$$

one sees that $\cup_m L^{p'}(O_m')$ is dense in $L^{p'}(E_2)$ because of

$$\cup_m O_m = E_2.$$

Since in $L^{p'}(O_m')$, $\|S_2(t)f\| = 0$ for $t > m$ then the local quasinilpotence subspace of $(S_2'(t))_{t\geq 0}$ is dense. This ends the proof because $\sigma(S_2(t)) = \sigma(S_2'(t))$ and $\sigma(T_2) = \sigma(T_2')$. ■

3.6 Spectra of the Third Reduced Advection (Semi)group

Theorem 16 *Let $\mathcal{S} := \sigma(T_3) \cap \mathbb{R}$ be the real spectrum of T_3. Then*

$$\sigma(T_3) = \mathcal{S} + i\mathbb{R} \text{ and } \sigma(S_3(t)) = e^{t\sigma(T_3)}.$$

Moreover, $supS = -\lambda_3^*$ *and* $\inf S = -\lambda_3^{**}$ *where*

$$\lambda_3^* = \lim_{t\to+\infty} \inf_{\{ s(y,-v)=\infty,\ s(y,v)=\infty\}} \frac{1}{t}\int_0^t \sigma(y+\tau v, v)d\tau$$

$$\lambda_3^* = \lim_{t\to+\infty} \sup_{\{ s(y,-v)=\infty,\ s(y,v)=\infty\}} \frac{1}{t}\int_0^t \sigma(y+\tau v, v)d\tau.$$

Proof The fact that $\sigma(T_3)$ is invariant by translation along the imaginary axis and that $e^{t\sigma(T_3)}$ is invariant under the rotations is a general feature of streaming semigroups in arbitrary geometry. The spectral mapping property for the real spectrum is due to the fact that $(S_3(t))_{t\in\mathbb{R}}$ is a positive C_0-group (see [27]). The type $-\lambda_3^*$ of $(S_3(t))_{t\geqslant 0}$ is obtained as for $(S_1(t))_{t\geqslant 0}$ or $(S_2(t))_{t\geqslant 0}$. Finally, λ_3^* is the spectral bound of the generator of $(S_3(-t))_{t\geqslant 0}$, (i.e. $-T_3$) and is obtained similarly. ■

Theorem 17 *If* $\sigma : \Omega \times V \to \mathbb{R}_+$ *is space-homogeneous then* $\mathcal{S} := \sigma(T_3) \cap \mathbb{R}$ *is nothing but the essential range of* $-\sigma(.)$.

See the details in [56]; in particular, $S := \sigma(T_3) \cap \mathbb{R}$ need *not* be connected. The description of $\sigma(T_3) \cap \mathbb{R}$ for general collision frequency $\sigma : \Omega \times V \to \mathbb{R}_+$ seems to be open. When $\Omega = \mathbb{R}^n$, the situation is well understood for bounded and compactly supported (in space) collision frequencies; see [28].

3.7 Reminders on Sun-Dual Theory

To study $\sigma(S_2(t))$ in L^1 spaces, we need to recall some material. Let X be a complex Banach space and let $(S(t))_{t\geqslant 0}$ be a C_0-semigroup on X with generator T. Let $(S'(t))_{t\geqslant 0}$ be the dual semigroup on the dual space X'. If X is not reflexive then a priori $(S'(t))_{t\geqslant 0}$ is *not* strongly continuous. Let

$$L^{\odot} := \left\{x' \in X';\ \left\|S'(t)x' - x'\right\| \to 0 \text{ as } t \to 0\right\}$$

the subspace of strong continuity of $(S'(t))_{t\geqslant 0}$. Then

- $L^{\odot}$ is a closed subspace of X' invariant under $(S'(t))_{t\geqslant 0}$.
- $L^{\odot} = \overline{D(T')}$ (the closure in X').

We denote by $(S^{\odot}(t))_{t\geqslant 0}$ the restriction of $(S'(t))_{t\geqslant 0}$ to $L^{\odot}$ (*sun-dual* C_0-*semigroup*). Its generator is given by

$$D(T^{\odot}) = \left\{x' \in D(T'),\ T'x' \in L^{\odot}\right\} \text{ and } T^{\odot}x' = T'x'$$

and we have $\sigma(T) = \sigma(T') = \sigma(T^{\odot})$ and $\sigma(S(t)) = \sigma(S'(t)) = \sigma(S^{\odot}(t))$; see e.g. [20, Chapter IV].

3.8 Sun-Dual Theory for Advection Semigroups

We consider $(S_2(t))_{t\geqslant 0}$ in $L^1(E_2)$ where

$$E_2 = \{(x, v) \in \Omega \times V;\ s(x, -v) = +\infty,\ s(x, v) < +\infty\}$$

and assume that

$$\sigma : \Omega \times V \to \mathbb{R}_+ \text{ is bounded.}$$

Since

$$\sigma(S_2(t)) = \sigma(S_2'(t)) = \sigma(S_2^{\odot}(t)),$$

it suffices to identify $\sigma(S_2^{\odot}(t))$. Because of the boundedness of σ,

$$L^{\odot} = \left\{ f \in L^{\infty}(E_2),\ \sup_{(x,v)} |f(x + tv, v) - f(x, v)| \to 0 \text{ as } t \to 0 \right\}.$$

Actually, we are going to work in the smaller closed subspace

$$L_0^{\odot} := \left\{ f \in L^{\odot},\ \sup_{\{(y,v),\ s(y,v)\geqslant r\}} |f(y, v)| \to 0 \text{ as } r \to \infty \right\}.$$

Lemma 4 $(S'(t))_{t\geqslant 0}$ *leaves invariant* $L_0^{\odot}$.

Proof

$$\left|\left(S'(t)f\right)(y, v)\right| \leq |f(y + tv, v)|$$

and $s(y+tv, v) = s(y, v)+t \to \infty$ if and only if $s(y, v) \to \infty$ so that $S'(t)f \in L_0^{\odot}$ if $f \in L_0^{\odot}$. ■

Let $\left(S_0^{\odot}(t)\right)_{t\geqslant 0}$ be the restriction of $\left(S_2^{\odot}(t)\right)_{t\geqslant 0}$ to $L_0^{\odot}$ and let $T_0^{\odot}$ be its generator. Then $\sigma_{ap}(S_0^{\odot}(t)) \subset \sigma_{ap}(S_2^{\odot}(t))$ and $\sigma_{ap}(T_0^{\odot}) \subset \sigma_{ap}(T_2^{\odot})$. In particular, $\sigma_{ap}(S_0^{\odot}(t)) \subset \sigma(S_2(t))$ and $\sigma_{ap}(T_0^{\odot}) \subset \sigma(T_2)$. Let

$$L_{00}^{\odot} := \left\{ f \in L^{\odot},\ \exists r > 0,\ f(y, v) = 0 \text{ for } s(y, v) \geqslant r \right\}.$$

Theorem 18 *$L_{00}^{\odot}$ is dense in $L_0^{\odot}$.*

Corollary 5 *The local quasinilpotence subspace of $\left(S_0^{\odot}(t)\right)_{t\geqslant 0}$ is dense in $L_0^{\odot}$.*

Proof of Corollary 5: The local quasinilpotence subspace of $\left(S_0^{\odot}(t)\right)_{t\geqslant 0}$ contains $L_{00}^{\odot}$. ■

Before proving Theorem 18, we need:

Lemma 5 *$L^{\odot}$ is an algebra.*

Proof For each $m \in \mathbb{N}$, let $\gamma_m : [0, +\infty[\to [0, 1]$ be smooth (say C^1) and such that

$$\gamma_m(s) = \begin{cases} 1 \; if \; s \leq m \\ 0 \; if \; s \geqslant 2m. \end{cases}$$

Lemma 6 *$\forall m \in \mathbb{N}$, $(x, v) \to \gamma_m(s(x, v))$ belongs to $L_{00}^{\odot}$.*

Proof We have just to show that $(x, v) \to \gamma_m(s(x, v))$ belongs to $L^{\odot}$. Since γ_m is Lipschitz then

$$\begin{aligned} |\gamma_m(s(x+tv, v)) - \gamma_m(s(x, v))| &= |\gamma_m(s(x, v) + t) - \gamma_m(s(x, v))| \\ &\leq Ct \;\; \forall (x, v). \end{aligned}$$

■

Proof of Theorem 18: Let $f \in L_0^{\odot}$ then $\forall m \in \mathbb{N}, (x, v) \to \gamma_m(s(x, v)) f(x, v)$ belongs to $L_{00}^{\odot}$.

$$\begin{aligned} |\gamma_m(s(x, v)) f(x, v) - f(x, v)| &= |(1 - \gamma_m(s(x, v))) f(x, v)| \\ &\leq \sup_{s(x,v)\geqslant 2m} |f(x, v)| \to 0 \text{ as } m \to \infty \end{aligned}$$

since $f \in L_0^{\odot}$. ■

By the general theory,

$$\sigma(T_0^{\odot}) = \left\{Re\lambda \leq \omega_0^{\odot}\right\}, \; \sigma(S_0^{\odot}(t)) = \left\{\mu;\; |\mu| \leq e^{\omega_0^{\odot} t}\right\}$$

where $\omega_0^{\odot}$ is the type of $\left(S_0^{\odot}(t)\right)_{t\geqslant 0}$. We have to *identify* $\omega_0^{\odot}$! The fact that $\left\|S_0^{\odot}(t)\right\| \leq \left\|S_2'(t)\right\| = \|S_2(t)\|$ implies

$$\omega_0^{\odot} \leq \omega_2 = \text{ type of } (S_2(t))_{t\geqslant 0}.$$

On the other hand,

$$g_m : (x, v) \to \gamma_m(s(x, v))$$

belongs to $L_0^{\odot}$ and $\|g_m\| \leq 1$ so that

$$\left\|S_0^{\odot}(t)\right\| \geqslant \left\|S_0^{\odot}(t)g_m\right\| = \sup_{(y,v)}\left(e^{-\int_0^t \sigma(y+\tau v,v)d\tau}\gamma_m(s(y+tv,v))\right)$$

$$= \sup_{(y,v)}\left(e^{-\int_0^t \sigma(y+\tau v,v)d\tau}\gamma_m(s(y,v)+t)\right) \quad \forall m \in \mathbb{N}.$$

But $\gamma_m(s(y,v)+t) = 1$ if $s(y,v) \leq m-t$ so

$$\left\|S_0^{\odot}(t)\right\| \geqslant \sup_{\{s(y,v)\leq m-t\}} e^{-\int_0^t \sigma(y+\tau v,v)d\tau} \quad \forall m \in \mathbb{N}.$$

Finally

$$\left\|S_0^{\odot}(t)\right\| \geqslant \sup_{\{s(y,v)<+\infty\}} e^{-\int_0^t \sigma(y+\tau v,v)d\tau} = \|S_2(t)\|$$

whence $\omega_2 \leq \omega_0^{\odot}$ and we are done. ■

A similar theory can be built for general vector fields. Indeed, consider $\mathcal{F} : \mathbb{R}^n \to \mathbb{R}^n$ a Lipschitz vector field and denote by $\Phi(x,t)$ the unique global solution to

$$\frac{d}{dt}X(t) = \mathcal{F}(X(t)), \ \ t \in \mathbb{R}$$
$$X(0) = x.$$

Let $\Omega \subset \mathbb{R}^n$ be an open set and let

$$s_{\pm}(x) := \inf\{s > 0;\ \Phi(x, \pm s) \notin \Omega\}$$

be the exit times from Ω (with the convention that $\inf \emptyset = +\infty$). We define a weighted shift semigroup

$$U(t) : f \to U(t)f$$

where

$$U(t)f = e^{-\int_0^t \nu(\Phi(x,-s))ds} f(\Phi(x,-t))\chi_{\{t<s_-(x)\}}(x).$$

We introduce the sets

$$\Omega_1 = \{x \in \Omega;\ s_+(x) < \infty\},\ \Omega_2 = \{x \in \Omega;\ s_+(x) = \infty,\ s_-(x) < \infty\}$$

and

$$\Omega_3 = \{x \in \Omega;\ s_+(x) = \infty,\ s_-(x) = \infty\}.$$

Then $L^p(\Omega_i)$ $(i = 1, 2, 3)$ are invariant under $(U(t))_{t\geqslant 0}$ and we can extend the previous spectral theory of advection semigroups, see [39].

4 Spectra of Perturbed Operators

This section deals with functional analytic results on stability of essential spectra for perturbed generators or perturbed semigroups on Banach spaces. Let X be a complex Banach space and $D \subset \mathbb{C}$ be open and connected. A compact operator valued meromorphic mapping

$$A : D \to \mathcal{C}(X)$$

($\mathcal{C}(X) \subset \mathcal{L}(X)$ is the closed subspace of compact operators) is called *essentially meromorphic* on D if A is holomorphic on D except at a discrete set of points $z_k \in D$ where A has poles with Laurent expansions

$$A(z) = \sum_{n=-m_k}^{\infty} (z - z_k)^m A_n(z_k) \quad (0 < m_k < \infty)$$

where $A_n(z_k)(n = -1, -2, \ldots, -m_k)$ are finite rank operators. We recall now a fundamental analytic Fredholm alternative:

Theorem 19 (*[65] Corollary II*) *Let X be a complex Banach space and $D \subset \mathbb{C}$ be open and connected. Let*

$$A : D \to \mathcal{C}(X)$$

be essentially meromorphic. Then

(i) Either $\lambda = 1$ is an eigenvalue of all $A(z)$
(ii) or $[I - A(z)]^{-1}$ exists except for a discrete set of points and $[I - A(z)]^{-1}$ is essentially meromorphic on D.

Let $T : D(T) \subset X \to X$ be a closed operator. We define its "essential resolvent set" as

$$\rho_e(T) = \rho(T) \cup \sigma_{discr}(T)$$

where $\sigma_{discr}(T)$ refers to the isolated eigenvalues of T with finite algebraic multiplicities. This set is open. We note that if $T \in \mathcal{L}(X)$ then the unbounded component of $\rho_e(T)$ coincides with the unbounded component of the Fredholm domain $\rho_F(T)$ (see [21, p. 204]). We give first a result from [77] and some of its consequences.

Theorem 20 *Let $T : D(T) \subset X \to X$ be a closed operator and let Ω be a connected component of $\rho_e(T)$. Let $B : D(T) \to X$ be T-bounded such that there exists $n \in \mathbb{N}$ and*

$$\left[B(\lambda - T)^{-1}\right]^n \text{ is compact } (\lambda \in \Omega \cap \rho(T)).$$

We assume that there exists some $\lambda \in \Omega \cap \rho(T)$ such that $I - \left[B(\lambda - T)^{-1}\right]^n$ is invertible (i.e. 1 is not an eigenvalue of $\left[B(\lambda - T)^{-1}\right]^n$). Then $\Omega \subset \rho_e(T + B)$ and

$$\left[B(\lambda - T - B)^{-1}\right]^n \text{ is compact } (\lambda \in \Omega \cap \rho(T + B)).$$

Proof We note that $(\lambda - T)^{-1}$ is essentially meromorphic on Ω. Then $B_\lambda := B(\lambda - T)^{-1}$ and $B_\lambda^n = \left[B(\lambda - T)^{-1}\right]^n$ are also essentially meromorphic on Ω. Since B_λ^n is operator compact valued then, by the analytic Fredholm alternative $(I - B_\lambda^n)^{-1}$ is also essentially meromorphic on Ω. On the other hand

$$I - B_\lambda^n = (I - B_\lambda)(I + B_\lambda + \ldots + B_\lambda^{n-1})$$

shows that

$$(I - B_\lambda)^{-1} = (I - B_\lambda^n)^{-1}(I + B_\lambda + \ldots + B_\lambda^{n-1})$$

is also essentially meromorphic on Ω and then so is

$$(\lambda - T - B)^{-1} = (\lambda - T)^{-1}(I - B_\lambda)^{-1}$$

i.e. $\Omega \subset \rho_e(T + B)$. Finally

$$\left[B(\lambda - T - B)^{-1}\right]^n = \left[B_\lambda(I - B_\lambda)^{-1}\right]^n = B_\lambda^n(I - B_\lambda)^{-n}$$

is also compact on $\Omega \cap \rho(T + B)$. ■

We give now a more precise version of the theorem above under an additional assumption.

Corollary 6 *Let $T : D(T) \subset X \to X$ be a closed operator and let Ω be a connected component of $\rho_e(T)$. Let $B : D(T) \to X$ be T-bounded such that there exists $n \in \mathbb{N}$ and*

$$\left[B(\lambda - T)^{-1}\right]^n \text{ is compact } (\lambda \in \Omega \cap \rho(T)).$$

We assume that there exists a sequence $(\lambda_j)_j \subset \Omega \cap \rho(T)$ *such that*

$$\left\| B(\lambda_j - T)^{-1} \right\| \to 0 \; (j \to +\infty).$$

Then $(\lambda_j)_j \subset \rho(T+B)$ *for* j *large enough and* $\left\| B(\lambda_j - T - B)^{-1} \right\| \to 0$ *as* $j \to +\infty$. *Furthermore* Ω *is a component of* $\rho_e(T+B)$.

Proof We note that for j large enough

$$\left\| B(\lambda_j - T - B)^{-1} \right\| = \left\| B_{\lambda_j}(I - B_{\lambda_j})^{-1} \right\| \leq \sum_{k=1}^{\infty} \left\| B_{\lambda_j} \right\|^k \to 0 \;\; (j \to +\infty)$$

Let Ω' be the component of $\rho_e(T+B)$ which contains Ω. We know that

$$\left[B(\lambda - T - B)^{-1} \right]^n$$

is compact $(\lambda \in \Omega \cap \rho(T+B))$. By analyticity, this extends to $\Omega' \cap \rho(T+B)$. By considering T as $(T+B)-B$ and reversing the arguments in the previous theorem one gets $\Omega' \subset \rho_e(T)$ and consequently $\Omega = \Omega'$. ■

Corollary 7 *Let* $T, B \in \mathcal{L}(X)$. *We assume that* $\left[B(\lambda - T)^{-1} \right]^n$ *is compact on the unbounded component of* $\rho(T)$, *i.e.*

$$\left[B(\lambda - T)^{-1} \right]^n \; \textit{is compact} \; (\lambda \in \Omega \cap \rho(T))$$

where Ω *is the unbounded connected component of* $\rho_e(T)$. *The unbounded components of* $\rho_e(T)$ *and* $\rho_e(T+B)$ *coincide and then*

$$r_e(T) = r_e(T+B).$$

Proof Since $\left\| (\lambda - T)^{-1} \right\| \to 0$ as $|\lambda| \to \infty$ we apply the corollary above. ■

Corollary 8 *Let* $T, B \in \mathcal{L}(X)$.

(i) If B *is compact then* $r_e(T) = r_e(T+B)$.
(ii) If $X = L^1(\mu)$ *and* B *is weakly compact then* $r_e(T) = r_e(T+B)$.

Proof The case (i) is clear with $n = 1$. In the case (ii) $B(\lambda - T)^{-1}$ is also weakly compact on $L^1(\mu)$ and consequently its square is compact. ■

4.1 Strong Integral of Operator Valued Mappings

Let (Ω, μ) be a *finite* measure space and let X, Y be two Banach spaces. Let

$$G : \omega \in \Omega \to G(\omega) \in \mathcal{L}(X, Y)$$

be bounded and *strongly* measurable in the sense that for each $x \in X$

$$\omega \in \Omega \to G(\omega)x \in Y$$

is (Bochner) measurable. We define the *strong integral* of G on Ω as the bounded operator

$$\int_\Omega G : x \in X \to \int_\Omega G(\omega)x\mu(d\omega) \in Y.$$

We note that strongly continuous mappings appear everywhere in semigroup theory!

Theorem 21 ([33, 80, 82]) *Under the conditions above, assume in addition that* $\forall\omega \in \Omega, G(\omega) \in \mathcal{C}(X, Y)$ *(i.e.* $G(\omega)$ *is a compact operator). Then*

$$\int_\Omega G \in \mathcal{C}(X, Y).$$

In the statement above, we can replace "compact" by "weakly compact" [69]. Direct proofs in Lebesgue spaces relying on Kolmogorov's compactness criterion and the Dunford-Pettis criterion of weak compactness are given in [50].

4.2 Spectra of Perturbed Generators

Theorem 22 *Let* X *be a complex Banach space. Let* $(U(t))_{t\geqslant 0}$ *be a* C_0*-semigroup with generator* T *and let* $K \in \mathcal{L}(X)$*. We denote by* $(V(t))_{t\geqslant 0}$ *the* C_0*-semigroup generated by* $T + K$*. We assume that* K *is* T*-power compact i.e. there exists* $n \in \mathbb{N}$ *such that (for* λ *in some right half-plane included in* $\rho(T + K)$*)*

$$\left[K(\lambda - T)^{-1}\right]^n \text{ is compact.}$$

Then the components of $\rho_e(T)$ *and* $\rho_e(T + K)$ *containing a right half-plane coincide. In particular*

$$\sigma(T + K) \cap \{Re\lambda > s(T)\}$$

consists at most of isolated eigenvalues with finite algebraic multiplicities.

The proof follows from Corollary 6 since $\left\|(\lambda - T)^{-1}\right\| \to 0$ as $Re\lambda$ goes to $+\infty$. This theorem is a refinement of a first version due to I. Vidav [75].

4.3 Dyson–Phillips Expansions

The perturbed semigroup $(V(t))_{t\geqslant 0}$ is related to the unperturbed one $(U(t))_{t\geqslant 0}$ by an integral equation (Duhamel equation)

$$V(t) = U(t) + \int_0^t U(t)KV(t-s)ds.$$

The integrals are interpreted in a strong sense i.e.

$$V(t)x = U(t)x + \int_0^t U(t)KV(t-s)xds \ \ (x \in X).$$

The Duhamel equation is solved by standard iterations

$$V_{j+1}(t)x = U(t)x + \int_0^t U(t)KV_j(t-s)xds \ (j \geqslant 0) \ \ V_0 = 0$$

and

$$\forall C > 0, \ \sup_{t\in[0,C]} \left\| V_j(t) - V(t) \right\|_{\mathcal{L}(X)} \to 0 \ \text{ as } j \to +\infty.$$

Finally, $(V(t))_{t\geqslant 0}$ is given by a Dyson–Phillips series

$$V(t) = \sum_0^{+\infty} U_j(t)$$

where

$$U_{j+1}(t) = \int_0^t U_0(t)KU_j(t-s)ds \ (j \geqslant 0) \ \ U_0(t) = U(t).$$

By remainder terms of Dyson–Phillips expansions we mean

$$R_m(t) := \sum_{j=m}^{+\infty} U_j(t) \ \ (m \geqslant 0).$$

For any strongly continuous mappings $f, g : \mathbb{R}_+ \to \mathcal{L}(X)$, we define their convolution by (the strong integral)

$$f * g = h(t) := \int_0^t f(s)g(t-s)ds$$

and note $[f]^n$ the n-fold convolution of f with the convention $[f]^1 = f$. Then we can express $U_j(.)$ for $j \geqslant 1$ as

$$U_j(.) = [UK]^j * U = U * [KU]^j \quad (j \geqslant 1).$$

Theorem 23 ([48] Chapter 2) *Let $n \in \mathbb{N}_*$ be given. Then $U_n(t)$ is compact for all $t \geqslant 0$ if and only if $R_n(t) := \sum_{j=n}^{+\infty} U_j(t)$ is compact for all $t \geqslant 0$. If $X = L^1(\mu)$ then we can replace "compact" by "weakly compact".*

Proof If $U_n(t)$ is compact for all $t \geqslant 0$ then

$$U_{n+1}(t) = \int_0^t U(s)KU_n(t-s)ds$$

is compact for all $t \geqslant 0$ as a strong integral of "compact operator" valued mappings. By induction $U_j(t)$ is compact (for all $t \geqslant 0$) for all $j \geqslant n$ and then $R_n(t)$ is compact for all $t \geqslant 0$ since the series converges in operator norm. Conversely, let $R_n(t)$ be compact for all $t \geqslant 0$. Then

$$\begin{aligned} R_{n+1}(t) = \sum_{j=n+1}^{+\infty} U_j(t) &= \sum_{j=n+1}^{+\infty} \left([UK]^j * U\right) \\ &= [UK] * \sum_{j=n+1}^{+\infty} \left([UK]^{j-1} * U\right) \\ &= [UK] * \sum_{j=n}^{+\infty} \left([UK]^j * U\right) \\ &= [UK] * R_n(t) = \int_0^t U(s)KR_n(t-s)ds \end{aligned}$$

shows that $R_{n+1}(t)$ is also compact for all $t \geqslant 0$ as a strong integral of "compact operator" valued mappings. Finally $U_n(t) = R_n(t) - R_{n+1}(t)$ is also compact for all $t \geqslant 0$. ■

The following result is given in [81] for unbounded perturbations. We give here a slightly different (and simpler) presentation of the proof thanks to the boundedness of K.

Theorem 24 *Let X be a complex Banach space. Let $(U(t))_{t\geqslant 0}$ be a C_0-semigroup with generator T and let $K \in \mathcal{L}(X)$. We denote by $(V(t))_{t\geqslant 0}$ the C_0-semigroup generated by $T + K$. If some remainder term*

$$R_n(t) := \sum_{j=n}^{+\infty} U_j(t)$$

is compact for t large enough then $\omega_e(V) \leq \omega_e(U)$ *where* $\omega_e(V)$ *(resp.* $\omega_e(U)$*) is the essential type of* $(V(t))_{t \geqslant 0}$ *(resp.* $(U(t))_{t \geqslant 0}$*). If* $X = L^1(\mu)$ *then we can replace "compact" by "weakly compact".*

Proof Let $\beta > \omega_e(U)$ then there exists a *finite range projection* P_β commuting with $(U(t))_{t \geqslant 0}$ such that for any $\beta' > \beta$

$$\left\| U(t)(I - P_\beta) \right\| \leq M_{\beta'} e^{\beta' t} \quad (t \geqslant 0).$$

(with $P_\beta = 0$ if $\beta > \omega(U)$ the type of $(U(t))_{t \geqslant 0}$). On the other hand, by stability of essential radius by compact (or weakly compact if $X = L^1(\mu)$) perturbation

$$r_e(V(t)) = r_e(\sum_{j=0}^{n-1} U_j(t)) = r_e(\sum_{j=0}^{n-1} [UK]^j * U)$$

for t large enough. We note that

$$\begin{aligned} [UK]^j &= \left[U(I - P_\beta + P_\beta) K (I - P_\beta + P_\beta) \right]^j \\ &= \left[U(I - P_\beta) K (I - P_\beta) \right]^j + C_j(t) \end{aligned}$$

where $C_j(t)$ is a sum of convolutions where each convolution involves at least one term of the form $U(I - P_\beta) K P_\beta$, $U P_\beta K P_\beta$ or $U P_\beta K (I - P_\beta)$. Such terms are compact (of finite rank) because of P_β so that the convolutions are compact for all time as strong integrals of "compact operator" valued mappings. Thus $C_j(t)$ is compact for all $t \geqslant 0$. Once again, the stability of essential radius by compact perturbation gives for t large enough

$$\begin{aligned} r_e(V(t)) &= r_e((I - P_\beta) U(t) (I - P_\beta) + \sum_{j=1}^{n-1} \left[U(I - P_\beta) K (I - P_\beta) \right]^j * U) \\ &\leq \left\| (I - P_\beta) U(t) (I - P_\beta) + \sum_{j=1}^{n-1} \left[U(I - P_\beta) K (I - P_\beta) \right]^j * U \right\| \\ &\leq \left\| (I - P_\beta) U(t) (I - P_\beta) \right\| + \sum_{j=1}^{n-1} \left\| \left[U(I - P_\beta) K (I - P_\beta) \right]^j * U \right\|. \end{aligned}$$

Observe that

$$\left[U(I - P_\beta) K (I - P_\beta) \right] * U = \left[\tilde{U} K \right] * \tilde{U}$$

where $(\tilde{U}(t))_{t \geqslant 0}$ is the semigroup $(U(t)(I - P_\beta))_{t \geqslant 0}$. More generally

$$[U(I - P_\beta)K(I - P_\beta)]^j * U = [\tilde{U}K]^j * \tilde{U}.$$

By using the estimate

$$\|\tilde{U}(t)\| \leq M_{\beta'} e^{\beta' t} \ (t \geqslant 0),$$

an elementary calculation shows that $\left\| [\tilde{U}K]^j * \tilde{U} \right\| \leq c_j t^j e^{\beta' t}$ so

$$r_e(V(t)) \leq p_n(t) e^{\beta' t}$$

where $p_n(t)$ is a polynomial of degree n. To end the proof, let $\beta'' > \beta'$. Then there exists a constant $M_{\beta''}$ such that

$$r_e(V(t)) \leq M_{\beta''} e^{\beta'' t}$$

for t large enough. Let $\omega_e(V)$ be the essential type of $(V(t))_{t \geqslant 0}$. The fact that

$$r_e(V(t)) = e^{\omega_e(V)t}$$

implies that $\omega_e(V) \leq \beta''$. Hence $\omega_e(V) \leq \omega_e(U)$ since $\beta' > \omega_e(U)$ and $\beta'' > \beta'$ are chosen arbitrarily. ■

Remark 1 A classical weaker estimate $\omega_e(V) \leq \omega(U)$ (where $\omega(U)$ is the type of $(U(t))_{t \geqslant 0}$) is due to I. Vidav [76]. The estimate $\omega_e(V) \leq \omega_e(U)$ is also derived in [70, 82] by using the properties of measure of noncompactness of strong integrals.

We have seen that if some remainder term $R_n(t) := \sum_{j=n}^{+\infty} U_j(t)$ is compact (or weakly compact when $X = L^1(\mu)$) for large t then $\omega_e(V) \leq \omega_e(U)$. We show now that if some remainder term is compact (or weakly compact when $X = L^1(\mu)$) for *all* $t \geqslant 0$ then $\omega_e(V) = \omega_e(U)$. We need a preliminary result:

Lemma 7 ([48] Chapter 2) *Let X be a complex Banach space. Let $(U(t))_{t \geqslant 0}$ be a C_0-semigroup with generator T and let $K \in \mathcal{L}(X)$. We denote by $(V(t))_{t \geqslant 0}$ the C_0-semigroup generated by $T + K$. Let $V(t) = \sum_0^{+\infty} U_j(t)$ be the Dyson–Phillips expansion of $(V(t))_{t \geqslant 0}$. Let $U(t) = \sum_0^{+\infty} V_j(t)$ be the Dyson–Phillips expansion of $(U(t))_{t \geqslant 0}$ considered as a perturbation of $(V(t))_{t \geqslant 0}$ (i.e. $T = (T + K) + (-K)$). For any $j \in \mathbb{N}_*$, $U_j(t)$ is compact for all $t \geqslant 0$ if and only if $V_j(t)$ is compact for all $t \geqslant 0$. If $X = L^1(\mu)$ then we can replace "compact" by "weakly compact".*

By reversing the rôle of $(U(t))_{t \geqslant 0}$ and $(V(t))_{t \geqslant 0}$ we obtain:

Corollary 9 *Let X be a complex Banach space. Let $(U(t))_{t \geqslant 0}$ be a C_0-semigroup with generator T and let $K \in \mathcal{L}(X)$. We denote by $(V(t))_{t \geqslant 0}$ the C_0-semigroup* *generated by $T + K$. Let*

$$U_{j+1}(t) = \int_0^t U_0(t) K U_j(t-s) ds \ (j \geqslant 0) \quad U_0(t) = U(t)$$

be the terms of the Dyson–Phillips expansion of $(V(t))_{t\geqslant 0}$. If for some $n \in \mathbb{N}_*$, $U_n(t)$ is compact (resp. weakly compact if $X = L^1(\mu)$) for all $t \geqslant 0$ then

$$\omega_e(V) = \omega_e(U).$$

Remark 2 We note that the stability of essential type appears also in [70, 81, 82] but under stronger assumptions.

4.4 Short Digression on Resolvent Approach

The following "resolvent characterization" is due to S. Brendle [9].

Theorem 25 *Let $n \in \mathbb{N}_*$. Then $U_n(t)$ is compact for all $t \geqslant 0$ if and only if:*

(i) $t \geqslant 0 \to U_n(t) \in \mathcal{L}(X)$ is continuous in operator norm
and
(ii) $\left[(\alpha + i\beta - T)^{-1} K\right]^n (\alpha + i\beta - T)^{-1}$ is compact for some $\alpha > \omega(U)$ and all $\beta \in \mathbb{R}$.

This is useful in some applications (for example for kinetic equations involving boundary operators relating the incoming and outgoing fluxes) where the unperturbed semigroup $(U(t))_{t\geqslant 0}$ is not explicit while the resolvent $(\lambda - T)^{-1}$ is "tractable"! The cost of the approach is that we need a priori that

$$t \geqslant 0 \to U_n(t) \in \mathcal{L}(X)$$

is continuous in operator norm. The following result gives sufficient conditions of continuity in operator norm.

Theorem 26 ([9]) *Let $n \in \mathbb{N}$. If X is a Hilbert space and if*

$$\left\| \left[(\alpha + i\beta - T)^{-1} K\right]^n (\alpha + i\beta - T)^{-1} \right\| \to 0 \ \text{as} \ |\beta| \to +\infty$$

then $t \geqslant 0 \to U_{n+2}(t) \in \mathcal{L}(X)$ is continuous in operator norm.

Note that the continuity of $t \geqslant 0 \to U_1(t) \in \mathcal{L}(X)$ (if we want to show the compactness of $V(t) - U(t)$) is out of reach of this theorem. We give now Sbihi's criterion of continuity in operator norm.

Theorem 27 ([67]) *Let X be a Hilbert space and let T be dissipative (i.e. $Re\,(Tx, x) \leq 0\ \forall x \in D(T)$). If*

$$\left\| K^*(\lambda - T)^{-1} K \right\| + \left\| K(\lambda - T)^{-1} K^* \right\| \to 0 \text{ as } |Im\lambda| \to +\infty$$

then $t \geqslant 0 \to U_1(t) \in \mathcal{L}(X)$ is continuous in operator norm.

Useful applications of this result are given in [34, 38, 67].

5 Collisional Transport Theory

In this section, we show how the previous functional analytic tools apply to neutron transport theory. We start with an unperturbed (advection) semigroup in $L^p(\Omega \times V;\ dx \otimes \mu(dv))$

$$U(t) : g \to e^{-\int_0^t \sigma(x - \tau v, v) d\tau} g(x - tv, v) 1_{\{t \leq s(x,v)\}}\ (t \geqslant 0)$$

(with generator T) where

$$s(x, v) = \inf\{s > 0;\ x - sv \notin \Omega\}$$

is the first exit time function from the spatial domain Ω. We regard the scattering operator

$$K : f \to \int_V k(x, v, v') f(x, v') \mu(dv')$$

as a bounded perturbation of T (we refer to [58] and references therein for generation results with unbounded scattering operators) and denote by $(V(t))_{t \geqslant 0}$ the perturbed neutron transport semigroup. We are faced with *two* main questions:

- When is $\left[(\lambda - T)^{-1} K\right]^n$ compact in $L^p(\Omega \times V)$ for some $n \in \mathbb{N}_*$?
- When is some remainder term $R_m(t)$ compact in $L^p(\Omega \times V)$?

We point out that the resolvent $(\lambda - T)^{-1}$ of T cannot be compact (e.g. in bounded geometries, $\sigma(T)$ is a half-plane when $0 \in V$!). The scattering operator

$$K : f \to \int_V k(x, v, v') f(x, v') \mu(dv')$$

is *local* with respect to the space-variable x so that K cannot be compact on $L^p(\Omega \times V)$. The good news is that compactness will emerge from subtle combinations of properties of T and those of K. For information, we recall some classical results: Under quite general assumptions on the scattering kernel $k(x, v, v')$, for bounded domains Ω and Lebesgue measure dv on $\mathbb{R}^n$ as velocity measure, the *second order* remainder term $R_2(t)$ is compact in $L^p(\Omega \times V)$ $(1 < p < \infty)$ or weakly compact in $L^1(\Omega \times V)$; see e.g. [25, 43, 45, 72, 79, 82].

We introduce now a useful class of scattering operators. Let $\Omega \subset \mathbb{R}^n$ $(n \geqslant 1)$ be an open subset and let μ be a positive Radon measure with support V. Let

$$X := L^p(\Omega \times V; dx \otimes \mu(dv))$$

with $1 \leq p < +\infty$. Let

$$k : (x, v, v') : \Omega \times V \times V \to k(x, v, v') \in \mathbb{R}_+$$

be measurable and such that

$$K : f \in L^p(\Omega \times V) \to \int_V k(x, v, v') f(x, v') dv' \in L^p(\Omega \times V)$$

is a bounded operator on $L^p(\Omega \times V)$. Since K is *local* in space variable, we may interpret it as a family of bounded operators on $L^p(V)$ indexed by the parameter $x \in \Omega$ i.e. a mapping

$$K : x \in \Omega \to K(x) \in \mathcal{L}(L^p(V)).$$

Then

$$\|K\|_{\mathcal{L}(L^p(\Omega \times V))} = \sup_{x \in \Omega} \|K(x)\|_{\mathcal{L}(L^p(V))} .$$

5.1 L^p Theory $(1 < p < \infty)$

In this section, we restrict ourselves to $1 < p < \infty$. A scattering operator K is called regular if

(i) $\{K(x);\ x \in \Omega\}$ is a set of *collectively compact* operators on $L^p(V)$, i.e. the set

$$\left\{K(x)\varphi;\ x \in \Omega,\ \|\varphi\|_{L^p(V)} \leq 1\right\}$$

is relatively compact in $L^p(V)$.

(ii) For any $\psi' \in L^{p'}(V)$, the set $\{K'(x)\psi';\ x \in \Omega\}$ is relatively compact in $L^{p'}(V)$ where p' is the conjugate exponent and $K'(x)$ is the dual operator of $K(x)$.

We note that the compactness of K with respect to velocities (at least in $L^2(V)$) is satisfied by most physical models [8].

Theorem 28 ([51]) *The class of regular scattering operators is the closure in the operator norm of $\mathcal{L}(L^p(\Omega \times V))$ of the class of scattering operators with kernels*

$$k(x, v, v') = \sum_{j \in J} \alpha_j(x) f_j(v) g_j(v')$$

where $\alpha_j \in L^\infty(\Omega)$, f_j, g_j continuous with compact supports and J finite.

The first part of the following lemma is given in ([48, Chapter 3, p. 32]) and the second part in [51].

Lemma 8 *Let μ be a finite Radon measure on $\mathbb{R}^n$.*

(i) If the hyperplanes (through the origin) have zero μ-measure then

$$\sup_{e \in S^{n-1}} \mu \{v;\ |v.e| \leq \varepsilon\} \to 0 \ \text{as}\ \varepsilon \to 0.$$

(ii) If the affine (i.e. translated) hyperplanes have zero μ-measure then

$$\sup_{e \in S^{n-1}} \mu \otimes \mu \left\{(v, v');\ \left|(v - v').e\right| \leq \varepsilon\right\} \to 0 \ \text{as}\ \varepsilon \to 0.$$

Theorem 29 ([51]) *We assume that Ω has finite Lebesgue measure and the scattering operator is regular in $L^p(\Omega \times V; dx \otimes \mu(dv))$ $(1 < p < \infty)$. If the hyperplanes have zero μ-measure then $(\lambda - T)^{-1}K$ and $K(\lambda - T)^{-1}$ are compact on $L^p(\Omega \times V; dx \otimes \mu(dv))$.*

Remark 3 The compactness of $K(\lambda - T)^{-1}$ can be expressed as an averaging lemma in open sets Ω with finite volume, i.e. if F is a bounded subset of $D(T)$ (for the graph norm) then

$$\{K\varphi;\ \varphi \in F\} \ \text{is relatively compact in}\ L^p(\Omega \times V).$$

We note that if

$$\sup_{e \in S^{n-1}} \mu \{v; |v.e| \leq \varepsilon\} \leq c\varepsilon^\alpha$$

and if Ω is bounded and convex then the compactness can be measured in terms of fractional Sobolev regularity [23], (see also [1]).

Corollary 10 *We assume that Ω has finite Lebesgue measure, the scattering operator is regular in $L^p(\Omega \times V; dx \otimes \mu(dv))$ $(1 < p < \infty)$ and the hyperplanes have zero μ-measure. Then*

$$\sigma_{ess}(T + K) = \sigma_{ess}(T).$$

In particular $\sigma(T+K)\cap\{Re\lambda > s(T)\}$ *consists at most of isolated eigenvalues with finite algebraic multiplicities where*

$$s(T) = -\lim_{t\to+\infty}\inf_{(y,v)}\frac{1}{t}\int_0^t \sigma(y+\tau v, v)d\tau.$$

Theorem 30 ([51]) *We assume that* Ω *has finite Lebesgue measure and the scattering operator is regular in* $L^p(\Omega\times V; dx\otimes\mu(dv))$ $(1<p<\infty)$*. If the hyperplanes have zero* μ*-measure then*

$$V(t)-U(t) \text{ is compact on } L^p(\Omega\times V) \quad (t\geqslant 0),$$

i.e. the first remainder $R_1(t)=\sum_{j=1}^{+\infty}U_j(t)$ *is compact on* $L^p(\Omega\times V)$*.*

Strategy of the proof:

- $R_1(t)=\sum_{j=m}^{+\infty}U_j(t)$ is compact for all $t\geqslant 0$ if and only if $U_1(t)=\int_0^t U(t-s)KU(s)ds$ is. So we deal with $U_1(t)$.
- $U_1(t)$ depends (linearly and) continuously on the scattering operator K. So, by approximation, we may assume that

$$k(x,v,v')=\sum_{j\in J}\alpha_j(x)f_j(v)g_j(v')$$

  where $\alpha_j\in L^\infty(\Omega)$, f_j, g_j continuous with compact supports and J is finite.
- By linearity, we may even choose

$$k(x,v,v')=\alpha(x)f(v)g(v')$$

  where $\alpha\in L^\infty(\Omega)$, f,g are continuous with compact supports. In this case, $U_1(t)$ operates on all $L^q(\Omega\times V)$ $(1\leq q\leq+\infty)$. So, by an interpolation argument, we may restrict ourselves to the case $p=2$.
- *Domination* arguments: In L^p spaces $(1<p<\infty)$, if O_i $(i=1,2)$ are two positive operators such that

$$O_1f\leq O_2f \quad \forall f\in L_+^p$$

  and if O_2 is compact then O_1 is also compact; see [3]. So we may assume that V is compact, $\alpha=f=g=1$ and $\sigma=0$.
- Because of $\sigma=0$

$$U(t)\varphi=\varphi(x-tv,v)1_{\{t\leq s(x,v)\}}$$

  where $s(x,v)=\inf\{s>0;\ x-sv\notin\Omega\}$. If $\tilde\varphi\in L^2(\mathbb{R}^n\times V)$ is the trivial extension of φ by zero outside $\Omega\times V$ then, for nonnegative φ

$$U(t)\varphi(x,v)\leq\tilde\varphi(x-tv,v) \ \ \forall(x,v)\in\Omega\times V.$$

so $U(t)\varphi \leq \mathcal{R}U_{\infty}(t)\mathcal{E}\varphi$ where $\mathcal{E} : L^2(\Omega \times V) \to L^2(\mathbb{R}^n \times V)$ is the trivial extension operator, $\mathcal{R} : L^2(\mathbb{R}^n \times V) \to L^2(\Omega \times V)$ is the restriction operator and

$$U_{\infty}(t) : \psi \in L^2(\mathbb{R}^n \times V) \to \psi(x - tv, v) \in L^2(\mathbb{R}^n \times V).$$

- Thus

$$\begin{aligned} U_1(t) &= \int_0^t \mathcal{R}U_{\infty}(t-s)\mathcal{E}K\mathcal{R}U_{\infty}(s)\mathcal{E}ds \\ &= \mathcal{R}\left(\int_0^t U_{\infty}(t-s)\mathcal{E}K\mathcal{R}U_{\infty}(s)ds\right)\mathcal{E} \\ &\leq \mathcal{R}\left(\int_0^t U_{\infty}(t-s)KU_{\infty}(s)ds\right)\mathcal{E}. \end{aligned}$$

and we are led to deal with the compactness of

$$\mathcal{R}\int_0^t U_{\infty}(t-s)KU_{\infty}(s)ds : L^2(\mathbb{R}^n \times V) \to L^2(\Omega \times V).$$

- Note that

$$\int_0^t U_{\infty}(t-s)KU_{\infty}(s)\psi ds = \int_0^t ds \int_V \psi(x-(t-s)v - sv', v')\mu(v').$$

For any $\psi(.,.) \in L^2(\mathbb{R}^n \times V)$, we denote by

$$\hat{\psi}(\zeta, v) = \lim_{M\to\infty} \frac{1}{(2\pi)^{\frac{n}{2}}} \int_{|\zeta|\leq M} \psi(x,v)e^{-i\zeta.x}d\zeta$$

its partial Fourier transform with respect to *space* variable where the limit holds in $L^2(\mathbb{R}^n \times V)$ norm. Then

$$\|\psi\|^2_{L^2(\mathbb{R}^n\times V)} = \int_V \int_{\mathbb{R}^n} \left|\hat{\psi}(\zeta, v)\right|^2 d\zeta\mu(dv)$$

and

$$\psi(x,v) = \lim_{M\to\infty} \frac{1}{(2\pi)^{\frac{n}{2}}} \int_{|\zeta|\leq M} \hat{\psi}(\zeta, v)e^{i\zeta.x}d\zeta$$

where the limit holds in $L^2(\mathbb{R}^n \times V)$ norm.

Hence

$$\int_0^t U_\infty(t-s)KU_\infty(s)\psi ds$$
$$= \int_0^t ds \int_V \psi(x-(t-s)v - sv', v')\mu(dv')$$
$$= \lim_{M\to\infty} \frac{1}{(2\pi)^{\frac{n}{2}}} \int_{|\zeta|\le M} e^{ix.\zeta} \int_0^t ds \int_V \hat{\psi}(\zeta, v) e^{-i((t-s)v+sv').\zeta} \mu(dv')$$
$$= \lim_{M\to\infty} \frac{1}{(2\pi)^{\frac{n}{2}}} \int_{|\zeta|\le M} \int_V e^{ix.\zeta}\hat{\psi}(\zeta, v) \left(\int_0^t dse^{-i((t-s)v+sv').\zeta} ds\right) \mu(dv')$$

where the limit holds in $L^2(\mathbb{R}^n \times V)$ norm. For each $M > 0$, let

$$O_M : L^2(\mathbb{R}^n \times V) \to L^2(\mathbb{R}^n \times V)$$

$$\psi \to \int_{|\zeta|\le M} \int_V e^{ix.\zeta}\hat{\psi}(\zeta, v) \left(\int_0^t dse^{-i((t-s)v+sv').\zeta} ds\right) \mu(dv').$$

We observe that $\mathcal{R}O_M$ is a Hilbert Shmidt operator because Ω has finite volume and V is compact (keep in mind that $x \in \Omega$!). It suffices to show that

$$O_M\psi \to \int_0^t ds \int_V \psi(x-(t-s)v - sv', v')\mu(dv')$$

in $L^2(\mathbb{R}^n \times V)$ *uniformly* in $\|\psi\|_{L^2(\mathbb{R}^n\times V)} \le 1$, i.e.

$$\int_{|\zeta|>M} e^{ix.\zeta} \int_V \hat{\psi}(\zeta, v') \left(\int_0^t e^{-i((t-s)v+sv').\zeta} ds\right) \mu(dv') \to 0$$

in $L^2(\mathbb{R}^n \times V)$ uniformly in $\|\psi\|_{L^2(\mathbb{R}^n\times V)} \le 1$. By the Parseval identity, this amounts to

$$\int_V \mu(dv) \int_{|\zeta|>M} d\zeta \left| \int_V \hat{\psi}(\zeta, v') \left(\int_0^t e^{-i((t-s)v+sv').\zeta} ds\right) \mu(dv')\right|^2 \to 0$$

uniformly in $\|\psi\|_{L^2(\mathbb{R}^n\times V)} \le 1$. Using Cauchy-Schwarz inequality, we majorize by

$$\sup_{|\zeta|>M} \int_{V\times V} \mu(dv')\mu(dv) \left|\int_0^t e^{-i((t-s)v+sv').\zeta} ds\right|^2 \int_{\mathbb{R}^n\times V} \left|\hat{\psi}(\zeta, v')\right|^2$$
$$\le \sup_{|\zeta|>M} \int_{V\times V} \mu(dv')\mu(dv) \left|\int_0^t e^{-i((t-s)v+sv').\zeta} ds\right|^2 \quad \forall \|\psi\| \le 1.$$

Now

$$\int_{V\times V}\mu(dv')\mu(dv)\left|\int_0^t e^{-i((t-s)v+sv').\zeta}ds\right|^2$$
$$=\int_{V\times V}\mu(dv')\mu(dv)\left|\int_0^t e^{is(v-v').\zeta}ds\right|^2$$
$$=\int_{V\times V}\mu(dv')\mu(dv)\left|\int_0^t e^{is|\zeta|(v-v').e}ds\right|^2$$

where $\zeta = |\zeta|\, e,\ (e \in S^{n-1})$ and

$$\int_{V\times V}\mu(dv')\mu(dv)\left|\int_0^t e^{is|\zeta|(v-v').e}ds\right|^2$$
$$=\int_{\{|(v-v').e|\le\varepsilon\}}\mu(dv')\mu(dv)\left|\int_0^t e^{is|\zeta|(v-v').e}ds\right|^2$$
$$+\int_{\{|(v-v').e|>\varepsilon\}}\mu(dv')\mu(dv)\left|\int_0^t e^{is|\zeta|(v-v').e}ds\right|^2$$
$$\le t^2\int_{\{|(v-v').e|\le\varepsilon\}}\mu(dv')\mu(dv)$$
$$+\int_{\{|(v-v').e|>\varepsilon\}}\mu(dv')\mu(dv)\left|\int_0^t e^{is|\zeta|(v-v').e}ds\right|^2.$$

The first term can made arbitrarily small for ε small enough (assumption on the velocity measure μ) and, for ε fixed, the second term goes to zero as $|\zeta| \to \infty$ because of $\int_0^t e^{is|\zeta|(v-v').e}ds$ (Riemann-Lebesgue lemma). This ends the proof. ∎

Corollary 11 *We assume that Ω has finite Lebesgue measure, the scattering operator is regular in $L^p(\Omega\times V; dx\otimes\mu(dv))$ $(1<p<\infty)$ and the hyperplanes have zero μ-measure. Then*

$$\sigma_{ess}(V(t)) = \sigma_{ess}(U(t)).$$

In particular $\omega_e(V) = \omega_e(U)$ and $\sigma(V(t))\cap\{\beta;\ |\beta| > e^{s(T)t}\}$ consists at most of isolated eigenvalues with finite algebraic multiplicities.

Here $\omega_e(U)$ denotes the essential type of $(U(t))_{t\geqslant 0}$ etc. The assumption on the velocity measure μ is "optimal":

Theorem 31 ([51]) *Let μ be finite, Ω bounded and*

$$K:\varphi\in L^2(\Omega\times V)\to\int_V\varphi(x,v)\mu(dv).$$

Let there exist a hyperplane $H = \{v; v.e = c\}$ $(e \in S^{n-1},\ c \in \mathbb{R})$ *with positive* μ*-measure. Then there exists* $\overline{t} > 0$ *such that* $V(t) - U(t)$ *is not compact on* $L^2(\Omega \times V)$ *for* $0 < t \leq \overline{t}$.

and

Theorem 32 ([51]) *In the general setting above, if every ball centred at zero contains at least a section (by a hyperplane) with positive* μ*-measure then* $V(t) - U(t)$ *is not compact on* $L^2(\Omega \times V)$ *for all* $t > 0$.

The assumption that the scattering operator is regular is "nearly optimal":

Theorem 33 ([51]) *Let* μ *be an arbitrary positive measure. We assume that its support* V *is bounded. If* $V(t) - U(t)$ *is compact on* $L^p(\Omega \times V)$ *for all* $t > 0$ *then, for any open ball* $B \subset \Omega$*, the strong integral*

$$\int_B K(x)dx$$

is a compact operator on $L^p(V)$.

Corollary 12 *Besides the conditions of Theorem 33, we assume that*

$$x \in \Omega \to K(x) \in \mathcal{L}(L^p(V))$$

is measurable (not simply strongly measurable!) e.g. is piecewise continuous in operator norm. Then $K(x)$ *is a compact operator on* $L^p(V)$ *for almost all* $x \in \Omega$.

Proof

$$\frac{1}{|B|}\int_B K(x)dx \to K(x) \text{ in } \mathcal{L}(L^p(V)) \text{ as } |B| \to 0$$

at the Lebesgue points x of $K : \Omega \to \mathcal{L}(L^p(V))$. ■

5.2 L^1 Theory

As noted previously, L^1 space is the physical setting for neutron transport because

$$\int_\Omega \int_V f(x, v, t)dx\mu(dv)$$

is the expected number of particles. The L^1 mathematical results are very different from those in L^p theory $(p > 1)$ and the analysis is much more involved! Weak compactness is a fundamental tool for spectral theory of neutron transport in L^1 spaces. To this end, we recall first some useful results. Let (E, m) be a σ-finite measure space. A bounded subset $B \subset L^1(E, m)$ is relatively weakly compact if

$$\sup_{f \in B} \int_A |f| \, dm \to 0 \text{ as } m(A) \to 0$$

and (if $m(E) = \infty$) there exists measurable sets $E_n \subset E$, $m(E_n) < +\infty$, $E_n \subset E_{n+1}$, $\cup E_n = E$ such that

$$\sup_{f \in B} \int_{E_n^c} |f| \, dm \to 0 \text{ as } n \to \infty.$$

A bounded subset $B \subset L^1(E, m)$ is relatively weakly compact if and only if B is relatively sequentially weakly compact. A bounded operator G on $L^1(E, m)$ is said to be weakly compact if G sends a bounded set into a relatively weakly compact one. If $G_i : L^1(E, m) \to L^1(E, m)$ $(i = 1, 2)$ are positive operators and $G_1 f \leq G_2 f$ $\forall f \in L^1_+(E, m)$ then G_1 is weakly compact if G_2 is; this follows easily from the above criterion of weak compactness. If G_i $(i = 1, 2)$ are two weakly compact operators on $L^1(E, m)$ then $G_1 G_2$ is a compact operator [18]. We are going to present (weak) compactness results for neutron transport operators and show their spectral consequences. We give here an overview of [52]. We treat first a model case where

$$\sigma(x, v) = 0, \ \ \mu \text{ is finite and } k(x, v, v') = 1.$$

We start with *negative* results:

Theorem 34 ([52]) *Let $\Omega \subset \mathbb{R}^n$ be an open set with finite Lebesgue measure and μ an arbitrary finite positive Borel measure on $\mathbb{R}^n$ with support V. Let $n \geqslant 2$ (or $n = 1$ and $0 \in V$). Let*

$$K : \varphi \in L^1(\Omega \times V) \to \int_V \varphi(x, v) \mu(dv) \in L^1(\Omega).$$

Then $K(\lambda - T)^{-1}$ is not weakly compact.

Proof We can assume without loss of generality that $0 \in \Omega$. Consider just the case $n \geqslant 2$. Let $(f_j)_j \subset C_c(\Omega \times V)$ a normalized sequence in $L^1(\Omega \times V)$ converging in the weak star topology of measures to the Dirac mass $\delta_{(0,\overline{v})}$. Then for any $\psi \in C_c(\Omega)$

$$\begin{aligned} \langle K(\lambda - T)^{-1} f_j, \psi \rangle &= \int_\Omega \psi(x) dx \int_V \mu(dv) \int_0^{s(x,v)} e^{-\lambda t} f_j(x - tv, v) dt \\ &= \int_\Omega \int_V f_j(y, v) \left[\int_0^{s(y,-v)} e^{-\lambda t} \psi(y + tv) dt \right] dy \mu(dv) \\ &\to \int_0^{s(0,-\overline{v})} e^{-\lambda t} \psi(t\overline{v}) dt \end{aligned}$$

i.e. $K(\lambda - T)^{-1} f_j$ tends, in the weak star topology of measures, to a (non-trivial) Radon measure

$$m : \psi \in C_c(\Omega) \to \int_0^{s(0,-\overline{v})} e^{-\lambda t} \psi(t\overline{v}) dt$$

with support included in a segment. Hence $m \notin L^1(\Omega)$ and $K(\lambda - T)^{-1}$ is not weakly compact. ■

We note that this property was noted for the first time in the whole space in [23]. We have also:

Theorem 35 ([52]) *Let $n \geqslant 3$ and let $\Omega \subset \mathbb{R}^n$ be an open set with finite Lebesgue measure. Let μ be an arbitrary finite positive Borel measure on $\mathbb{R}^n$ with support V and*

$$K : \varphi \in L^1(\Omega \times V) \to \int_V \varphi(x, v)\mu(dv) \in L^1(\Omega).$$

Then

(i) $(\lambda - T - K)^{-1} - (\lambda - T)^{-1}$ is not weakly compact.
(ii) $V(t) - U(t)$ is not weakly compact.

We note that this theorem is false for $n = 1$ while the case $n = 2$ is open, see [52]. We recall now a necessary condition on μ.

Theorem 36 ([52]) *We assume that the velocity measure is invariant under the symmetry $v \to -v$. Let there exist $m \in \mathbb{N}$ such that*

$$\left[K(\lambda - T)^{-1}\right]^m$$

is compact on $L^1(\Omega \times V)$. Then the hyperplanes (through zero) have zero μ-measure.

We note that for any $m \in \mathbb{N}$

$$\left[K(\lambda - T)^{-1}K\right]^m \leq \mathcal{R}\left[K(\lambda - T_\infty)^{-1}K\right]^m \mathcal{E}$$

where

$$\mathcal{E} : L^1(\Omega \times V) \to L^1(\mathbb{R}^n \times V)$$

is the trivial extension operator,

$$\mathcal{R} : L^1(\mathbb{R}^n \times V) \to L^1(\Omega \times V)$$

is the restriction operator and T_∞ is the generator of

$$U_\infty(t) : \psi \in L^1(\mathbb{R}^n \times V) \to \psi(x - tv, v) \in L^1(\mathbb{R}^n \times V).$$

We start with a fundamental observation:

Lemma 9 *Let μ be an arbitrary finite positive measure on $\mathbb{R}^n$ with support V and*

$$K : \varphi \in L^1(\Omega \times V) \to \int_V \varphi(x, v)\mu(dv) \in L^1(\Omega).$$

For any $\lambda > 0$ there exists a finite positive measure β on $\mathbb{R}^n$ (depending on λ) such that

$$K(\lambda - T_\infty)^{-1}K\varphi = \beta * K\varphi.$$

Moreover,

$$\hat{\beta}(\zeta) = \frac{1}{(2\pi)^{\frac{n}{2}}} \int_{\mathbb{R}^n} \frac{\mu(dv)}{\lambda + i\zeta.v}.$$

Proof

$$\begin{aligned}
K(\lambda - T_\infty)^{-1}K\varphi &= \int_{\mathbb{R}^n} \mu(dv) \int_0^\infty e^{-\lambda t}(K\varphi)(x - tv)dt \\
&= \int_0^\infty e^{-\lambda t}dt \int_{\mathbb{R}^n} (K\varphi)(x - tv)\mu(dv) \\
&= \int_0^\infty e^{-\lambda t}dt \int_{\mathbb{R}^n} (K\varphi)(x - z)\mu_t(dz) \\
&= \int_0^\infty e^{-\lambda t}\left[\mu_t * K\varphi\right]dt
\end{aligned}$$

where μ_t is the image of μ under the dilation $v \to tv$. So

$$K(\lambda - T_\infty)^{-1}K\varphi = \beta * K\varphi$$

where $\beta := \int_0^\infty e^{-\lambda t}\mu_t dt$ (strong integral). Finally $\hat{\beta}(\zeta)$ is given by

$$\begin{aligned}
\int_0^\infty e^{-\lambda t}\hat{\mu_t}(\zeta)dt &= \frac{1}{(2\pi)^{\frac{n}{2}}} \int_0^\infty e^{-\lambda t}\left[\int_{\mathbb{R}^n} e^{-i\zeta.v}\mu_t(dv)\right]dt \\
&= \frac{1}{(2\pi)^{\frac{n}{2}}} \int_0^\infty e^{-\lambda t}\left[\int_{\mathbb{R}^n} e^{-it\zeta.v}\mu(dv)\right]dt \\
&= \frac{1}{(2\pi)^{\frac{n}{2}}} \int_{\mathbb{R}^n} \frac{\mu(dv)}{\lambda + i\zeta.v}.
\end{aligned}$$

This ends the proof. ■

Remark 4 $\hat{\beta}(\zeta) = \int_{\mathbb{R}^n} \frac{\mu(dv)}{\lambda + i\zeta.v} \to 0$ for $|\zeta| \to \infty$ if and only if the hyperplanes (through zero) have zero μ-measure; see [52]. We are going to show that the compactness results rely on how fast $\hat{\beta}(\zeta)$ goes to zero as $|\zeta| \to \infty$.

Theorem 37 ([52]) *We assume that $\Omega \subset \mathbb{R}^n$ is an open set with finite Lebesgue measure. Let the velocity measure μ be finite and such that*

$$\int_{\mathbb{R}^n} \left|\hat{\beta}(\zeta)\right|^{2m} d\zeta < +\infty$$

for some $m \in \mathbb{N}$. Then $\left[K(\lambda - T)^{-1}K\right]^m$ is weakly compact in $L^1(\Omega \times V)$ and $\left[K(\lambda - T)^{-1}K\right]^{m+1}$ is compact in $L^1(\Omega \times V)$.

Proof We start from

$$\left[K(\lambda - T_\infty)^{-1}K\right]^m \varphi = \beta^{(m)} * K\varphi$$

where

$$\beta^{(m)} = \beta * \ldots. * \beta \ \ (m \text{ times}).$$

Since

$$\hat{\beta^{(m)}}(\zeta) = \left(\hat{\beta}(\zeta)\right)^m$$

then our assumption amounts to $\hat{\beta^{(m)}} \in L^2(\mathbb{R}^n)$. In particular $\beta^{(m)} \in L^2(\mathbb{R}^n)$ ($\beta^{(m)}$ is now a function!). It follows that

$$\left[K(\lambda - T_\infty)^{-1}K\right]^m \varphi = \beta^{(m)} * K\varphi \in L^2(\mathbb{R}^n)$$

and then $\left[K(\lambda - T_\infty)^{-1}K\right]^m$ maps continuously$L^1(\mathbb{R}^n \times V)$ into $L^2(\mathbb{R}^n)$. Hence

$$\mathcal{R}\left[K(\lambda - T_\infty)^{-1}K\right]^m : L^1(\mathbb{R}^n \times V) \to L^1(\Omega)$$

is weakly compact because the imbedding of $L^2(\Omega)$ into $L^1(\Omega)$ is weakly compact since Ω has finite Lebesgue measure (a bounded subset of $L^2(\Omega)$ is equi-integrable). Finally $\left[K(\lambda - T)^{-1}K\right]^m$ is also weakly compact by a domination argument. It follows that $\left[K(\lambda - T)^{-1}K\right]^{2m}$ is compact as a product of two weakly compact operators. Actually

$$\left[K(\lambda - T)^{-1}K\right]^{m+1} = K(\lambda - T)^{-1}K\left[K(\lambda - T)^{-1}K\right]^m$$

is compact because $K(\lambda - T)^{-1}$ is a *Dunford-Pettis* operator, see below. ■

We give now a geometrical condition on μ implying the compactness results.

Theorem 38 ([52]) *Let $\Omega \subset \mathbb{R}^n$ be an open set with finite Lebesgue measure. Let the velocity measure μ be finite and there exist $\alpha > 0$, $c > 0$ such that*

$$\sup_{e \in S^{n-1}} \mu\{v; |v.e| \leq \varepsilon\} \leq c\varepsilon^{\alpha}.$$

Then $\left[K(\lambda - T)^{-1}K\right]^{m+1}$ is compact in $L^1(\Omega \times V)$ for $m > \frac{n(\alpha+1)}{2\alpha}$.

Proof Note that $\hat{\beta}(\zeta)$ is a continuous function. According to the preceding theorem, we need just check the integrability of $\left|\hat{\beta}(\zeta)\right|^{2m}$ *at infinity.* Up to a factor $(2\pi)^{-\frac{n}{2}}$

$$\left|\hat{\beta}(\zeta)\right| = \left|\int_{\mathbb{R}^n} \frac{\mu(dv)}{\lambda + i\zeta.v}\right| \leq \int_{\mathbb{R}^n} \frac{\mu(dv)}{\sqrt{\lambda^2 + |\zeta|^2 |e.v|^2}}$$

where $e = \frac{\zeta}{|\zeta|}$. So, for any $\varepsilon > 0$,

$$\left|\hat{\beta}(\zeta)\right| \leq \int_{\{|e.v|<\varepsilon\}} \frac{\mu(dv)}{\sqrt{\lambda^2 + |\zeta|^2 |e.v|^2}} + \int_{\{|e.v|\geqslant\varepsilon\}} \frac{\mu(dv)}{\sqrt{\lambda^2 + |\zeta|^2 |e.v|^2}}$$

$$\leq \lambda^{-1}\mu\{|e.v| < \varepsilon\} + \frac{\|\mu\|}{|\zeta|\,\varepsilon} \leq \lambda^{-1}c\varepsilon^{\alpha} + \frac{\|\mu\|}{|\zeta|\,\varepsilon}.$$

Optimizing with respect to ε yields

$$\left|\hat{\beta}(\zeta)\right|^{2m} \leq \frac{C}{|\zeta|^{\frac{2m\alpha}{\alpha+1}}}$$

for some positive constant C depending on λ. We are done if $\frac{2m\alpha}{\alpha+1} > n$ i.e. if $m > \frac{n(\alpha+1)}{2\alpha}$. ■

In the same spirit (but with more involved estimates) we can show:

Theorem 39 ([52]) *Let $\Omega \subset \mathbb{R}^n$ be an open set with finite Lebesgue measure. Let the velocity measure μ be finite and there exist $\alpha > 0$, $c > 0$ such that*

$$\sup_{e \in S^{n-1}} \mu \otimes \mu\left\{(v, v');\ \left|(v - v').e\right| \leq \varepsilon\right\} \leq c\varepsilon^{\alpha}.$$

Then $U_m(t)$ is weakly compact in $L^1(\Omega \times V)$ for all $t \geqslant 0$ and for $m \geqslant m_0$ where m_0 is the smallest odd integer greater than $\frac{n(\alpha+1)}{2\alpha} + 1$.

We point out that in Theorems 38 and 39, the condition on m *does not depend* on the constant c in the statement; this fact is fundamental if we want to pass from model cases to more general models. We show now how to treat by approximation more general velocity measures and scattering kernels. Indeed, in the approximation procedure, a general (a priori infinite) velocity measure μ is approximated, by truncation, by a sequence of finite measures μ_j such that

$$\sup_{e\in S^{n-1}} \mu_j \otimes \mu_j \left\{(v,v');\, \left|(v-v').e\right| \leq \varepsilon\right\} \leq c_j \varepsilon^{\alpha}$$

where α is *independent* of j.

A scattering operator K in $L^1(\Omega \times V)$ is said to be regular if $\{K(x);\ x \in \Omega\}$ is a set of *collectively weakly compact operators* on $L^1(V)$, i.e. the set

$$\left\{K(x)\varphi;\ x \in \Omega,\ \|\varphi\|_{L^1(V)} \leq 1\right\}$$

is relatively weakly compact in $L^1(V)$. This class of scattering kernels appears (with Lebesgue measure dv on $\mathbb{R}^n$) in P. Takak [72] and L. W. Weis [82]. See B. Lods [37] for the extension of P. Takak's construction to abstract velocity measures μ.

Theorem 40 ([37]) *Let K be a regular scattering operator in $L^1(\Omega \times V)$. Then there exists a sequence $(K_j)_j$ of scattering operators such that:*

(i) $0 \leq K_j \leq K$
(ii) $\left\|K - K_j\right\|_{\mathcal{L}(L^1(\Omega\times V))} \to 0$ *as* $j \to +\infty$.
(iii) For each K_j there exists $f_j \in L^1(V)$ such that

$$K_j\varphi \leq f_j(v) \int \varphi(x,v')\mu(dv')\ \ \forall \varphi \in L^1_+(\Omega \times V).$$

We are ready to show:

Theorem 41 ([52]) *Let $\Omega \subset \mathbb{R}^n$ be an open set with finite Lebesgue measure and let K be a regular scattering operator in $L^1(\Omega \times V)$. We assume that the velocity measure μ is such that: There exists $\alpha > 0$ such that for any $c_1 > 0$ there exists $c_2 > 0$ such that*

$$\sup_{e\in S^{n-1}} \mu\{v; |v| \leq c_1,\ |v.e| \leq \varepsilon\} \leq c_2 \varepsilon^{\alpha}.$$

Then the components of $\rho_e(T)$ and $\rho_e(T+K)$ containing a right half-plane coincide. In particular

$$\sigma(T+K) \cap \{Re\lambda > s(T)\}$$

consists at most of isolated eigenvalues with finite algebraic multiplicities where

$$s(T) = -\lim_{t\to+\infty} \inf_{(y,v)} \frac{1}{t}\int_0^t \sigma(y+\tau v, v)d\tau.$$

Proof Let us show that $\left[K(\lambda - T)^{-1}K\right]^{m+1}$ is weakly compact in $L^1(\Omega \times V)$ for $m > \frac{n(\alpha+1)}{2\alpha}$. We fix $m > \frac{n(\alpha+1)}{2\alpha}$. It suffices to show the weak compactness of $\left[K_j(\lambda - T)^{-1}K_j\right]^{m+1}$ for all j (K_j from Theorem 40). By domination, we may replace K_j by $\hat{K}_j$ where

$$\widehat{K_j}\varphi = f_j(v)\int \varphi(x, v')\mu(dv').$$

By approximation again we may suppose that f_j is continuous with compact support. Let j be fixed and denote by V_j the support of f_j. We note that $\left[\widehat{K_j}(\lambda - T)^{-1}\widehat{K_j}\right]^m$ leaves invariant $L^1(\Omega \times V_j)$ and $\widehat{K_j}$ maps $L^1(\Omega \times V)$ into $L^1(\Omega \times V_j)$. By replacing $f_j(v)$ by its supremum, the model case (dealt with previously) in $L^1(\Omega \times V_j)$ insures that $\left[K_j(\lambda - T)^{-1}K_j\right]^m$ is weakly compact on $L^1(\Omega \times V_j)$ so

$$\left[K_j(\lambda - T)^{-1}K_j\right]^{m+1} = \left[K_j(\lambda - T)^{-1}K_j\right]^m \left[K_j(\lambda - T)^{-1}K_j\right]$$

is weakly compact on $L^1(\Omega \times V)$. Finally some power of $K(\lambda - T)^{-1}$ is compact on $L^1(\Omega \times V)$ and we conclude by the general theory. ■

Theorem 42 ([52]) *Let $\Omega \subset \mathbb{R}^n$ be an open set with finite Lebesgue measure and let K be a regular scattering operator in $L^1(\Omega \times V)$. We assume that the velocity measure μ is such that: There exists $\alpha > 0$ such that for any $c_1 > 0$ there exists $c_2 > 0$ such that*

$$\sup_{e \in S^{n-1}} \mu \otimes \mu \left\{(v, v'); |v|, |v'| \leq c_1, \left|(v - v').e\right| \leq \varepsilon\right\} \leq c_2 \varepsilon^\alpha.$$

Then $(V(t))_{t \geqslant 0}$ and $(U(t))_{t \geqslant 0}$ have the same essential type. In particular

$$\sigma(V(t)) \cap \left\{\alpha \in \mathbb{C};\ |\alpha| > e^{s(T)t}\right\}$$

consists at most of isolated eigenvalues with finite algebraic multiplicities.

Proof Let us show that $R_{m+1}(t)$ is weakly compact in $L^1(\Omega \times V)$ for all $t \geqslant 0$ and for $m \geqslant m_0$ where m_0 is the smallest odd integer greater than $\frac{n(\alpha+1)}{2\alpha} + 1$. We fix $m \geqslant m_0$. It suffices to show that $R^j_{m+1}(t)$ is weakly compact in $L^1(\Omega \times V)$ for all j where $R^j_{m+1}(t)$ is the remainder term of order $m+1$ corresponding to a perturbation K_j in place of K (K_j from Theorem 40). By domination, we may replace K_j by $\widehat{K_j}$ where

$$\widehat{K_j}\varphi = f_j(v)\int \varphi(x, v')\mu(dv').$$

By approximation again we may suppose that f_j is continuous with compact support. Let j be fixed and denote by V_j the support of f_j. We note that $R_m^j(t)$ leaves invariant $L^1(\Omega \times V_j)$ and $\widehat{K_j}$ *maps* $L^1(\Omega \times V)$ *into* $L^1(\Omega \times V_j)$. Note also that

$$R_{m+1}^j = R_m^j * \left[\widehat{K_j}U\right] = \int_0^t R_m^j(t-s)\widehat{K_j}U(s)ds.$$

By dominating f_j by its supremum, the previous model case (and a domination argument) shows that $R_m^j(t-s)$ is weakly compact in $L^1(\Omega \times V_j)$. Thus $R_m^j(t-s)\widehat{K_j}U(s)$ is weakly compact in $L^1(\Omega \times V)$ and then so is R_{m+1}^j as a strong integral of "weakly compact operator" valued mapping. We conclude by the general theory. ■

Remark 5 The conditions on μ in Theorems 41 and 42 are satisfied e.g. for Lebesgue measure on $\mathbb{R}^n$ or on spheres (multigroup models).

5.3 *Dunford-Pettis Operators in Transport Theory*

A bounded operator $G \in \mathcal{L}(L^1(\nu))$ is called a Dunford-Pettis (or a completely continuous) operator if G maps a weakly compact subset into a (norm) compact subset. For instance, a weakly compact operator on $L^1(\nu)$ is Dunford-Pettis; this explains why the product of two weakly compact operators on $L^1(\nu)$ is compact. More generally, if G_2 is weakly compact on $L^1(\nu)$ and G_1 is Dunford-Pettis on $L^1(\nu)$ then G_1G_2 is compact on $L^1(\nu)$. We have seen that various relevant operators for neutron transport are *not* weakly compact in $L^1(\Omega \times V)$, for instance:

$$K(\lambda - T)^{-1},\ (\lambda - T - K)^{-1} - (\lambda - T)^{-1} \text{ and } V(t) - U(t).$$

We can show however that they are *all* Dunford-Pettis, (see [50] for more information). This explains why we claimed in the proof of Theorem 37 that

$$\left[K(\lambda - T)^{-1}K\right]^{m+1} = K(\lambda - T)^{-1}K\left[K(\lambda - T)^{-1}K\right]^m$$

is compact since $\left[K(\lambda - T)^{-1}K\right]^m$ is weakly compact and $K(\lambda - T)^{-1}$ is Dunford-Pettis. We restrict ourselves to:

Theorem 43 ([50]) *Let $\Omega \subset \mathbb{R}^n$ be an open set with finite Lebesgue measure and let K be a regular scattering operator in $L^1(\Omega \times V)$. If the affine hyperplanes have zero μ-measure then $V(t) - U(t)$ is a Dunford-Pettis operator on $L^1(\Omega \times V)$.*

Proof We note first that $V(t) - U(t) = \sum_1^\infty U_j(t)$ is Dunford-Pettis for all $t \geqslant 0$ if and only if $U_1(t)$ is [50]. By approximation, we may assume that their exists f

continuous with support in $\{|v| < \overline{c}\}$ such that

$$K\varphi \leq f(v) \int \varphi(x, v')\mu(dv') \ \ \forall \varphi \in L^1_+(\Omega \times V).$$

Let $E \subset L^1(\Omega \times V)$ be relatively weakly compact. In particular

$$\sup_{\varphi \in E} \int_{|v| \geqslant c} \mu(dv) \int_{\Omega} |\varphi(x, v)| \, dx \to 0 \ \text{ as } \ c \to +\infty.$$

We decompose $\varphi \in E$ as

$$\varphi = \varphi \chi_{\{|v| < c\}} + \varphi \chi_{\{|v| \geqslant c\}}$$

so

$$\left\| U_1(t)(\varphi \chi_{\{|v| \geqslant c\}}) \right\| \leq \| U_1(t) \| \left\| \varphi \chi_{\{|v| \geqslant c\}} \right\| \to 0 \ \text{ as } \ c \to +\infty$$

uniformly in $\varphi \in E$. On the other hand $\varphi \chi_{\{|v| < c\}}$ is zero for $|v| > c$ and (for $c > \overline{c}$) for any $\psi \in L^1(\Omega \times V)$, $K\psi$ is zero for $|v| > c$. So we may assume from the beginning that V is bounded and then K maps also $L^2(\Omega \times V)$ into itself. We decompose $\varphi \in E$ as

$$\varphi = \varphi^1_\alpha + \varphi^2_\alpha := \varphi \chi_{\{|\varphi| < \alpha\}} + \varphi \chi_{\{|\varphi| \geqslant \alpha\}}$$

and note that

$$\int |\varphi| \geqslant \int_{\{|\varphi| \geqslant \alpha\}} |\varphi| \geqslant \alpha \, (dx \otimes \mu \, \{|\varphi| \geqslant \alpha\})$$

so

$$dx \otimes \mu \, \{|\varphi| \geqslant \alpha\} \to 0 \ \text{ as } \alpha \to +\infty.$$

The equi-integrability of E implies that

$$\left\| \varphi^2_\alpha \right\|_{L^1} \to 0 \ \text{ as } \alpha \to +\infty$$

uniformly in $\varphi \in E$ and finally $\left\| U_1(t) \varphi^2_\alpha \right\|_{L^1} \to 0$ as $c \to +\infty$ uniformly in $\varphi \in E$. Since, $\{\varphi^1_\alpha\}$ is bounded in L^1 and in L^∞ then the interpolation inequality

$$\left\| \varphi^1_\alpha \right\|_{L^2} \leq \left\| \varphi^1_\alpha \right\|^{\frac{1}{2}}_{L^1} \left\| \varphi^1_\alpha \right\|^{\frac{1}{2}}_{L^\infty}$$

shows that $\{\varphi_\alpha^1\}$ is also bounded in $L^2(\Omega \times V)$. We know (from the L^2 theory) that $U_1(t)$ is compact in $L^2(\Omega \times V)$ so $\{U_1(t)\varphi_\alpha^1\}$ is relatively compact in $L^2(\Omega \times V)$ and then relatively compact in $L^1(\Omega \times V)$ since $\Omega \times V$ has a finite measure. Thus $\{U_1(t)\varphi;\ \varphi \in E\}$ is as close to a relatively compact subset of $L^1(\Omega \times V)$ as we want and finally $\{U_1(t)\varphi;\ \varphi \in E\}$ is relatively compact. ∎

6 Comments

6.1 *Measure Convolution Operators in Transport Theory*

The compactness results for neutron transport operators in L^p spaces with $1 < p < \infty$ (see Subsection 5.1) can also be derived from the analysis of just two particular measure convolution operators on $\mathbb{R}^n$, see [59].

6.2 *Unbounded Geometries*

The compactness results given in this lecture in spatial domains Ω with finite Lebesgue measure need no be true in general domains, e.g. the results are false in the whole space (i.e. $\Omega = \mathbb{R}^n$) and *space homogeneous* cross-sections. However, under suitable assumptions on the cross-sections, we can recover the compactness results above in unbounded geometries [60]. Actually, for general geometries and cross-sections, the relevant perturbation theory does not concern the essential spectra (and the essential types) but rather the *critical* spectra (and the critical types); we refer to [10, 53, 54, 59, 62, 67] for the abstract theory and how to use it in the context of neutron transport theory.

6.3 *Leading Eigenvalue of Neutron Transport*

The time asymptotic behaviour of neutron transport semigroup is meaningful if the latter has a spectral gap or equivalently if its generator has a leading eigenvalue. This topic relies on peripheral spectral analysis of neutron transport. The first relevant question concerns irreducibility criteria of neutron transport semigroups for which we refer e.g. to [26, 47, 78] and [48, Chapter 5]. The second relevant question concerns the effective existence of leading eigenvalues; besides the isotropic case dealt with by [29], we refer to ([48, Chapter 5]) for general tools. Variational characterizations of the leading eigenvalue in L^p spaces ($InfSup$ or $SupInf$ criteria) and lower bounds of this eigenvalue are given in [55]. The criticality eigenvalue problem is dealt with in [41,66] and [48, Chapter 5]. Finally, we refer to [40] for variational characterizations of the criticality eigenvalue.

6.4 Partly Elastic Scattering Operators

Most of the literature on spectral theory of neutron transport is devoted either to one speed models or to inelastic models. Despite their apparent difference, these two models can be covered by a unique general formalism as we did in Sect. 5. However, more complex models which take into account of both elastic and inelastic scatterings appear e.g. in [35]:

$$\frac{\partial f}{\partial t} + v.\frac{\partial f}{\partial x} + \sigma(x,v) f(t,x,v) = K_e f + K_i f$$

where

$$K_i f = \int_{\mathbb{R}^3} k(x,v,v') f(x,v') dv' \text{ (inelastic operator)}$$

and

$$K_e f = \int_{S^2} k(x,\rho,\omega,\omega') f(x,\rho\omega') dS(\omega') \text{ (elastic operator)}$$

where $v = \rho\omega$. The peculiarity of the elastic scattering operator is that it is *not* compact "in velocities" in contrast to usual inelastic scattering operators. This explains the complexity of $\sigma(T + K_e)$ which consists of a half-plane and various *"curves"*[35]. We find in [68] various compactness results (due to K_i) and spectral results. In particular the semigroups generated by $T + K_e$ and by $T + K_e + K_i$ have the same essential type.

6.5 Generalized Boundary Conditions

We point out that "zero incoming flux" is the natural (i.e. physical) boundary condition for neutron transport. However, various boundary conditions (relating e.g. the incoming and outgoing fluxes) were also considered in kinetic theory [11] or in structured cell population models (see e.g. [5]). The literature on the subject is considerable and we do not try to comment on it. We just note that the corresponding advection semigroup is no longer explicit and, for this reason, spectral analysis of such kinetic models with non local boundary conditions is much more technical. We refer for instance to [38] and references therein.

References

1. V.I. Agoshkov, Spaces of functions with differential-difference characteristics and smoothness of solutions of the transport equations. Sov. Math. Dokl. **29**, 662–666 (1987)
2. S. Albertoni, B. Montagnini, On the spectrum of neutron transport equations in finite bodies. J. Math. Anal. Appl. **13**, 19–48 (1966)
3. D. Aliprantis, O. Burkinshaw, Positive compact operators on Banach lattices. Math. Z **174** 289–298 (1980)
4. N. Angelescu, V. Protopopescu, On a problem in linear transport theory. Rev. Roum. Phys. **22**, 1055–1061 (1977)
5. O. Arino, Some spectral properties for the asymptotic behavior of semigroups connected to population dynamics. SIAM Rev. **34**(4), 445–476 (1992)
6. C. Bardos, R. Santos, R. Sentis, Diffusion approximation and computation of the critical size. Trans. Am. Math. Soc. **284**(2), 617–649 (1984)
7. G.I. Bell, Stochastic theory of neutron transport. J. Nucl. Sci. Eng. **21**, 390–401 (1965)
8. M. Borysiewicz, J. Mika, Time behaviour of thermal neutrons in moderating media. J. Math. Anal. Appl. **26**, 461–478 (1969)
9. S. Brendle, On the asymptotic behaviour of perturbed strongly continuous semigroups. Math. Nachr. **226**, 35–47 (2001)
10. S. Brendle, R. Nagel, J. Poland, On the spectral mapping theorem for perturbed strongly continuous semigroups. Archiv Math. **74**, 365–378 (2000)
11. C. Cercignani, *The Boltzmann Equation and Its Applications* (Springer, New York, 1988)
12. M. Cessenat, Théorèmes de trace L^p pour des espaces de fonctions de la neutronique. C. R. Acad. Sci. Paris Ser. I **299**, 831–834 (1984)
13. M. Cessenat, Théorèmes de trace pour les espaces de fonctions de la neutronique. C. R. Acad. Sci. Paris Ser. I **300**, 89–92 (1985)
14. R. Dautray, J.L. Lions (eds.), *Analyse mathématique et calcul numérique pour les sciences et les techniques*. Tome 1 (Masson, Paris, 1985)
15. E.-B. Davies, *One-Parameter Semigroups* (Academic Press, London, 1980)
16. M.L. Demeru, B. Montagnini, Complete continuity of the free gas scattering operator in neutron thermalization theory. J. Math. Anal. Appl. **12**, 49–57 (1965)
17. J.J. Duderstadt, W.R. Martin, *Transport Theory* (John Wiley & Sons, Inc, New York, 1979)
18. N. Dunford, J.T. Schwartz, *Linear Operators*, Part I (John Wiley & Sons, Inc, Interscience New York, 1958)
19. D.E. Edmunds, W.D. Evans, *Spectral Theory and Differential Operators* (Clarendon Press, Oxford, 1989)
20. K.-J. Engel, R. Nagel, *One-Parameter Semigroups for Linear Evolution Equations* (Springer, New York, 2000)
21. I. Gohberg, S. Goldberg, M.A. Kaashoek, *Classes of Linear Operators*, vol. 1 (Birkhauser, Basel, 1990)
22. J.A. Goldstein, *Semigroups of Linear Operators and Applications*. Oxford Mathematical Monographs (Oxford University Press, Oxford, 1985)
23. F. Golse, P.L. Lions, B. Perthame, R. Sentis, J. Funct. Anal. **76**, 110–125 (1988)
24. G. Greiner, Zur Perron-Frobenius Theorie stark stetiger Halbgruppen. Math. Z. **177**, 401–423 (1981)
25. G. Greiner, Spectral properties and asymptotic behavior of the linear transport equation. Math. Z. **185**, 167–177 (1984)
26. G. Greiner, An irreducibility criterion for the linear transport equation, in *Semesterbericht Funktionalanalysis* (University of Tubingen, 1984)
27. G. Greiner, A spectral decomposition of strongly continuous group of positive operators. Q. J. Math. Oxf. **35**, 37–47 (1984)
28. A. Huber, Spectral properties of the linear multiple scattering operator in L^1 Banach lattice. Int. Equ. Opt. Theory **6**, 357–371 (1983)

29. T. Hiraoka, S. Ukaï, Eigenvalue spectrum of the neutron transport operator for a fast multiplying system. J. Nucl. Sci. Technol. **9**(1), 36–46 (1972)
30. K. Jarmouni-Idrissi, M. Mokhtar-Kharroubi, A class of non-linear problems arising in the stochastic theory of neutron transport. Nonlinear Anal. **31**(3–4), 265–293 (1998)
31. K. Jorgens, Commun. Pure. Appl. Math. **11**, 219–242 (1958)
32. T. Kato, *Perturbation Theory of Linear Operators* (Springer, Berlin, 1984)
33. M. Kunze, G. Schluchtermann, Strongly generated Banach spaces and measures of non-compactness. Math. Nachr. **191**, 197–214 (1998)
34. K. Latrach, B. Lods, Spectral analysis of transport equations with bounce-back boundary conditions. Math. Methods Appl. Sci. **32**, 1325–1344 (2009)
35. E.W. Larsen, P.F. Zweifel, On the spectrum of the linear transport operator. J. Math. Phys. **15**, 1987–1997 (1974)
36. J. Lehner, M. Wing, On the spectrum of an unsymmetric operator arising in the transport theory of neutrons. Commun. Pure. Appl. Math. **8**, 217–234 (1955)
37. B. Lods, On linear kinetic equations involving unbounded cross-sections. Math. Methods Appl. Sci. **27**, 1049–1075 (2004)
38. B. Lods, M. Sbihi, Stability of the essential spectrum for 2D-transport models with Maxwell boundary conditions. Math. Methods Appl. Sci. **29**, 499–523 (2006)
39. B. Lods, M. Mokhtar-Kharroubi, M. Sbihi, Spectral properties of general advection operators and weighted translation semigroups. Commun. Pure. Appl. Anal. **8**(5), 1–24 (2009)
40. B. Lods, Variational characterizations of the effective multiplication factor of a nuclear reactor core. Kinet. Relat. Models **2**, 307–331 (2009)
41. J. Mika, Fundamental eigenvalues of the linear transport equation. J. Quant. Spectr. Radiat. Transf. **11**, 879–891 (1971)
42. M. Mokhtar-Kharroubi, La compacité dans la théorie du transport des neutrons. C. R. Acad. Sci. Paris Ser. I **303**, 617–619 (1986)
43. M. Mokhtar-Kharroubi, Les équations de la neutronique: Positivité, Compacité, Théorie spectrale, Comportement asymptotique en temps. Thèse d'Etat (French habilitation), Paris (1987)
44. M. Mokhtar-Kharroubi, Compactness properties for positive semigroups on Banach lattices and applications. Houston J. Math. **17**(1), 25–38 (1991)
45. M. Mokhtar-Kharroubi, Time asymptotic behaviour and compactness in transport theory. Eur. J. Mech. B Fluids **11**(1), 39–68 (1992)
46. M. Mokhtar-Kharroubi, On the stochastic nonlinear neutron transport equation. Proc. R. Soc. Edinb. **121**(A), 253–272 (1992)
47. M. Mokhtar-Kharroubi, Quelques applications de la positivité en théorie du transport. Ann. Fac. Toulouse **11**(1), 75–99 (1990)
48. M. Mokhtar-Kharroubi, *Mathematical Topics in Neutron Transport Theory, New Aspects*, vol. 46 (World Scientific, Singapore, 1997)
49. M. Mokhtar-Kharroubi, L. Thevenot, On the diffusion theory of neutron transport on the torus. Asymp. Anal. **30**(3–4), 273–300 (2002)
50. M. Mokhtar-Kharroubi, On the convex compactness property for the strong operator topology and related topics. Math. Methods Appl. Sci. **27**(6), 687–701 (2004)
51. M. Mokhtar-Kharroubi, Optimal spectral theory of the linear Boltzmann equation. J. Funct. Anal. **226**, 21–47 (2005)
52. M. Mokhtar-Kharroubi, On L1-spectral theory of neutron transport. Differ. Int. Equ. **18**(11), 1221–1242 (2005)
53. M. Mokhtar-Kharroubi, M. Sbihi, Critical spectrum and spectral mapping theorems in transport theory. Semigroup Forum **70**(3), 406–435 (2005)
54. M. Mokhtar-Kharroubi, M. Sbihi, Spectral mapping theorems for neutron transport, L^1-theory. Semigroup Forum **72**, 249–282 (2006)
55. M. Mokhtar-Kharroubi, On the leading eigenvalue of neutron transport models. J. Math. Anal. Appl. **315**, 263–275 (2006)

56. M. Mokhtar-Kharroubi, Spectral properties of a class of positive semigroups on Banach lattices and streaming operators. Positivity **10**(2), 231–249 (2006)
57. M. Mokhtar-Kharroubi, F. Salvarani. Convergence rates to equilibrium for neutron chain fissions. Acta. Appl. Math. **113**, 145–165 (2011)
58. M. Mokhtar-Kharroubi, New generation theorems in transport theory. Afr. Mat. **22**, 153–176 (2011)
59. M. Mokhtar-Kharroubi, On some measure convolution operators in neutron transport theory. Acta Appl. Math. (2014). doi:10.1007/s10440-014-9866-3.
60. M. Mokhtar-Kharroubi, Compactness properties of neutron transport operators in unbounded geometries (work in preparation)
61. R. Nagel (ed.), *One-Parameter Semigroups of Positive Operators*. Lecture Notes in Mathematics, vol. 1184 (Springer, Berlin, 1986)
62. R. Nagel, J. Poland, The critical spectrum of a strongly continuous semigroup. Adv. Math **152**(1), 120–133 (2000)
63. B. de Pagter, Irreducible compact operators. Math. Z. **192**, 149–153 (1986)
64. A. Pazy, P. Rabinowitz, A nonlinear integral equation with applications to neutron transport theory. Arch. Rat. Mech. Anal. **32**, 226–246 (1969)
65. M. Ribaric, I. Vidav, Analytic properties of the inverse $A(z)^{-1}$ of an analytic linear operator valued function $A(z)$. Arch. Ration. Mech. Anal. **32**, 298–310 (1969)
66. R. Sanchez, The criticality eigenvalue problem for the transport with general boundary conditions. Transp. Theory Stat. Phys. **35**, 159–185 (2006)
67. M. Sbihi, A resolvent approach to the stability of essential and critical spectra of perturbed C_0-semigroups on Hilbert spaces with applications to transport theory. J. Evol. Equ. **7**(1), 35–58 (2007)
68. M. Sbihi, Spectral theory of neutron transport semigroups with partly elastic collision operators. J. Math. Phys. **47**, 123502 (2006).
69. G. Schluchtermann, On weakly compact operators. Math. Ann. **292**, 263–266 (1992)
70. G. Schluchtermann, Perturbation of linear semigroups, in *Recent Progress in Operator Theory, (Regensburg 1995)*. Oper. Theory Adv. Appl., vol. 103 (Birkhauser, Basel, 1998), pp. 263–277
71. Y. Shizuta, On the classical solutions of the Boltzmann equation. Commun. Pure. Appl. Math. **36**, 705–754 (1983)
72. P. Takak, A spectral mapping theorem for the exponential function in linear transport theory. Transp. Theory Stat. Phys. **14**(5), 655–667 (1985)
73. A.E. Taylor, D.C. Lay, *Introduction to Functional Analysis* (Krieger Publishing Company, Melbourne, 1980)
74. S. Ukaï, Eigenvalues of the neutron transport operator for a homogeneous finite moderator. J. Math. Anal. Appl. **30**, 297–314 (1967)
75. I. Vidav, Existence and uniqueness of nonnegative eigenfunctions of the Boltzmann operator. J. Math. Anal. Appl. **22**, 144–155 (1968)
76. I. Vidav, Spectra of perturbed semigroups with application to transport theory. J. Math. Anal. Appl. **30**, 264–279 (1970)
77. J. Voigt, A perturbation theorem for the essential spectral radius of strongly continuous semigroups. Monatsh. Math. **90**, 153–161 (1980)
78. J. Voigt, Positivity in time dependent linear transport theory. Acta Appl. Math. **2**, 311–331 (1984)
79. J. Voigt, Spectral properties of the neutron transport equation. J. Math. Anal. Appl. **106**, 140–153 (1985)
80. J. Voigt, On the convex compactness property for the strong operator topology. Note di Math. **12**, 259–269 (1992)
81. J. Voigt, Stability of essential type of strongly continuous semigroups. Proc. Steklov Inst. Math. **3**, 383–389 (1995)
82. L.W. Weis, A generalization of the Vidav-Jorgens perturbation theorem for semigroups and its application to transport theory. J. Math. Anal. Appl. **129** 6–23 (1988)

Reaction-Diffusion-ODE Models of Pattern Formation

Anna Marciniak-Czochra

1 Introduction

This chapter is devoted to analysis of a class of reaction-diffusion type models arising from mathematical biology. We focus on mechanisms pattern formation in reaction-diffusion equations coupled to ordinary differential equations. Such systems are applied to modelling of interactions between cellular processes and diffusing signalling factors.

After introducing the classical Turing concept of pattern formation in two reaction-diffusion equations, we present recent results concerning reaction-diffusion-ODE models accounting for:

1. Diffusion-driven instability of all regular stationary solutions (including spatially heterogenous ones).
2. Mass concentration and dynamical spike patterns in the systems with autocatalysis of the non-diffusing component.
3. Stable discontinuous patterns in the systems with hysteresis in the quasi-stationary ODE subsystem.

The material is illustrated by examples from mathematical biology. The theoretical concepts presented in this paper can be applied to construct and analyse models of biological pattern formation.

The chapter is organised as follows: Sect. 2 is devoted to the Turing mechanism of pattern formation in classical models consisting of two linear reaction-diffusion equations. We provide a systematic mathematical analysis of the phenomenon of the

A. Marciniak-Czochra (✉)
Institute of Applied Mathematics, Interdisciplinary Center of Scientific Computing (IWR), University of Heidelberg, Heidelberg, Germany
e-mail: anna.marciniak@iwr.uni-heidelberg.de

© Springer International Publishing Switzerland 2015

J. Banasiak, M. Mokhtar-Kharroubi (eds.), *Evolutionary Equations with Applications in Natural Sciences*, Lecture Notes in Mathematics 2126,
DOI 10.1007/978-3-319-11322-7_8

diffusion-driven instability (DDI) and characterise Turing patterns. Then, we focus on the models with a degenerated diffusion, i.e. reaction-diffusion-ODE systems. Our aim is to show novelty and mathematical challenges arising in such models. In Sect. 3 we summarise results on the existence and regularity of solutions of such models and the corresponding linearised stability principle. In Sect. 4 we focus on systems coupling one reaction-diffusion equation with one ODE and exhibiting DDI. We investigate Turing-type phenomenon and show that autocatalysis of non-diffusing component is a necessary and sufficient condition for DDI in such systems. However, the same mechanism which destabilises constant solutions, destabilises also all continuous spatially heterogenous stationary solutions. We provide a rigorous result on nonlinear (in)stability analysis, which is extended to discontinuous patterns for a specific class of nonlinearities. Simulations, supplemented by numerical analysis, indicate a novel pattern formation phenomenon based on non-stationary structures tending asymptotically to the sum of Dirac deltas. Finally, Sect. 5 is devoted to a systematic description of hysteresis-driven pattern formation. The emerging structures may be monotone, periodic or irregular. We show that bistability without a hysteresis effect is not sufficient for the existence of stable patterns and provide a criterion for nonlinear stability of the discontinuous stationary solutions in case of a basic reaction-diffusion-ode model.

2 Models with Diffusion-Driven Instability

2.1 Turing Instability

Classical concept for pattern formation in reaction-diffusion equations originates from the seminal paper of Allan Turing [66]. The Turing hypothesis can be stated as follows: *When two chemical species with different diffusion rates react with each other, the spatially homogeneous state may become unstable, thereby leading to a nontrivial spatial structure.*

The idea looks counter-intuitive, since diffusion is expected to lead to the uniform distribution of the particles. In case of a scalar reaction-diffusion equation in a convex domain, all stationary spatially heterogenous solutions are unstable and the only stable equilibria are spatially homogenous, see e.g. [7].

Mathematical analysis of reaction-diffusion equations provides explanation for the phenomenon postulated by Turing. The mechanism is related to a local behaviour of solutions of a reaction-diffusion system in the neighbourhood of a constant stationary solution that is destabilised through diffusion. Patterns arise through a bifurcation, called *diffusion-driven instability (DDI)* or *Turing instability*, which is defined as follows:

Definition 1 (Turing Instability) A system of reaction-diffusion equations exhibits diffusion-driven instability (DDI; Turing instability), if and only if there exists a constant stationary solution, which is stable to spatially homogenous perturbations, but unstable to spatially heterogenous perturbations.

The emerging patterns can be spatially monotone corresponding to the gradients in positional information or spatially periodic. They are located around the destabilised constant equilibrium (*close to equilibrium* patterns).

The original concept was presented by Turing on the example of two linear reaction-diffusion equations describing dynamics of two substances (chemical or biochemical) of the form

$$\begin{aligned} \frac{\partial w}{\partial t} &= D\Delta w + Aw && \text{in } \Omega, \\ \partial_n w(t,0) &= 0 && \text{on } \partial\Omega, \\ w(0,x) &= w_0(x), && \end{aligned} \tag{1}$$

where $w \in \mathbb{R}^2$ is a vector of two variables $w = (u, v)$, D is a diagonal matrix with nonnegative coefficients d_u, d_v on the diagonal, and $\partial_n = \frac{\partial}{\partial n}$, where n denotes the unit outer normal vector to $\partial\Omega$ (no-flux condition), and Ω is a bounded region.

Following Turing [66], we formulate the following result on DDI

Theorem 1 (Conditions for Turing Instability) *Assume that*

$$\begin{aligned} \operatorname{tr} A &< 0, \\ \det A &> 0, \end{aligned} \tag{2}$$

and $d_v > 0$. *There exists* $d_u > 0$ *(small enough) such that the constant steady state* $(0,0)$ *is unstable for the reaction-diffusion equation* (1).

The proof is based on a spectral decomposition of the Laplace operator with homogenous Neumann boundary conditions and calculation of eigenvalues of obtained finite dimensional operator.

Proposition 1 (Spectral Decomposition) *If the domain* Ω *is* C^1, *connected, open and bounded, then there exists a spectral basis* $(\lambda_k, \phi_k)_{k\geq 1}$ *such that*

$$\begin{aligned} -\Delta\phi_k &= \lambda_k\phi_k \quad \textit{in} \quad \Omega, \\ \partial_n\phi_k &= 0 \quad \textit{on} \quad \partial\Omega. \end{aligned}$$

1. λ_k *is a nondecreasing sequence* $0 = \lambda_1 < \lambda_2 \leq \ldots \leq \lambda_k \to \infty$,
2. $(\phi_k)_{k\geq 1}$ *is an orthonormal basis in* $L^2(\Omega)$,
3. $\phi_1(x) = \frac{1}{|\Omega|1/2} > 0$ *and* $(\phi_k)_{k>1}$ *change the sign on* Ω.

Remark 1 Connectivity of Ω guarantees that the first eigenvalue is simple and the corresponding eigenfunction is positive on Ω.

Remark 2 The eigenvalues of the Laplacian λ_k are called *wave numbers* related to ϕ_k.

Proof (of Theorem 1) First we consider a system of two ordinary differential equations

$$\frac{du}{dt} = a_{11}u + a_{12}v,$$
$$\frac{dv}{dt} = a_{21}u + a_{22}v \tag{3}$$

fulfilling stability conditions (2). Consequently, the matrix

$$A = \begin{pmatrix} a_{11} & a_{12} \\ a_{21} & a_{22} \end{pmatrix}. \tag{4}$$

has two eigenvalues with a negative real part (or one negative eigenvalue and the Jordan form) and the constant stationary solution $(\bar{u}, \bar{v}) = (0, 0)$ is asymptotically stable.

Then, we consider a corresponding system of reaction-diffusion equations on a bounded domain Ω,

$$\frac{\partial u}{\partial t} = d_u \Delta u + a_{11}u + a_{12}v,$$
$$\frac{\partial v}{\partial t} = d_v \Delta v + a_{21}u + a_{22}v, \tag{5}$$

with zero-flux boundary conditions on $\partial\Omega$,

$$\partial_n u = 0,$$
$$\partial_n v = 0,$$

and initial conditions $u(0, x) = u_0(x)$, $v(0, x) = v_0(x)$.

In the remainder of this section, we assume positive diffusion coefficients $d_u > 0$ and $d_v > 0$.

To investigate the spectrum of $D\Delta + A$, we apply spectral decomposition of the Laplace operator with Neumann homogeneous boundary conditions provided by Proposition 1. We use the orthonormal basis of eigenfunctions $(\phi_k)_{k\geq 1}$ associated with positive eigenvalues to decompose the solutions, which can be written in the form

$$u(t, x) = \Sigma_{k=1}^{\infty} \alpha_k(t)\phi_k(x),$$
$$v(t, x) = \Sigma_{k=1}^{\infty} \beta_k(t)\phi_k(x), \tag{6}$$

Projecting the system on the finite dimensional space spanned by those eigenfunctions, we obtain a system of linear ordinary differential equations with coefficients depending on the wave numbers.

For simplicity of presentation, we restrict further calculations to one-dimensional domain $\Omega = [0, 1]$. The eigenelements $((\phi_k)_{k\geq 1}, \lambda_k)$ can be directly calculated

$$\begin{aligned}\lambda_k &= k^2\pi^2,\\ \phi_k &= \cos(k\pi x).\end{aligned}$$

We search for solutions with exponential growth in time, i.e. having a form

$$\begin{pmatrix} u(t,x) \\ barv(t,x) \end{pmatrix} = e^{\lambda t}\begin{pmatrix} c_1 \cos k\pi x \\ c_2 \cos k\pi x \end{pmatrix}. \tag{7}$$

Inserting (7) to (5) yields

$$\begin{aligned}\lambda e^{\lambda t} c_1 \cos(k\pi x) &= -d_u e^{\lambda t} c_1 \cos(k\pi x)(k\pi)^2\\ &\quad + a_{11} e^{\lambda t} c_1 \cos(k\pi x) + a_{12} e^{\lambda t} c_2 \cos(k\pi x),\\ \lambda e^{\lambda t} c_2 \cos(k\pi x) &= -d_v e^{\lambda t} c_2 \cos(k\pi x)(k\pi)^2 + a_{21} e^{\lambda t} c_1 \cos(k\pi x)\\ &\quad + a_{22} e^{\lambda t} c_2 \cos(k\pi x)\end{aligned}$$

and we obtain

$$\begin{aligned}\lambda c_1 &= -d_u c_1 (k\pi)^2 + a_{11} c_1 + a_{12} c_2,\\ \lambda c_2 &= -d_v c_2 (k\pi)^2 + a_{21} c_1 + a_{22} c_2.\end{aligned}$$

This is a system of linear algebraic equations for (c_1, c_2) and it has a nonzero solution if and only if its determinant vanishes. The number λ is then an eigenvalue of the matrix

$$\tilde{A} = \begin{pmatrix} a_{11} - d_u(k\pi)^2 & a_{12} \\ a_{21} & a_{22} - d_v(k\pi)^2 \end{pmatrix}$$

and it depends on the model parameters and on the wave numbers $\lambda_k = (k\pi)^2$.

Definition 2 The dependence $\lambda(\lambda_k)$ is called a *dispersion relation*.

The question arises whether there exists λ with a positive real part and how it depends on the wave numbers, see Fig. 1.

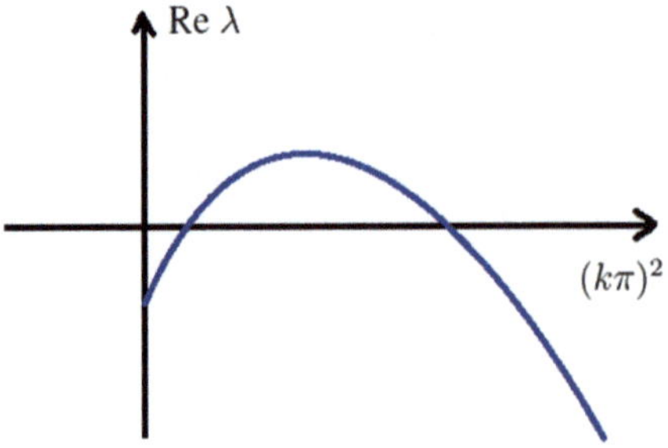

Fig. 1 Dispersion relation for model (5)

Since $\mathrm{tr}\tilde{A}$ is always negative, only one (if any) eigenvalue can change the sign. The necessary condition for that is $\det \tilde{A} < 0$. Hence, we obtain a necessary and sufficient condition for existence of an eigenvalue λ of the matrix $\tilde{A}$ such that $\mathrm{Re}\lambda > 0$

$$(a_{11} - d_u(k\pi)^2)(a_{22} - d_v(k\pi)^2) - a_{12}a_{21} < 0. \tag{8}$$

The left hand-side of (8) is a second order polynomial of $\lambda_k = (k\pi)^2$.
Since d_u and d_v are positive, we may rewrite the above inequality as

$$\lambda_k^2 - \lambda_k \frac{d_u a_{22} + d_v a_{11}}{d_u d_v} + \frac{a_{11}a_{22} - a_{12}a_{21}}{d_u d_v} < 0.$$

The estimates $\lambda_k > 0$ and $(a_{11}a_{22} - a_{12}a_{21})/(d_u d_v) > 0$ yield that the polynomial can take negative values only if $(d_u a_{22} + d_v a_{11})/(d_u d_v) > 0$ large enough and $(a_{11}a_{22} - a_{12}a_{21})/(d_u d_v)$ small enough.

We take $\theta = d_u/d_v$ and calculate roots of the above polynomial

$$\Lambda_\pm = \frac{1}{2d_v\theta}(a_{22}\theta + a_{11} \pm \sqrt{(a_{22}\theta + a_{11})^2 - 4\theta(a_{11}a_{22} - a_{12}a_{21})}),$$

$$\Lambda_\pm = \frac{a_{22}\theta + a_{11}}{2d_v\theta}(1 \pm \sqrt{1 - \frac{4\theta(a_{11}a_{22} - a_{12}a_{21})}{(a_{22}\theta + a_{11})^2}}).$$

It remians to show that there exists $\lambda_k \in [\Lambda_-, \Lambda_+]$. For this purpose, we restrict ourselves to the regime of small θ. Using Taylor expansion we obtain

$$\Lambda_\pm \approx \frac{a_{11}}{2d_v\theta}(1 \pm (1 - \frac{2\theta(a_{11}a_{22} - a_{12}a_{21})}{(a_{22}\theta + a_{11})^2})).$$

Therefore, $\Lambda_- \approx \frac{a_{11}a_{22} - a_{12}a_{21}}{a_{11}d_v} = O(1)$ and $\Lambda_+ \approx \frac{a_{11}}{\theta d_v}$. For small d_u and for d_v of order 1, the interval $[\Lambda_-, \Lambda_+]$ becomes large and hence we know that there exist some λ_k which are from that interval.

Since $\lim_{k\to\infty}\lambda_k = +\infty$, there exist only a finite number of unstable modes λ_k. □

Turing patterns corresponding to different unstable wave numbers are presented in Fig. 2.

Remark 3 (Two-Dimensional Domain) In a rectangle $(0, L_1) \times (0, L_2)$, the family of eigenelements has the form

$$\lambda_{kl} = (\frac{\pi k}{L_1})^2 + (\frac{\pi l}{L_2})^2,$$
$$w_{kl} = \cos\frac{\pi k x}{L_1}\cos\frac{\pi l y}{L_2}.$$

In a narrow domain, i.e. $L_2 \approx 0$ and L_1large, the condition $\lambda_{kl} \in [\Lambda_-, \Lambda_+]$ yields $l = 0$. Otherwise λ_{kl} is very large and it does not belong to the required interval.

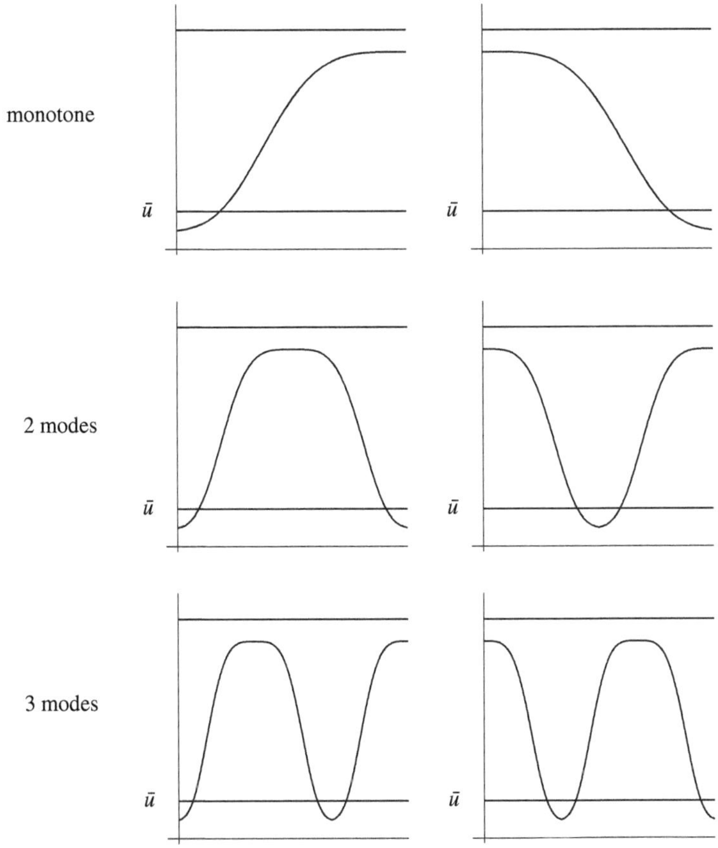

Fig. 2 Example of Turing patterns corresponding to different unstable wave numbers

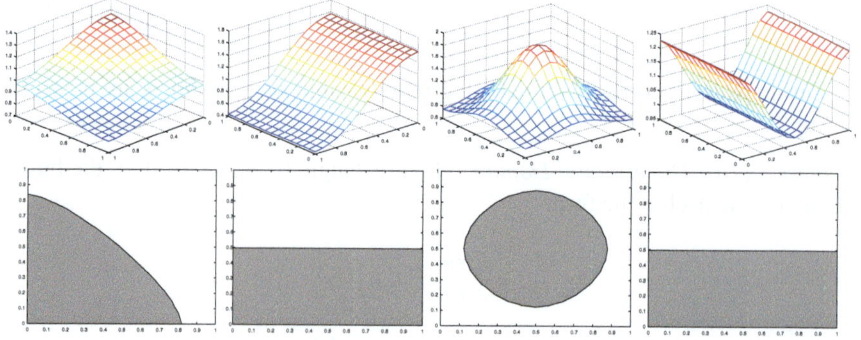

Fig. 3 Spatial patterns obtained in numerical simulation of a reaction-diffusion system with Schnakenberg kinetics on the unit square with Neumann boundary conditions. *Upper panel* solutions (values of the activator u) of the model; *lower panel* a pattern obtained by assigning a *grey color* to all points of domain with the value of activator being above a certain threshold (Courtesy of Dirk Hartmann)

The corresponding eigenfunctions are constant in y (bands parallel to y axis). If $L_1 \approx L_2$, we obtain equal repartition in both directions (spots patterns) (Fig. 3)

Due to the local character of Turing instability, the notion can be extended in a natural way to nonlinear equations using linearisation around a constant positive steady state. We consider a system of equations

$$\begin{aligned}\frac{\partial u}{\partial t} &= D_u \Delta u + f(u, v),\\ \frac{\partial v}{\partial t} &= D_v \Delta v + g(u, v)\end{aligned} \tag{9}$$

with homogeneous Neumann boundary conditions on $\partial\Omega$

$$\partial_n u = 0,$$
$$\partial_n v = 0$$

and initial conditions $u(0, x) = u_0(x)$, $v(0, x) = v_0(x)$, for $x \in \Omega \subset \mathbb{R}^n$, Ω open, bounded.

Let $(\bar{u}, \bar{v}) \in \mathbb{R}^2$ be a constant stationary solution, i.e.

$$f(\bar{u}, \bar{v}) = 0,$$
$$g(\bar{u}, \bar{v}) = 0.$$

We study the behaviour of the solution in a neighbourhood of $(\bar{u}, \bar{v})$ using the linearisation

$$f(u,v) \approx f_u(\bar{u},\bar{v})u + f_v(\bar{u},\bar{v})v,$$
$$g(u,v) \approx g_u(\bar{u},\bar{v})u + g_v(\bar{u},\bar{v})v.$$

For the principle of linearised stability in reaction-diffusion equations we refer to [60] and [16, Theorem 5.1.3]. The case with a degenerated matrix of diffusion coefficients, i.e. allowing some of the coefficients to be equal to zero, will be studied in details in the next section.

Remark 4 The nonlinearities are typically satisfying the inequalities $f_u > 0$ and $g_v < 0$. Then, the DDI phenomenon requires the condition $f_u g_v - f_v g_u > 0$, and therefore, f_v and g_u satisfy also the following *compensation condition*

$$g_u(\bar{u},\bar{v})\, f_v(\bar{u},\bar{v}) < 0. \tag{10}$$

Consequently, the signs of f_v and g_u should be opposite and there are only two possibilities

$$\begin{pmatrix} f_u & f_v \\ g_u & g_v \end{pmatrix} = \quad \text{either} \quad \begin{pmatrix} + & - \\ + & - \end{pmatrix} \quad \text{or} \quad \begin{pmatrix} + & + \\ - & - \end{pmatrix},$$

what is called the *activator-inhibitor* system, and the *resource-consumer* system, respectively.

We refer to the Murray book [47] and to the review article [62] and references therein for more information on DDI in the case of two component reaction-diffusion systems and to the paper [58] in the case of several component systems.

Following all these observations, we define *Turing patterns* in the following way:

Definition 3 (Turing Patterns) By Turing patterns we call the solutions of reaction-diffusion equations that are

- stable,
- stationary,
- continuous,
- spatially heterogenous and
- arise due to the Turing instability (DDI) of a constant steady state. See Fig. 4

A system of reaction-diffusion equations may also exhibit a Turing-type Hopf bifurcation, which leads to spatio-temporal oscillations [24]. In general, we can distinguish two types of Turing patterns (depending on the imaginary part of the eigenvalue with positive real part):

- Stationary patterns, when a single eigenvalue becomes positive and the bifurcating solution is a nonconstant steady state. Long-time solutions are stationary, spatially heterogeneous structures.
- Wave patterns, when 2 complex conjugate eigenvalues cross the imaginary axis. It is a supercritical Hopf bifurcation from a homogeneous solution to a stable periodic and nonconstant solution.

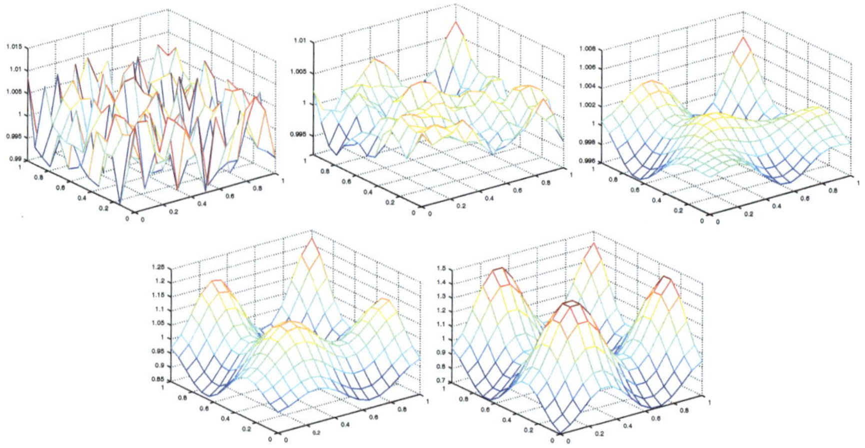

Fig. 4 Growth of patterns in two-dimensional domain after a small spatially random perturbation of a constant stationary solution exhibiting DDI. Each panel presents the solution at a different time step as the time progresses. We observe evolution of a regular, spatially periodic pattern from a randomly perturbed constant state (Courtesy of Dirk Hartmann)

Remark 5 Stationary Turing patterns are possible in a 2-equation system, while for wave bifurcation 3 variables are necessary.

2.2 A Prominent Example: Activator-Inhibitor Model of Gierer and Meinhardt

The most famous realisation of Turing's idea in a mathematical model of biological pattern formation and the first numerical simulation of a system exhibiting Turing patterns is the activator-inhibitor model proposed by Gierer and Meinhardt in 1972 [13]. The model, defined on one-dimensional domain $(0, 1)$ consists of two reaction-diffusion equations

$$\begin{aligned}\frac{\partial}{\partial t}a &= D_a\frac{\partial^2}{\partial x^2}a + \rho_a\frac{a^2}{h} + \sigma_a - \mu_a a,\\ \frac{\partial}{\partial t}h &= D_h\frac{\partial^2}{\partial x^2}h + \rho_h a^2 + \sigma_h - \mu_h h,\end{aligned} \tag{11}$$

completed by homogeneous Neumann boundary conditions and initial conditions.

Variables a and h denote concentrations of two morphogens, called *activator* and *inhibitor*, respectively. The parameters σ_a and σ_h describe *de novo* production, μ_a and μ_h are the rates of degradation and ρ_a and ρ_h the parameters of the activator-inhibitor interactions.

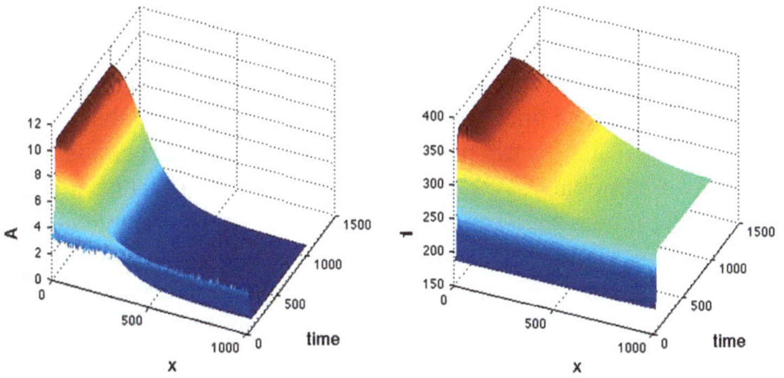

Fig. 5 Simulations of the activator-inhibitor model. Formation of a gradient-like pattern of activator A and inhibitor I from a randomly perturbed homogeneous steady state

The model and several its modifications have been applied to various problems from developmental biology, see e.g. [41–43, 47] and references therein. It aims to explain symmetry breaking and *de novo* pattern formation due to a coupling of a local activation and a long-range inhibition process. The activator promotes the differentiation process and stimulates its own production. The inhibitor acts as a suppressant against the self-enhancing activator to prevent the system from unlimited growth. *Gradients of morphogens* are formed by the DDI mechanism, see Fig. 5 for numerical simulation of gradient-like patterns.

Due to its interesting mathematical features and emerging singularities, the model has attracted also a lot of attention from the side of mathematical analysis. Existence and boundedness of solutions of (11) and its generalisations have been studied by Masuda and Takahashi [40], Li et al. [26], Jiang [19] and Suzuki [61], while existence of spatially heterogenous stationary solutions have been analysed by Takagi [65] in one-dimensional domain, and by Ni and Takagi [53] in multi-dimensional domains with certain symmetries, see also [50–52]. Additionally, the location, profiles, and stability of spike-layer patterns have been investigated, see [48, 69] for a survey. The model leads also to interesting phenomena such as *a collapse of patterns* studied in [64] and stable periodic solutions proven in [20].

2.3 Extensions and Limitations

The majority of theoretical studies in theory of pattern formation due to DDI focus on the analysis of systems of two reaction-diffusion equations. The Turing theory has been also extended to systems of n reacting and diffusing chemicals [58]. However, it consists only in conditions for DDI, while long term behaviour of the model solutions is determined by the character of nonlinearities.

The mechanism of pattern selection, by which one mode is chosen from many admissible modes to grow to heterogeneous steady state, is a complex issue. The dispersion relation, algebraic equation for the growth rate, indicates which modes can grow to determine the final long term pattern. All wave numbers in the region where Re $\lambda(\lambda_k) > 0$ are linearly unstable. The best-known dispersion relation (i.e. known from two-component systems), such as in Fig. 1, determines the bounded range of unstable modes the size of which depends on the bifurcation parameter d_u. Then a certain range of d_u can be determined for which there exists only one unstable mode. This feature of the Turing type systems is usually used in the applications. It allows to select (changing the scaling parameter d_u) the mode which grows to the long-term heterogeneous pattern. If the range of unstable modes is finite but includes more values, then there is a competition between patterning modes. Murray [47] suggests that the mechanism of initiation of pattern formation may determine the mode selected, but so far there is no systematic analysis of this problem. Interestingly, in case of systems with degenerated diffusion, there exists an infinite range of unstable modes [15, 29].

Furthermore, it may happen in a reaction-diffusion system with DDI that all Turing patterns are unstable. A solution may converge to another constant solution, to some discontinuous structures or even exhibit unbounded growth and blow-up. In summary, Turing instability is neither sufficient nor necessary condition for emergence of stable patterns in reaction-diffusion type systems.

Another important problem, related to pattern selection, arises in the models with multiple constant solutions. In such cases, there may exist heterogenous *far from equilibrium* structures and the global behaviour of the solutions cannot be predicted by the properties of the linearised system, e.g. [33, 63].

We observe a variety of possible dynamics depending on the type of nonlinearities. Tight control of the initial conditions is needed to select the desired pattern. In particular, in case of models with DDI, the scaling parameter (corresponding to the domain size and diffusion coefficients) must have proper relative magnitudes.

The concept of Turing instability can be extended also to degenerated systems such as reaction-diffusion-ODE models or integro-differential equations, for example reaction-diffusion equations with nonlocal terms such as the nonlocal Fisher/KPP equation [4] or shadow systems obtained through reduction of the reaction-diffusion model [36]. We will address these issues in the next section.

3 Reaction-Diffusion-ODE Models

In this section we focus on models consisting of a single reaction-diffusion equation coupled to an ordinary differential equation (ODE).

$$u_t = f(u, v), \qquad \text{for} \quad x \in \overline{\Omega}, \quad t > 0, \tag{12}$$

$$v_t = d_v \Delta v + g(u, v) \qquad \text{for} \quad x \in \Omega, \quad t > 0 \tag{13}$$

in a bounded domain $\Omega \subset \mathbb{R}^N$ for $N \geq 1$, with a sufficiently regular boundary $\partial\Omega$, supplemented with the Neumann boundary condition

$$\partial_n v = 0 \qquad \text{for} \quad x \in \partial\Omega, \quad t > 0, \tag{14}$$

and initial data

$$u(x,0) = u_0(x), \qquad v(x,0) = v_0(x). \tag{15}$$

The nonlinearities $f = f(u, v)$ and $g = g(u, v)$ are assumed to be locally Lipschitz continuous functions.

Such systems of equations arise, e.g., from modelling of interactions between cellular or intracellular processes and diffusing growth factors and have already been employed in various biological contexts, see e.g. [17, 21, 29, 32, 34, 54, 67]. In some cases they can be obtained as a homogenisation limit of the models describing coupling of cell-localised processes with cell-to-cell communication via diffusion in a cell assembly [31, 35]. Other initial-boundary value problems, where one reaction-diffusion equation is coupled with a system of ordinary differential equations are discussed e.g. in [8, 10, 39, 46, 68] and in references therein. To understand the role of non-diffusive components in pattern formation process, we focus on systems involving a single reaction-diffusion equation coupled to ODEs. It is an interesting case, since a scalar reaction-diffusion equation cannot exhibit stable spatially heterogenous patterns [7].

In the remainder of this section we present a summary of results on existence, regularity and stability of solutions of reaction-diffusion-ode models. We focus on two-equation system (12)–(13) with nonlinearities $f = f(u, v)$ and $g = g(u, v)$ being arbitrary C^2-functions that satisfy certain growth properties. Since we can rescale the time variable and the functions f and g so that $d_v = 1$, we assume diffusivity to be equal to one.

In these notes, we consider only *non-degenerate stationary solutions*, i.e. such that

$$f_u(\bar{u},\bar{v}) + g_v(\bar{u},\bar{v}) \neq 0, \quad \det\begin{pmatrix} f_u(\bar{u},\bar{v}) & f_v(\bar{u},\bar{v}) \\ g_u(\bar{u},\bar{v}) & g_v(\bar{u},\bar{v}) \end{pmatrix} \neq 0, \quad \text{and} \quad f_u(\bar{u},\bar{v}) \neq 0. \tag{16}$$

The first two conditions in (16) allow us to study the asymptotic stability of a constant stationary solution $(\bar{u}, \bar{v})$ as a solution to the corresponding *kinetic system* of ordinary differential equations

$$\frac{du}{dt} = f(u,v), \qquad \frac{dv}{dt} = g(u,v), \tag{17}$$

by analyzing eigenvalues of the corresponding linearisation matrix. The last condition in (16) guarantees that the equation $f(U, V) = 0$ can be uniquely solved with respect to U in the neighbourhood of $(\bar{u}, \bar{v})$.

3.1 Existence of Solutions

We begin our study of the initial-boundary value problem (12)–(15) by recalling the results on local-in-time existence and uniqueness of solutions for all bounded initial conditions.

Theorem 2 (Local-in-Time Solution) *Assume that* $u_0, v_0 \in L^\infty(\Omega)$. *Then, there exists* $T = T(\|u_0\|_\infty, \|v_0\|_\infty) > 0$ *such that the initial-boundary value problem* (12)–(15) *has a unique local-in-time mild solution* $u, v \in L^\infty\big([0,T], L^\infty(\Omega)\big)$.

We recall that a *mild solution* of problem (12)–(15) is a couple of measurable functions $u, v : [0,T]\times\overline{\Omega} \mapsto \mathbb{R}$ satisfying the following system of integral equations

$$u(x,t) = u_0(x) + \int_0^t f\big(u(x,s), v(x,s)\big)\, ds, \tag{18}$$

$$v(x,t) = S(t)v_0(x) + \int_0^t S(t-s)g\big(u(x,s), v(x,s)\big)\, ds, \tag{19}$$

where S is the semigroup of linear operators generated by Laplacian with homogeneous Neumann boundary conditions.

Since our nonlinearities $f = f(u,v)$ and $g = g(u,v)$ are locally Lipschitz continuous, to construct a local-in-time unique solution of system (18)–(19), it suffices to apply the Banach fixed point theorem. Details of this approach and the proof of Theorem 2 for more general systems of reaction-diffusion equations can be found in the book of Rothe [56, Theorem 1, p. 111], see also [37, Chap. 3] for a construction of nonnegative solutions of a particular reaction-diffusion-ODE problems.

Remark 6 For more regular initial conditions $v_0 \in H^2_N(\Omega) = \{v \in H^2(\Omega) \mid \partial_n v = 0 \text{ on } \partial\Omega\}$, we obtain more regular solutions $(u, v) \in C\big(\overline{\Omega}; H^2_N(\Omega) \times L^\infty(\Omega)\big)$. It can be proven using theory of strongly continuous semigroups [6].

Remark 7 If $u_0 \in C^\alpha(\overline{\Omega})$, $v_0 \in C^{2+\alpha}(\overline{\Omega})$ for some $\alpha \in (0,1)$, and the compatibility condition holds $\partial_n v_0 = 0$ on $\partial\Omega$, then the mild solution of problem (12)–(15) is smooth and satisfies $u \in C^{1,\alpha}\big([0,T]\times\overline{\Omega}\big)$ and $v \in C^{1+\alpha/2, 2+\alpha}\big([0,T]\times\overline{\Omega}\big)$. We refer to [56, Theorem 1, p. 112] as well as to [12] for studies of general reaction-diffusion-ODE systems in Hölder spaces.

3.2 Linearised Stability Principle for Reaction-Diffusion-ODE Problems

The next goal of this section is to provide a linearised stability principle for reaction-diffusion-ODE problems. Here, we are interested in showing that instability of the solutions of linearised system imply instability in the nonlinear system.

We recall the approach applied recently to reaction-diffusion-ODE models by Marciniak-Czochra et al. [38], and based on theory developed originally for analysis of the Euler equation and other fluid dynamics models [11, 25]. This classical method has been also applied by Mulone and Solonnikov [46] to specific reaction-diffusion-ODE problems, however under assumptions which are not satisfied by pattern formation models considered by us.

We formulate our problem as a general evolution equation having a form

$$w_t = \mathscr{L} w + \mathscr{N}(w), \qquad w(0) = w_0 \tag{20}$$

where $\mathscr{L}$ (in our case the Laplace operator with Neumann homogeneous boundary conditions) is a generator of a C_0-semigroup of linear operators $\{e^{t\mathscr{L}}\}_{t\geq 0}$ on a Banach space

$$Z = L^2(\Omega) \times L^2(\Omega),$$

and $\mathscr{N}$ is a nonlinear operator such that $\mathscr{N}(0) = 0$.

Now we recall the notion of nonlinear stability of a trivial solution of equation (20).

Definition 4 (Spectral Gap) $\mathscr{L}$ has a *spectral gap* if there exists a subset of the spectrum $\sigma(\mathscr{L})$, which has a positive real part, separated from zero.

Definition 5 (Nonlinear Stability and Instability) Let (X, Z) be a pair of Banach spaces such that $X \subset Z$ with a dense and continuous embedding. A solution $w \equiv 0$ of the Cauchy problem (20) is *nonlinearly stable* if for every $\varepsilon > 0$, there exists $\delta > 0$ such for $w(0) \in X$ and $\|w(0)\|_Z < \delta$ it holds:

1. There exists a global in time solution to (20) such that $w \in C([0, \infty); X)$;
2. $\|w(t)\|_Z < \varepsilon$ for all $t \in [0, \infty)$.

An equilibrium $w \equiv 0$ that is not stable in the above sense is called *Lyapunov unstable*.

Theorem 3 (Criterion for Nonlinear Instability) *Assume that*

- *The semigroup of linear operators $\{e^{t\mathscr{L}}\}_{t\geq 0}$ on Z satisfies the spectral gap condition, i.e. for every $t > 0$, the spectrum σ of the linear operator $e^{t\mathscr{L}}$ can be decomposed. $\sigma = \sigma(e^{t\mathscr{L}}) = \sigma_- \cup \sigma_+$ with $\sigma_+ \neq \emptyset$, where*

$$\sigma_- \subset \{z \in \mathbb{C} : e^{\kappa t} < |z| < e^{\mu t}\} \quad \textit{and} \quad \sigma_+ \subset \{z \in \mathbb{C} : e^{Mt} < |z| < e^{\Lambda t}\}$$

and

$$-\infty \leq \kappa < \mu < M < \Lambda < \infty \qquad \textit{for some} \quad M > 0.$$

- *The nonlinear term $\mathscr{N}$ satisfies*

$$\|\mathscr{N}(w)\|_Z \leq C_0 \|w\|_X \|w\|_Z \qquad \textit{for all} \quad w \in X \quad \textit{with} \quad \|w\|_X < \rho \tag{21}$$

for some constants $C_0 > 0$ and $\rho > 0$.

Then, the trivial solution $w_0 \equiv 0$ of equation (20) *is nonlinearly unstable.*

The proof of this theorem follows the one by Friedlander et al. [11, Theorem 2.1], with a modification concerning the estimate from below for the constant κ. This extension to infinite κ is important to deal with the operator $\mathscr{L}$ related to our reaction-diffusion problem generating a semigroup of linear operators with unbounded sequence of eigenvalues.

In the next step, we apply this result to problem (12)–(15). Let (U, V) be a bounded stationary solution of problem (12)–(15). Substituting

$$u = U + \tilde{u} \qquad \text{and} \qquad v = V + \tilde{v}$$

into (12)–(13), we obtain an initial-boundary value problem for the perturbation from stationary solution $(\tilde{u}, \tilde{v})$ having a form

$$\frac{\partial}{\partial t}\begin{pmatrix}\tilde{u}\\ \tilde{v}\end{pmatrix} = \mathscr{L}\begin{pmatrix}\tilde{u}\\ \tilde{v}\end{pmatrix} + \mathscr{N}\begin{pmatrix}\tilde{u}\\ \tilde{v}\end{pmatrix}, \tag{22}$$

where the linear operator $\mathscr{L}$ is given by

$$\mathscr{L}\begin{pmatrix}\tilde{u}\\ \tilde{v}\end{pmatrix} = \begin{pmatrix}0\\ \Delta\tilde{v}\end{pmatrix} + \begin{pmatrix} f_u(U,V) & f_v(U,V)\\ g_u(U,V) & g_v(U,V)\end{pmatrix}\begin{pmatrix}\tilde{u}\\ \tilde{v}\end{pmatrix} \tag{23}$$

with homogeneous Neumann boundary condition $\partial_\nu \tilde{v} = 0$.

Lemma 1 (Spectral Mapping Theorem) *The linear operator $\mathscr{L}$ given by* (23) *with $D(\mathscr{L}) = L^2(\Omega) \times W^{2,2}(\Omega)$ generates an analytic semigroup $\{e^{t\mathscr{L}}\}_{t\geq 0}$ of linear operators on $L^2(\Omega) \times L^2(\Omega)$, which satisfies*

$$\sigma(e^{t\mathscr{L}}) \setminus \{0\} = e^{t\sigma(\mathscr{L})} \qquad \textit{for every} \quad t \geq 0. \tag{24}$$

Proof Operator $\mathscr{L}$ is a bounded perturbation of the linear operator

$$\mathscr{L}_0\begin{pmatrix}\tilde{u}\\ \tilde{v}\end{pmatrix} \equiv \begin{pmatrix}0\\ \Delta\tilde{v}\end{pmatrix}$$

with homogeneous Neumann boundary condition for $\tilde{v}$ and domain of definition $D(\mathscr{L}_0) = L^2(\Omega) \times W^{2,2}(\Omega)$, which generates an analytic semigroup on $L^2(\Omega) \times L^2(\Omega)$, and therefore also $\mathscr{L}$ generates an analytic semigroup, for details see e.g. [6, 9].

Every analytic semigroup of linear operators is eventually norm-continuous (see [9, Chap. IV, Corollary 3.12]) and property (24) is satisfied for eventually norm-continuous semigroups (see [9, Chap. IV.3.10]), what completes the proof.

□

Now we show that the nonlinearity in equation (22) satisfies the assumption (21) in Theorem 3.

Lemma 2 *Let* (U, V) *be a bounded stationary solution of* (12)–(15). *The nonlinear operator* $\mathscr{N}$ *given by*

$$\mathscr{N}\begin{pmatrix}\tilde{u}\\ \tilde{v}\end{pmatrix} \equiv \begin{pmatrix} f(U+\tilde{u}, V+\tilde{v}) - f(U,V) \\ g(U+\tilde{u}, V+\tilde{v}) - g(U,V)\end{pmatrix} - \begin{pmatrix} f_u(U,V) & f_v(U,V) \\ g_u(U,V) & g_v(U,V)\end{pmatrix}\begin{pmatrix}\tilde{u}\\ \tilde{v}\end{pmatrix}$$

satisfies

$$\begin{aligned}\|\mathscr{N}(\tilde{u},\tilde{v})\|_{L^2\times L^2} &\le C(\rho, \|U\|_{L^\infty}, \|V\|_{L^\infty})\big(\|\tilde{u}\|_{L^\infty} + \|\tilde{v}\|_{L^\infty}\big)\big(\|\tilde{u}\|_{L^2} + \|\tilde{v}\|_{L^2}\big)\\ &= C(\rho, \|U\|_{L^\infty}, \|V\|_{L^\infty})\|(\tilde{u},\tilde{v})\|_{L^\infty\times L^\infty}\|(\tilde{u},\tilde{v})\|_{L^2\times L^2}\end{aligned}$$

for all $\tilde{u}, \tilde{v} \in L^\infty$ *such that* $\|\tilde{u}\|_{L^\infty} < \rho$ *and* $\|\tilde{v}\|_{L^\infty} < \rho$, *for an arbitrary constant* $\rho > 0$.

Proof Applying Taylor formula to C^2-nonlinearities $f = f(u,v)$ and $g = g(u,v)$.

□

4 Reaction-Diffusion-ODE Models with Turing Instability

4.1 DDI in Two-Component Reaction-Diffusion-ODE Systems

In this section, we provide a systematic analysis of the DDI phenomenon in models of one reaction-diffusion equation coupled to one ODE. The presented material is based on two recent papers by Marciniak-Czochra et al. [37, 38] and a paper by Härting and Marciniak-Czochra [15]. DDI in the case of three component systems with some diffusion coefficients equal to zero has been considered in the recent work [1].

Theorem 4 (DDI in a Reaction-Diffusion-ODE Model) *Let* $(\bar{u}, \bar{v})$ *be a constant stationary solution of problem* (12) *such that*

$$\mathrm{tr} A < 0, \qquad \det A > 0, \tag{25}$$

with

$$A = \begin{pmatrix} f_u(\bar{u},\bar{v}) & f_v(\bar{u},\bar{v}) \\ g_u(\bar{u},\bar{v}) & g_v(\bar{u},\bar{v}) \end{pmatrix}.$$

Then, the autocatalysis condition, i.e.

$$a_{11} = f_u\big(\bar{u},\bar{v}\big) > 0 \tag{26}$$

is a necessary and sufficient condition for DDI of $(\bar{u},\bar{v})$.

Remark 8 Inequality (26) can be interpreted as a self-enhanced growth of u at the steady state $(\bar{u},\bar{v})$.

Proof The proof follows the same lines as the proof of Theorem 1. Conditions (25) provide local stability of $(\bar{u},\bar{v})$ to spatially homogenous perturbations. Then, applying spectral decomposition of the Laplace operator with Neumann homogeneous boundary conditions leads to

$$\tilde{A} = \begin{pmatrix} f_u(\bar{u},\bar{v}) & f_v(\bar{u},\bar{v}) \\ g_u(\bar{u},\bar{v}) & g_v(\bar{u},\bar{v}) - \lambda_k \end{pmatrix}.$$

The resulting characteristic polynomial has solutions with a positive real part if and only if there exists unstable wave numbers, i.e. λ_k such that

$$-\lambda_k a_{11} + \det A < 0.$$

It provides the assertion of the theorem, since $\det A > 0$ and $\lambda_k \to \infty$. □

Remark 9 We observe that the degeneration of the system by setting one diffusion coefficient to zero leads to completely different shape of dispersion relation. In particular, the autocatalysis condition, if satisfied, provides DDI independently on the size of diffusion. Moreover, DDI is related to infinitely many unstable modes. Changing the size of diffusion leads to a shift of the minimal unstable mode (see Fig. 6).

4.2 Nonhomogenous Stationary Solutions and Their Instability

We search for Turing patterns of problem (12)–(14), i.e. solutions of the boundary value problem

$$f(U,V) = 0 \quad \text{for} \quad x \in \overline{\Omega}, \tag{27}$$

$$\Delta V + g(U,V) = 0 \quad \text{for} \quad x \in \Omega, \tag{28}$$

$$\partial_\nu V = 0 \quad \text{for} \quad x \in \partial\Omega. \tag{29}$$

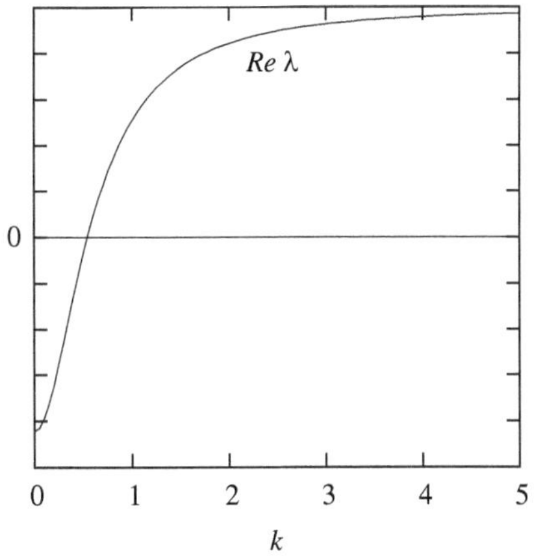

Fig. 6 Dispersion relation for the operator resulting from a linearisation of (12) at $(\bar{u}, \bar{v})$. There exists an infinite range of unstable modes

Solutions of (27)–(29) satisfy the following elliptic problem

$$\begin{aligned} \Delta V + h(V) &= 0 \quad \text{for} \quad x \in \Omega, \\ \partial_\nu V &= 0 \quad \text{for} \quad x \in \partial\Omega, \end{aligned} \tag{30}$$

where

$$h(V) = g\big(k(V), V\big) \qquad \text{and} \qquad U(x) = k(V(x)). \tag{31}$$

Equations of this type, so called *systems with one degree of freedom* have been intensively studied in classical mechanics. For a deeper analysis of the system on one-dimensional domain we refer to the book of Arnold [2, Sect. 12]. We recall this classical approach and apply it to a system in one-dimensional domain in Sect. 5. In case of a multidimensional domain and corresponding solutions to elliptic boundary-value problems we refer to the review paper and the book of Ni [48, 49].

Here, we skip a proof of existence of regular stationary solutions and focus on their properties. In particular, we show that all regular stationary solutions are unstable. We start by showing that *a non-constant solution* (U, V) *touches a constant solution* $(\bar{u}, \bar{v})$.

Lemma 3 *Assume that* (U, V) *is a non-constant regular solution of stationary problem* (27). *Then, there exists* $x_0 \in \overline{\Omega}$, *such that vector* $(\bar{u}, \bar{v}) \equiv \big(U(x_0), V(x_0)\big)$ *is a constant solution of problem* (27).

Proof Integrating the elliptic equation over Ω and using the Neumann boundary condition, we obtain $\int_\Omega g\big(U(x), V(x)\big)\, dx = 0$. Due to continuity of U and V, we conclude that there exists $x_0 \in \overline{\Omega}$ such that $g\big(U(x_0), V(x_0)\big) = 0$ and hence, $f\big(U(x_0), V(x_0)\big) = 0$. □

Now we can formulate our main result for problem (12)–(15) under the autocatalysis condition required for DDI.

Theorem 5 (Instability of Regular Solutions) *Let (U, V) be a regular solution of problem* (27) *satisfying autocatalysis condition*

$$f_u\big(U(x), V(x)\big) > 0 \quad \textit{for all } x \in \overline{\Omega}. \tag{32}$$

Then, (U, V) is an unstable solution of problem (12)–(15).

The instability results from Theorem 4 and Theorem 5 can be summarised in the following way: *In case of a single reaction-diffusion equation coupled to ODE the same mechanism which destabilizes constant solutions, destabilizes also all non-constant regular stationary solutions. Consequently, no Turing patterns can be stable.*

In the proof of Theorem 5 the following general result on a family of compact operators is applied.

Theorem 6 (Analytic Fredholm Theorem) *Assume that H is a Hilbert space and denote by $L(H)$ the Banach space of all bounded linear operators acting on H. For an open connected set $D \subset \mathbb{C}$, let $f : D \to L(H)$ be an analytic operator-valued function such that $f(z)$ is compact for each $z \in D$. Then, either*

(a) *$(I - f(z))^{-1}$ exists for no $z \in D$, or*
(b) *$(I - f(z))^{-1}$ exists for all $z \in D \setminus S$, where S is a discrete subset of D (i.e. a set which has no limit points in D).*

A proof of this analytic Fredholm theorem can be found in the book by Reed and Simon [55, Theorem VI.14].

We present below a sketch of the proof. For details of the proof we refer to [38].

Proof (of Theorem 5) Let $(U(x), V(x))$ be a regular stationary solution to problem (12)–(15). To show instability of $(U(x), V(x))$ using Theorem 3 combined with Lemmas 1 and 2, it suffices to study the spectrum $\sigma(\mathscr{L})$ of the linear operator $\mathscr{L}$ defined by formula (23) with $D(\mathscr{L}) = L^2(\Omega) \times W^{2,2}(\Omega)$.

To show that $\mathscr{L}$ has a *spectral gap*, we prove that the spectrum $\sigma(\mathscr{L}) \subset \mathbb{C}$ consists of two parts:

1. All numbers from the interval $[\lambda_0, \Lambda_0]$, where

$$\lambda_0 = \inf_{x\in\overline{\Omega}} f_u\big(U(x), V(x)\big) > 0 \qquad \text{and} \qquad \Lambda_0 = \sup_{x\in\overline{\Omega}} f_u\big(U(x), V(x)\big) > 0, \tag{33}$$

   where positivity of λ_0 is a consequence of the autocatalysis condition (26);
2. A set of eigenvalues of $(\mathscr{L}, D(\mathscr{L}))$ which are isolated points in $\mathbb{C}$, see Fig. 7.

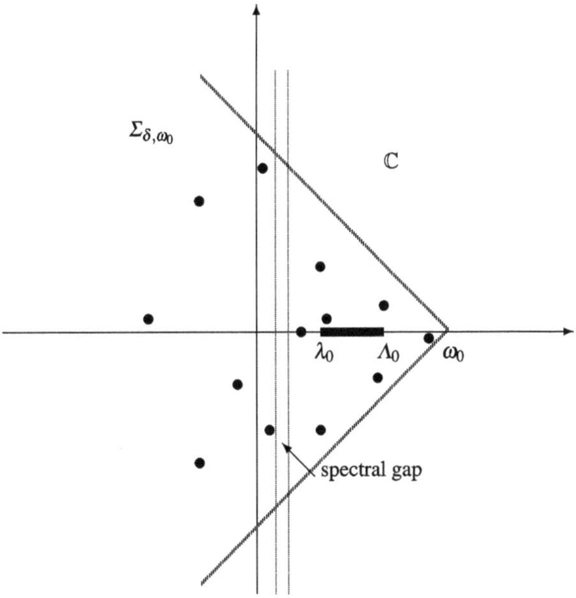

Fig. 7 The spectrum $\sigma(\mathscr{L})$ is marked by *thick dots* and by the interval $[\lambda_0, \Lambda_0]$ in the sector Σ_{δ,ω_0}. The spectral gap is represented by the strip $\{\lambda \in \mathbb{C} : \mu \leq \mathrm{Re}\,\lambda \leq M\}$ without elements of $\sigma(\mathscr{L})$

Part I: Interval $[\lambda_0, \Lambda_0]$. We show that for each $\lambda \in [\lambda_0, \Lambda_0]$ the operator

$$\mathscr{L} - \lambda I : L^2(\Omega) \times W^{2,2}(\Omega) \to L^2(\Omega) \times L^2(\Omega)$$

defined by formula

$$(\mathscr{L} - \lambda I)(\varphi, \psi) = \big((f_u - \lambda)\varphi + f_v\psi,\ \Delta\psi + g_u\varphi + (g_v - \lambda)\psi\big),$$

where the derivatives are evaluated at the stationary solution $\big(U(x), V(x)\big)$, cannot have a bounded inverse.

We show it by contradiction using eliptic regularity of our problem: Suppose that $(\mathscr{L}-\lambda I)^{-1}$ exists and is bounded. Then, it holds for a constant $K = \|(\mathscr{L}-\lambda I)^{-1}\|$ that

$$\begin{aligned} &\|\varphi\|_{L^2(\Omega)} + \|\psi\|_{W^{2,2}(\Omega)} \\ &\quad \leq K\big(\|(f_u - \lambda)\varphi + f_v\psi\|_{L^2(\Omega)} + \|\Delta\psi + g_u\varphi + (g_v - \lambda)\psi\|_{L^2(\Omega)}\big). \end{aligned} \tag{34}$$

A contradiction is obtained by showing that inequality (34) does not hold for all $(\varphi, \psi) \in L^2(\Omega) \times W^{2,2}(\Omega)$. Consequently, we obtain that each $\lambda \in [\lambda_0, \Lambda_0]$ belongs to $\sigma(\mathscr{L})$.

Part II: Eigenvalues. To show that the remainder of the spectrum of $\big(\mathscr{L}, D(\mathscr{L})\big)$ consists of a discrete set of eigenvalues $\{\lambda_n\}_{n=1}^{\infty} \subset \mathbb{C} \setminus [\lambda_0, \Lambda_0]$, we consider the resolvent equations

$$(f_u - \lambda)\varphi + f_v\psi = F \qquad \text{in} \quad \overline{\Omega} \tag{35}$$

$$\Delta\psi + g_u\varphi + (g_v - \lambda)\psi = G \qquad \text{in} \quad \Omega \tag{36}$$

$$\partial_\nu \psi = 0 \qquad \text{on} \quad \partial\Omega, \tag{37}$$

with arbitrary $F, G \in L^2(\Omega)$. Solving equation (35) with respect to φ and substituting the resulting expression

$$\varphi = (F - f_v\psi)/(f_u - \lambda) \in L^2(\Omega)$$

into (36), we obtain for any $\lambda \in \mathbb{C} \setminus [\lambda_0, \Lambda_0]$ the boundary value problem

$$\Delta\psi + q(\lambda)\psi = p(\lambda) \qquad \text{for} \quad x \in \Omega, \tag{38}$$

$$\partial_\nu \psi = 0 \qquad \text{for} \quad x \in \partial\Omega, \tag{39}$$

where

$$q(\lambda) = q(x, \lambda) = -\frac{g_u f_v}{f_u - \lambda} + g_v - \lambda,$$

$$p(\lambda) = p(x, \lambda) = G - \frac{g_u F}{f_u - \lambda}. \tag{40}$$

For a fixed $\lambda \in \mathbb{C}\setminus[\lambda_0, \Lambda_0]$, by Theorem 6, either the inhomogeneous problem (38)–(39) has a unique solution or else the homogeneous boundary value problem

$$\Delta\psi + q(\lambda)\psi = 0 \qquad \text{for} \quad x \in \Omega, \tag{41}$$

$$\partial_\nu \psi = 0 \qquad \text{for} \quad x \in \partial\Omega, \tag{42}$$

has a nontrivial solution ψ. Hence, it suffices to consider those $\lambda \in \mathbb{C} \setminus [\lambda_0, \Lambda_0]$, for which problem (41)–(42) has a nontrivial solution.

It can be proven that the set $\sigma(\mathscr{L}) \setminus [\lambda_0, \Lambda_0]$ consists only of isolated eigenvalues of $\mathscr{L}$ by rewriting (41)–(42) in the form

$$\psi = G\big[-(q(\lambda) + \ell)\psi\big] \equiv R(\lambda)\psi, \tag{43}$$

and applying Theorem 6 to the resulting operator $R(\lambda) : L^2(\Omega) \to L^2(\Omega)$.

$G = (\Delta - \ell I)^{-1}$ in (43) is an operator supplemented with Neumann homogeneous boundary conditions, $\ell \in \mathbb{R}$ is a fixed number different from eigenvalues of Laplacian with Neumann homogeneous boundary condition. For each $\lambda \in \mathbb{C} \setminus [\lambda_0, \Lambda_0]$, $R(\lambda)$ is a compact operator as the superposition of the compact operator G and of the continuous multiplication operator with the function $q(\lambda) + \ell \in L^\infty(\Omega)$. The mapping $\lambda \mapsto R(\lambda)$ from an open set $\mathbb{C} \setminus [\lambda_0, \Lambda_0]$ into the Banach space of linear compact operators is analytic, which can be easily seen using the explicit form of $q(\lambda)$ in (40). Showing invertibility of $I - R(\lambda)$ for some $\lambda \in \mathbb{C} \setminus [\lambda_0, \Lambda_0]$, we exclude the case (a) in Theorem 6. Finally, we conclude that the set $\sigma(\mathscr{L}) \setminus [\lambda_0, \Lambda_0]$ consists of isolated points.

Part III: Spectral gap. By Lemma 1, there exists a number $\omega_0 \geq 0$ such that the operator $\big(\mathscr{L} - \omega_0 I, D(\mathscr{L})\big)$ generates a bounded analytic semigroup on $L^2(\Omega) \times L^2(\Omega)$, hence, this is a sectorial operator, see [9, Chap. II, Theorem 4.6]. In particular, there exists $\delta \in (0, \pi/2]$ such that $\sigma(\mathscr{L}) \subset \Sigma_{\delta,\omega_0} \equiv \{\lambda \in \mathbb{C} : |\arg\ (\lambda - \omega_0)| \geq \pi/2 + \delta\}$, see Fig. 7. The part of the spectrum $\sigma(\mathscr{L})$ in $\Sigma_{\delta,\omega_0} \cap \{\lambda \in \mathbb{C} : \operatorname{Re} \lambda > 0\}$ consists of all numbers from the interval $[\lambda_0, \Lambda_0]$ with $\lambda_0 > 0$ and of a discrete sequence of eigenvalues with accumulation points restricted to the interval $[\lambda_0, \Lambda_0]$. Thus, there exist infinitely many $0 \leq \mu < M \leq \lambda_0$, for which the spectrum $\sigma(\mathscr{L})$ can be decomposed as required in Theorem 3 using (24). Due to the relation $|e^z| = e^{\operatorname{Re} z}$ for every $z \in \mathbb{C}$, the spectral gap condition holds if for every $\lambda \in \sigma(\mathscr{L})$, either $\operatorname{Re} \lambda \in (\kappa, \mu)$ or $\operatorname{Re} \lambda \in (M, \Lambda)$. □

Interestingly, in case of several models from applications the derivative $f_u(U(x), V(x))$ is constant along regular stationary solutions. In such case the spectrum $\sigma(\mathscr{L})$ can be closer characterised for systems satisfying additionally a *compensation condition*, which is an extension of condition (10) to x-dependent solutions.

Definition 6 A regular solution (U, V) of problem (27) satisfy the *compensation condition* if

$$g_u\big(U(x), V(x)\big) f_v\big(U(x), V(x)\big) < 0 \qquad \text{for all} \quad x \in \overline{\Omega}. \tag{44}$$

Corollary 1 *Let (U, V) be a regular solution of problem* (27). *Assume that there exists a constant $\lambda_0 > 0$ such that*

$$0 < \lambda_0 = f_u\big(U(x), V(x)\big) \qquad \textit{for all} \quad x \in \overline{\Omega}, \tag{45}$$

and that the compensation condition (44) *holds at the stationary solution (U, V). Then, the spectrum $\sigma(\mathscr{L})$ contains λ_0, which is an element of the continuous spectrum of $\mathscr{L}$ and a sequence of real eigenvalues $\{\lambda_n\}_{n=1}^\infty$ of $\mathscr{L}$ converging towards λ_0.*

For the proof of this result we refer to [38].

4.3 Discontinuous Stationary Solutions

The initial-boundary value problem (12)–(15) may also have non-regular steady states in the case when the equation $f(U,V) = 0$ is not uniquely solvable. Choosing different branches of solutions of the equation $f\big(U(x), V(x)\big) = 0$, we obtain the relation $U(x) = k\big(V(x)\big)$ with a discontinuous, piecewise C^1-function k.

Definition 7 A couple $\big(U, V\big) \in L^\infty(\Omega) \times W^{1,2}(\Omega)$ is a *weak solution* of problem (27)–(29) if the equation $f\big(U(x), V(x)\big) = 0$ is satisfied for almost all $x \in \Omega$ and if

$$-\int_\Omega \nabla V(x) \cdot \nabla \varphi(x)\, dx + \int_\Omega g\big(U(x), V(x)\big)\varphi(x)\, dx = 0$$

for all test functions $\varphi \in W^{1,2}(\Omega)$.

For existence of such discontinuous solutions we refer to classical works [3, 45, 57] as well as to a recent paper [37, Theorem 2.9] for explanation how to construct such solutions to one dimensional model of cancerogenesis presented in Sect. 4.4 using the phase portrait analysis. In Sect. 5 we will apply the method to a reaction-diffusion-ODE model with hysteresis.

As next, we aim to find a counterpart of the autocatalysis condition (26), which leads to instability of weak (including discontinuous) stationary solutions.

The following two corollaries can be proven in the same way as Theorem 5.

Corollary 2 *Assume that (U, V) is a weak bounded solution of problem* (27)–(29) *satisfying the following counterpart of the autocatalysis condition*

$$\text{Range } f_u(U,V) \equiv \{f_u(U(x), V(x)) \,:\, x \in \overline{\Omega}\} \subset [\lambda_0, \Lambda_0] \tag{46}$$

for some constants $0 < \lambda_0 \le \Lambda_0 < \infty$. Suppose, moreover, that there exists $x_0 \in \Omega$ such that $f_u(U, V)$ is continuous in a neighbourhood of x_0. Then, (U, V) is an unstable solution the initial-boundary value problem (12)–(15).

The following corollary covers the case in which the generalised autocatalysis condition (46) is not satisfied.

Corollary 3 (Instability of Weak Solutions) *Assume that the nonlinear term in equation* (12) *satisfies $f(0, v) = 0$ for all $v \in \mathbb{R}$. Suppose that (U, V) is a weak bounded solution of problem* (27)–(29) *with the following property: there exist constants $0 < \lambda_0 < \Lambda_0 < \infty$ such that*

$$\lambda_0 \le f_u\big(U(x), V(x)\big) \le \Lambda_0 \qquad \textit{for all} \quad x \in \Omega, \quad \textit{where} \quad U(x) \ne 0. \tag{47}$$

Suppose, moreover, that there exists $x_0 \in \Omega$ such that $U(x_0) \ne 0$ and that the functions $U = U(x)$ as well as $f_u(U, V)$ are continuous in a neighbourhood of x_0. Then, (U, V) is an unstable solution the initial-boundary value problem (12)–(15).

Remark 10 A typical nonlinearity satisfying the assumptions of Corollary 3 has the form $f(u, v) = r(u, v)u$ and appears in the models where the unknown variable u evolves according to the Malthusian law with a growth rate r depending on u and other variables of the model.

4.4 Examples of Reaction-Diffusion-ODE Models with Autocatalysis

We present here examples of models from mathematical biology, to which the presented theory applies. We show that both, the autocatalysis and the compensation, conditions are satisfied. Consequently, we obtain instability of all positive regular stationary solutions of the considered models.

4.4.1 Activator-Inhibitor System with a Nondiffusing Activator

Let us start with a generic version of activator-inhibitor model (11) introduced in Sect. 2, i.e. with zero source terms but general activation/inhibition nonlinearities, i.e.

$$f(u, v) = -u + \frac{u^p}{v^q},$$

$$g(u, v) = -v + \frac{u^r}{v^s},$$

where the exponents satisfy $p > 1, q, r > 0$ and $s \geq 0$.

Remark 11 Such generic activator-inhibitor model can be characterised by two numbers essential for its dynamics.

- The *net self-activation index* $\rho_A \equiv (p-1)/r$ comparing the strength of self-activation of the activator with the cross-activation of the inhibitor. If ρ_A is large, then the net growth rate of the activator is large in spite of the inhibitor.
- The *net cross-inhibition index* $\rho_I \equiv q/(s+1)$ comparing how strongly the inhibitor suppresses the production of the activator with that of itself. The large value of ρ_I means that the production of the activator is strongly suppressed by the inhibitor.

Guided by biological interpretation of the model, it is assumed that

$$0 < \frac{p-1}{r} < \frac{q}{s+1}, \tag{48}$$

In the remainder of this section, we assume no diffusion of activator and consider the following system of equations

$$u_t = f(u,v) \quad \text{for} \quad x \in \overline{\Omega},\ t > 0, \tag{49}$$

$$v_t = \Delta v + g(u,v) \quad \text{for} \quad x \in \Omega,\ t > 0, \tag{50}$$

supplemented with positive initial data and with the zero-flux boundary condition for $v = v(x,t)$.

Every positive regular stationary solution satisfies the relation

$$U^{p-1} = V^q,$$

and hence, the function $V = V(x)$ is a solution of the boundary value problem

$$\Delta V - V + V^Q = 0 \quad \text{for} \quad x \in \Omega, \tag{51}$$

$$\partial_\nu V = 0 \quad \text{for} \quad x \in \partial\Omega, \tag{52}$$

where $Q = qr/(p-1) - s$.

This problem with $Q > 1$ was considered in a series of papers, e.g. [27, 51, 52, 65]. In particular, for $1 < Q < \frac{N+2}{N-2}$ for $N \geq 3$, and $1 < Q < \infty$ for $N = 1, 2$, existence of a positive solution V for a sufficiently large domain Ω has been proven, see the review article [48] for further comments and references.

Since derivative f_u, evaluated at a stationary solution, does not depend on x, we can easily check its sign

$$f_u(U,V) = -1 + p\frac{U^{p-1}}{V^q} = -1 + p > 0 \qquad \text{for} \quad U^{p-1} = V^q.$$

It follows that $\lambda_0 = f_u\big(U(x), V(x)\big)$ does not depend on x and the autocatalysis condition (26) is satisfied at all regular stationary solutions of system (49)–(50).

Also the compensation assumption (44) can be checked directly

$$f_v(U,V)\cdot g_u(U,V) = \left(-q\frac{U^p}{V^{q+1}}\right)\left(r\frac{U^{r-1}}{V^s}\right) = -rq\frac{U^{p+r-1}}{V^{s+q+1}} < 0 \quad \text{for all} \quad x \in \overline{\Omega},$$

for every positive stationary solution (U, V).

Applying Corollary 3, we conclude about instability of all positive regular stationary solutions of the activator-inhibitor model with a non-diffusing activator.

4.4.2 Gray-Scott Model

As a next example we consider a model proposed to describe pattern formation in chemical reactions [14] with non-diffusing activator

$$u_t = -(B+k)u + u^2 v \quad \text{for} \quad x \in \overline{\Omega},\ t > 0,$$
$$v_t = \Delta v - u^2 v + B(1-v) \quad \text{for} \quad x \in \Omega,\ t > 0,$$

where B and k are positive constants, with the zero-flux boundary condition for v and with nonnegative initial conditions.

Regular positive stationary solutions (U, V) satisfy the following boundary value problem

$$U = (B+k)/V,$$
$$\Delta V - BV - \frac{(B+k)^2}{V} + B = 0 \quad \text{for} \quad x \in \Omega,$$
$$\partial_\nu V = 0. \quad \text{for} \quad x \in \partial\Omega.$$

Similarly as in the previous example, $f_u\big(U(x), V(x)\big)$ evaluated at a stationary solution, i.e. satisfying the condition $U = (B+k)/V$, is independent of x. It holds

$$\lambda_0 = f_u\big(U(x), V(x)\big) = -(B+k)+2U(x)V(x) = B+k > 0 \qquad \text{for all} \quad x \in \Omega.$$

and the autocatalysis assumption (32) is satisfied.

Since the compensation assumption (44) is also satisfied at a positive solution (U, V) is also valid due to the following calculation

$$f_v(U,V) \cdot g_u(U,V) = -2U^3 V < 0$$

for all positive stationary solutions, we can apply Corollary 5 to conclude about instability of all Turing patterns.

4.4.3 Model of Early Carcinogenesis

Finally, we introduce an example of a reaction-diffusion-ODE system describing growth of a spatially-distributed cell population which proliferation is controlled by diffusing growth factors. We consider a system having a form of one ordinary differential equation coupled to one reaction-diffusion equation

$$u_t = \Big(\frac{auw}{1+uw} - d_c\Big)u \qquad \text{for } x \in \overline{\Omega},\ t > 0, \tag{53}$$

$$w_t = D\Delta w - d_g w - \frac{d_b}{d_b + d} u^2 w + \kappa_0 \qquad \text{for } x \in \Omega,\ t > 0. \tag{54}$$

supplemented with zero-flux boundary conditions for the function w

$$\partial_n w = 0 \qquad \text{for} \quad x \in \partial\Omega \quad \text{and} \quad t > 0 \tag{55}$$

and with nonnegative initial conditions.

$$u(x,0) = u_0(x), \quad w(x,0) = w_0(x). \tag{56}$$

All model parameters $a, d_c, d_b, d_g, d, D, \kappa_0$ are positive constants.

This system is a reduction of a three-equation model introduced and applied to study various biological scenarios related to early lung cancer development in the series of papers [32–34]. For the proof of model reduction using quasi-steady state approximation we refer to [38]. The three-equation model was analysed in [37], and the reduced model was investigated in [15]. A stochastic version of the model has been investigated in [5].

The autocatalysis assumption (45) and the compensation assumption (44) are satisfied by simple calculations, which are analogous to those in previous examples. As a consequence, all nonnegative stationary solutions of system (53)–(54) are unstable due to Theorem 5 and Corollary 3. A corresponding result for the original three-equation model has been proven in [37].

4.5 Dynamical Spike Patterns in Reaction-Diffusion-ODE Models with DDI

In the previous sections we showed that in reaction-diffusion-ODE models with a single diffusing species, all Turing patterns are unstable. Then, beside a possibility of pattern collapse, two scenarios may be observed: (1) either the model solutions tend to discontinuous *far from equilibrium* patterns, or (2) the emerging spatially heterogeneous structures have a dynamical character. The mechanism leading to emergence of discontinuous patterns will be presented in the next section. Dynamical patterns have been observed in the models for which the instability result holds also for discontinuous patterns, as in Corollary 3. Simulations of different models of this form indicate formation of dynamical, multimodal and apparently irregular structures, the shape of which depends strongly on initial conditions [15, 33, 54].

An example of such dynamics is given by the cancerogenesis model (53)–(54), see Fig. 8 for simulation of the model showing emergence of one or two spike solutions depending on the size of diffusion (scaling coefficient). Simulations suggest that the number of peaks corresponds to the minimal unstable mode for

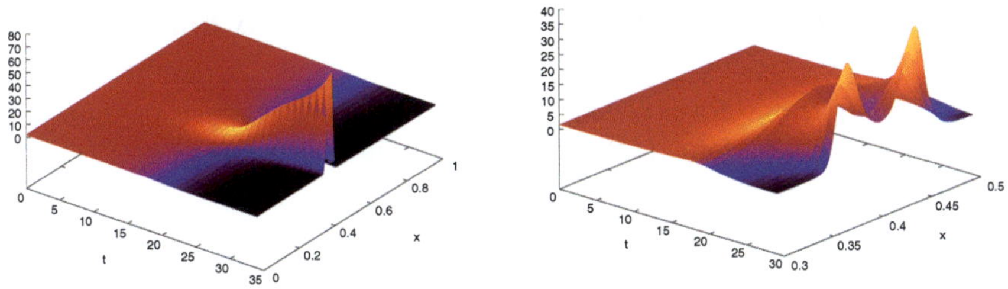

Fig. 8 Numerical simulation of reaction-diffusion-ode model (53)–(54) defined on one-dimensional domain. We observe growth of dynamical spike patterns in the non-diffusing component u. The two solutions are obtained for the same model parameters and initial data but different diffusion coefficient. We observe a shape of final pattern corresponding to the smallest unstable mode (Courtesy of Steffen Härting)

initial data in being a small perturbation of the constant steady state. However, for large perturbations a variety of patterns may be induced depending of the shape of initial data. Systematic numerical study of this phenomenon is provided in [15].

Analysis of the shadow-type limit of the model shows possibility of unbounded growth and mass concentration in [36]. Since solutions of the system remain uniformly bounded in absence of diffusion, we call this phenomenon *diffusion-driven unbounded growth*. Interestingly, in some cases DDI may even lead to a finite time blow-up of solutions (*diffusion-driven blow-up*) [36].

5 Reaction-Diffusion-ODE Models with Bistability and Hysteresis

In this section we show a different mechanism of pattern formation in reaction-diffusion-ODE models, which leads to formation of *far from equilibrium* patterns. The phenomenon is based on existence of multiple quasi-stationary solutions in the ODE subsystem. Diffusion tries to average different states and is the cause of spatio-temporal patterns.

These concepts proved to be basis for the explanation of the morphogenesis of *Hydra* [30], dorso-ventral patterning in *Drosophila* [67] as well as for modelling of formation of growth patterns in populations of micro-organisms [18]. The patterns observed in such models are not Turing patterns and the system does not need to exhibit DDI. In most cases its constant steady states do not change stability.

The material of this section is based on the recent results by Köthe and Marciniak-Czochra [22, 23] and Marciniak-Czochra et al. [39].

5.1 Configuration of the Bistability and Hysteresis Kinetics

We present hysteresis-driven pattern formation on example of a reaction-diffusion-ODE system of the form (12)–(15) with diffusion coefficient $d_v = \frac{1}{\gamma}$ and nonlinearities

$$f(u,v) = v - p(u),$$
$$g(u,v) = \alpha u - \beta v, \tag{57}$$

where α, β are positive constants and

$$p(u) = a_2 u^3 + a_1 u^2 + a_0 u$$

is a polynomial of degree three. The parameters are chosen so that the nullclines $f(u,v) = 0$ and $g(u,v) = 0$ have three intersection points with nonnegative coordinates.

$$S_0 = (0,0), \qquad S_1 = (u_1, v_1) \quad \text{and} \quad S_2 = (u_2, v_2). \tag{58}$$

For simplicity of presentation, we restrict our considerations in this section to one-dimensional domain $(0,1)$.

Our aim is to understand the role of *bistability* and *hysteresis* in pattern formation. Therefore, we analyze two cases of the kinetic functions of the form (57) with monotone and non-monotone dependence $v = p(u)$:

Case 1 Monotone increasing $p(u)$ (see Fig. 9 left panel).
Case 2 Non-monotone S-shaped $p(u)$ (see Fig. 9 right panel).

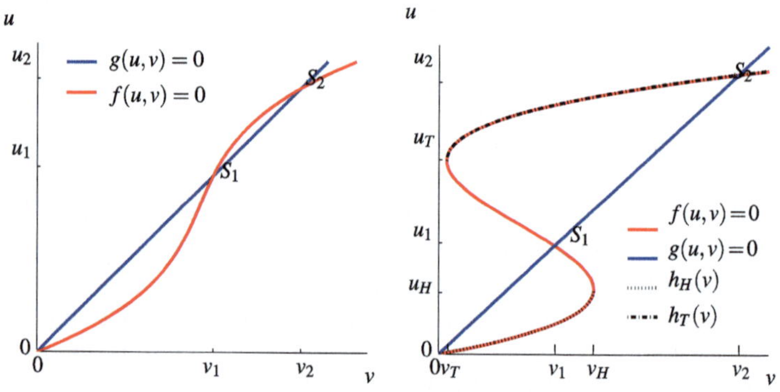

Fig. 9 Configurations of the zero sets of kinetic functions with monotone (on the *left hand-side*) and non-monotone (*hysteresis*) (on the *right hand-side*) dependence $v = p(u)$ corresponding to Case (1) and (2), respectively. In both cases there exist three intersection points $S_0 = (0,0)$, S_1 and S_2 giving three constant stationary solutions of the model

In both cases we obtain bistability in the kinetic system. Case 2 exhibits additionally the *hysteresis* effect. In this case we denote by $H = (v_H, u_H) = (p(u_H), u_H)$ the local maximum of $u \mapsto p(u)$ and by $T = (v_T, u_T) = (p(u_T), u_T)$ the local minimum of $u \mapsto p(u)$. Moreover, we assume that the coordinates of H and T are positive and $\lim_{u \to +\infty} p(u) = +\infty$, see Fig. 9.

We start our investigations with analysis of the *kinetic system.*

Lemma 4 *Consider the kinetic system*

$$u_t = f(u, v), \qquad v_t = g(u, v). \tag{59}$$

with nonlinearities given by (57). *The stationary solutions S_0 and S_2, as defined in* (58), *are asymptotically stable and S_1 is a saddle.*

Proof We calculate Jacobian matrix A at a steady state $(\bar{u}, \bar{v})$:

$$A = \begin{pmatrix} \partial_u f(\bar{u}, \bar{v}) & \partial_v f(\bar{u}, \bar{v}) \\ \partial_u g(\bar{u}, \bar{v}) & \partial_v g(\bar{u}, \bar{v}) \end{pmatrix} = \begin{pmatrix} -p'(\bar{u}) & 1 \\ \alpha & -\beta \end{pmatrix}.$$

As α/β is the slope of $g(u, v) = 0$ solved with respect to u, we obtain

$$p'(0) > \alpha/\beta > 0,$$
$$p'(u_2) > \alpha/\beta > 0.$$

Hence, $\det A(S_0) > 0$ and $\det A(S_2) > 0$, whereas the trace is negative at both states. Therefore, linearisation of the model at S_0 and S_2 has only negative eigenvalues.

For S_1 we have to distinguish between the two cases. In Case 2, it holds $p'(u_1) < 0$, hence $\det A(S_1) < 0$. In Case 1 it holds $p'(u_1) > 0$, but $p'(u_1) < \alpha/\beta$, thus $\det A(S_1) < 0$. Consequently, linearisation of the system at S_1 has one positive and one negative eigenvalue. □

Remark 12 There exists a stable manifold W^s providing a separatrix for the kinetic system (59). Consequently, the solutions tend either to S_0 or S_2, depending on the position of initial data in respect to W^s. For details we refer to [22, 39].

Remark 13 Model (12)–(15) with nonlinearities (57) does not satisfy autocatalysis condition neither in S_0 nor in S_2. Hence, Theorem 4 yields that the model does not exhibit DDI and does not belong to the class of pattern formation models considered in the previous section.

5.2 *Instability of all Spatially Heterogenous Solutions in the Model Without Hysteresis*

When discussing systems with multistability, we face a question whether the bistability without a hysteresis effect is sufficient for creation of stable patterns. Now, following [23], we show that the model with monotone kinetics behaves in this respect as a scalar reaction-diffusion equation, i.e. there exist no stable stationary spatially heterogenous solutions.

Theorem 7 *There exists no stable spatially heterogeneous stationary solution* (U, V) *of the generic model* (12)–(15) *with nonlinearities* (57) *satisfying Case 1.*

First, we recall the Sturm comparison principle, which we will apply in the proof of this theorem.

Proposition 2 (Sturm Comparison Principle) *Let* ϕ_1 *and* ϕ_2 *be nontrivial solutions of the equations*

$$Dw_{xx} + q_1(x)w = 0 \quad \text{and}$$
$$Dw_{xx} + q_2(x)w = 0,$$

respectively, for $x \in (0, 1)$ *and the diffusion coefficient* $D > 0$. *We assume that the functions* q_1 *and* q_2 *are continuous on* $[0, 1]$ *and that the inequality*

$$q_1(x) \leq q_2(x)$$

holds for all $x \in [0, 1]$. *Then between any two consecutive zeros* x_1 *and* x_2 *of* ϕ_1, *there exists at least one zero of* ϕ_2 *unless* $q_1(x) \equiv q_2(x)$ *on* $[0, 1]$.

Proof (of Theorem 7) We consider system (12)–(15), with nonlinearities (57) and diffusion coefficient $d_v = \frac{1}{\gamma}$, linearised at $(U(x), V(x))$

$$\begin{pmatrix} \tilde{u}_t \\ \tilde{v}_t \end{pmatrix} = \begin{pmatrix} 0 \\ \frac{1}{\gamma}\tilde{v}_{xx} \end{pmatrix} + \begin{pmatrix} -p'(U(x)) & 1 \\ \alpha & -\beta \end{pmatrix} \begin{pmatrix} \tilde{u} \\ \tilde{v} \end{pmatrix} =: \mathscr{L} \begin{pmatrix} \tilde{u} \\ \tilde{v} \end{pmatrix}$$

with boundary conditions $\tilde{v}_x(0) = \tilde{v}_x(1) = 0$.

The eigenvalue equation

$$\mathscr{L} \begin{pmatrix} \varphi \\ \psi \end{pmatrix} = \lambda \begin{pmatrix} \varphi \\ \psi \end{pmatrix}$$

with boundary condition $\psi_x(0) = \psi_x(1) = 0$ reads

$$\psi - (p'(U(x)) + \lambda)\varphi = 0 \tag{60}$$

$$\frac{1}{\gamma}\psi_{xx} - (\beta + \lambda)\psi + \alpha\varphi = 0. \tag{61}$$

We will show that all nonconstant stationary solutions are linearly unstable.

We define a depending on λ operator

$$A(\lambda) : \psi \mapsto \frac{1}{\gamma}\psi_{xx} + r(\lambda, x)\psi$$

with

$$r(\lambda, x) := \frac{\alpha}{\lambda + p'(U(x))} - \beta,$$

and domain of definition

$$\mathscr{D}\big(A(\lambda)\big) = \{\psi \in C^2\big([0,1]\big) \mid \psi_x(0) = \psi_x(1) = 0\}.$$

By assumptions of Case 1, polynomial p is monotone increasing. Denoting

$$K = \min_{x\in[0,1]} p'(U(x)) > 0,$$

we observe that for $\lambda > -K$, the denominator $\lambda + p'(U(x))$ is positive and linearly growing in λ. Therefore, $r(\lambda, x)$ is decreasing in λ and it holds

$$\frac{\alpha}{\lambda + p'(U(x))} - \beta \leq \frac{\alpha}{p'(U(x))} - \beta = r(0, x) \leq \frac{\alpha}{K} - \beta =: C.$$

C is positive, because otherwise for $\alpha/\beta < K$ the graph of p would be entirely on one side of $g = 0$, which contradicts the assumption of the existence of three intersection points.

Hence, we obtain that there exists $C > 0$ independent of $\lambda \geq 0$ such that

$$\big|r(\lambda, x)\big| \leq C$$

for all $x \in [0, 1]$. Consequently, the eigenvalue problem (60)–(61) is equivalent to the equation

$$A(\lambda)\psi = \lambda\psi.$$

To show instability of all patterns, we denote by $\mu_0(\lambda)$ the largest eigenvalue of the Neumann problem

$$A(\lambda)\psi = \mu(\lambda)\psi \quad \psi_x(0) = \psi_x(1) = 0,$$

and by $\nu_0(\lambda)$ the largest eigenvalue of the Dirichlet problem

$$A(\lambda)\psi = \nu(\lambda)\psi \quad \psi(0) = \psi(1) = 0.$$

Using Proposition 2 we can show that

$$\mu_0(\lambda) > \nu_0(\lambda)$$

holds. Indeed, $\mu_0(\lambda) \leq \nu_0(\lambda)$ yields

$$q_2(x) := r(\lambda, x) - \mu_0(\lambda) \geq r(\lambda, x) - \nu_0(\lambda) =: q_1(x).$$

By definition, the principal eigenfunction of the Dirichlet problem does not change sign and it has its only zeros at $x = 0$ and $x = 1$. Hence, the Sturm comparison principle yields that the principal eigenfunction corresponding to the Neumann problem has a zero in $(0, 1)$, what leads to a contradiction, since the principal eigenfunction of the Neumann problem does not change sign (see Proposition 1).

Further, calculating the derivative of equation (13) leads to

$$0 = \frac{1}{\gamma}(V_{xx})_x + \alpha h'(V)V_x - \beta V_x = \frac{1}{\gamma}(V_x)_{xx} + \left(\alpha \frac{1}{p'(U)} - \beta\right) V_x = A(0)V_x.$$

Hence, we obtain that $V_x(x)$ is an eigenfunction for the eigenvalue 0 of the Dirichlet problem with $\lambda = 0$. Hence, $\nu(0) \geq 0$ and, therefore, $\mu(0) > 0$.

Furthermore, $\mu(\lambda)$ depends continuously on λ and it can be expressed as

$$\mu(\lambda) = \sup_{\psi \in W^{1,2}(0,1), \|\psi\|_2 = 1} \left(-\frac{1}{\gamma}\langle \psi_x, \psi_x \rangle + \langle r(\lambda, x)\psi, \psi \rangle \right),$$

where $\langle \cdot, \cdot \rangle$ denotes the L^2-scalar product.

We obtain

$$-\int_0^1 \frac{1}{\gamma}\psi_x\psi_x \mathrm{d}x + \int_0^1 r(\lambda, x)\psi^2 \mathrm{d}x \leq -\int_0^1 \frac{1}{\gamma}\psi_x\psi_x \mathrm{d}x + \int_0^1 C\psi^2 \mathrm{d}x \leq C.$$

Therefore, $\mu(\lambda)$ is bounded and, hence, there exists a value $\hat{\lambda} > 0$ fulfilling $\mu(\hat{\lambda}) = \hat{\lambda}$. It yields existence of an eigenfunction $\hat{\psi} \neq 0$ satisfying

$$A(\hat{\lambda})\hat{\psi} = \mu(\hat{\lambda})\hat{\psi} = \hat{\lambda}\hat{\psi} \quad \text{with} \quad \hat{\psi}_x(0) = \hat{\psi}_x(1) = 0,$$

which proves the existence of a positive eigenvalue of problem (60)–(61). □

5.3 Existence of Monotone Discontinuous Patterns

In this section, we show how to construct discontinuous stationary solutions in a model with *hysteresis* (Case 2). We focus on monotone increasing solutions having a form of transition layers or boundary layers.

The stationary problem corresponding our problem can be reduced to a boundary value problem for a single reaction-diffusion equation with discontinuous nonlinearity. Construction of transition layer solutions for such systems was undertaken in Mimura et al. [45] by using a shooting method. The result was applied by Mimura [44] to show the existence of discontinuous patterns in a model with density dependent diffusion. While in those models, the transition layer solution was unique, in the system face the problem of the existence of infinite number of solutions with changing connecting point. To deal with this difficulty we apply the approach proposed recently in [39] to construct all monotone stationary solutions having either a transition layer or a boundary layer. The analysis presented in this section follows [22].

In Case 2, the polynomial equation $f(U, V) = 0$ cannot be solved uniquely with respect to U. We denote $U = h_H(V)$ the solution branch connecting $(0, 0)$ with the turning point (v_H, u_H) and $U = h_T(V)$ the branch connecting (v_T, u_T) with S_2 (see Fig. 9 right panel).

Remark 14 We do not analyse the third solution branch $U = h_0(V)$ (the middle one), since this branch contains the unstable solution S_1. We may check later that, in fact, this branch does not satisfy the stability condition.

Solving stationary problem (27)–(29) (as described in Sect. 4.2), we obtain a boundary value problem

$$0 = \frac{1}{\gamma} V_{xx} + g(V, h_i(V)), \qquad \text{for } x \in (0, 1) \tag{62}$$

with zero-flux boundary condition $V_x(0) = V_x(1) = 0$, for any of the two branches $h_i, i = H, T$ of stationary solutions of the algebraic equation $f(U, V) = 0$.

However, in this case phase plane analysis provides no solution of this problem neither for $i = H$ nor $i = T$.

Lemma 5 *Neither* (62) *with* $i = T$*, nor* (62) *with* $i = H$ *has a nonconstant solution fulfilling zero-flux boundary conditions.*

Proof We rewrite the equations as systems of first order ODEs

$$\begin{aligned} V_x &= W \\ W_x &= -g_i(V) \quad \text{for} \quad i = H, T, \end{aligned}$$

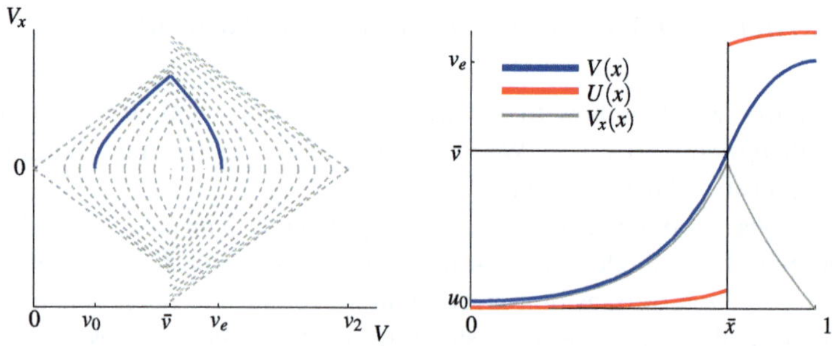

Fig. 10 Phase plane analysis for equation $V_{xx} + q_{\bar{v}}(V) = 0$: the *blue* trajectory $(V(x), V_x(x))$ connects the points $(v_0, 0)$ and $(v_e, 0)$ and it provides a solution of the boundary value problem with $V_x(0) = V_x(L) = 0$ for some L. The solution is given by $V_x = \sqrt{2(Q_{\bar{v}}(v_0) - Q_{\bar{v}}(V))}$. $V(x)$ is C^1, the derivative $V_x(x)$ is continuous, but not differentiable, whereas $U(x)$ is discontinuous at $x = \bar{x}$ (Courtesy of Alexandra Köthe)

where $g_i(V) = g(V, h_i(V))$. By definition, $g_H(0) = 0$ and $-g_H(V) > 0$ for $v \in (0, v_H)$. It means that the flux at the point $(v_0, 0)$ always points upwards and to the right. Therefore, a solution starting at $(v_0, 0)$ for $0 < v_0$ will never reach the W-axis again. Similarly, $g_T(v_2) = 0$ and $-g_T(V) < 0$ for $v \in (v_T, v_2)$. Consequently, all orbits ending at $(v_e, 0)$ with $v_e < v_2$ have started at some point with positive W-component, compare Fig. 10 (left panel). □

We conclude that there exist no stable patterns which are continuous in u. Next, we notice that the phase planes associated to (62) for $i = H$ and $i = T$ overlap for $v \in (v_T, v_H)$. Heuristically, to construct a solution, we select a value $\bar{v} \in (v_T, v_H)$ and "glue" the phase planes together at $\bar{v}$.

Definition 8 We denote by $q_{\bar{v}}$ the function with discontinuity at $\bar{v}$ defined by

$$q_{\bar{v}}(v) = \begin{cases} g(v, h_H(v)) & \text{when} \quad v \leq \bar{v} \\ g(v, h_T(v)) & \text{when} \quad v > \bar{v}. \end{cases}$$

Equation (62) can be then rewritten as

$$\frac{1}{\gamma} V_{xx} + q_{\bar{v}}(V) = 0 \qquad \text{for } x \in (0, 1) \tag{63}$$

$$V_x(0) = V_x(1) = 0.$$

Definition 9 A pair of functions (U, V) is called a *solution with jump at* $\bar{v}$, if $V \in C^1([0, 1])$ is a *weak solution* (as in Definition 7) of problem (63). The function $U \in L^\infty(0, 1)$ is given for almost all $x \in [0, 1]$ by

$$U(x) = \begin{cases} h_H(V(x)) & \text{if} \quad V(x) \leq \bar{v} \\ h_T(V(x)) & \text{if} \quad V(x) > \bar{v}. \end{cases}$$

Definition 10 The value $0 < \bar{x} < 1$ with $V(\bar{x}) = \bar{v}$ is called the *layer position* of the solution (U, V).

Remark 15 For x such that $V(x) < \bar{v}$ the function $V(x)$ is a classical solution of $\frac{1}{\gamma}V_{xx} + q_H(V) = 0$ and, therefore, it is a C^2 function. Similarly, $V(x)$ is a classical solution of $\frac{1}{\gamma}V_{xx} + q_T(V) = 0$ for x such that $V(x) > \bar{v}$. The branches can be connected satisfying $V \in C^1([0,1])$. The function U has a discontinuity at the layer position $\bar{x}$.

Theorem 8 *For all diffusion coefficients* $\frac{1}{\gamma}$ *with* $\gamma > 0$, *stationary problem* (63) *has a unique monotone increasing solution* (U, V) *with jump at* $\bar{v} \in (v_T, \min(v_H, v_2))$.

To analyse equation (63), we apply the method of phase plane analysis [2] and the analysis of time-maps [59].

Changing variables $x \mapsto Lx$, where $L = \sqrt{\gamma}$, we transform problem (63) into

$$V_{xx} + q_{\bar{v}}(V) = 0 \quad \text{for } x \in (0, L) \tag{64}$$

with boundary condition

$$V_x(0) = V_x(L) = 0. \tag{65}$$

This leads to a boundary value problem for a system of first order equations

$$\begin{aligned} V_x &= W, \\ W_x &= -q_{\bar{v}}(V). \end{aligned} \tag{66}$$

The first integral of (64) is given by

$$\frac{V_x^2(x)}{2} + Q_{\bar{v}}(V(x)) = E, \tag{67}$$

where $Q_{\bar{v}}(v) = \int_0^u q_{\bar{v}}(\tilde{v})d\tilde{v}$ is called the *potential*. The constant $E \in \mathbb{R}$ is arbitrary and corresponds to the *total energy* of the system.

A monotone increasing solution of problem (64) is given by a solution of system (66) such that

$$\big(V(0), W(0)\big) = (v_0, 0) \quad \text{and} \quad \big(V(L), W(L)\big) = (v_e, 0), \tag{68}$$

where $v_0, v_e \in \mathbb{R}$ satisfy $v_0 < v_e$ and $W(x) > 0$ for all $x \in (0, L)$. Such solution describes a curve in the phase plane, see Fig. 10 (left panel).

Calculating first integral yields (67)

$$W = V_x = \sqrt{2(E - Q_{\bar{v}}(V))} \tag{69}$$

The Neumann homogeneous boundary condition at $x = 0$ and $x = L$ provides

$$E = Q_{\bar{v}}(v_0) = Q_{\bar{v}}(v_e). \tag{70}$$

Moreover, (69) leds to inequality $Q_{\bar{v}}(V(x)) \leq E$ for a solution V and $x \in [0, L]$.

By definition $Q_{\bar{v}}$ satisfies

$$Q'_{\bar{v}}(v) = q_{\bar{v}}(v)$$

and

$$Q_{\bar{v}}(0) = 0.$$

It is continuous for all v but not differentiable at $\bar{v}$. Equalities

$$q_{\bar{v}}(0) = f(h_H(0), 0) = f(0, 0) = 0$$

and

$$q_{\bar{v}}(v_2) = f(h_H(v_2), v_2) = f(u_2, v_2) = 0$$

guarantee existence of local maxima of the potential at $v = 0$ and at $v = v_2$, respectively. It has a local minimum at $v = \bar{v}$, requiring that $\bar{v} < v_2$. Therefore, for $\bar{v} \in \big(v_T, \min(v_H, v_2)\big)$, there exist $0 < v_0 < \bar{v} < v_e < v_2$ satisfying (70) and such that $Q_{\bar{v}}(v) < Q_{\bar{v}}(v_0)$ for all $v \in (v_0, v_e)$.

Solving equation (69) provides L such that a trajectory starting at $\big(V(0), W(0)\big) = (v_0, 0)$ for a fixed value $v_0 \in (0, \bar{v})$ satisfies $W(L) = 0$ and $V(L) = v_e(v_0)$ for the first time.

$$L = \frac{1}{\sqrt{2}} \int_0^L \frac{V_x(x)\mathrm{d}x}{\sqrt{E - Q_{\bar{v}}(V(x))}} = \frac{1}{\sqrt{2}} \int_{v_0}^{v_e} \frac{\mathrm{d}v}{\sqrt{E - Q_{\bar{v}}(v)}}.$$

For further calculation, we split the integral at the minimum of the potential $\bar{v}$, and denote

$$T^1_{\bar{v}}(v_0) = \frac{1}{\sqrt{2}} \int_{v_0}^{\bar{v}} \frac{\mathrm{d}v}{\sqrt{Q_{\bar{v}}(v_0) - Q_{\bar{v}}(v)}}, \qquad T^2_{\bar{v}}(v_e) = \frac{1}{\sqrt{2}} \int_{\bar{v}}^{v_e} \frac{\mathrm{d}v}{\sqrt{Q_{\bar{v}}(v_e) - Q_{\bar{v}}(v)}}.$$

$T^1_{\bar{v}}(v_0)$ is classically interpreted as "time" x, for which a forward orbit in the phase plane, associated to system (66) with $q_{\bar{v}}(v) = f(v, h_H(v))$, starting at $(0, v_0)$ needs to reach the $v = \bar{v}$ axis for the first time. Analogously, $T^2_{\bar{v}}(v_e)$ is a "time" x, for which a backward orbit in the phase plane, associated to system (66) with $q_{\bar{v}}(v) = f(v, h_T(v))$, starting at $(v_e, 0)$ needs to reach the $v = \bar{v}$ axis for the first time.

Thus, we can calculate the total "time" to connect $(v_0, 0)$ with $(v_e, 0)$

$$T_{\bar{v}}(v_0) := T_{\bar{v}}^1(v_0) + T_{\bar{v}}^2(v_e(v_0)),$$

where v_e depends on v_0 by relation (70). A monotone increasing solution of (64) starts at $V(0) = v_0$ which satisfies

$$T_{\bar{v}}(v_0) = L.$$

To show existence of a monotone solution for a given diffusion coefficient $\frac{1}{\gamma}$, we find a corresponding time map with $T_{\bar{v}}(v_0) = \sqrt{\gamma}$. We characterise the time maps with following lemmas.

Lemma 6 *The maps* $T_{\bar{v}}^1 : (0, \bar{v}) \to (0, \infty)$ *and* $T_{\bar{v}}^2 : (\bar{v}, v_2) \to (0, \infty)$ *are well-defined, continuous and surjective.*

Proof Follows [37, 59].

To show surjectivity we investigate behaviour of the time-maps for v_0 and v_e in the limits at the borders of their domain of definition.

Taylor expansion at 0 and existence of a local maximum of $Q_{\bar{v}}$ at 0 yield the estimate

$$Q_{\bar{v}}(v_0) - Q_{\bar{v}}(v) \le Q_{\bar{v}}(0) - \left[Q_{\bar{v}}(0) + q_H(0)(v) + \frac{q_H'(\eta)}{2}v^2\right] \le Cv^2,$$

where $C = \max_{\eta \in (0,\bar{v})} \frac{q_H'(\eta)}{2}$. Hence, we obtain

$$\lim_{v_0 \to 0} T_{\bar{v}}^1(v_0) \ge \lim_{v_0 \to 0} \frac{1}{\sqrt{2C}} \int_{v_0}^{\bar{u}} \frac{\mathrm{d}v}{\sqrt{v^2}} = \lim_{v_0 \to 0} \frac{1}{\sqrt{2C}} \big(\ln(\bar{v}) - \ln(v_0)\big) = \infty.$$

For v_0 and v close to $\bar{v}$, we use the Taylor expansion of $Q_{\bar{v}}$ at v_0 and obtain

$$Q_{\bar{v}}(v_0) - Q_{\bar{v}}(v) = Q_{\bar{v}}(v_0) - \big(Q_{\bar{v}}(v_0) + q_H(\eta)(v - v_0)\big) = -q_H(\eta)(v - v_0).$$

with $\eta \in (v_0, v)$. Thus, we calculate the limit

$$\begin{aligned}
\lim_{v_0 \to \bar{v}} T_{\bar{v}}^1(v_0) &= \lim_{v_0 \to \bar{v}} \frac{1}{\sqrt{2}} \int_{v_0}^{\bar{v}} \frac{\mathrm{d}v}{\sqrt{-q_H(\eta)(v - v_0)}} \\
&\ge \lim_{v_0 \to \bar{v}} \left(\min_{\eta \in (v_0, \bar{v})} \frac{1}{\sqrt{2|q_H(\eta)|}} \right) \int_{v_0}^{\bar{v}} \frac{\mathrm{d}v}{\sqrt{v - v_0}} \\
&= \frac{1}{\sqrt{2|q_H(\bar{v})|}} \lim_{v_0 \to \bar{v}} 2\sqrt{\bar{v} - v_0} = 0.
\end{aligned}$$

The corresponding results for $T_{\bar{v}}^2(v_e)$ are obtained in a similar way. □

The maps $T^1_{\bar{v}}$ and $T^2_{\bar{v}}$ are defined for all $v_0 \in (0, \bar{v})$ and $v_e \in (\bar{v}, v_2)$, respectively. By definition of $T_{\bar{v}}$, they must satisfy (70). Thus, they depend on the local maxima of the potential. Therefore, for $Q_{\bar{v}}(v_2) \geq Q_{\bar{v}}(0) = 0$, we denote $v_{\min} = 0$ and $\bar{v} < v_{\max} \leq v_2$ is the solution of $Q_{\bar{v}}(v_{\max}) = 0$. For $Q_{\bar{v}}(v_2) \leq 0$, we denote $v_{\max} = v_2$ and $0 \leq v_{\min} < \bar{v}$ is the solution of $Q_{\bar{v}}(v_{\min}) = Q_{\bar{v}}(v_2)$.

To show existence of a monotone increasing solution of problem (63) for every diffusion coefficient $\frac{1}{\gamma} > 0$, we prove:

Lemma 7 *The map $T_{\bar{v}} : (v_{\min}, \bar{v}) \to (0, \infty)$ is well-defined, continuous and surjective.*

Proof $T_{\bar{v}}$ is continuous as sum and composition of continuous functions. Continuity of $Q_{\bar{v}}$ yields that $u_0 \to \bar{v}$ implies $v_e \to \bar{v}$ and, therefore,

$$\lim_{v_0 \to \bar{v}} T_{\bar{v}}(v_0) = \lim_{v_0 \to \bar{v}} T^1_{\bar{v}}(v_0) + \lim_{v_e \to \bar{v}} T^2_{\bar{v}}(v_e) = 0.$$

Finally, we know that either $v_{\min} = 0$ or $v_{\max} = u_2$, which yields

$$\lim_{v_0 \to v_{\min}} T_{\bar{v}}(v_0) = \lim_{v_0 \to v_{\min}} T^1_{\bar{v}}(v_0) + \lim_{v_e \to v_{\max}} T^2_{\bar{v}}(v_e) = \infty,$$

because either the first or the second limit is infinite. □

In the next step, we show uniqueness of this solution, i.e. monotonicity of the time-map $T_{\bar{v}}$. For this we use the following representation of derivatives of the time-maps.

Lemma 8 *The time-maps are differentiable and their derivatives satisfy*

$$\frac{\mathrm{d}}{\mathrm{d}v_0} T^1_{\bar{v}}(v_0) = \frac{-\frac{1}{\sqrt{2}} q_H(v_0)}{Q_{\bar{v}}(v_0) - Q_{\bar{v}}(\bar{v})} \int_{v_0}^{\bar{v}} \left[\frac{(Q_{\bar{v}}(v) - Q_{\bar{v}}(\bar{v})) q'_H(v)}{q_H(v)^2} - \frac{1}{2} \right] \frac{\mathrm{d}v}{\sqrt{Q_{\bar{v}}(v_0) - Q_{\bar{v}}(v)}} \tag{71}$$

and

$$\frac{\mathrm{d}}{\mathrm{d}v_e} T^2_{\bar{v}}(v_e) = \frac{-\frac{1}{\sqrt{2}} q_T(v_e)}{Q_{\bar{v}}(v_e) - Q_{\bar{v}}(\bar{v})} \int_{\bar{v}}^{v_e} \left[\frac{(Q_{\bar{v}}(v) - Q_{\bar{v}}(\bar{v})) q'_T(v)}{q_T(v)^2} - \frac{1}{2} \right] \frac{\mathrm{d}u}{\sqrt{Q_{\bar{v}}(v_e) - Q_{\bar{v}}(v)}}. \tag{72}$$

Proof Under the assumption $q_{\bar{v}} \in C^0$ with piecewise continuous $q'_{\bar{v}}$ and $q_{\bar{v}}(\bar{u}) = 0$, this formula has been proven by Loud in [28]. Our problem does not satisfy Loud's hypotheses. However, the formula holds, what can be shown by adapting the proof given in [28]. □

Proposition 3 *Derivatives of $T^1_{\bar{v}}$ and $T^2_{\bar{v}}$ satisfy*

$$\frac{\mathrm{d}}{\mathrm{d}v_0}T^1_{\bar{v}}(v_0) < 0 \quad \frac{\mathrm{d}}{\mathrm{d}v_e}T^2_{\bar{v}}(v_e) > 0.$$

Therefore, for all jumps $\bar{v} \in \big(v_T, \min(v_H, v_2)\big)$, it holds

$$\frac{\mathrm{d}}{\mathrm{d}v_0}T_{\bar{v}}(v_0) < 0.$$

Proof To show the negativity of $\frac{\mathrm{d}}{\mathrm{d}v_0}T^1_{\bar{v}}(v_0)$, we rewrite the integral representation (71) of the derivative of the time-map as

$$\frac{\mathrm{d}}{\mathrm{d}v_0}T^1_{\bar{v}}(v_0) = -\frac{q_H(v_0)}{\sqrt{2}(Q_{\bar{v}}(v_0) - Q_{\bar{v}}(\bar{v}))} \cdot \int_{v_0}^{\bar{v}} l_H(v)\frac{\mathrm{d}v}{\sqrt{Q_{\bar{v}}(v_0) - Q_{\bar{v}}(v)}},$$

with a function l_H defined by

$$l_H(v) = \frac{\big(Q_{\bar{v}}(v) - Q_{\bar{v}}(\bar{v})\big)q'_H(v)}{q_H(v)^2} - \frac{1}{2}.$$

We observe that $-\frac{1}{\sqrt{2}}q_H(v_0)(Q_{\bar{v}}(v_0) - Q_{\bar{v}}(\bar{v}))^{-1}$ is positive, as $q_H(v_0) < 0$ and $Q_{\bar{v}}$ is decaying on $(0, \bar{v})$. Thus, it remains to show that l_H is negative for $v \in [v_0, \bar{v}]$.

Multiplying l_H by the square of q_H and calculating the derivative

$$\frac{\mathrm{d}}{\mathrm{d}v}\left(q_H(v)^2 l_H(v)\right) = (Q_{\bar{v}}(v) - Q_{\bar{v}}(\bar{v}))q''_H(v)$$

and using

$$q_H(v)^2 l_H(v)|_{v=\bar{v}} = -\frac{1}{2}q_H(\bar{v})^2 =: -C_H < 0,$$

we obtain the representation

$$\begin{aligned} q_H(v)^2 l_H(v) &= -C_H + \int_{\bar{v}}^{v} (Q_{\bar{v}}(\tilde{v}) - Q_{\bar{v}}(\bar{v}))q''_H(\tilde{v})\mathrm{d}\tilde{v} \\ &= -C_H - \int_{v}^{\bar{v}} (Q_{\bar{v}}(\tilde{v}) - Q_{\bar{v}}(\bar{v}))q''_H(\tilde{v})\mathrm{d}\tilde{v}. \end{aligned}$$

To show negativity of this expression, we notice that the second derivative $q''_H(v) = \alpha h''_H(v) - -\alpha p''(h_H(v))/p'(h_H(v))^3$ is positive. Indeed, for all $v \in (v_0, \bar{v})$, we have $p'(h_H(v)) > 0$ as well as $p''(h_H(v)) < 0$. Finally, we obtain $l_H(v) < 0$, which proves $\frac{\mathrm{d}}{\mathrm{d}v_0}T^1_{\bar{v}}(v_0) < 0$. Similarly, the we can prove that $\frac{\mathrm{d}}{\mathrm{d}v_e}T^2_{\bar{v}}(v_e) > 0$.

To accomplish the proof, we deduce

$$\frac{\mathrm{d}}{\mathrm{d}v_0}v_e(v_0) = \frac{q_H(v_0)}{q_T(v_e)} < 0,$$

by differentiating (70) with respect to u_0. It yields

$$\frac{\mathrm{d}}{\mathrm{d}v_0}T_{\bar{v}}(v_0) = \frac{\mathrm{d}}{\mathrm{d}v_0}T^1_{\bar{v}}(v_0) + \frac{\mathrm{d}}{\mathrm{d}v_e}T^2_{\bar{v}}(v_e)\cdot\frac{\mathrm{d}}{\mathrm{d}v_0}v_e(v_0) < 0.$$

□

5.4 Stability of Discontinuous Patterns

In this section we provide a stability condition for monotone discontinuous stationary solutions of system (12)–(15) with nonlinearities (57) satisfying Case 2. Due to discontinuity of patterns, linearised stability analysis cannot be directly applied. To cope with this difficulty, an approach using a special topology which allows for disclosure of the discontinuity points was proposed by Weinberger in [70] and applied in [3] in analysis of a model with density dependent diffusion. We present here analysis based on direct estimates as proposed in [22] for this particular model.

5.5 L^∞ Versus L^2 Perturbations

As shown in the previous section, a generic model with hysteresis admits an infinite number of monotone increasing stationary solutions. Simulations indicate that the resulting pattern strongly depends on the initial conditions. More precisely, the stationary solution depends on the position of the initial condition relative to the separatrix of the initial condition. For this reason it is suitable to consider stability in $L^\infty(0,1)$ sense. A perturbation which is small in $L^\infty(0,1)$ cannot change the position of points on the stationary solution $\big(U,V\big)$ from one side of the separatrix to the other side. In contrast, a small $L^2(0,1)$ perturbation may lead to large changes of values on some small interval, leading to another stationary solution.

Example 1 We consider a model with hysteresis and nonlinearities satisfying $g(u,v) = 1.4u - v$ and $p(u) = u^3 - 6.3u^2 + 10u$. Let $\big(U(x),V(x)\big)$ be a monotone increasing stationary solution with layer position at $\bar{x} = 0.4$ (see Fig. 11, upper left panel).

We observe that after a random perturbation of the stationary solution $\big(U(x),V(x)\big)$ with $\big(\varphi_0(x),\psi_0(x)\big)$ such that $\|\varphi_0\|_{L^\infty(0,1)} = \|\psi_0\|_{L^\infty(0,1)} = 0.4$, the solution with initial condition $\big(U(x)+\varphi_0(x), V(x)+\psi_0(x)\big)$ converges back to $\big(U(x),V(x)\big)$, see Fig. 11 (upper right panel).

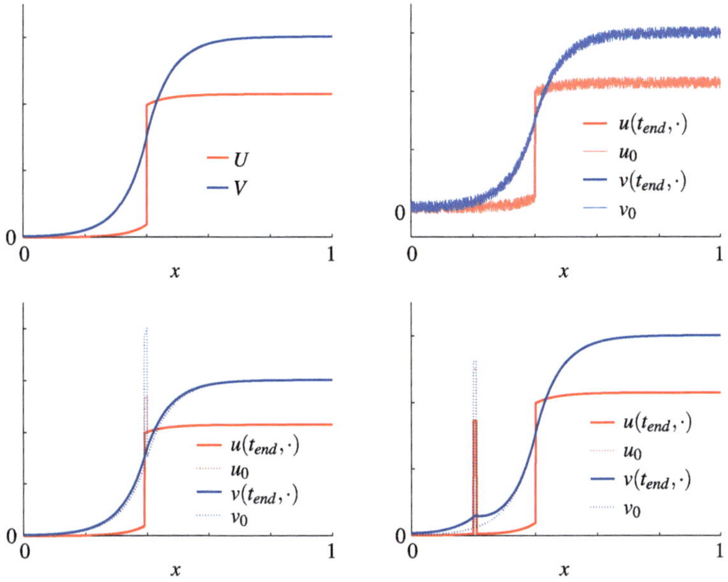

Fig. 11 Simulations of the generic model in the hysteresis case for different types of perturbations of a stationary solution. The plots show the initial condition (*dotted lines*) and the approached stationary solution (*continuous lines*) after a sufficiently large time t_{end}. *Upper left panel* a monotone increasing stationary solution $(U(x), V(x))$ with layer position at $\bar{x} = 0.4$. *Upper right panel* a perturbation which is small in $L^\infty(0,1)$ norm does not change the stationary solution. *Lower left panel* a perturbation which is small in $L^2(0,1)$, but not in $L^\infty(0,1)$ norm shifts the layer position. *Lower right panel* a perturbation which is small in $L^2(0,1)$, but not in $L^\infty(0,1)$ norm leads to a nonmonotone stationary solution

Next, we apply a perturbation with a step function

$$\varphi_0(x) = \psi_0(x) = \begin{cases} 5 & \text{for } x \in [0.39, 0.4] \\ 0 & \text{else.} \end{cases}$$

It is small in $L^2(0,1)$, but large in $L^\infty(0,1)$, and we observe a simulated pattern with a shift of the layer position, see Fig. 11 (lower left panel). The solution converges to a stationary transition layer with $\bar{x} = 0.39$.

Perturbations which are small in $L^2(0,1)$ but large in $L^\infty(0,1)$ may also lead to a stationary solution which is not monotone anymore, see Fig. 11 (lower right panel) for the solution of the model with

$$\varphi_0(x) = \psi_0(x) = \begin{cases} 5 & \text{for } x \in [0.19, 0.2] \\ 0 & \text{else.} \end{cases}$$

5.6 *Stability Analysis of a Reaction-Diffusion-Ode Model with Hysteresis*

Let $(U(x), V(x))$ be a nonhomogeneous stationary solution of system (12)–(15) with nonlinearities (57) satisfying Case 2. In the remainder of this section we assume that the following condition holds

$$\text{ess inf}_{x\in[0,1]}\, p'(U(x)) := K > \frac{\alpha}{\beta}. \tag{73}$$

We call (73) the *stability condition.*

We consider a small nonlinear perturbation of the stationary solution in the $L^\infty(0,1)$-norm, i.e.

$$\begin{aligned} u(t,x) &= U(x) + \varphi(t,x), \\ v(t,x) &= V(x) + \psi(t,x) \end{aligned} \tag{74}$$

with $(\varphi, \psi) \in C\big((0,\infty); L^\infty(0,1) \times L^\infty(0,1)\big)$.

Lemma 9 *The nonlinear perturbation*

$$(\varphi, \psi) \in C\big((0,\infty); L^\infty(0,1) \times L^\infty(0,1)\big),$$

with bounded initial data, is a solution of the system

$$\varphi_t(t,x) = \psi(t,x) - p'(U(x))\varphi(t,x) + R(x)\varphi^2(t,x), \quad t > 0, x \in [0,1] \tag{75}$$

$$\psi_t(t,x) = \frac{1}{\gamma}\psi_{xx}(t,x) + \alpha\varphi(t,x) - \beta\psi(t,x), \qquad t > 0, x \in (0,1) \tag{76}$$

with homogenous Neumann boundary conditions and a bounded function $R(x)$.

Proof Using the linearity of $g(u,v)$, we directly obtain (76) from (74). For (75), we calculate

$$\varphi_t = u_t = V+\psi-p(U+\varphi) = \psi+p(U)-p(U+\varphi) = \psi-p'(U(x))\varphi-\frac{1}{2}p''(\theta(x))\varphi^2,$$

where we use the Taylor expansion of $p(U(x))$ for x fixed with $\theta(x)$ being the rest. Since $\theta(x) \in (0, u_2)$ and $p''(u) = 6a_1u + 2a_2$ is bounded for u being in some interval, we conclude the proof by setting $R(x) := p''(\theta(x))$ which has the desired property. □

Proposition 4 *The nonlinear system (75)–(76) with bounded initial condition $\varphi(0,x) = \varphi_0(x) \in L^\infty(0,1)$ and $\psi(0,x) = \psi_0(x) \in L^\infty(0,1)$ under boundary condition $\psi_x(t,0) = \psi_x(t,1) = 0$ has a unique bounded solution*

$$(\varphi,\psi) \in C\big((0,\infty); L^\infty(0,1) \times L^\infty(0,1)\big).$$

given by

$$\varphi(t,x) = e^{-p'(U(x))t}\varphi_0(x) + \int_0^t e^{-(t-s)p'(U(x))}\psi(s,x)ds$$
$$+ \int_0^t e^{-(t-s)p'(U(x))}R(x)\varphi^2(s,x)ds, \tag{77}$$

$$\psi(t,x) = S(t)\psi_0(x) + \alpha\int_0^t S(t-s)\varphi(s,x)ds, \tag{78}$$

where $S(t)$ is the semigroup generated by the operator

$$D(A) = \{u \in H^2(0,1) \mid v_x(0) = v_x(1) = 0\},$$
$$A := \frac{1}{\gamma}v_{xx} - \beta v \quad \textit{for } v \in D(A),$$

which fulfils the estimate

$$\|S(t)\psi_0\|_\infty \le e^{-t\beta}\|\psi_0\|_\infty. \tag{79}$$

Remark 16 By smoothing property of the analytic semigroup $[S(t)]_{t\ge0}$ generated by $\gamma^{-1}d^2/dx^2 - \beta$ on $L^p(0,1)$, for $\psi_0 \in L^\infty(0,1) \subset \cap_{1<p<\infty}L^p(0,1)$ we obtain that $S(t)\psi_0$ is a smooth function on $0 \le x \le 1$ whenever $t > 0$. Hence, $S(t)\psi_0 \in C^0[0,1] \subset L^\infty(0,1)$ for $t > 0$. If in addition, $\|S(t)\psi_0 - \psi_0\|_\infty \to 0$ as $t \downarrow 0$, then it must necessarily hold $\psi_0 \in C^0[0,1]$, because the continuous functions converge uniformly to a continuous limit. It yields that $[S(t)]_{t\ge0}$ cannot be interpreted as a strongly continuous semigroup on $L^\infty(0,1)$. Nevertheless, L^∞ is a convenient space to handle nonlinearities. To circumvent this difficulty, we make use of the fact that $\|S(t)\psi_0\|_\infty \le e^{-t\beta}\|\psi_0\|_\infty$ for initial data $\psi_0 \in L^\infty(0,1)$ and $t > 0$. Then, for $g \in L^\infty((0,T) \times [0,1])$ the function $s \to S(t-s)g(s,\cdot)$ is meaningful as a function from $0 \le s \le t$ into $L^2(0,1)$ and we obtain a continuous function $t \to \int_0^t S(t-s)g(s,\cdot)ds$ from $0 \le s \le t$ to $L^2(0,1)$.

Theorem 9 (Nonlinear Stability) *If φ_0 and ψ_0 are initial perturbations of a stationary solution fulfilling assumption (73) with $\|\varphi_0\|_\infty$ and $\|\psi_0\|_\infty$ sufficiently small, then $\|\varphi(t,\cdot)\|_\infty$ and $\|\psi(t,\cdot)\|_\infty$ stay small for all times t. More precisely, let $R_0 = \operatorname{ess\,sup}_{x\in[0,1]} R(x)$ and let the initial perturbations fulfil*

$$\|\psi_0\|_\infty + K\|\varphi_0\|_\infty \le \frac{1}{4R_0}\Big(K - \frac{\alpha}{\beta}\Big)^2, \tag{80}$$

and

$$\|\varphi_0\|_\infty \leq \frac{1}{2R_0}\Big\{(K-\frac{\alpha}{\beta}) - \sqrt{(K-\frac{\alpha}{\beta})^2 - 4R_0(\|\psi_0\|_\infty + K\|\varphi_0\|_\infty)}\Big\} \tag{81}$$

then the following estimates hold

$$\|\varphi(t,\cdot)\|_\infty < \frac{1}{2R_0}\Big(K-\frac{\alpha}{\beta}\Big), \tag{82}$$

$$\|\psi(t,\cdot)\|_\infty < \|\psi_0\|_\infty + \frac{\alpha}{\beta}\frac{1}{2R_0}\Big(K-\frac{\alpha}{\beta}\Big), \tag{83}$$

for all $t \in [0,\infty)$.

Proof We define increasing continuous functions

$$\Phi(t) := \sup_{0\leq\tau\leq t} \|\varphi(\tau,\cdot)\|_\infty \quad \Psi(t) := \sup_{0\leq\tau\leq t} \|\psi(\tau,\cdot)\|_\infty.$$

Using the integral representation (78), the estimate of the semigroup (79) and taking the supremum over $\tau \in [0,t]$, we obtain

$$\Psi(t) \leq \|\psi_0\|_\infty + \alpha\int_0^t e^{-\beta(t-s)}\Phi(s)\mathrm{d}s \tag{84}$$

and similarly using (77), we obtain

$$\Phi(t) \leq \|\varphi_0\|_\infty + \int_0^t e^{-K(t-s)}\Psi(s)\mathrm{d}s + R_0\int_0^t e^{-K(t-s)}\Phi^2(s)\mathrm{d}s. \tag{85}$$

$\Phi(t)$ and $\Psi(t)$ are increasing by definition, thus, estimates (84) and (85) yield

$$\begin{aligned}\Phi(t) &\leq \|\varphi_0\|_\infty + \Psi(t)\int_0^t e^{-K(t-s)}\mathrm{d}s + R_0\Phi^2(t)\int_0^t e^{-K(t-s)}\mathrm{d}s,\\ &\leq \|\varphi_0\|_\infty + \frac{1}{K}\Psi(t) + \frac{1}{K}R_0\Phi^2(t).\end{aligned} \tag{86}$$

$$\Psi(t) \leq \|\psi_0\|_\infty + \alpha\Phi(t)\int_0^t e^{-\beta(t-s)}\mathrm{d}s \leq \|\psi_0\|_\infty + \frac{\alpha}{\beta}\Phi(t). \tag{87}$$

Substituting (87) into (86) and multiplying by K, we obtain the estimate

$$0 \leq K\|\varphi_0\|_\infty + \|\psi_0\|_\infty + \Big(\frac{\alpha}{\beta} - K\Big)\Phi(t) + R_0\Phi^2(t). \tag{88}$$

The graph of

$$y \mapsto \|\varphi_0\|_\infty + K\|\psi_0\|_\infty + \left(\frac{\alpha}{\beta} - K\right)y + R_0 y^2 \tag{89}$$

is a parabola which is positive at $y = 0$. As the linear term is negative by the stability condition (73), this parabola intersects the y-axis if the constant term $K\|\psi_0\|_\infty\|\varphi_0\|_\infty$ is small enough.

Thus, because $\Phi(t)$ is nonnegative for all times, it is bounded by the smallest zero of the parabola, provided $\Phi(0) = \|\varphi_0\|_\infty$ is smaller than this zero.

To be more precise, we calculate the discriminant of the parabola (89)

$$D = \left(K - \frac{\alpha}{\beta}\right)^2 - 4R_0(\|\psi_0\|_\infty + K\|\varphi_0\|_\infty).$$

Therefore, under the condition (81) the discriminant D is positive, which yields that the parabola has two zeros, given by

$$y_1 = \frac{1}{2R_0}\left(K - \frac{\alpha}{\beta} - \sqrt{D}\right), \qquad y_2 = \frac{1}{2R_0}\left(K - \frac{\alpha}{\beta} + \sqrt{D}\right).$$

The smallest zero is necessarily less than $\frac{1}{2R_0}(K - \frac{\alpha}{\beta})$, which leads to the bound for $\Phi(t)$. Together with the estimate

$$\Psi(t) \leq \|\psi_0\|_\infty + \frac{\alpha}{\beta}\Phi(t).$$

we obtain the bound for $\Psi(t)$. □

Theorem 10 (Asymptotic Stability) *Let $\big(U(x), V(x)\big)$ be a stationary solution fulfilling the stability condition (73). If φ_0 and ψ_0 are initial perturbations, which fulfil the estimate (81), then for the nonlinear perturbation (φ, ψ) holds*

$$\big(\|\varphi(t,\cdot)\|_\infty, \|\psi(t,\cdot)\|_\infty\big) \to (0,0)$$

for $t \to \infty$.

Proof We define the values

$$\Phi^{NL} := \limsup_{t\to\infty} \|\varphi(t,\cdot)\|_\infty \quad \text{and} \quad \Psi^{NL} := \limsup_{t\to\infty} \|\psi(t,\cdot)\|_\infty,$$

which are finite due to Proposition 4.

Using the integral representation (78) for ψ yields

$$\Psi^{NL} \leq \limsup_{t\to\infty} e^{-\beta t}\|\psi_0\|_\infty + \alpha\Phi^{NL} \limsup_{t\to\infty} \int_0^t e^{-\beta(t-s)} ds.$$

Together with (79), we deduce the inequality

$$\Psi^{NL} \leq \frac{\alpha}{\beta}\Phi^{NL}. \tag{90}$$

From (77) we deduce in the same way

$$\Phi^{NL} \leq \limsup_{t\to\infty} e^{-Kt}\|\varphi_0\|_\infty + \Psi^{NL}\limsup_{t\to\infty}\int_0^t e^{-K(t-s)}\mathrm{d}s + R_0(\Phi^{NL})^2\limsup_{t\to\infty}\int_0^t e^{-K(t-s)}\mathrm{d}s$$

and, thus,

$$\Phi^{NL} \leq \frac{1}{K}\Psi^{NL} + \frac{R_0}{K}(\Phi^{NL})^2. \tag{91}$$

The inequalities (91) and (90) lead to

$$0 \leq \Big(\frac{\alpha}{\beta} - K\Big)\Phi^{NL} + R_0(\Phi^{NL})^2.$$

Assuming that $0 \leq \big(\alpha/\beta - K\big) + R_0\Phi^{NL}$ yields $\Phi^{NL} \geq \frac{1}{R_0}\big(K - \alpha/\beta\big)$. Theorem 9 provides for initial perturbations fulfilling (81) that $\Phi^{NL} = \lim_{t\to\infty}\Phi(t) \leq \frac{1}{2R_0}\big(K - \alpha/\beta\big)$, which is a contradiction. Finally, we obtain $\Phi^{NL} = 0$. □

At the first glance, the stability condition seems to be difficult to check, because it depends on the stationary solution. However, it can be translated into a condition for the jump. For this purpose, we denote by v_H^{cr} the zero of q_H' in $[0, v_H]$ and by v_T^{cr} the zero of q_T' in $[v_T, v_2]$. These values exist and they are unique. Their relative position depends on the kinetic functions.

Corollary 4 *We consider the model* (12)–(15) *with nonlinearities* (57) *such that* $v_T^{cr} < v_H^{cr}$. *Let* $\big(U(x), V(x)\big)$ *be a stationary solution with jump* $v_T^{cr} < \bar{v} < v_H^{cr}$, *then* $\big(U(x), V(x)\big)$ *is asymptotically stable.*

Proof For $x \in [0, 1]$ fulfilling $V(x) \leq \bar{v} < v_H^{cr}$, it holds $q_H'(V(x)) < 0$ by definition of v_H^{cr}.

$$q_H'(V(x)) = \alpha\frac{1}{p'(h_H(V(x)))} - \beta = \alpha\Big(\frac{1}{p'(h_H(V(x)))} - \frac{\beta}{\alpha}\Big) < 0.$$

Moreover, for those x the function $U(x)$ is given by $h_H(V(x))$ and we obtain that the stability condition $p'(h_H(V(x))) > \frac{\alpha}{\beta}$ is fulfilled.

Similarly, for $x \in [0,1]$ fulfilling $V(x) > \bar{v} > v_T^{cr}$ it holds

$$q_T'(V(x)) = \alpha \frac{1}{p'(h_T(V(x)))} - \beta = \alpha\Big(\frac{1}{p'(h_T(V(x)))} - \frac{\beta}{\alpha}\Big) < 0.$$

Thus, $p'(h_T(V(x))) > \frac{\alpha}{\beta}$ is fulfilled. □

Acknowledgements A. M-C was supported by European Research Council Starting Grant No 210680 "Multiscale mathematical modelling of dynamics of structure formation in cell systems" and Emmy Noether Programme of German Research Council (DFG).

References

1. A. Anma, K. Sakamoto, T. Yoneda, Unstable subsystems cause Turing instability. Kodai Math. J. **35**, 215–247 (2012)
2. V.I. Arnold, *Ordinary Differential Equations* (Springer, Berlin, 1992)
3. D.G. Aronson, A. Tesei H. Weinberger, A density-dependent diffusion system with stable discontinuous stationary solutions. Ann. Mat. Pura Appl. **152**, 259–280 (1988)
4. H. Berestycki, G. Nadin, P. Perthame, L. Ryzhik, The non-local Fisher-KPP equation: traveling waves and steady states. Nonlinearity **22**, 2813–2844 (2009)
5. R. Bertolusso, M. Kimmel, Modeling spatial effects in early carcinogenesis: stochastic versus deterministic reaction-diffusion systems. Math. Modell. Nat. Phenomena **7**, 245–260 (2012)
6. H. Brezis, *Functional Analysis, Sobolev Spaces and Partial Differential Equations* (Springer, New York, 2010)
7. R. Caste, C. Holland, Instability results for reaction-diffusion equations with Neumann boundary conditions. J. Differ. Equ. **27**, 266–273 (1978)
8. H. Chuan Le, T. Tsujikawa, A. Yagi, Asymptotic behaviour of solutions for forest kinematic model. Funkcial. Ekvac. **49**, 427–449 (2006)
9. K.-L. Engel, R. Nagel, *One-Parameter Semigroups for Linear Evolution Equations*. Graduate Texts in Mathematics, vol 194 (Springer, New York, 2000)
10. J.W. Evans, Nerve axon equations. IV. The stable and the unstable impulse. Indiana Univ. Math. J. **24**, 1169–1190 (1974/1975)
11. S. Friedlander, W. Strauss, M. Vishik, Nonlinear instability in an ideal fluid. Ann. Inst. H. Poincaré Anal. Non Linéaire **14**, 187–209 (1997)
12. M.G. Garroni, V.A. Solonnikov, M.A. Vivaldi, Schauder estimates for a system of equations of mixed type. Rend. Mat. Appl. **29**, 117–132 (2009)
13. A. Gierer, H. Meinhardt, A theory of biological pattern formation. Kybernetik **12**, 30–39 (1972)
14. P. Gray, S.K. Scott, Autocatalytic reactions in the isothermal continuous stirred tank reactor: isolas and other forms of multistability. Chem. Eng. Sci. **38**, 29–43 (1983)
15. S. Härting, A. Marciniak-Czochra, Spike patterns in a reactiondiffusion ODE model with Turing instability. Math. Methods Appl. Sci. 1–15 (2013). doi:10.1002/mma.2899
16. D. Henry, *Geometric Theory of Semilinear Parabolic Equations* (Springer, New York, 1981)
17. S. Hock, Y. Ng, J. Hasenauer, D. Wittmann, D. Lutter, D. Trümbach, W. Wurst, N. Prakash, F.J. Theis, Sharpening of expression domains induced by transcription and microRNA regulation within a spatio-temporal model of mid-hindbrain boundary formation. BMC Syst. Biol. **7**, 48 (2013)
18. F. Hoppensteadt, W. Jäger, C. Pöppe, in *A Hysteresis Model for Bacterial Growth Patterns*, ed. by S. Levin, Lecture Notes in Biomathematics: Modelling of Patterns in Space and Time (Springer, Heidelberg, 1983)

19. H. Jiang, Global existence of solutions of an activator-inhibitor system. Discrete Contin. Dyn. Syst. **14**, 737–751 (2006)
20. G. Karali, T. Suzuki, Y. Yamada, Global-in-time behaviour of the solution to a Gierer-Meinhardt system. Discrete Contin. Dyn. Syst. **33**, 2885–2900 (2013)
21. V. Klika, R.E. Baker, D. Headon, E.A. Gaffney, The influence of receptor-mediated interactions on reaction-diffusion mechanisms of cellular self-organisation. Bull. Math. Biol. **74**, 935–957 (2012)
22. A. Köthe, Hysteresis-driven pattern formation in reaction-diffusion-ODE models. Ph.D. thesis, University of Heidelberg, 2013
23. A. Köthe, A. Marciniak-Czochra, Multistability and hysteresis-based mechanism of pattern formation in biology, in *Pattern Formation in Morphogenesis-Problems and Their Mathematical Formalisation*, eds. by V. Capasso, M. Gromov, N. Morozova (Springer, New York, 2012)
24. S. Krömker, Model and Analysis of Heterogeneous Catalysis with Phase Transition. Ph.D. thesis, University of Heidelberg, 1997
25. O.A. Ladyzenskaja, V.A. Solonnikov, The linearisation principle and invariant manifolds for problems of magnetohydrodynamics, Boundary value problems of mathematical physics and related questions in the theory of functions, 7. Zap. Naucn. Sem. Leningrad. Otdel. Mat. Inst. Steklov. (LOMI) **38**, 46–93 (1973) (in Russian)
26. M.D. Li, S.H. Chen, Y.C. Qin, Boundedness and blow up for the general activator-inhibitor model. Acta Math. Appl. Sinica (English Ser.), **11**, 59–68 (1995)
27. C.-S. Lin, W.-M. Ni, I. Takagi, Large amplitude stationary solutions to a chemotaxis system. J. Differ. Equ. **72**, 1–27 (1988)
28. W. Loud, Periodic solutions of $x'' + cx' + g(x) = f(t)$. Mem. Am. Math. Soc. **31**, 58 (1959)
29. A. Marciniak-Czochra, Receptor-based models with diffusion-driven instability for pattern formation in hydra. J. Biol. Sys. **11**, 293–324 (2003)
30. A. Marciniak-Czochra, Receptor-based models with hysteresis for pattern formation in hydra. Math. Biosci. **199**, 97–119 (2006)
31. A. Marciniak-Czochra, Strong two-scale convergence and corrector result for the receptor-based model of the intercellular communication. IMA J. Appl. Math. (2012). doi:10.1093/imamat/hxs052
32. A. Marciniak-Czochra, M. Kimmel, Dynamics of growth and signalling along linear and surface structures in very early tumors. Comput. Math. Methods Med. **7**, 189–213 (2006)
33. A. Marciniak-Czochra, M. Kimmel, Modelling of early lung cancer progression: influence of growth factor production and cooperation between partially transformed cells. Math. Models Methods Appl. Sci. **17**, 1693–1719 (2007)
34. A. Marciniak-Czochra, M. Kimmel, Reaction-diffusion model of early carcinogenesis: the effects of influx of mutated cells. Math. Model. Nat. Phenom. **3**, 90–114 (2008)
35. A. Marciniak-Czochra, M. Ptashnyk, Derivation of a macroscopic receptor-based model using homogenisation techniques. SIAM J. Mat. Anal. **40**, 215–237 (2008)
36. A. Marciniak-Czochra, S. Härting, G. Karch, K. Suzuki, Dynamical spike solutions in a nonlocal model of pattern formation (2013). Preprint available at http://arxiv.org/abs/1307.6236
37. A. Marciniak-Czochra, G. Karch, K. Suzuki, Unstable patterns in reaction-diffusion model of early carcinogenesis. J. Math. Pures Appl. **99**, 509–543 (2013)
38. A. Marciniak-Czochra, G. Karch, K. Suzuki, Unstable patterns in autocatalytic reaction-diffusion-ODE systems (2013). Preprint available at http://arxiv.org/abs/1301.2002
39. A. Marciniak-Czochra, M. Nakayama, I. Takagi, Pattern formation in a diffusion-ODE model with hysteresis (2013). Preprint available at http://arXiv:1311.1737
40. K. Masuda, K. Takahashi, Reaction-diffusion systems in the Gierer–Meinhardt theory of biological pattern formation. Jpn J. Appl. Math. **4**, (1987) 47–58
41. H. Meinhardt, A model for pattern formation of hypostome, tentacles and foot in hydra: how to form structures close to each other, how to form them at a distance. Dev. Biol. **157**, 321–333 (1993)

42. H. Meinhardt, Turing's theory of morphogenesis of 1952 and the subsequent discovery of the crucial role of local self-enhancement and long-range inhibition. Interface Focus (2012). doi:10.1098/rsfs.2011.0097
43. H. Meinhardt, Modeling pattern formation in hydra—a route to understand essential steps in development. Int. J. Dev. Biol. (2012). doi:10.1387/ijdb.113483hm
44. M. Mimura, Stationary pattern of some density-dependent diffusion system with competitive dynamics. Hiroshima Math. J. **11**, 621–635 (1981)
45. M. Mimura, M. Tabata, Y. Hosono, Multiple solutions of two-point boundary value problems of Neumann type with a small parameter. SIAM J. Math. Anal. **11**, 613–631 (1980)
46. G. Mulone, V.A. Solonnikov, Linearisation principle for a system of equations of mixed type. Nonlinear Anal. **71**, 1019–1031 (2009)
47. J.D. Murray, *Mathematical Biology II. Spatial Models and Biomedical Applications*. Interdisciplinary Applied Mathematics, 3rd edn, vol 18 (Springer, New York, 2003)
48. W.-M. Ni, Qualitative properties of solutions to elliptic problems, in *Handbook of Differential Equations: Stationary Partial Differential Equations*, eds. by M. Chipot, P. Quittner, vol 1 (North-Holland, Amsterdam, 2004) pp. 157–233
49. W.-M. Ni, *The Mathematics of Diffusion* (Cambridge University Press, Cambridge, 2013)
50. W.-M. Ni, I. Takagi, On the Neumann problem for some semilinear elliptic equations and systems of activator-inhibitor type. Trans. Am. Math. Soc. **297**, 351–368 (1986)
51. W.-M. Ni, I. Takagi, On the shape of least energy solution to a semilinear Neumann problem. Commun. Pure Appl. Math. **44**, 819–851 (1991)
52. W.-M. Ni, I. Takagi, Locating the peaks of least-energy solutions to a semilinear Neumann problem, Duke Math. J. **70**, 247–281 (1993)
53. W.-M. Ni, I. Takagi, Point condensation generated by a reaction-diffusion system in axially symmetric domains. Jpn. J. Indust. Appl. Math. **12**, 327–365 (1995)
54. K. Pham, A. Chauviere, H. Hatzikirou, X. Li, H.M. Byrne, V. Cristini, J. Lowengrub, Density-dependent quiescence in glioma invasion: instability in a simple reaction-diffusion model for the migration/proliferation dichotomy. J. Biol. Dyn. **6**, 54–71 (2011)
55. M. Reed, B. Simon, *Methods of Modern Mathematical Physics. I. Functional Analysis*, 2nd edn (Academic Press Inc., New York, 1980)
56. F. Rothe, *Global Solutions of Reaction-Diffusion Systems*. Lecture Notes in Mathematics, vol 1072 (Springer, Berlin, 1984)
57. K. Sakamoto, Construction and stability analysis of transition layer solutions in reaction-diffusion systems. Tohoku Math. J. **42**, 17–44 (1990)
58. R.A. Satnoianu, M. Menzinger, P.K. Maini, Turing instabilities in general systems. J. Math. Biol. **41**, 493–512 (2000)
59. R. Schaaf, *Global Solution Branches of Two Point Boundary Value Problems*. Lecture Notes in Mathematics (Springer, New York, 1990)
60. J. Smoller, *Shock Waves and Reaction-Diffusion Equations*, Grundlehren der Mathematischen Wissenschaften, 2nd edn, vol 258 (Springer, New York, 1994)
61. K. Suzuki, Existence and behaviour of solutions to a reaction-diffusion system modelling morphogenesis. Ph.D. thesis, Tohoku University, Sendai, Japan, 1996
62. K. Suzuki, Mechanism generating spatial patterns in reaction-diffusion systems. Interdiscip. Inform. Sci. **17** 131–153 (2011)
63. K. Suzuki, I. Takagi, Collapse of patterns and effect of basic production terms in some reaction-diffusion systems. GAKUTO Internat. Ser. Math. Sci. Appl. **32**, 168–187 (2010)
64. K. Suzuki, I. Takagi, On the role of basic production terms in an activator-inhibitor system modelling biological pattern formation. Funkcial. Ekvac. **54**, 237–274 (2011)
65. I. Takagi, Point-condensation for a reaction-diffusion system. J. Differ. Equ. **61**, 208–249 (1986)
66. A.M. Turing, The chemical basis of morphogenesis. Philos. Trans. R. Soc. B **237**, 37–72 (1952)
67. D.M. Umulis, M. Serpe, M.B. O'Connor, H.G. Othmer, Robust, bistable patterning of the dorsal surface of the Drosophila embryo. Proc. Natl. Acad. Sci. **103**, 11613–11618 (2006)

68. L. Wang, H. Shao, Y. Wu, Stability of travelling front solutions for a forest dynamical system with cross-diffusion. IMA J. Appl. Math. **78**, 494–512 (2013)
69. J. Wei, Existence and stability of spikes for the Gierer–Meinhardt system, in *Handbook of Differential Equations: Stationary Partial Differential Equations*, vol V, Handbook of Differential Equations (Elsevier/North-Holland, Amsterdam, 2008) pp. 487–585
70. H. Weinberger, A simple system with a continuum of stable inhomogeneous steady states. Nonlinear Partial Differ. Equ. Appl. Sci. **81**, 345–359 (1983)

Nonlinear Hyperbolic Systems of Conservation Laws and Related Applications

Mapundi Kondwani Banda

1 Introduction

In this chapter, we present an overview of evolution equations defined by nonlinear hyperbolic conservation laws. This overview is not comprehensive rather it is a bird's eye view of the ideas that govern the analysis of nonlinear hyperbolic conservation equations. Various mathematical concepts as well as some ideas related to their discrete algorithms will be presented. Thus basic ideas of viscous regularisation, entropy, monotonicity, total variation bounds and the Riemann problem are discussed. These are also the underlying ideas in the development of discrete solutions as well as discrete theory. Finite volume methods such as the relaxation schemes will also be introduced. This is not a recommendation for the reader to use the method but rather a bias due to the author's previous work in the field. Some examples of applications of the nonlinear conservation laws in networks as well as optimal control leading to stabilisation of the system of interacting equations will be presented.

By a nonlinear conservation law in one space dimension, we imply a first-order partial differential equation (PDE) of the form:

$$\frac{\partial \rho}{\partial t} + \frac{\partial (\rho u)}{\partial x} = 0. \tag{1}$$

In this equation ρ represents a conserved quantity, for example, the mass density. The variables t and x are independent variables representing time and space,

M.K. Banda (✉)
Department of Mathematics and Applied Mathematics, University of Pretoria, Private Bag X20, Hatfield 0028, South Africa
e-mail: mapundi.banda@up.ac.za

© Springer International Publishing Switzerland 2015

J. Banasiak, M. Mokhtar-Kharroubi (eds.), *Evolutionary Equations with Applications in Natural Sciences*, Lecture Notes in Mathematics 2126,
DOI 10.1007/978-3-319-11322-7_9

respectively, and u is a flow velocity variable. The term $(\rho u)(x,t)$ is the mass flux of the conserved quantity through a cross-section of the flow domain in the normal direction at a point x and t. Equation (1) is popularly referred to as a continuity equation [50].

Examples of such equations include the density of gas flowing in a pipe considering that significant changes in the flow are one-dimensional, the continuity equation represents the flow of density of a non-viscous (inviscid) gas in which the flux represents a flow rate of gas through the cross-section of the pipe at a point x and t. To get a closed relation between ρ and u, one needs a constitutive relation or additional conservation laws. If u is known, the equation is referred to as an advection equation. Another popular example is the traffic flow model on a section of a single lane highway. Assume that on a highway section the car density is defined by ρ, cars are a conserved quantity giving Eq. (1). To close the equation, there is need for a relation between ρ and u, for example [50]

$$u = u(\rho) = u_{\max}\left(1 - \frac{\rho}{\rho_{\max}}\right)$$

i.e. the denser the traffic, the slower the cars will move. Here $u_{\max}$ is the maximum velocity and $\rho_{\max}$ is the maximum traffic density, i.e. when the cars are so-to-say bumper-to-bumper. This defines the flux as

$$f(\rho) = \rho u_{\max}\left(1 - \frac{\rho}{\rho_{\max}}\right).$$

The two equations discussed are prototypes of hyperbolic scalar conservation laws in one space dimension.

A popular protoype for a system of conservation laws is the inviscid Euler Equations of gas dynamics model, a system of hyperbolic conservation laws. In this case density (mass), momentum and energy are conserved. Let the flow of a gas be defined by its density ρ, velocity v, energy E and pressure p. To model the complete system, there is need for additional conservation laws (i.e. a system). The model takes the form:

$$\rho_t + (\rho v)_x = 0; \quad \text{conservation of mass,} \tag{2a}$$

$$(\rho v)_t + (\rho v^2 + p)_x = 0; \quad \text{conservation of momentum,} \tag{2b}$$

$$E_t + (v(E + p))_x = 0; \quad \text{conservation of energy.} \tag{2c}$$

The subscripts in this case represent partial differentiation with respect to the variable in the subscript. To close the system there is need for constitutive relations. Assume T is the absolute temperature of a gas then:

$$p = \rho T; \quad \text{the ideal gas law,}$$

$$E = \frac{\rho v^2}{2} + c_v \rho T; \quad \text{representing kinetic and thermal (internal) energy.}$$

In this case a system of three equations for three conserved quantities is obtained. To close the system a relation involving variable p which depends on ρ, v, E is assumed based on empirical considerations. In this case the hyperbolic conservation laws can be written in the familiar form with

$$u = \begin{pmatrix} \rho \\ \rho v \\ E \end{pmatrix} \quad \text{and} \quad F(u) = \begin{pmatrix} \rho v \\ \rho v^2 + p(u) \\ v(E + p(u)) \end{pmatrix}$$

giving a hyperbolic system of conservation laws in one space dimension:

$$\partial_t u + \partial_x f(u) = 0. \tag{3}$$

A further simplification of the above system would be the isothermal Euler equations of gas dynamics: assume temperature, T, is constant then

$$p = a^2 \rho$$

where a is the speed of sound in the gas. Hence we obtain a system that takes the form:

$$\begin{aligned} \rho_t + (\rho v)_x &= 0; \text{ conservation of mass,} \\ (\rho v)_t + (\rho v^2 + a^2 \rho)_x &= 0; \text{ conservation of momentum.} \end{aligned}$$

The flow is referred to as isothermal flow.

The above are but a few examples of conservation laws. In general the Cauchy problem for the hyperbolic conservation laws consists of Eq. (3) with initial conditions

$$u(x,0) = u_0(x). \tag{4}$$

In this chapter, a lot of discussion will centre around the Riemann problem which is (3) with initial conditions of the form:

$$u(x,0) = \begin{cases} u_L, & \text{for } x < 0; \\ u_R, & \text{for } x \geq 0. \end{cases} \tag{5}$$

The solution for a nonlinear Riemann problem can either be a shock propagating, a contact discontinuity or a rarefaction wave.

In this introductory review, we give an overview of the mathematical methods used in investigating the well-posedness and constructing of solutions of the scalar nonlinear conservation laws (1). This discussion will be extended to systems in one space dimension (3). This will be undertaken in Sect. 2. For further reading, we recommend [46, 63, 71]. For the theory of numerical methods, the reader may consult [33, 50, 64]. A brief introduction of numerical methods based on the Relaxation schemes will be presented in Sect. 3. Applications of these methods to networked flow or evolution equations can be found in 4. The interested reader is also encouraged to consult the following references: [62] for a classical treatment of systems; [48,59] for more recent aspects of nonlinear hyperbolic conservation laws; [31] for conservation laws on networks; [14,30,42] for careful studies and advances in the theory of low-order numerical approximation of hyperbolic conservation laws.

2 Mathematical Modelling and Analysis with Hyperbolic Conservation Laws

In this section the basic formulation of conservation laws will be discussed. These are derived from the basic physical principles of conservation, for example, conservation of mass or momentum, as alluded to in Sect. 1. Examples of conservations laws have been presented. Such equations are also popularly referred to as evolution equations. To be more specific, the special case of hyperbolic systems of conservation laws will be discussed in detail. These are first-order hyperbolic partial differential equations (PDE). They also play a prominent role in modelling flow and transport processes. It needs to be noted that interesting cases of this class of equations are usually nonlinear. It can be said that due to their nonlinear nature these models present special difficulties: formation of shocks (jump discontinuities) even though the initial data is continuous. As a consequence a great deal about their mathematical structure is not yet known. A popular alternative to approximate the solutions of these models is to apply numerical approximations which are also able to resolve such jump discontinuities.

We would like to point out that systems of conservation laws in a single space variable have been well-studied. For example, Euler equations of compressible flow, the inviscid gas dynamics equations in Eq. (2), are an important example of a hyperbolic system of conservation laws. The second popularly discussed model is the linear wave equation. In such phenomena signals propagate with finite speed. Singularities propagate along characteristics, such singularities arise spontaneously, leading to formation of shocks or jump discontinuities. Therefore, in nonlinear cases, time is not reversible as for linear equations thus future and past are different [46]. There is loss of information as time moves forward, which can be interpreted as an increase in entropy. Basic existence theory of solutions of hyperbolic conservation laws in single space variables was discussed in [32]. In general, apart from isolated results, no comparable theory exists for more space dimensions.

In this chapter we will, therefore, discuss the case of one spatial dimensional problems. In general, we seek a weak solution $u : \mathbb{R} \times [0,T] \to \mathbb{R}^k$ to the Cauchy problem, see also Eq. (3):

$$\partial_t u + \partial_x f(u) = 0; \quad u(x,0) = u_0(x)$$

with $f : \mathbb{R}^k \to \mathbb{R}^k$, which is a system of k conservation laws in $\mathbb{R} \times [0,T]$ with flux functions f. In this case T represents the final time. This system can also be written in quasilinear form as:

$$\partial_t u + A(u)\partial_x u = 0; \quad u(x,0) = u_0(x). \tag{6}$$

where

$$A(u) = \left(\frac{\partial f_i}{\partial u_j}(u)\right)_{1 \le i,j \le k}$$

is the Jacobian matrix of $f = (f_1, f_2, \ldots, f_k)^T$ i.e. $A(u) = D_u f$.

For smooth solutions, the solutions for Eqs. (3) and (6) are equivalent. In cases where it is not possible to obtain classical or smooth solutions, we seek weak solutions which will be discussed in Sect. 2.1. For basic knowledge on conservation laws, we recommend the following literature [26,33,46,52,63]. The peculiarities of the conservation laws include:

1. the evolution of shock discontinuities which require weak (in the distributional sense) solutions of (3).
2. non-uniqueness of weak solutions of (3).
3. identification of unique 'physically relevant' weak solutions of (3). In general we seek a solution, $u = u(x,t)$, which can be defined as a viscosity limit solution, $u = \lim_{\epsilon \to 0} u^\epsilon$ such that [15,26,33]

$$\partial_t u^\epsilon + \partial_x f(u^\epsilon) = \epsilon \partial_x (\nu \partial_x u^\epsilon), \qquad \epsilon\nu > 0.$$

4. The entropy condition which requires that for all convex entropy functions, $\eta(u)$, the following holds:

$$\eta(u)_t + \psi(u)_x \le 0.$$

The viscosity limit solution is somehow related to the entropy solution, u.

At this point, we would like to give a formal definition of a hyperbolic system in Definition 1.

Definition 1 Consider the quasilinear system in Eq. (6). The system is referred to as (strictly) hyperbolic if and only if the matrix $A(u)$ has only real (pairwise disjoint) eigenvalues (characteristic speeds) and the eigenvectors form a basis of $\mathbb{R}^k$.

Examples such as the inviscid Euler equations of gas dynamics have been given in Sect. 1. For the one-dimensional scalar conservation laws, the equation is always strictly hyperbolic. For Eq. (6) one needs to investigate the matrix $A(u)$. Below we give some examples of hyperbolic conservation laws.

One-Dimensional Isentropic Gas Dyamics Equations Consider a one-dimensional isentropic gas dynamics model in Lagrangian coordinates [33]:

$$\frac{\partial v}{\partial t} - \frac{\partial u}{\partial x} = 0;$$
$$\frac{\partial u}{\partial t} + \frac{\partial}{\partial x} p(v) = 0;$$

v is specific volume, u is the velocity, and the pressure $p(v)$. For a polytropic isentropic ideal gas: $p(v) = \alpha v^{-\gamma}$ for some constant $\alpha = \alpha(s) > 0$ (depending on entropy), $\gamma > 1$. The above system in quasilinear form is of the form:

$$\frac{\partial w}{\partial t} + A(w)\frac{\partial w}{\partial x} = 0;$$

where $A(w)$ is the Jacobian matrix with two real distinct eigenvalues

$$\lambda_1 = -\sqrt{(-p'(v))} < \lambda_2 = \sqrt{(-p'(v))}$$

i.e. $p'(v) < 0$. Hence the system is strictly hyperbolic provided $p'(v) < 0$. This is also referred to as the p-system. The p-system is the simplest nontrivial example of a nonlinear system of conservation laws.

Nonlinear Wave Equation Consider any nonlinear wave equation [33]

$$\frac{\partial^2 g}{\partial t^2} - \frac{\partial}{\partial x}\Big(\sigma\big(\frac{\partial g}{\partial x}\big)\Big) = 0.$$

This equation can be re-written in the form of a p-system as follows: set

$$u = \frac{\partial g}{\partial t}, \quad v = \frac{\partial g}{\partial x}, \quad p(v) = -\sigma(v).$$

Isothermal Flow Consider the isothermal Euler equation [33, 50]:

$$u = \begin{pmatrix} \rho \\ q \end{pmatrix}; \quad f(u) = \begin{pmatrix} q \\ \frac{q^2}{\rho} + a^2\rho \end{pmatrix}; \quad A(u) = \begin{pmatrix} 0 & 1 \\ -\frac{q^2}{\rho^2} + a^2 & \frac{2q}{\rho} \end{pmatrix} \tag{7}$$

where the momentum $q = \rho v$. The matrix has eigenvalues

$$\lambda_1 = v + a; \qquad \lambda_2 = v - a$$

where a is the speed of sound in the flow.

2.1 *Weak Solutions*

In this section, the idea of a weak solution will be presented. Two one-dimensional examples will be presented: the Burgers equation with a smooth initial condition and periodic boundary conditions and secondly the inviscid Euler equations of gas dynamics with jump initial conditions. The first example demonstrates the evolution of a jump discontinuity even though the initial conditions are continuous. The latter example will introduce the three types of solutions expected: propagating shocks, rarefaction waves as well as contact discontinuities.

2.1.1 Burgers Equation

Consider the Burgers equation with the given initial condition:

$$\partial_t u + \partial_x\left(\frac{u^2}{2}\right) = 0; \quad u(x,0) = 0.5 + \sin(x), \quad x \in [0, 2\pi].$$

According to Fig. 1, as time evolves the solution develops steep gradients which eventually become a discontinuity which propagates with time. Such discontinuous

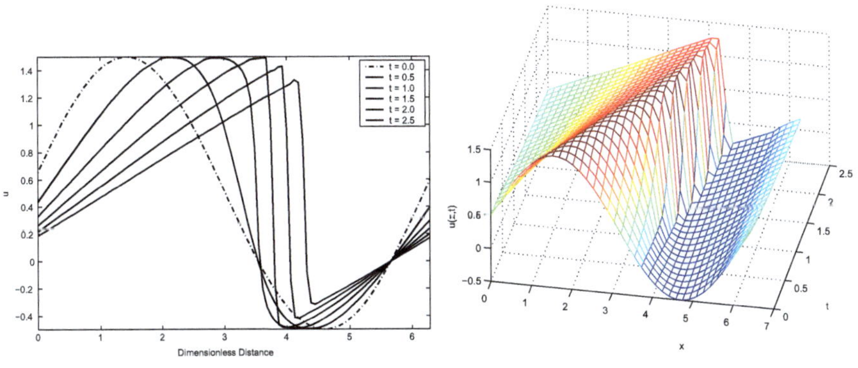

Fig. 1 Evolution of the solution for the Burgers equation at different times up to $T = 2.5$ to demonstrate the development of a jump discontinuity

solutions do not satisfy the PDE in the classical sense at all points. Therefore, there is need for a weaker definition of a solution of the PDE.

2.1.2 Sod's Problem

In this example, a nonlinear system of inviscid Euler Equations of gas dynamics, Eq. (2), is considered. The system is considered subject to the following initial conditions, a Riemann problem, with boundary conditions.

$$u(x,0) = \begin{cases} u_L, \, x < 0; \\ u_R, \, x > 0; \end{cases} \text{ and transparent boundary conditions,}$$

where $u_L = (1,0,2.5)^T$ and $u_R = (0.125,0,0.25)^T$. This can be considered as a case of a gas in a long pipe in which the two halves of the pipe are initially separated by a membrane. The gas on the left side of the membrane is subjected to a higher pressure (density) and as the experiment starts, the membrane is broken. In the solution in Fig. 2 in which different numerical schemes were used to resolve the solution, it can be seen that on the left hand side of the pipe a rarefaction wave (smooth solution) moving leftwards has developed. Around the position $x = 6$ a contact discontinuity has developed in which there is a jump in density but not in pressure and velocity. Lastly around $x = 8$ one observes a shock wave travelling to the right. This is a jump discontinuity.

The two examples demonstrate that there is need to re-think what we mean by a solution. Thus we seek weak solutions rather than classical ones. The idea of weak solutions will be discussed below. Before we discuss weak solutions, we would

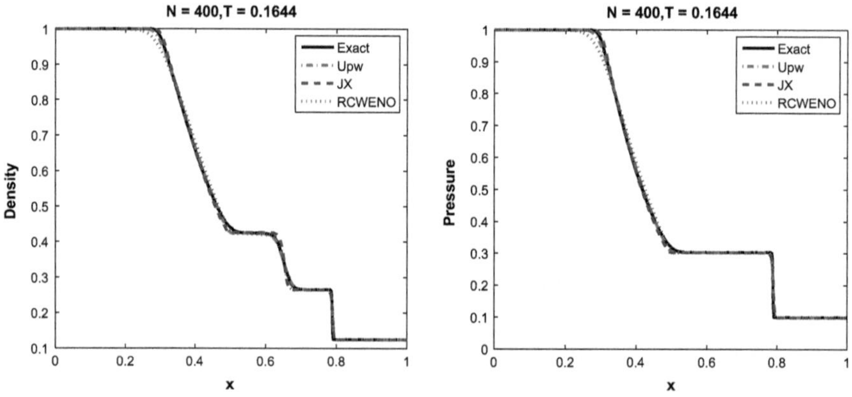

Fig. 2 Solution of the Sod's problem at time $t = 0.1644$ obtained using three methods: the upwind scheme (Upw), the second-order relaxation scheme (JX) and the third order relaxation scheme (RCWENO) [39] based on central weighted essentially non-oscillatory (CWENO) interpolation [44] compared with the exact solution (Exact)

like to demonstrate how discontinuities actually develop by using the method of characteristics in the next section.

2.2 *Method of Characteristics*

In this section a Cauchy problem for scalar conservation laws is considered:

$$u_t + f(u)_x = 0; \qquad u : \mathbb{R} \times \mathbb{R}^+ \to \mathbb{R}, \qquad f : \mathbb{R} \to \mathbb{R}, \tag{8}$$

$$u(x,0) = u_0(x), \qquad x \in \mathbb{R}. \tag{9}$$

To solve the problem, the method of characteristics will be discussed. For simplicity, consider an advection equation:

$$u_t + au_x = 0, \qquad a \in \mathbb{R}, \tag{10}$$

or the Burgers Equation:

$$u_t + \left(\frac{u^2}{2}\right)_x = 0. \tag{11}$$

The solution of the Cauchy problem for the advection equation is:

$$u(x,t) = u_0(x - at),$$

i.e. the initial data travels unchanged with velocity a. The left side of Eq. (10) can be interpreted as a directional derivative with:

$$\frac{du}{dt} = 0, \qquad \frac{dx}{dt} = a.$$

The equation $\dfrac{dx}{dt} = a, x(0) = x_0$ is the characteristic line which cuts through x_0 in the (x,t)-plane, i.e.

$$x = x_0 + at.$$

It can be shown that along the characteristics the solution is constant [33,50]. Hence suppose (x,t) is such that $x = x_0 + at$ holds, then

$$u(x,t) - u_0(x_0) - u_0(x - at)$$

is constant.

On the other hand Eq. (8) can be written in the form

$$u_t + f'(u)u_x = 0$$

for smooth solution u. The term $f'(u)$ is referred to as the characteristic velocity, the velocity with which information propagates. To find characteristics one needs to solve:

$$x'(t) = f'(u(x(t),t)), \qquad x(0) = x_0.$$

Along the solution $x(t)$, $u(x,t)$ is constant, i.e.:

$$\frac{d}{dt}u(x(t),t) = \frac{\partial u(x(t),t)}{\partial t} + \frac{\partial}{\partial x}u(x(t),t)x'(t) = u_t + f'(u)u_x = 0.$$

Therefore, $x'(t)$ is constant since $f'(u(x(t),t)$ is constant. Thus characteristics are straight lines given by:

$$x = x(t) = x_0 + f'(u(x_0,0))t = x_0 + f'(u_0(x_0))t. \tag{12}$$

If the above can be solved for x_0 for all (x,t), then the solution of the conservation law takes the form:

$$u(x,t) = u(x_0,0) = u_0(x_0)$$

where x_0 is implicitly given as:

$$x = x_0 + f'(u_0(x_0))t.$$

Unfortunately, Eq. (12) above cannot always be uniquely solved—the characteristics can intersect after some time. For example, in the case of the Burgers equation $f(u) = \frac{u^2}{2}$. Let the initial value be given by $u_0(x) = -x$. Then the characteristics are

$$x = x(t) = x_0 - x_0 t$$

and this gives

$$t = \frac{x_0 - x}{x_0} = 1 - \frac{x}{x_0}$$

and also

$$u(x,t) = u_0(x_0) = u_0\left(\frac{x}{1-t}\right).$$

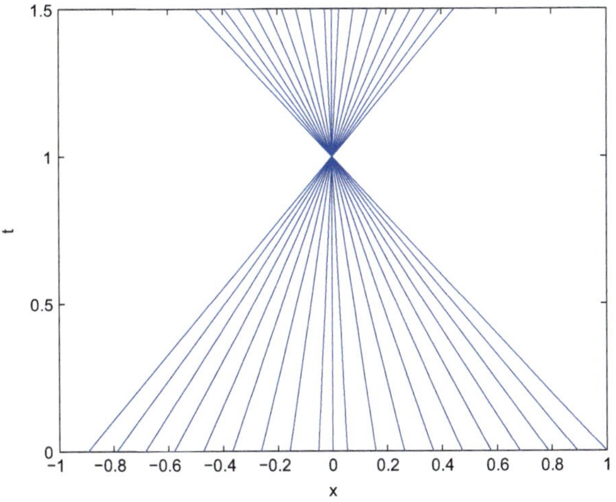

Fig. 3 Characteristics showing intersection resulting in shock development

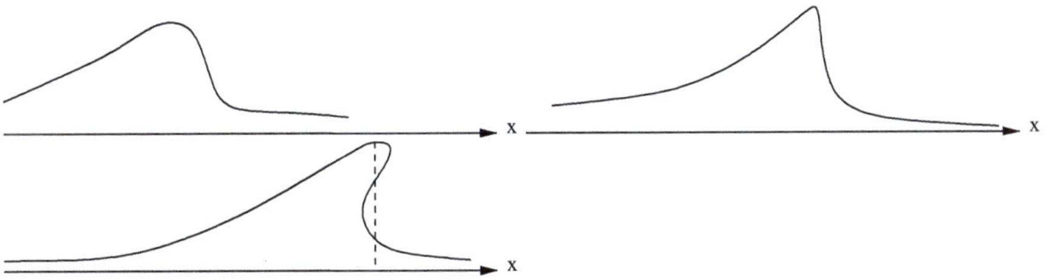

Fig. 4 Evolution of waves—development of a multi-valued function

The characteristics intersect at the point $(x, t) = (0, 1)$. In this case there is no unique solution for (x, t) anymore, see Figs. 3 and 4

In these figures, it can be seen that one attains a multivalued solution of the equation. Physically, for example, the density of a gas cannot be multivalued. In this case the discontinuity propagates as a shock. This can now be mathematically accurately treated with the help of weak (non-differentiable) solutions. In this case we seek solutions in L^1.

Definition 2 A function $u = u(x, t)$ is called a weak solution of the Cauchy–Problem, Eq. (8), if

$$\int_{\mathbb{R}} \int_0^{\infty} [u\phi_t + f(u)\phi_x]\, dt\, dx + \int_{\mathbb{R}} u_0(x)\, \phi(x, 0)\, dx = 0, \quad \forall \phi \in C_0^1(\mathbb{R} \times \mathbb{R}).$$

If u is a smooth solution of the equation, then

$$\int_{\mathbb{R}}\int_0^{\infty} [\phi u_t + \phi f(u)_x]\, dt\, dx = 0, \quad \forall \phi \in C_0^1(\mathbb{R}\times\mathbb{R}). \tag{13}$$

Integration by parts gives

$$\int_{\mathbb{R}}\left[-\int_0^{\infty} \phi_t u dt + \phi u\Big|_0^{\infty}\right] dx + \int_0^{\infty}\left[-\int_{\mathbb{R}} \phi_x f(u) dx + \phi f(u)\Big|_{-\infty}^{\infty}\right] = 0.$$

So smooth solutions are also weak solutions. Indeed, if the Riemann problem in Eq. (5) is considered, a weak solution is, for example, the shock wave

$$u(x,t) = \begin{cases} u_l\,, & x < st \\ u_r\,, & x > st, \end{cases}$$

where s is the shock speed:

To find s, the shock speed, the Rankine–Hugoniot conditions

$$s = \frac{f(u_l) - f(u_r)}{u_l - u_r} =: \frac{[f]}{[u]}$$

are applied [33, 50].

In addition the differential equation can be integrated to obtain:

$$\int_{t_1}^{t_2}\int_{x_1}^{x_2} \left\{\partial_t u(x,t) + \partial_x f(u(x,t))\right\} dx\, dt = 0; \quad u(x,0) = u_0(x)$$

and further giving the following integral forms:

1. $\displaystyle\int_{x_1}^{x_2} u(x,t_2)dx = \int_{x_1}^{x_2} u(x,t_1)dx + \int_{t_1}^{t_2} f(u(x_1,t))dt - \int_{t_1}^{t_2} f(u(x_2,t))dt$
2. $\displaystyle\frac{d}{dt}\int_{x_1}^{x_2} u(x,t)dx = f(u(x_1,t)) - f(u(x_2,t)).$

These are more difficult to work with than the differential equation. As a consequence, the "weak" form of the PDE is introduced. The weak form allows discontinuous solutions and is easier to work with. This form is also fundamental in development and analysis of numerical methods. Integrating (13) gives:

$$\int_0^{\infty}\int_{-\infty}^{\infty} \left\{\phi_t u(x,t) + \phi_x f(u(x,t))\right\} dx dt = \int_{-\infty}^{\infty} \phi(x,0) u(x,0) dx$$

$$\forall \phi \in C_0^1(\mathbb{R}\times\mathbb{R})$$

which is integration over a bounded domain. This approach introduces a new problem: non-uniqueness of solutions with same initial data—hence a criterion for finding the physically relevant solution is necessary. As an example, consider Burgers equation

$$f(u) = \frac{u^2}{2}, \quad s = \frac{1}{2} \cdot \frac{u_l^2 - u_r^2}{u_l - u_r} = \frac{1}{2}(u_l + u_r). \tag{14}$$

If the characteristics run into the shock, then it is called stable otherwise unstable. The mean value theorem implies:

$$s = \frac{f(u_l) - f(u_r)}{u_l - u_r} = f'(\xi), \quad \xi \in (u_l, u_r) \text{ or } \xi \in (u_r, u_l).$$

If f is convex, i.e. f' is monotone increasing, then

$$u_l > u_r \Rightarrow f'(u_l) > f'(\xi) > f'(u_r),$$

one has a stable shock, see Fig. 5, if on the other hand

$$u_l < u_r \Rightarrow f'(u_l) < f'(\xi) < f'(u_r),$$

then one has an unstable shock [33, 46, 50].

Another weak solution for $u_l < u_r$ is given by the so called rarefaction wave

$$u(x,t) = \begin{cases} u_l, & \frac{x}{t} \leq f'(u_l) \\ v(\frac{x}{t}), & f'(u_l) \leq \frac{x}{t} \leq f'(u_r) \\ u_r, & \frac{x}{t} \geq f'(u_r) \end{cases}$$

with $f'(v(\xi)) = \xi$ (Figs. 6 and 7).

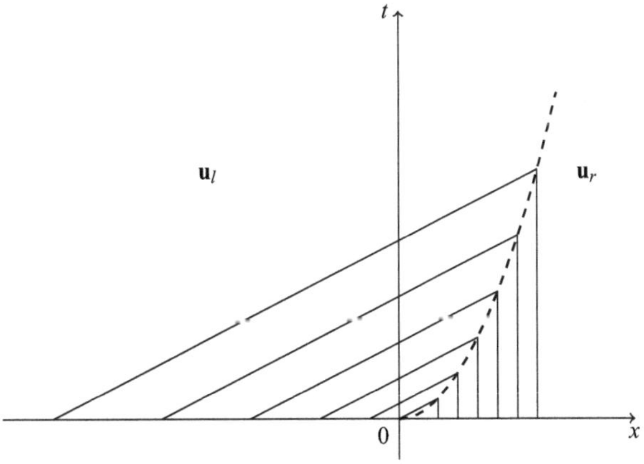

Fig. 5 Characteristics for $u_l > u_r$

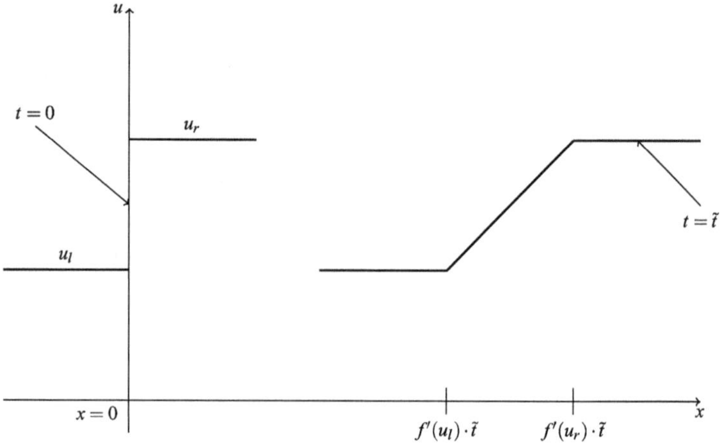

Fig. 6 A rarefaction wave

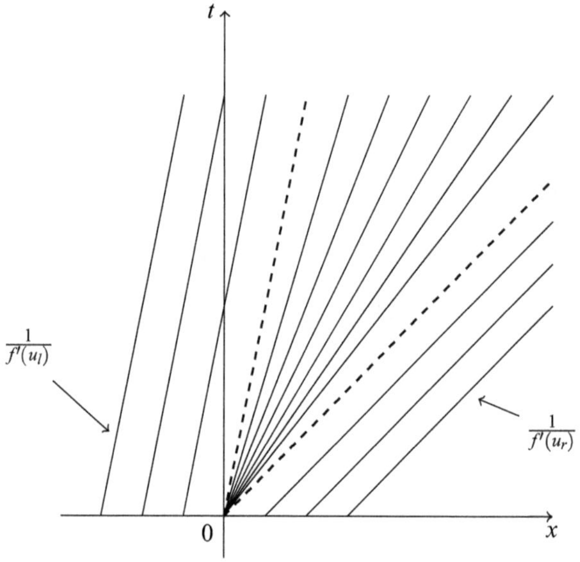

Fig. 7 Characteristics for rarefaction wave

For the Burgers equation ($f'(u) = u, v(\xi) = \xi$) then

$$u(x,t) = \begin{cases} u_l, & x \leq u_l t \\ \frac{x}{t}, & u_l \leq \frac{x}{t} \leq u_r \\ u_r, & x \geq u_r t \end{cases},$$

i.e. for $u_l < u_r$, we obtain two weak solutions, not a unique solution. But only one can be physically correct. Supplement the PDE by additional "jump conditions" that are satisfied across discontinuities. Rankine-Hugoniot conditions give the shock speed, s. Unfortunately, even if equations have the same smooth solutions, it does not necessarily mean that their weak solutions will remain the same in cases where a discontinuity develops. For example, multiplying the Burgers equation (11) by $2u$ gives

$$\partial_t(u^2) + \partial_x\left(\frac{2}{3}u^3\right) = 0; \tag{15}$$

which is also a conservation law for u^2. Equation (11) and (15) possess the same smooth solution. Considering a Riemann Problem with $u_l > u_r$, one can observe that Eq. (15) has shocks travelling at speed

$$s_2 = \frac{[\frac{2}{3}u^3]}{[u^2]} = \frac{2}{3}\left(\frac{u_r^3 - u_l^3}{u_r^2 - u_l^2}\right).$$

A comparison of the shock speeds for Eqs. (11) and (15) shows that

$$s_2 - s = \frac{1}{6}\frac{(u_l - u_r)^2}{(u_l + u_r)}$$

hence $s_2 \neq s$ when $u_l \neq u_r$. Hence, we obtain different discontinuous solutions [50]. Also note that to obtain equation (15), careful manipulation of the smooth solution had to be undertaken. To obtain a unique solution, it is possible to use viscosity regularisation or to supplement the conservation law with an entropy condition. In the following section, we will discuss the entropy condition which is a concept that finds its roots in thermodynamics.

2.3 Energy and Entropy

In this section a discussion of entropy will be undertaken. Entropy is usually at the centre of defining solutions of hyperbolic conservation laws in the weak sense (non-classical solutions). It is usually used to identify weak solutions that are of physical relevance. The entropy used as such is referred to as mathematical entropy. Before the concept of entropy is discussed in subsequent sections, a short discussion of the physical entropy will be undertaken.

In this section the presentation in [38] will be followed closely. It is known that the energy content of a system is measured by its internal energy per unit mass e. In a fluid, the total energy considered in the conservation of energy equation is the sum of its internal energy and its kinetic energy per unit mass:

$$E = e + \frac{v^2}{2}$$

where E denotes the total energy. Using the first law of thermodynamics the sources of variation of the total energy is due to work of forces acting on the system as well as heat transmitted to the system. Hence two fluxes are considered: the convective flux F_C:

$$F_C = \rho v(e + \frac{v^2}{2}) = \rho v E$$

as well as the diffusive flux F_D:

$$F_D = -\gamma \rho \kappa \nabla e$$

where κ is the thermal diffusivity coefficient and is defined empirically while γ is the ratio of specific heat coefficients under constant pressure and constant volume:

$$\gamma = c_p / c_v.$$

It can be noted that the diffusion flux represents the diffusion of heat in a medium at rest due to molecular conduction. Under Fourier's law of heat conduction it can be written as

$$F_D = -k \nabla T$$

where T is the absolute temperature and k is the thermal conductivity coefficient.

Furthermore, one can also consider energy variations due to two sources: the volume sources which are due to work of the volume forces f_e and heat source other than conduction (radiation, chemical reactions) denoted by q_H giving $Q_v = \rho f_e \cdot v + q_H$. There are also surface energy sources due to work done by internal shear stresses acting on the surface of the volume $Q_s = \sigma \cdot v = -pv + \tau \cdot v$ where p is the pressure, τ are the viscous shear stresses. Accounting for all the energy contributions, the conservation of energy takes the form:

$$\frac{\partial}{\partial t}(\rho E) + \nabla \cdot (\rho v E) = \nabla \cdot (k \nabla T) + \nabla \cdot (\sigma \cdot v) + W_t + q_H$$

where W_t is the work of the external volume forces $W_t = \rho f_e \cdot v$.

From the equation above, one can obtain an equation of internal energy of the form:

$$\frac{\partial}{\partial t}(\rho e) + \nabla \cdot (\rho v h) = (v \cdot \nabla) p + \varepsilon_v + q_H + \nabla \cdot (k \nabla T) \tag{16}$$

where $h = e + \frac{p}{\rho}$ and ε_v is the dissipation term. For details the reader may consult [38]. In addition the dissipation term ε_v takes the form

$$\varepsilon_v = (\tau \cdot \nabla) \cdot v = \frac{1}{2\mu}(\tau \otimes \tau^T) = \tau_{ij}\frac{\partial v_i}{\partial x_j}$$

where μ is the dynamic viscosity.

Introducing the continuity equation, Eq. (1), in Eq. (16) above, one obtains:

$$\rho\frac{de}{dt} = -p(\nabla \cdot v) + \varepsilon_v + \nabla \cdot (k\nabla T) + q_H \tag{17}$$

where: $\dfrac{d}{dt} = \dfrac{\partial}{\partial t} + v \cdot \nabla$ is the material or convective derivative, $p(\nabla \cdot v)$ is the reversible work of the pressure forces (vanishes in incompressible flows). Other terms are considered as heat additions: ε_v dissipation term—acting as an irreversible heat source; $F_C = -k\nabla T$ is flux due to heat conduction (Fourier's Law); q_H are heat sources as discussed above.

To clarify the entropy contribution, the entropy s per unit mass will be introduced through the thermodynamic relation

$$T\ ds = de + p\ d\left(\frac{1}{\rho}\right) = dh - \frac{dp}{\rho} \tag{18}$$

where h is the enthalpy. Thus it is now possible to separate reversible and irreversible heat additions using:

$$T\ ds = dq + dq'$$

where dq is a reversible heat addition and dq' is an irreversible heat addition. From the Second Principle of Thermodynamics: $dq' \geq 0$, i.e. in adiabatic flow ($dq = 0$). Hence entropy will always increase. Introducing Eq. (18) in (17) the following equation is obtained:

$$\rho T\frac{ds}{dt} = \varepsilon_v + \nabla \cdot (k\nabla T) + q_H \tag{19}$$

where the last two terms are reversible heat addition by conduction and other sources. For $q_H = 0$ and $k = 0$ the non-negative dissipation term ε_v is a non-reversible heat source. Equation (19) is the entropy equation of the flow. In conclusion it can be noted that this equation is important but not independent from the energy equation. Therefore, either this equation or the energy conservation equation needs to be added to the mass and momentum conservation equations. In addition entropy can not be classified as a 'conservative' quantity in the sense discussed above.

Thus mathematical expression of second principle of thermodynamics for an adiabatic flow without heat conduction or heat sources can be expressed as:

$$\rho T \frac{ds}{dt} = \varepsilon_v;$$

$$\rho T \Big(\frac{\partial s}{\partial t} + v \cdot \nabla s \Big) = \varepsilon_v.$$

The viscous dissipation is positive i.e. $\varepsilon_v \geq 0$ which implies that solutions of the inviscid Euler equations, which also physically imply a vanishing viscosity limit, need to satisfy the following entropy condition

$$\rho T \Big(\frac{\partial s}{\partial t} + v \cdot \nabla s \Big) \geq 0.$$

In other words any solution of Euler equations with a physical meaning satisfies the entropy condition above. Therefore, entropy is used as an additional condition, where necessary, to exclude non-physical solutions for uniqueness.

To conclude, it must be noted also that non-viscous, non-heat-conducting fluid, flow in the limit of vanishing viscosity admits both continuous and discontinuous solutions. Therefore, entropy variations for continuous flow variations give the entropy equation:

$$\rho T \Big(\frac{\partial s}{\partial t} + v \cdot \nabla s \Big) = 0$$

i.e. entropy constant along flow path. In absence of discontinuities, inviscid Euler equation describes isentropic flows where the value of entropy only varies from one flow path to another. From the inviscid Euler equations classical solutions also satisfy the entropy equation which can be derived as:

$$\frac{\partial(\rho S)}{\partial t} + \frac{\partial(v\rho S)}{\partial x} = 0$$

where S is the specific entropy. For a strict convex entropy function, for example, of the form $U = -\rho S$ [33] it can be shown that the limit of the viscosity solution will also satisfy, in the weak sense, the entropy inequality

$$\frac{\partial(\rho S)}{\partial t} + \frac{\partial(v\rho S)}{\partial x} \leq 0.$$

This is an instance of the Clausius-Duhem inequality [26].

Nature has its own way of solving the same problems. Equations are models of reality and some physical effects are ignored. For example, fluid flow always has viscous effects: strong near discontinuity i.e. a discontinuity can be thought of as a thin region with very steep gradients. Thus, one needs further conditions, in order to eliminate the non-physical weak solutions. Such a simple additional assumption is the entropy-condition [33, 46, 50, 63]. We observe, geometrically, for the Burgers equation that

- for convex f and $u_l > u_r$ we need to get a stable shock.
- For $u_l < u_r$ the shock solution disappears. In this case the rarefaction wave is the correct physical solution.

For convex f the entropy conditions discussed above are sufficient, but we need a more general solution, in which case we appeal to viscosity regularisation:

1. introduce a diffusive term in equation: consider linear advection equation:

$$\partial_t u + a\partial_x u = 0; \quad u(x,0) = u_0(x) \quad \text{gives } u(x,t) = u_0(x - at),$$

2. to obtain an advection-diffusion equation:

$$\partial_t u + a\partial_x u = Du_{xx}; \quad u(x,0) = u_0(x) \quad \text{where the flux is } f = f(u, u_x).$$

   The parabolic, conservation law, always has a smooth solution for $t > 0$ even if u is discontinuous. As the reader may notice, for $D \ll 1$, it is a good approximation for a linear advection equation.
3. Letting $D \to 0$—gives a "vanishing viscosity" method which is useful for analysis, in general, but practically not optimal [50].

For practical purposes, the entropy condition has been well accepted. Let η be a convex function and let ψ exist, for which we have for all smooth solutions u of the conservation law:

$$\eta(u)_t + \psi(u)_x = 0,$$

i.e. $\eta'(u)u_t + \psi'(u)u_x = 0$. For smooth solutions $u_t + f'(u)u_x = 0$ or $\eta'(u)u_t + \eta'(u)f'(u)u_x = 0$ giving

$$\eta'(u)f'(u) = \psi'(u). \tag{20}$$

Equation (20) needs to be solvable for the case $n \geq 2$. Such functions (η, ψ) are called entropy–entropy flux pairs. The entropy condition is sufficient for existence and uniqueness of weak solutions. The function $u(x,t)$ is the entropy solution if, for all convex entropy functions, $\eta(u)$, and corresponding entropy fluxes, $\psi(u)$, the inequality

$$\eta(u)_t + \psi(u)_x \leq 0,$$

is satisfied in the weak sense: i.e.

$$\int_{\mathbb{R}}\int_0^\infty (\phi_t\eta(u) + \phi_x\psi(u))\,dt\,dx + \int_{\mathbb{R}} \phi(x,0)\eta(u)(x,0)\,dx \leq 0 \tag{21}$$

$$\forall\phi \in C_0^1(\mathbb{R}\times\mathbb{R}), \quad \phi \geq 0.$$

Now we are able to define what we mean by an entropy solution:

Definition 3 Let u be a weak solution of the conservation law. Moreover, suppose u fulfills (21) for any entropy–entropy flux pairs (η, ψ), then the function u is called an entropy solution.

For scalar conservation laws every convex function η leads to an entropy, i.e. there exist infinite many entropies. It is also interesting to note that for multi-dimensional systems, entropy weak solutions are not unique. Recent results show that the usual concept of entropy solutions is inadequate for uniqueness of compressible and incompressible Euler equations [18, 27, 29].

At this point it is convenient to introduce more function spaces that are useful in analysing nonlinear conservation laws. These spaces are also well exploited in designing numerical schemes for approximating solutions for these equations. Here the notions of monotonicity, total variation bounds are significant.

Definition 4 Let $u \in L^\infty(\Omega)$, $\Omega \subset \mathbb{R}^n$ be open. Then the total variation of u is defined by

$$\mathrm{TV}(u) = \limsup_{\varepsilon \to 0} \frac{1}{\varepsilon} \int_\Omega |u(x+\varepsilon) - u(x)|\, dx.$$

The space of bounded variation is

$$\mathrm{BV}(\Omega) := \{u \in L^\infty(\Omega) : \mathrm{TV}(u) < \infty\}.$$

If $u' \in L^1(\Omega)$ holds, then

$$\mathrm{TV}(u) = \int_\Omega |u'|\, dx.$$

Theorem 1 (Kruskov[43]) *The scalar Cauchy-Problem*

$$\begin{aligned} u_t + (f(u))_x &= 0, \qquad f \in C^1(\mathbb{R}); \\ u(x,0) &= u_0(x), \quad u_0 \in L^\infty(\mathbb{R}) \end{aligned}$$

has a unique entropy solution $u \in L^\infty(\mathbb{R} \times \mathbb{R}^+)$, having the following properties:

(i) $\|u(\cdot,t)\|_{L^\infty} \le \|u_0(\cdot)\|_{L^\infty}, \quad t \in \mathbb{R}^+$;
(ii) $u_0 \ge v_0 \Rightarrow u(\cdot,t) \ge v(\cdot,t), \quad t \in \mathbb{R}^+$;
(iii) $u_0 \in BV(\mathbb{R}) \Rightarrow u(\cdot,t) \in BV(\mathbb{R})$ *and* $TV(u(\cdot,t)) \le TV(u_0)$;
(iv) $u_0 \in L^1(\mathbb{R}) \Rightarrow \int_{\mathbb{R}} u(x,t)\, dx = \int_{\mathbb{R}} u_0(x)\, dx, \quad t \in \mathbb{R}^+$;

(i)–(iv) are called L^∞-stability, monotonicity, TV-stability, conservativity.

The theorem can be extended to several dimensions $x \in \mathbb{R}^d$, $d > 1$. The theorem cannot be extended to the general case of systems ($k > 1$). Until now, there is no general proposition proved for the system case.

At this point some general remarks are in order [67]:

1. The family of admissible entropies in the scalar case consists of all convex functions, and the envelop of this family leads to Kruzkov's entropy pairs[43]

$$\eta(u; c) = |u-c|, \qquad \psi(u, c) = sgn(u-c)(f(u)-f(c)), \qquad c \in \mathbb{R}. \qquad (22)$$

2. L^1-contraction. If u^1, u^2 are two entropy solutions of the scalar conservation law, then

$$\|u^2(\cdot,t) - u^1(\cdot,t)\|_{L^1(x)} \leq \|u_0^2(\cdot) - u_0^1(\cdot)\|_{L^1(x)}$$

This implies that the entropy solution operator related to scalar conservation laws in L^1 is non-expansive or contractive and by Crandall-Tartar lemma [24], it is monotone

$$u_0^2(\cdot) \geq u_0^1(\cdot) \Rightarrow u^2(\cdot,t) \geq u^1(\cdot,t)$$

3. Semi-group: Let $\{S_t, t \geq 0\}$ be a one-parameter family of operators which form a semi-group of constant-preserving, monotone operators. S_t preserves constants if

$$S_t[u \equiv \text{Const}] = \text{Const} \qquad (23)$$

and its monotonicity implies

$$u_0^2(\cdot) \geq u_0^1(\cdot) \Rightarrow u^2(\cdot,t) \geq u^1(\cdot,t), \qquad \forall t \geq 0. \qquad (24)$$

Assuming $\{S_t\}$ satisfies basic semi-group relations:

$$S_{t+s} = S_t S_s, \qquad S_0 = I,$$

and it has an infinitesimal generator,

$$\partial_x \Lambda(u) = \lim_{\Delta t \downarrow 0} (\Delta t)^{-1}(S_t(u) - u).$$

Thus $S_t u_0 = u(t)$ may be identified as the solution of the abstract Cauchy problem

$$u_t + \partial_x f(u(t)) - 0 \qquad (25)$$

subject to initial conditions $u(0) = u_0$. For an L^1-setup for quasilinear evolution equations, the reader may consult [23]. In addition monotone, constant preserving solution operators of the Cauchy problem (25) are uniquely identified by the following entropy condition [67].

Theorem 2 (Kruzkov's Entropy Condition) *Assume* $\{u(t), t \geq 0\}$ *is a family of solutions for the Cauchy problem* (25) *which is constant-preserving,* (23)*, and satisfies the monotonicity condition* (24)*. Then the following entropy inequality holds:*

$$\partial_t \left|u(t) - c\right| + \partial_x \{sgn(u - c)(f(u) - f(c))\} \leq 0, \qquad \forall c.$$

Hence one observes that monotonicity and constant preserving properties recover Kruzkov entropy pairs.

2.4 Analysis of Systems

At this point we would like to delve into the ideas that guide the construction of solutions for one-dimensional systems. The building block for such constructions is the Riemann problem discussed in Sect. 2.1. We will firstly consider linear problems and then close with a brief discussion of the nonlinear systems. In this case we will only concentrate on the strict hyperbolic case.

Consider systems of conservation laws in one-space-dimension in (3) with $x \in \mathbb{R}, \quad u, f \in \mathbb{R}^k, \quad f = (f_1, \ldots, f_k)$. In quasilinear form, Eq. (6),

$$A(u) = f'(u) = \frac{\partial f^{(i)}(u)}{\partial u_j}, \qquad 1 \leq i, j \leq k.$$

Hence the linear systems are denoted as

$$u_t + Au_x = 0, \quad A \in \mathbb{R}^{k \times k}, \quad u(x, 0) = u_0(x).$$

For a hyperbolic equation A is diagonalisable with eigenvalues $\lambda_1, \ldots, \lambda_k$ and eigenvectors $r_1, \ldots, r_k$. Let $R = (r_1| \ldots |r_k), \quad AR = RD, \quad A = RDR^{-1}$. Using this, we can diagonalise the system as follows: Let $v = R^{-1}u$ (characteristic variables)

$$Rv_t + RDR^{-1}Rv_x = 0 \quad \text{or} \quad v_t + Dv_x = 0,$$

since R is constant. We obtain k scalar problems for $(v_1, \ldots, v_p, \ldots, v_k)$ with solutions

$$v_p(x, t) = v_p(x - \lambda_p t, 0).$$

Given $v(x, 0) = R^{-1}u_0(x)$, we obtain

$$u(x,t) = Rv(x,t) = \sum_{p=1}^{k} v_p(x,t)r_p = \sum_{p=1}^{k} v_p(x - \lambda_p t, 0)r_p.$$

The curves $x = x_0 + \lambda_p t$ are called characteristics of the p-th family ($x_p'(t) = \lambda_p$). The characteristics of the p-th family are given by $x_p(t)$ and take the form

$$x_p'(t) = \lambda_p, \quad x_p(0) = x_0, \quad p = 1, \dots, k.$$

Note that in general problems depend on u and are strongly coupled due to $R = R(u)$. In terms of the Riemann problem, we consider systems of conservation laws in one-space-dimension:

$$u_t + Au_x = 0,$$

where A is a constant matrix and initial conditions as given in Eq. (4). Assume we can express the initial conditions as

$$u_L = \sum_{p=1}^{k} \alpha_p r_p, \quad u_R = \sum_{p=1}^{k} \beta_p r_p, \quad v_p(x,0) = \begin{cases} \alpha_p, & x < 0; \\ \beta_p, & x > 0. \end{cases}$$

Now

$$v_p(x,t) = \begin{cases} \alpha_p, & x - \lambda_p t < 0; \\ \beta_p, & x - \lambda_p t > 0. \end{cases}$$

Hence

$$u(x,t) = \sum_{p=1}^{k_0} \beta_p r_p + \sum_{p=k_0+1}^{k} \alpha_p r_p.$$

where k_0 is the minimal value of p with $x - \lambda_p t > 0$, $(\lambda_1 \leq \dots \leq \lambda_k)$. State u can also be expressed in terms of jump discontinuities as follows:

$$\begin{aligned} u(x,t) &= \sum_{p=1}^{k} \alpha_p r_p + \sum_{p=1}^{k_0} \beta_p r_p - \sum_{p=1}^{k_0} \alpha_p r_p \\ &= u_L + \sum_{p=1}^{k_0} (\beta_p - \alpha_p) r_p \\ &= u_R - \sum_{p=k_0+1}^{k} (\beta_p - \alpha_p) r_p \end{aligned}$$

and

$$u_R - u_L = \sum_{p=1}^{k} (\beta_p - \alpha_p) r_p$$

Hence the solution of a Riemann problem can be considered as a splitting of difference $u_R - u_L$ in a sum of jumps, which move with velocity λ_p in the direction r_p in the phase space. For $k = 2$ a phase plot can be used to determine the intermediate states between u_L and u_R. If u_m is the intermediate state:

$$u_R - u_L = (\beta_1 - \alpha_1) r_1 + (\beta_2 - \alpha_2) r_2 = u_m - u_L + u_R - u_m$$

The jump $u_m - u_L$ moves with velocity λ_1 in the direction r_1 and the jump $u_R - u_m$ with λ_2 in the direction r_2. Now $\lambda_1 \leq \lambda_2$, which implies velocity of $u_m - u_L$ is smaller, thus this must be the first jump (recall, entropy conditions).

In the general nonlinear case, we consider Eq. (3). Rewriting the equation in quasilinear form, we obtain:

$$u_t + A(u) u_x = 0$$

for which $A(u)$ has eigenvalues $\lambda_1(u), \ldots, \lambda_k(u)$ and eigenvectors $r_1(u), \ldots, r_k(u)$. In this case hyperbolicity needs to be understood in the sense that $A(u)$ has a complete real eigensystem. For the ensuing discussion strict hyperbolicity is assumed, i.e. distinct eigenvalues $\lambda_i(u) \neq \lambda_j(u)$ for $i \neq j$.

We first start with introducing the Lax entropy conditions:

Definition 5 (Lax Entropy Condition) A discontinuity with left state u_L and right state u_R, moving with speed s for a conservation law with a convex flux function is entropy satisfying if

$$f'(u_L) > s > f'(u_R).$$

This implies that the characteristics are ingoing into the shock as time evolves. We also note that if f is convex then the correct weak solution is the limit as $\epsilon \to 0$ of the viscous problem if and only if the Lax entropy condition holds [63].

Definition 6 The p-th characteristic field is genuinely nonlinear, if

$$\nabla \lambda_p(u) \cdot r_p(u) \neq 0, \qquad \forall u$$

or it is linear degenerate, if

$$\nabla \lambda_p(u) \cdot r_p(u) = 0, \qquad \forall u.$$

At this point we observe that in the linear case all fields are linear degenerate.

The building block for constructing solutions in the one-dimensional case is the solution of the Riemann problem. A solution for (3) is sought subject to initial conditions (4). The solution is composed of k simple waves. Each of these waves is associated with one eigenpair $(\lambda_p(u), r_p(u)),\quad 1 \leq p \leq k$. The simple waves come in three forms as pointed out in the Sod's problem: if the p-th field is genuinely nonlinear, the waves are either p-shock or p-rarefaction waves. On the other hand if the p-th field is linearly degenerate, we obtain a contact wave. These simple waves are centred and depend on $\xi = \frac{x}{t}$. This is how the simple waves are determined:

1. A p-shock discontinuity of the form

$$u(\xi) = \begin{cases} u_L, & \xi < s \\ u_R, & \xi > s. \end{cases}$$

   As usual s denotes the shock speed which can be determined by the Rankine-Hugoniot condition and to be entropy satisfying $\lambda_p(u_L) > s > \lambda_p(u_R)$.
2. A p-rarefaction wave, $u(\xi)$, directed along the p-eigenvector, $\dot{u}(\xi) = r_p(u(\xi))$. The eigenvector is normalised such that $r_p \cdot \nabla \lambda_p \equiv 1$. This ensures that the gap between $\lambda_p(u_L) < \lambda_p(u_R)$ is filled with a fan of the form

$$\lambda_p(u(\xi)) = \begin{cases} \lambda_p(u_L), & \xi < \lambda_p(u_L) \\ \xi, & \lambda_p(u_L) < \xi < \lambda_p(u_R) \\ \lambda_p(u_R), & \lambda_p(u_R) < \xi. \end{cases}$$

3. A p-contact discontinuity of the form

$$u(\xi) = \begin{cases} u_L, & \xi < s \\ u_R, & \xi > s. \end{cases}$$

   As usual s denotes the shock speed which can be determined by the Rankine-Hugoniot condition such that $\lambda_p(u_L) = s = \lambda_p(u_R)$.

The admissibility of systems has been discussed in [46]. The theorem below summarises the admissibility of systems:

Theorem 3 (Lax Solution of Riemann Problems [15]) *The strictly hyperbolic admissible system* (3), *subject to Riemann initial data* (4) *with* $u_L - u_R$ *sufficiently small, admits a weak entropy solution, which consists of shock-, rarefaction- and contact-waves.*

For more discussion on the solution of Riemann problems, the reader may consult [16]. An extension to generalised Riemann problems subject to piecewise-linear initial data can be found in [13, 49].

In conclusion to this section we would like to briefly mention Glimm's theorem [32] which is celebrated as essential for existence theorems which are designed for general one-dimensional systems, see also [53, 60]

Theorem 4 ([67]) *There exists a weak entropy solution, $u(\cdot,t) \in L^{\infty}[BV \cap L^{\infty}(\mathbb{R}), [0,T]]$, of strictly hyperbolic systems (3), subject to initial conditions with sufficiently small variation, $\|u_o(\cdot)\|_{BV\cap L^{\infty}(\mathbb{R})} \leq \epsilon$.*

In the next section, we will consider numerical approximations for solutions of nonlinear hyperbolic equations. The relaxation scheme will be considered. Its simplicity and ease of implementation has been a motivation for presenting this scheme.

3 Numerical Approximation

In this section numerical approximations of solutions of nonlinear hyperbolic conservation laws will be presented. We will start by presenting the simplest and basic scheme in the finite volume framework. In the latter part of the section derivation of the relaxation scheme will be presented and numerical results on inviscid Euler gas dynamics equations will be presented.

3.1 Design Ideas of Numerical Schemes: The Godunov Upwind Scheme and the Lax-Friedrichs Scheme

Consider a scalar equation of the form in Eq. (3) with a linear flux function $f(u) = au$ in which $a =$ const, also referred to as a wave propagation speed. To approximate the solution on the spatial domain which is a subinterval of $\mathbb{R}$, we discretise the interval to approximate the solution at discrete points by taking a uniform grid with mesh-size Δx. In addition we similarly discretise the time variable t using Δt. Denote u_i^n as an approximation of $u(x_i, t^n)$ at the point $x_i = i\Delta x$ which is the midpoint of the cell $[x_{i-1/2}, x_{i+1/2}]$, $t^n = n\Delta t$. Thus we need explicit conservation schemes to approximate the equation in (3) in the form:

$$u_i^{n+1} = u_i^n + \frac{\Delta t}{\Delta x}\big[f_{i-1/2} - f_{i+1/2}\big], \quad i \in \mathbb{Z}, \quad n \geq 0 \tag{26}$$

where $f_{i+1/2}$ is the intercell numerical flux and u_i^0 for which the continuous version is given in (4). For more details on discrete solution methods, the reader may consult [33, 50, 67]. A first-order scheme referred to as the upwind scheme can be defined as follows: for $a > 0$:

$$u_i^{n+1} = u_i^n - \frac{\Delta t}{\Delta x}\left[au_i - au_{i-1}\right]$$

and for $a < 0$:

$$u_i^{n+1} = u_i^n - \frac{\Delta t}{\Delta x}\left[au_{i+1} - au_i\right].$$

The above schemes can be generalised by defining the discrete flux function as

$$f_{i+1/2} = \frac{1}{2}\left(f(u_{i+1}) + f(u_i)\right) - \frac{1}{2}|a|\left(u_{i+1} - u_i\right).$$

This is the simplest case of the famous Godunov methods [33, 50, 67]. It is also referred to as the Rusanov Scheme.

In general,

$$f_{i+1/2} = F(u_{i-k+1}^n, \ldots, u_{i+k}^n) \tag{27}$$

where F is a continuous numerical flux. The Lax-Friedrichs Method takes the form:

$$u_i^{n+1} = \frac{1}{2}(u_{i+1}^n + u_{i-1}^n) - \frac{a\Delta t}{2\Delta x}(u_{i+1}^n - u_{i-1}^n).$$

One has to also consider stability issues when designing the method [33,50,67]. For example, the Lax-Friedrichs Method is stable for

$$\left|\frac{a\Delta t}{\Delta x}\right| \leq 1/2.$$

As a consequence the methods are versatile but due to the fact that waves of different families are averaged together at each computational cell, their resolution is very diffuse.

Alternatively, one discretises space leaving time continuous to obtain a system of ordinary differential equations. This is referred to as a semi-discretisation or a method of lines. A stability analysis can be carried out on the system of ordinary differential equations. For Cauchy-Problems consider a subinterval e.g. $[0, 1]$ and prescribe a boundary condition. Prescription of boundary conditions depends on the direction of transport (i.e. wind direction). For example, for $a > 0$, we need a condition at $x = 0$. On the other hand, periodical boundary conditions can be given as

$$u(0, t) = u(1, t), \quad \forall t \geq 0.$$

To construct finite difference or finite volume schemes of the form discussed above, we will discuss some properties that the numerical flux needs to satisfy. Let the numerical flux be defined as in Eq. (27) which is an approximation of $f(u)$, a flux across the interface between neighbouring cells. The essential feature of the numerical scheme in Eq. (26) is their conservation form. This implies that the discrete flux over any spatial domain needs to depend on the discrete flux across the boundaries of the domain.

Definition 7 Now consider the hyperbolic conservation law in discretised form given in Eq. (26). The scheme is said to be consistent if

$$F(u, u, \dots, u) = f(u), \quad \forall u.$$

This is a $(2k + 1)$-point scheme, for $k = 1$ it is a three-point scheme.

Definition 8 Consider Eq. (26) where $f_{i+1/2}$ is the intercell numerical flux and u_i^0 is given. The scheme (26) is also referred to as a conservative scheme and it is in conservative form.

Theorem 5 (Lax and Wendrof [47]) *Consider the conservative difference scheme (26), with consistent numerical flux. Let* $\Delta t \downarrow 0$ *with fixed grid-ratios* $\lambda_j = \dfrac{\Delta t}{\Delta x_j} \equiv const_j$, *and let* $u^{\Delta t} = \{u_i^n\}$ *denote the corresponding solution (parameterised with respect to the vanishing grid-size). Assume that* $u^{\Delta t}$ *"converges" strongly to a function u (in some sensible way), the limit u is a weak solution of Eq.* (3).

It can be pointed out that the Lax-Wendroff theorem is fundamental in designing the so called shock-capturing schemes. In these schemes, instead of tracking jump discontinuities (evolving smooth pieces of the approximate solution on both sides of the discontinuity), conservative schemes capture a discrete shock discontinuity.

To study stability of the scheme, use the (discrete) L^p-norms for the sequences $u^n = (u_j^n)$. Other notions which are used in understanding the convergence behavior of discrete sequences of solutions of conservation laws are monotonicity/TVD. These will be considered in the scalar case:

Definition 9
- A scheme is monotone if given two sequences $v^0 = (v_j^0)$ and $w^0 = (w_j^0)$, $v^0 \geq w^0$ then $v^1 \geq w^1$, where $v \geq w$ means for all j, $v_j \geq w_j$ and $v_j^1 = (v_j^1)$
- A scheme is Total Variation Diminishing (TVD) if $\forall v^0 = (v_j^0)$,

$$\mathrm{TV}(v^n) \leq \mathrm{TV}(v^0), \tag{28}$$

where

$$\mathrm{TV}(v) = \sum_{j \in \mathbb{Z}} |v_{j+1} - v_j|.$$

In general, monotonicity and thus the TVD property is ensured if

$$\min_k\{u_k^n\} \le u_j^{n+1} \le \max_k\{u_k^n\}.$$

Hence it transforms a monotone sequence, say non-decreasing one, into a monotone (non-decreasing) sequence. Therefore, oscillations can not occur. Numerical solutions of conservation laws that satisfy the TVD property (28), also possess the following properties [67]:

- Convergence: the piecewise-constant numerical solution, $u^{\Delta x}(x,t^n) = \sum_i u_i^n \mathscr{X}_i(x)$, where $\mathscr{X}_i$ is a characteristic function defined on the cell $[x_{i+1/2}, x_{i-1/2}]$, converges strongly to a limit function, $u(x,t^n)$, as the spatial grid is refined. This with equicontinuity in time (in the L^1-norm) and the Lax-Wendroff theorem, give a weak solution, $u(x,t)$, of the conservation law (3).
- Spurious solutions can not occur due to the TVD condition, also the BV condition in Sect. 2.
- Accuracy: one can derive schemes of accuracy higher than first-order. Monotone schemes are at most first-order accurate [36]. TVD schemes are instead not restricted to first-order at least in one-dimension. This means we replace L^1-contractive solutions with (the weaker) condition of bounded variation solutions.

Thus to apply the schemes, the discrete equations must be written in conservation form which is able to capture the correct speeds of discontinuities. To obtain $f_{i+1/2}$, one needs an extrapolation of the solution in the cell to the boundary of the cell i.e. need $u_{i+1/2}$. The Godunov first-order upwind method uses piecewise constant data to extrapolate the solution to cell edges. A higher-order method, the Modified Upwind Scheme for Conservation Law (MUSCL), modifies the piecewise constant data, replaces the Godunov approach by some monotone first-order centred scheme, avoiding explicitly solving the Riemann Problem as follows: Consider piecewise constant data $\{u_i^n\}$, replace constant states u_i^n, understood as integral averages in cells $I_i = [x_{i-1/2}, x_{i+1/2}]$, by piecewise linear functions $u_i(x)$:

$$u_i(x) = u_i^n + \frac{(x - x_i)}{\Delta x}\Delta_i, \qquad x \in [0, \Delta x]$$

where Δ_i is a suitably chosen slope of $u_i(x)$ in cell I_i. For a discussion of slope choices, we refer to [33, 50], in which slope limiters based on TVD constraints are employed to remove possible oscillations. Thus one chooses $\bar{\Delta}_i = \phi\Delta_i$ where ϕ is a slope limiter function, for example, minmod, van Leer [69], or Superbee [66]. The centre of x_i in local coordinates is $x = \frac{1}{2}\Delta x$ and $u_i(x_i) = u_i^n$. Thus the boundary extrapolated values are:

$$u_i^L = u_i(0) = u_i^n - \frac{1}{2}\bar{\Delta}_i;$$

$$u_i^R = u_i(\Delta x) = u_i^n + \frac{1}{2}\bar{\Delta}_i.$$

The values u_i^L and u_i^R are the new arguments of the numerical flux function, $f_{i+1/2}(u_{i+1}^L, u_i^R)$. Thus the problem reduces to a flux estimate at each interface. For ideas on different flux approximations, we refer to [45, 68].

The characteristic variables play an essential role: upwind schemes are generally derived for a scalar equation hence for systems, the quantities that are being convected are the characteristic variables. Therefore, to prevent spurious oscillations, characteristic variables must also be numerically transported.

3.2 The Relaxation Schemes

Below a scheme, which is simple to implement since it does not actively consider the direction of the flow, is presented [2, 5, 39]. The scheme is developed with the spirit of central schemes like the Rusanov or Lax-Friedrichs but based on kinetic considerations of the flow [1, 55]. The conservation law in Eq. (3) is re-written as a balance law in which the right-hand side is a stiff relaxation term as follows:

$$u_t + v_x = 0; \tag{29}$$

$$v_t + a^2 u_x = -\frac{1}{\varepsilon}(v - f(u)); \tag{30}$$

where a is a value yet to be determined. Equation (29)–(30) can be rewritten in the form:

$$\begin{pmatrix} u \\ v \end{pmatrix}_t + \begin{pmatrix} 0 & 1 \\ a^2 & 0 \end{pmatrix} \begin{pmatrix} u \\ v \end{pmatrix}_x = \frac{1}{\varepsilon} g(u).$$

The eigenvalues of this Jacobian matrix are $\pm a$. The characteristic variables take the form: $\begin{pmatrix} v + au \\ v - au \end{pmatrix} =: W$ from which we obtain a semi-linear system coupled by the source term that takes the form:

$$\frac{\partial W}{\partial t} + \Lambda \frac{\partial W}{\partial x} = \frac{1}{\varepsilon} G.$$

3.2.1 Spatial Discretisation

Let $\Delta x = x_{i+1/2} - x_{i-1/2}$, $\Delta t = t_{n+1} - t_n$, and $\omega_{i+1/2}^n := \omega(x_{i+1/2}, t_n)$ with the cell average

$$\omega_i^n = \frac{1}{\Delta x} \int_{x-1/2}^{x+1/2} \omega(x, t_n)\, dx.$$

Then the semi-discrete relaxation system takes the form:

$$\frac{du_i}{dt} + D_x v_i = 0; \tag{31}$$

$$\frac{dv_i}{dt} + a^2 D_x u_i = -\frac{1}{\varepsilon}(v_i - f(u_i)); \tag{32}$$

where D_x is the discrete equivalent of ∂_x. Now the flux at cell boundaries can be approximated as follows: consider interval $I_i = [x_{i-1/2}, x_{i+1/2}]$, denote an approximating polynomial on cell I_i by $p_i(x,t)$ then

$$\tilde{u}(x,t) = \sum_i p_i(x,t;u)\mathscr{X}_i(x);$$

where $\mathscr{X}$ is a characteristic function defined on cell I_i. Denote the values of u at cell boundary point between cell I_i and I_{i+1}, $x_{i+1/2}$, as:

$$u^R(x_{i+1/2};u) = p_{i+1}(x_{i+1/2};u);$$
$$u^L(x_{i+1/2};u) = p_i(x_{i+1/2};u).$$

To apply the MUSCL approach, characteristic variables are used in the reconstruction

$$(v-au)_{i+1/2} = (v-au)^R_{i+1/2} = p_{i+1}(x_{i+1/2}; v-au);$$
$$(v+au)_{i+1/2} = (v+au)^L_{i+1/2} = p_i(x_{i+1/2}; v+au);$$

giving:

$$u_{i+1/2} = \frac{1}{2a}\Big(p_i(x_{i+1/2}; v+au) - p_{i+1}(x_{i+1/2}; v-au)\Big);$$
$$v_{i+1/2} = \frac{1}{2}\Big(p_i(x_{i+1/2}; v+au) + p_{i+1}(x_{i+1/2}; v-au)\Big).$$

The first-order schemes are derived by choosing the polynomial: $p_i(x,u) = u_i$ giving $(v+au)_{i+1/2} = (v+au)_i$ and $(v-au)_{i+1/2} = (v-au)_{i+1}$ and

$$u_{i+1/2} = \frac{u_i + u_{i+1}}{2} - \frac{v_{i+1} - v_i}{2a};$$
$$v_{i+1/2} = \frac{v_i + v_{i+1}}{2} - a\frac{u_{i+1} - u_i}{2}.$$

A second-order polynomial with slope limiters can also be used, giving:

$$u_{i+1/2} = \frac{u_i + u_{i+1}}{2} - \frac{v_{i+1} - v_i}{2a} + \frac{\sigma_i^+ + \sigma_{i+1}^-}{4a};$$

$$v_{i+1/2} = \frac{v_i + v_{i+1}}{2} - a\frac{u_{i+1} - u_i}{2} + \frac{\sigma_i^+ - \sigma_{i+1}^-}{4}.$$

Using Sweby's notation: slopes of $v \pm au$ can be defined as:

$$\sigma^\pm = (v_{i+1} \pm au_{i+1} - v_i \mp au_i)\phi(\theta_i^\pm);$$

$$\theta_i^\pm = \frac{v_i \pm au_i - v_{i-1} \mp au_{i-1}}{v_{i+1} \pm au_{i+1} - v_i \mp au_i}.$$

Below, some popular slope limiters are presented: Minmod Slope Limiter:

$$\phi(\theta) = \max(0, \min(1, \theta))$$

and van Leer slope limiter:

$$\phi(\theta) = \frac{|\theta| + \theta}{1 + |\theta|}.$$

It should be observed that if $\sigma_i^\pm = 0$ or $\phi = 0$, we obtain the first-order discretisation.

To complete the description of the relaxation systems initial conditions are approximated as: $u(x,0) = u_0(x), \quad v(x,0) = v_0(x) = f(u_0(x))$. It should also be pointed out that the choice of a, the free parameter, must be made such that the problem is well-defined. For this the regularisation properties of the relaxation framework are exploited in the choice of a as $\varepsilon \to 0$. The Chapman-Enskog asymptotic analysis [39] gives the sub-characteristic condition:

$$-a \le f'(u) \le a; \quad \forall u. \tag{33}$$

The limit $\varepsilon \to 0$ also gives the discretization of the original conservation law given in (3). In practice, for the cases where $\varepsilon \to 0$, the constraint $v = f(u)$ is used in Eqs. (29)–(30) (in which case the first-order scheme becomes the Rusanov scheme). This scheme is also referred to as the relaxed scheme. Otherwise one chooses a small value of ε, for example, $\varepsilon = 10^{-6}$, and solves (29)–(30) numerically. This scheme is referred to as the relaxing scheme.

3.2.2 Time Discretization

In general the time discretisation schemes can be derived using TVD Runge-Kutta schemes [34]. Consider Eq. (31). In this case, it can be observed that the strategy is as follows:

- Treat space discretizations separately using a MUSCL-type formulation.
- Treat time by an ordinary differential equation (ODE) solver (method of lines), for example, Implicit-Explicit (IMEX) schemes since the source term (RHS) is stiff.
- Choice of a^2 is wide as long as the sub-characteristic condition (33) is satisfied: a_i's can be chosen as global characteristic speeds, local speeds, depending on the stability and accuracy needs of implementation.
- Source terms are incorporated accordingly.

The semi-discrete formulation above is a system of ODEs which can be integrated using Implicit-Explicit (IMEX) Runge-Kutta approaches as follows: consider the semi-discrete equation (31) in a general form as follows [56–58]:

$$\frac{d\mathscr{Y}}{dt} = \mathscr{F}(\mathscr{Y}) - \frac{1}{\varepsilon}\mathscr{G}(\mathscr{Y}),$$

In general the following strategy is applied in solving the system:

1. Treat non-stiff stage, $\mathscr{F}$, with an explicit Runge-Kutta scheme.
2. Treat the stiff stage, $\mathscr{G}$, with a diagonally implicit Runge-Kutta (DIRK) scheme.
3. Scheme must be asymptotic-preserving.
4. The limiting Scheme i.e. as $\varepsilon \to 0$ must be Strong Stability Preserving (SSP) [34]

$$\|\mathscr{Y}^{n+1}\| \leq \|\mathscr{Y}^{n}\|.$$

Here $\mathscr{Y}^n$ is the approximate solution at $t = n\Delta t$. Examples of these schemes can be represented using the so called Butcher tables [58]:

$$\begin{array}{c|ccccc} \tilde{c}_1 & \tilde{a}_{11} & \tilde{a}_{12} & \tilde{a}_{13} & \cdots & \tilde{a}_{1s} \\ \tilde{c}_2 & \tilde{a}_{21} & \tilde{a}_{22} & \tilde{a}_{23} & \cdots & \tilde{a}_{2s} \\ \tilde{c}_3 & \tilde{a}_{31} & \tilde{a}_{32} & \tilde{a}_{33} & \cdots & \tilde{a}_{3s} \\ \vdots & \vdots & \vdots & \vdots & \vdots & \vdots \\ \tilde{c}_s & \tilde{a}_{s1} & \tilde{a}_{s2} & \tilde{a}_{s3} & \cdots & \tilde{a}_{ss} \\ \hline & \tilde{b}_1 & \tilde{b}_2 & \tilde{b}_3 & \cdots & \tilde{b}_s \end{array} \qquad \begin{array}{c|ccccc} c_1 & a_{11} & a_{12} & a_{13} & \cdots & a_{1s} \\ c_2 & a_{21} & a_{22} & a_{23} & \cdots & a_{2s} \\ c_3 & a_{31} & a_{32} & a_{33} & \cdots & a_{3s} \\ \vdots & \vdots & \vdots & \vdots & \vdots & \vdots \\ c_s & a_{s1} & a_{s2} & a_{s3} & \cdots & a_{ss} \\ \hline & b_1 & b_2 & b_3 & \cdots & b_s \end{array}$$

For practical purposes, the IMEX method is implemented as follows:

- For $l = 1, \ldots, s$,

  1. Evaluate $\mathbf{K}_l^*$ as:

$$\mathbf{K}_l^* = \mathscr{Y}^n + \Delta t \sum_{m=1}^{l-1} \tilde{a}_{lm}\mathscr{F}(\mathbf{K}_m) - \frac{\Delta t}{\varepsilon} \sum_{m=1}^{l-1} a_{lm}\mathscr{G}(\mathbf{K}_m).$$

2. Solve for $\mathbf{K}_l$:

$$\mathbf{K}_l = \mathbf{K}_l^* - \frac{\Delta t}{\varepsilon} a_{ll} \mathscr{G}(\mathbf{K}_l).$$

- Update $\mathscr{Y}^{n+1}$ as:

$$\mathscr{Y}^{n+1} = \mathscr{Y}^n + \Delta t \sum_{l=1}^{s} \tilde{b}_l \mathbf{F}(\mathbf{K}_l) - \frac{\Delta t}{\varepsilon} \sum_{l=1}^{s} b_l \mathscr{G}(\mathbf{K}_l).$$

Below, we present some examples of the Butcher tables for the second-order and third-order schemes:

Second-Order Scheme

$$\begin{array}{c|cc} 0 & 0 & 0 \\ 1 & 1 & 0 \\ \hline & \frac{1}{2} & \frac{1}{2} \end{array} \qquad \begin{array}{c|cc} -1 & -1 & 0 \\ 2 & 1 & 1 \\ \hline & \frac{1}{2} & \frac{1}{2} \end{array}$$

Third-Order Scheme

$$\begin{array}{c|ccc} 0 & 0 & 0 & 0 \\ \gamma & \gamma & 0 & 0 \\ 1-\gamma & \gamma-1 & 2-2\gamma & 0 \\ \hline & 0 & \frac{1}{2} & \frac{1}{2} \end{array} \qquad \begin{array}{c|ccc} 0 & 0 & 0 & 0 \\ \gamma & 0 & \gamma & 0 \\ 1-\gamma & 0 & 1-2\gamma & \gamma \\ \hline & 0 & \frac{1}{2} & \frac{1}{2} \end{array}$$

where $\gamma = \frac{3+\sqrt{3}}{6}$. The advantage of this approach is that numerically neither linear algebraic nor nonlinear source terms can arise. As $\varepsilon \longrightarrow 0$ the time integration procedure tends to an SSP time integration scheme of the limit equation (3). The only restriction is the usual CFL condition

$$\text{CFL} = \max\left(\frac{\Delta t}{\Delta x}, \lambda \frac{\Delta t}{\Delta x}\right) \le 1.$$

In the following section, we present some numerical examples for which the relaxation scheme have been applied. We also wish to point out that the results in Figs. 1 and 2 were also produced by the relaxation schemes of different accuracy. In the case below, we extend the application of the scheme to two dimensional Euler equations.

3.2.3 Some Numerical Results

In this section, examples of results produced using the above defined schemes are presented. The one-dimensional examples have been presented in Sect. 2 above. A dimension-by-dimension extension can be undertaken to solve problems in higher space dimensions [4, 5, 7, 61]. It must be mentioned here that one advantage of the finite volume formalism is the ability to deal with unstructured meshes in multi-dimensional problems while still applying the one-dimensional numerical schemes. Here, two examples of two dimensional inviscid Euler problems are presented:

$$\partial_t \rho + \partial_x m + \partial_y n = 0; \tag{35a}$$

$$\partial_t m + \partial_x(\rho u^2 + p) + \partial_y(\rho uv) = 0; \tag{35b}$$

$$\partial_t n + \partial_x(\rho uv) + \partial_y(\rho v^2 + p) = 0; \tag{35c}$$

$$\partial_t E + \partial_x(u(E + p)) + \partial_y(v(E + p)) = 0; \tag{35d}$$

$$p = (\gamma - 1)(E - \frac{\rho}{2}(u^2 + v^2)). \tag{35e}$$

Double Mach Reflection Problem [72]

Consider flow in the domain $\Omega = [0, 4] \times [0, 1]$, the bottom wall is a reflecting wall in interval: $[1/6, 4]$. A Mach 10 shock is introduced at $x = 1/6$, $y = 0$, 60^0 angle with x-axis. Further, exact post-shock condition at bottom boundary $[0, 1/6]$ are imposed. The rest of the boundary is a reflective boundary. On the top boundary exact motion of a Mach 10 shock is used. Results are displayed at $t = 0.2$ in Fig. 8.

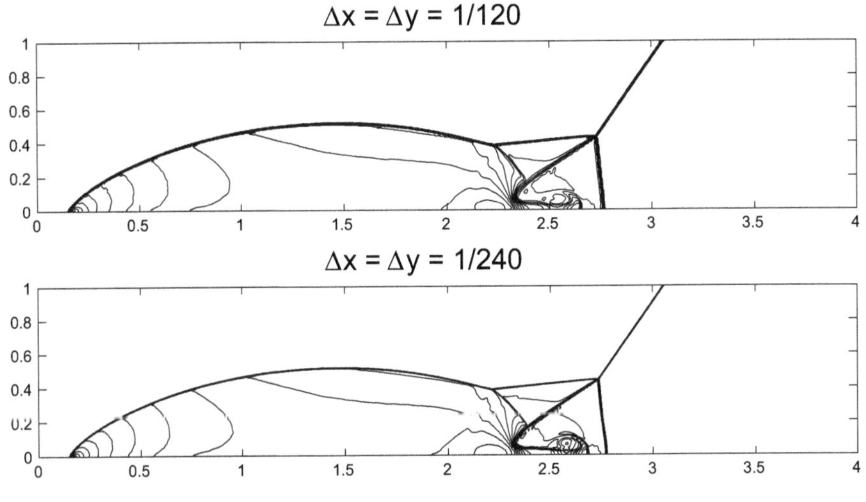

Fig. 8 Solution of the double Mach reflection problem obtained using the fifth-order relaxation scheme [5]

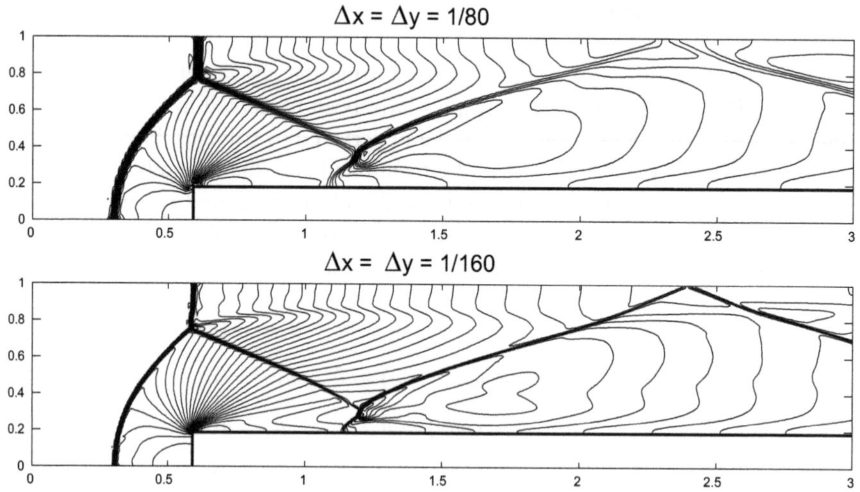

Fig. 9 Solution of forward-facing step problem obtained using the fifth-order relaxation scheme [5]

Forward-Facing Step Problem [72]

The next experiment involves a right-going Mach 3 uniform flow entering a wide tunnel 1 unit by 3 units long. The step is 0.2 units high located 0.6 units from left hand end of tunnel. As initial conditions a uniform right-going Mach 3 flow is considered. The boundary conditions are defined by a reflecting boundary condition along walls, inflow and outflow boundary conditions are applied at entrance and exit of tunnel. Results at $t = 4.0$ are given in Fig. 9.

3.2.4 Other Applications of the Relaxation Schemes

The relaxation approach has been applied in numerical methods for equations that avail themselves to being expressed in a kinetic formulation. In this formulation the $u(x,t)$ can be represented as an average of a 'microscopic' density function, $f(x,v,t)$. The formulation is based on classical kinetic models such as the Boltzmann equation. The useful tool is the velocity averaging which yields equations of moments of the kinetic model. These moments are solutions of the classical fluid flow equations like the incompressible Navier-Stokes equations. The most popular numerical approach for this is the Lattice Boltzmann Model (LBM). The relaxation approach can also be derived from the Boltzmann model by way of asymptotic analysis based on the Hilbert expansion in the small Knudsen and Mach number. Without going into details, this approach has been applied in simulating incompressible Navier-Stokes equations [7, 11, 40, 41], turbulent flow [6], the incompressible flow models with radiative transfer [10].

4 Applications in Networked Flows

Consider applications which have an inherent network and transport structure: traffic flow, gas or water transportation networks, telecommunication, blood flow or production systems (supply chain networks). The network is considered as a directed graph with arcs (edges/links) and vertices (nodes). Flow or Transport phenomena along each arc are sufficiently described by a one-dimensional hyperbolic equation, for example, a conservation law or a balance law (conservation law with source terms). To define flow or transport at the vertex, physical coupling conditions at each vertex are described by an algebraic condition. The mathematical problem can be considered as a coupled system of hyperbolic balance or conservation laws. In this section we discuss or develop an appropriate model for the arcs then define coupling conditions at nodes. To close the section, we will also investigate optimal control problems such as boundary stabilisation.

4.1 Modelling and Simulation of Networked Flow

In network flow problems, the mathematical model of the underlying physical process is formulated by an interconnected hierarchy of dynamics on the arcs/edges using mathematical modelling principles. These dynamics are coupled and interact with other processes by transmission conditions at the vertices/nodes of the graph/network. The transmission conditions might obey other independent dynamic laws defined based on physical, chemical or engineering considerations, the reader may refer to Banda et al. [8, 9] and references therein.

Arising mathematical issues deal with modelling and the interaction of different dynamics on arcs and the influence of the transmission conditions on the local and global dynamics of the network. This in particular involves the interaction between different dynamic and/or discrete models. Other problems arise due to the underlying complex physics and governing principles inside the vertices which are usually not known exactly, but need to be formulated in a concise mathematical formulation for further treatment.

From a practical point of view, network problems are usually large-scale and contain additional complexity due to their geometry. Hence, efficient numerical methods and new techniques are required to compute solutions for given accuracy and within a reasonable time. In most cases real-time solutions would be the desired goal.

To address such problems, one considers modelling of the dynamics on the arcs and development or application of efficient and accurate numerical methods for resolving these dynamics. Already developed models are available and can be applied. Where necessary a careful derivation of simplified dynamics which yields a trade-off between accuracy and computability can be considered. The second step is the modelling of transmission or coupling conditions informed by the underlying physical processes and mathematical analysis of the derived conditions. In particular, the compatibility of the coupling with both mathematical and engineering needs are of interest. For example, in the case of gas dynamics in pipelines, results are already available. The details of mathematical proofs of existence of solutions to a coupled system of dynamics on arcs and vertices are presented in [8, 9]. Here we give a summary of the results that are known from literature:

1. Gas Networks: conditions of conservation of mass and equal pressure at the nodes are well-posed.
2. Compressor condition and conservation of mass through compressor are well-posed.
3. Assumption of gas with minor losses at the junction was not taken into consideration (i.e. pressure drop factor—depends on geometry, flow and density). In addition results on three-dimensional flow, which is more realistic have not been considered.

4.1.1 Flow in a Single Pipe

To present flow in a network of pipes, it is necessary to consider flow in a single pipe. Thereafter extensions of single pipe flow to coupled pipes in a network will be discussed. The link between the two are algebraic relations which are defined using physical considerations.

The flow model is a non-linear partial differential equation (PDE) presented in Eq. (7). Some simplifications assumed on the Euler equations are as follows. For pipelines the cross-section of the pipes is very small compared to the length of the pipeline segments (pipes). Such being the case, the flow is assumed to be one-dimensional. Hence, pressure and velocity variations across the cross-sectional area are assumed negligible. The temperature of the gas is assumed to be constant. This is the case since most of the pipes in the real-world are buried underground and the soil is assumed to be a large heat sink. Therefore, the flow is assumed to be in thermodynamic equilibrium. Only two forces acting on the gas are considered significant: internal friction of the pipe and inclination of the pipe due to topography.

With these assumptions the isothermal Euler equations in one space dimension augmented with non-linear source terms are obtained [8, 9]:

$$\frac{\partial \rho}{\partial t} + \frac{\partial q}{\partial x} = 0; \tag{36a}$$

$$\frac{\partial (\rho u)}{\partial t} + \frac{\partial}{\partial x}\Big(\frac{q^2}{\rho} + p(\rho)\Big) = s_1(\rho, u) + s_2(x, \rho). \tag{36b}$$

As described in Sect. 1, the first equation defines the conservation of mass. The second equation is based on Newton's laws of motion and describes the conservation of momentum (mass flux of the gas) $q = \rho u$ where $\rho(x,t)$ is the mass density of the gas, $u(x,t)$ is the gas velocity and $p(\rho)$ is a pressure law. The pressure law satisfies the following properties:

(P) $p \in C^2(\mathbb{R}^+;\mathbb{R}^+)$ with $p(0) = 0$ and $p'(\rho) > 0$ and $p'' \geq 0$ for all $\rho \in \mathbb{R}^+$.

With the above assumptions, $p(\rho)$ is often chosen as

$$p(\rho) = \frac{Z\mathscr{R}T}{M_g}\rho = a^2\rho, \tag{37}$$

where Z is the natural gas compressibility factor, $\mathscr{R}$ the universal gas constant, T the absolute gas temperature, and M_g the gas molecular weight. Since temperature has been assumed constant, one considers the constant a to be the speed of sound in the gas. Also note that a depends on the type of gas as well as the temperature.

For the sake of simplicity, further assumptions are made [28, 54, 65]: pipe wall expansion or contraction under pressure loads is negligible; hence, pipes have constant cross-sectional area. The diameter, D, of the pipe is constant.

The source term modelling the influence of friction is modelled by assuming steady state friction for all pipes [54, 73]. The friction factor f_g is calculated using Chen's equation [17]:

$$\frac{1}{\sqrt{f_g}} := -2\log\Big(\frac{\varepsilon/D}{3.7065} - \frac{5.0452}{N_{Re}}\log\big(\frac{1}{2.8257}\big(\frac{\varepsilon}{D}\big)^{1.1098} + \frac{5.8506}{0.8981 N_{Re}}\big)\Big) \tag{38}$$

where N_{Re} is the Reynolds number $N_{Re} = \rho u D/\mu$, μ the gas dynamic viscosity and ϵ the pipeline roughness, which are again assumed to be the same for all pipes. With the above assumptions, the friction term takes the form

$$s_1(\rho, u) = -\frac{f_g}{2D}\rho u|u|.$$

The pipe inclination takes the form

$$s_2(x,\rho) = -g\rho\sin\alpha(x)$$

where g is acceleration due to gravity and $\alpha(x)$ is the slope of the pipe.

In addition, two additional assumptions need to be made:

A1. There are no vacuum states, i.e. $\rho > 0$.
A2. All flow states are subsonic, i.e. $\frac{q}{\rho} < a$.

These assumptions are backed by physical considerations. It is reasonable to assume that atmospheric pressure is the lower bound for the pressure in the pipes. Lower pressure can occur due to waves travelling through the pipe. Waves which would create vacuum states are untenable since they would cause the pipe to explode or implode. Generally pipelines are operated at high pressure (40–60 bar) and the gas velocity is very low (< 10 m/s). As the speed of sound in natural gas is around

370 m/s, the second assumption also makes sense. In pipeline networks, all states are a reasonable distance from the sonic states so that travelling waves cannot create sonic or supersonic states.

Briefly, some mathematical properties of the above system will be discussed [50]. These mathematical properties are used to discuss the well-posedness of the flow problem in a single pipe. With the above assumptions Eq. (36) is also strictly hyperbolic. To solve Riemann problems, characteristic fields based on Lax-curves are employed [26, 50]. The discussion of the Riemann problems has been given in Sect. 2.

4.1.2 Flow in Networks or Coupling Pipes and Nodes

To complete the discussion of flows in networks it is necessary to discuss coupling of different pipes at a node. Coupling of pipes is believed to be the major part of gas networks [65]. Approaches in the engineering community are exploited.

Definition 10 A network is a finite directed graph $(\mathscr{I}, \mathscr{V})$. The arcs, $I_j \in \mathscr{I}$, are connected together by vertices or nodes, $v \in \mathscr{V}$.

We need to study a Cauchy problem on the whole network. This depends on the solution at the vertices.

Applying assumption **A2** and the notation in [8, 35], a set of edges is denoted by $\mathscr{J}$ and a set of nodes is denoted by $\mathscr{V}$. Both sets are taken to be non-empty. Each edge $j \in \mathscr{J}$ corresponds to a pipe parametrised by an interval $I_j := [x_j^a, x_j^b]$. Each node $v \in \mathscr{V}$ corresponds to a single intersection of pipes. For each node $v \in \mathscr{V}$, the set of all indices of pipes $j \in \mathscr{J}$ ingoing and outgoing to the node can be separated into two sets δ_v^- and δ_v^+, respectively. The set of all pipes intersecting at a node $v \in \mathscr{V}$ can be denoted as $\delta_v = \delta_v^- \cup \delta_v^+$. In addition, the degree of a vertex $v \in \mathscr{V}$ is the number of pipes connected to the node.

Further, the nodes $\mathscr{V}$ can also be classified according to their physical use. Any node of degree one, i.e. $|\delta_v^- \cup \delta_v^+| = 1$, is either an inflow ($\delta_v^- = \emptyset$) or an outflow ($\delta_v^+ = \emptyset$) boundary node for the network. These nodes can be interpreted as suppliers or consumers of gas and the sets of such nodes can be denoted as $\mathscr{V}_I$ or $\mathscr{V}_O$, respectively. Some nodes of degree two are controllable nodes (for example, compressor stations or valves). This subset of nodes is denoted by $\mathscr{V}_C \subset \mathscr{V}$. The rest of the nodes, $\mathscr{V}_P = \mathscr{V} \backslash (\mathscr{V}_I \cup \mathscr{V}_O \cup \mathscr{V}_C)$, can be considered the standard pipe-to-pipe intersections.

In addition to the above assumptions, the following can also be imposed:

A3. All pipes have the same diameter D. The cross-sectional area is given by $A = \dfrac{D^2}{4}\pi$. Similar to a single pipe, the walls do not expand or contract due to pressure load.

A4. The friction factor, f_g, is the same for all pipes.

In general, the value at a node v depends only on the flow in the ingoing and outgoing pipes and where needed some (possibly) time-dependent controls. Thus for the network, on each edge $l \in \mathcal{J}$, assume that the dynamics is governed by the isothermal Euler equation (36) for all $x \in [x_l^a, x_l^b]$ and $t \in [0, T]$ supplemented with initial data U_l^0. In addition, on each vertex $v \in \mathcal{V}$ systems of the type (36) are coupled by suitable coupling conditions:

$$\Psi(\rho^{(l_1)}, q^{(l_1)}, \ldots, \rho^{(l_n)}, q^{(l_n)}) = \Pi(t), \qquad \delta_v = \{l_1, \ldots, l_n\}.$$

Hence the network model for gas flow in pipelines consisting of m pipes takes the form

$$\frac{\partial \rho^{(l)}}{\partial t} + \frac{\partial q^{(l)}}{\partial x} = 0; \tag{39a}$$

$$\frac{\partial q^{(l)}}{\partial t} + \frac{\partial}{\partial x}\Big(\frac{(q^{(l)})^2}{\rho} + a^2 \rho^{(l)}\Big) = s(x, \rho^{(l)}, q^{(l)}), \qquad l \in \{1, \ldots, n\}; \tag{39b}$$

$$\Psi(\rho^{(l_1)}, q^{(l_1)}, \ldots, \rho^{(l_m)}, q^{(l_n)}) = \Pi(t). \tag{39c}$$

Now let us consider different junctions and the resulting coupling conditions.

Inflow and Outflow Nodes

The inflow and outflow nodes $v \in \mathcal{V}_I \cup \mathcal{V}_O$ can be identified with boundary conditions for the Eq. (36), modelling time-dependent inflow pressure or the outgoing mass flux of the gas. This formulation is well known and widely discussed in the literature [50].

Standard Junctions

On junctions $v \in \mathcal{V}_P$, the amount of gas is conserved at pipe-to-pipe intersections:

$$\sum_{l \in \delta_v^+} q^{(l)} = \sum_{\bar{l} \in \delta_v^-} q^{(\bar{l})}$$

In addition, pressure is assumed equal for all pipes at the node [25, 28, 70], i.e.

$$p(\rho^{(l)}) = p(\rho^{(\bar{l})}), \qquad \forall\, l, \bar{l} \in \delta_v^- \cup \delta_v^+.$$

An introduction of such coupling conditions was undertaken in [9] and analysed in [8]. The conservation of mass at the nodes is unanimously agreed upon in the literature. The pressure conditions, however provide room for debate. The engineering community apply pressure tables but it is still not clear how these can be represented in mathematical terms. Different conditions can be applied [19]. Here we assume there is 'good mixing' at each vertex so that the condition of equal pressure can be imposed. Hence at the vertex the pipes are coupled using the following conditions:

$$\Psi(\rho^{(l_1)}, q^{(l_1)}, \dots, \rho^{(l_n)}, q^{(l_n)}) = \begin{pmatrix} \sum_{l \in \delta_v^+} q^{(l)} - \sum_{\bar{l} \in \delta_v^-} q^{(\bar{l})} \\ p(\rho^{(l_1)}) - p(\rho^{(l_2)}) \\ \dots \\ p(\rho^{(l_1)}) - p(\rho^{(l_n)}) \end{pmatrix} = 0 \tag{40}$$

for $\delta_v^- \cup \delta_v^+ = \{l_1, \dots, l_n\}$.

To solve the problem at the vertex, ideas from solutions of the standard Riemann problem are adopted [15, 26, 50] also presented in Sect. 2. Half-Riemann problems are solved instead [37]. In addition to that a few remarks on the process are in order. Given a node with n pipes coupled through coupling conditions (40), in each of the n pipes assume a constant subsonic state $(\bar{\rho}^{(l)}, \bar{q}^{(l)})$. The states $(\bar{\rho}^{(1)}, \bar{q}^{(1)}), \dots, (\bar{\rho}^{(n)}, \bar{q}^{(n)}))$ need not satisfy the coupling conditions. Hence, similar to the Riemann problem for a simple domain, we construct the intermediate states that develop at the node as in the state u_m in Sect. 2.4. Unfortunately, solutions to this problem are only realisable if the waves generated travel with negative speed in incoming pipes and with positive speed in outgoing pipes, which means the coupling conditions do not provide feasible solutions to all choices of states $(\bar{\rho}^{(1)}, \bar{q}^{(1)}), \dots, (\bar{\rho}^{(n)}, \bar{q}^{(n)}))$.

A solution of a Riemann problem is a vector $(u^{(1)}(x,t), \dots, u^{(n)}(x,t))$ of functions $u^{(l)} : I_l \times (0, \infty) \to \Omega$ satisfying:

- for every $l \in \{1, \dots, n\}$, $u^{(l)}$ is a restriction to $I_l \times (0, \infty)$ of the solution to the classical Riemann Problem:

$$\begin{aligned} (u^{(l)})_t + f(u^{(l)})_x &= 0, \quad t > 0, x \in \mathbb{R}; \\ u^{(l)}(x,0) &= \bar{u}^{(l)}, \qquad x < 0; \\ u^{(l)}(x,0) &= v^{(l)}(0), \quad x > 0 \end{aligned}$$

  Coupling or transmission conditions in Eq. (40) guarantee that we obtain a well-posed model. Of course, we have to prove that we indeed yield a well-posed mathematical model for the junction. See [8, 9, 12] for details. In addition, the derivation of $v^{(l)}(0)$ is a result of the process of solving Riemann Problems at the junction [8, 9].
- for every $l \in \{1, \dots, n\}$

$$\lim_{t \to 0+} u^{(l)}(\cdot, t) = \bar{u}^{(l)}$$

  with respect to the L^1_{loc} topology.

The coupling conditions as presented above are well-posed. This result was discussed in [8, 37]. Below a numerical example will be presented. In addition, this work has been extended further to prove well-posedness for multi-phase flows. The special case of the drift-flux model has been investigated in [12].

In summary, the following is the general result: consider the Cauchy Problem and a Riemann solver. Under some strict technical assumptions on the initial conditions as the case might be, there exists a unique weak solution at vertex v: $(u^{(1)}(x,t), u^{(2)}(x,t), \ldots, u^{(n)}(x,t))$ such that

1. for every $l \in \{1, \ldots, n\}$, $u^{(l)}(x, 0) = u_0^{(l)}(x)$ for a.e. $x \in I_l$;
2. for a.e. $t > 0$ a solution for the Riemann solver exists.

Moreover, for up to a 2×2 system, the solution depends in a Lipschitz continuous way on the initial conditions. As pointed out above, the main assumption is the initial data needs to be sub-sonic and small TV-norm of initial data and

$$\det\Big(D_1\Psi(\bar{u}), \ldots, D_n\Psi(\bar{u})\Big) \neq 0 \quad \text{where} \quad D_j\Psi(\bar{u}) = \frac{\partial}{\partial u_j}\Psi(\bar{u}).$$

A solution may contain shock waves just as in the Cauchy problem.

4.1.3 Numerical Results

At this point an example taken from [8] is presented. A time-splitting approach can be applied:

$$\partial_t \begin{pmatrix} \rho^{(l)} \\ q^{(l)} \end{pmatrix} + \partial_x \begin{pmatrix} q^{(l)} \\ q^{(l)2}/\rho^{(l)} + P^{(l)}(\rho^{(l)}) \end{pmatrix} = \begin{pmatrix} 0 \\ 0 \end{pmatrix};$$

$$\partial_t \begin{pmatrix} \rho^{(l)} \\ q^{(l)} \end{pmatrix} = \begin{pmatrix} 0 \\ s(\rho^{(l)}, q^{(l)}) \end{pmatrix}.$$

1. A high resolution scheme is applied to the homogeneous system.
2. An exact solution can be computed for the ordinary differential equations.

A One-to-One Network Example

Consider an incoming wave on pipe $l = 1$, which cannot freely pass the intersection, due to a given low flux profile on the outgoing pipe $l = 2$. Friction in the pipes is taken to be $f_g = 10^{-3}$. The sound speeds are $a_l = 360$ m/s and the pipe diameter is $D = 0.1$ m. An inflow of $q_{in} = 70$ kg m^{-2} s^{-1} at $x = x_1^a$ and an outflow

Fig. 10 Snapshots of the solution ρ_l to the problem of two connected pipes at different times t [8]

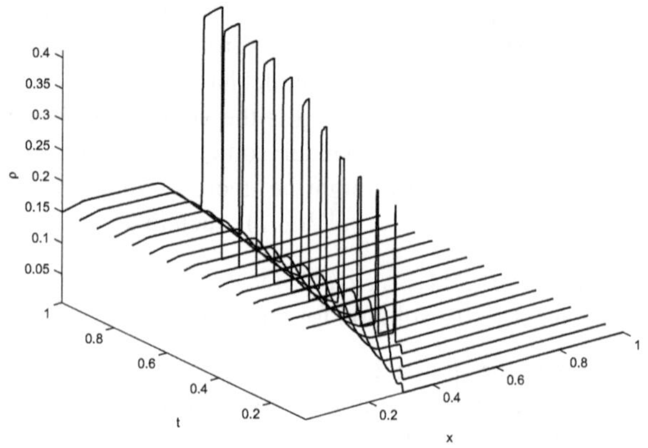

of $q_{out} = 1$ at $x = x_2^b$ is prescribed. Both pipes have the same initial condition $U_j^0 = [\frac{1}{360}, 1]$. In Fig. 10, snapshots of the density evolution are shown. Take note that the inflow profile moves on pipe 1 until it reaches the intersection. Since the maximal flow on the outgoing pipe is $q^* = 1$, a backwards moving shock wave on pipe 1 develops and increases the density near the intersection.

A Five-Pipe Network

In the following example, we consider a small network with five pipes, refer to Fig. 11. The initial conditions are defined as follows: on pipes 1,3,5 we set the density and flux as: $(4, 2)$, $(4, 4)$, $(4, 6)$.

The initial conditions on pipe 2,4 are a realisation of a pressure increase and decrease as follows:

$$U_2^0(x) = \begin{cases} (4, 2), & x < \frac{1}{2}; \\ (4 + \frac{1}{2}\sin(\pi(2x-1)), 2), & x > \frac{1}{2}; \end{cases}$$

$$U_4^0(x) = \begin{cases} (4 + \frac{1}{2}\sin(4\pi(x - \frac{1}{4})), 2), & \frac{1}{2} < x < \frac{3}{4} \\ (4, 2), & \text{else}. \end{cases}$$

The flow in the pipes 1, 3, 5 is displayed in Fig. 12. In the figure it can be seen that the pressure equality has been achieved. The evolution of the shocks due to the pressure variations is also visible.

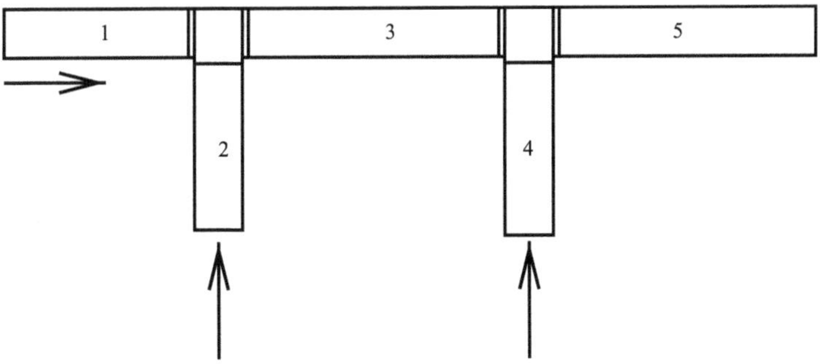

Fig. 11 Network setup

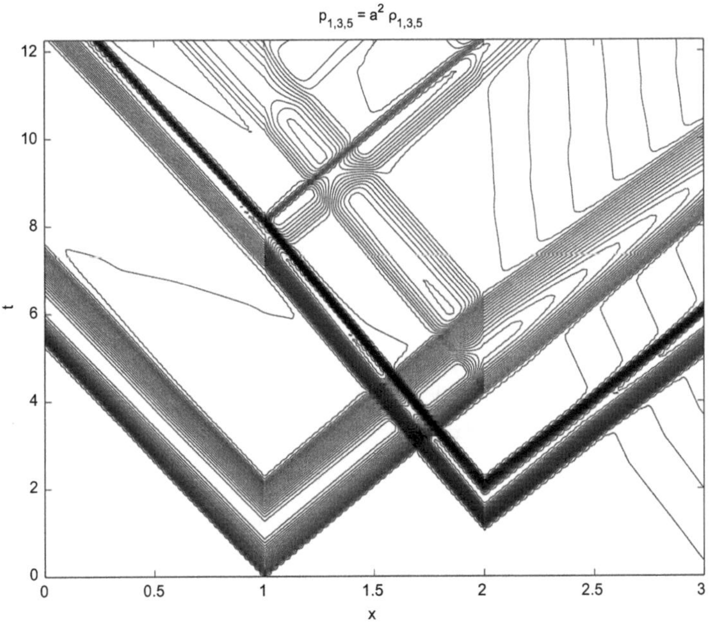

Fig. 12 Pressure distribution in time for pipes 1, 3, 5

4.2 Boundary Stabilisation of a Dynamic System of Hyperbolic Conservation Equations

Consider dynamic systems described by hyperbolic partial differential equations. The problem now is to find methods for designing certain control action at the boundary of a domain in order to attain a desired stable equilibrium status. This kind of need can be observed in daily activities such as the use of traffic lights on traffic flow networks or the installation of gates for hydraulic networks. For such networks to guarantee navigation, a critical issue is to stabilise the water level in the reaches taking into consideration the variations of the water flow rate of the river. The stability of a boundary control system is crucial to attain a desired equilibrium status. Convergence of the solution towards the desired equilibrium is due to boundary damping only.

The significant tool that has been introduced in the study of boundary stabilisation problems are suitable Lyapunov functions. For stability of the solution to the partial differential equation, the exponential decay of such a Lyapunov function can be established in a variety of cases—see [22], for example. For a more detailed discussion, the reader may refer to [21] and the references therein. Our focus was the analysis of the numerical discretisation analogous to the continuous results. Therefore, the conditions under which, in a numerical scheme, an exponential decay of the discrete solution to the hyperbolic system can be observed were derived [3]. Mathematical proofs for explicit decay rates for general three-point finite volume schemes, using the discrete counterpart of the Lyapunov function, are presented in [3]. In general, for a numerical discretisation the exponential decay of discrete L^2-Lyapunov functions up to a given finite terminal time only has been proved.

4.2.1 The Continuous Results

The following results are (probably) the most generally available continuous stabilisation results [20, 22]. Consider the non-linear hyperbolic partial differential equation

$$\frac{\partial u}{\partial t} + \frac{\partial f(u)}{\partial x} = 0, \qquad u(x,0) = u_0(x), \tag{41}$$

where $t \in [0, +\infty)$, $x \in [0, L]$, $u : [0, L] \times [0, +\infty) \to \mathbb{R}^k$ and $f : \mathbb{R}^k \to \mathbb{R}^k$ denotes a possibly non-linear smooth flux function. The gas dynamics model discussed above is a special case of this with $k = 2$. The flux function f is assumed to be strictly hyperbolic. If the solution, u, is smooth, solving Eq. (41) is equivalent to solving

$$\frac{\partial u}{\partial t} + A(u)\frac{\partial u}{\partial x} = 0, \; x \in [0, L], t \in [0, +\infty), u \in \mathbb{R}^k \tag{42}$$

with $A(u) = D_u f(u)$ a $k \times k$ real matrix and

$$u(x,0) = u^0(x). \tag{43}$$

It must be emphasised that here solution in H^2 are considered, and not general discontinuous solutions as in the previous sections. Consider the case where $L = 1$ and hence $x \in [0, 1]$. In addition, $\|a\|_q$ denotes the q-norm of the vector $a \in \mathbb{R}^k$, $k > 1$, and $\|x\|$ denotes the absolute value of a real number $x \in \mathbb{R}$.

Feedback boundary conditions [21, 22, 51] for (42) can be prescribed as follows:

$$\begin{pmatrix} u_+(0,t) \\ u_-(1,t) \end{pmatrix} = G \begin{pmatrix} u_+(1,t) \\ u_-(0,t) \end{pmatrix}, \quad t \in [0, +\infty) \tag{44}$$

where $G : \mathbb{R}^k \to \mathbb{R}^k$ is a possible non-linear function. The variables u_+ and u_- will be defined at a later stage below. Denote by Λ_i for $i = 1, \dots, k$ the eigenvalues of $A(0)$. As in [22], $A(0)$ may be assumed to be diagonal (otherwise an appropriate state transformation is applied). Due to the strict hyperbolicity the eigenvalues are distinct, i.e. $\Lambda_i \neq \Lambda_j$ for $i \neq j$. In addition, assume that $\Lambda_i \neq 0$ for all $i = 1, \dots, k$. The eigenvalues are ordered such that $\Lambda_i > 0$ for $i = 1, \dots, m$ and $\Lambda_i < 0$ for $i = m+1, \dots, k$. For any $u \in \mathbb{R}^k$, define $u_+ \in \mathbb{R}^m$ and $u_- \in \mathbb{R}^{k-m}$ by requiring

$$u = \begin{pmatrix} u_+ \\ u_- \end{pmatrix}.$$

Hence, $u_\pm$ are the components of the vector $u \in \mathbb{R}^k$ corresponding to the positive and negative eigenvalues of the diagonal matrix $A(0)$. Thus u is now defined in the eigenvector basis of $A(0)$ corresponding to ordered eigenvalues. Similarly, for A (and G), define for $A_+ : \mathbb{R}^k \to \mathbb{R}^{m \times k}$, $G_+ : \mathbb{R}^k \to \mathbb{R}^m$ and $A_- : \mathbb{R}^k \to \mathbb{R}^{k-m \times k}$, $G_- : \mathbb{R}^k \to \mathbb{R}^{k-m}$ by

$$A(u) = \begin{pmatrix} A_+(u) \\ A_-(u) \end{pmatrix} \text{ and } G(u) = \begin{pmatrix} G_+(u) \\ G_-(u) \end{pmatrix}.$$

Again, assume that $A_\pm(u)$ and $G_\pm(u)$ are diagonal, for all u. Let $G_+(u) = G_+((u_+, u_-)^T)$. The partial derivatives with respect to u_+ and u_- will be denoted by $G'_{+,u_+}(u)$ and $G'_{+,u_-}(u)$, respectively and analogously for G_-. This notation is as in [22]. Then, the definition of stability of the equilibrium solution $u \equiv 0$ is given by:

Definition 11 (Definition 2.2[22]) The equilibrium solution $u \equiv 0$ of the non-linear system (42)–(43) is exponentially stable (in the H^2-norm) if there exists $\epsilon > 0$, $\nu > 0$, $C > 0$ such that, for every $u^0 \in H^2((0,1); \mathbb{R}^k)$ satisfying $\|u^0\|_{H^2((0,1);\mathbb{R}^k)} \leq \epsilon$ and the compatibility conditions

$$\begin{pmatrix} u_+^0(0) \\ u_-^0(1) \end{pmatrix} = G \begin{pmatrix} u_+^0(1) \\ u_-^0(0) \end{pmatrix};$$

$$A_+(u^0(0))u_x^0(0) = G'_{+,u_+} \begin{pmatrix} u_+^0(1) \\ u_-^0(0) \end{pmatrix} A_+(u^0(1))u_x^0(1) + G'_{+,u_-} \begin{pmatrix} u_+^0(1) \\ u_-^0(0) \end{pmatrix} A_-(u^0(0))u_x^0(0);$$

$$A_-(u^0(1))u_x^0(1) = G'_{-,u_+} \begin{pmatrix} u_+^0(1) \\ u_-^0(0) \end{pmatrix} A_+(u^0(1))u_x^0(1) + G'_{-,u_-} \begin{pmatrix} u_+^0(1) \\ u_-^0(0) \end{pmatrix} A_-(u^0(0))u_x^0(0);$$

the classical solution u to the Cauchy problem (42) and (43) with boundary conditions (44) is defined for all $t \in [0, +\infty)$ and satisfies

$$\|u(\cdot, t)\|_{H^2((0,1);\mathbb{R}^k)} \le C \exp(-\nu t)\|u^0\|_{H^2((0,1);\mathbb{R}^p)}. \tag{45}$$

For any real $k \times k$ matrix A, $\rho_1(A) := \inf\{\|\Delta A \Delta^{-1}\|_2 : \Delta \in \mathscr{D}_{k,+}\}$, where $\mathscr{D}_{k,+}$ is the set of all real $k \times k$ diagonal matrices with strictly positive diagonal elements. Then, the following general result holds:

Theorem 6 (Theorem 2.3[22]) *Assume $A(0)$ is diagonal with distinct and non-zero eigenvalues and A is of class $C^2(\mathbb{R}^k; \mathbb{R}^k)$ in a neighbourhood of zero. Take $A(0) = diag(\Lambda_i)_{i=1}^k$ and $\Lambda_i > 0$ for $i = 1, \dots, m$ and $\Lambda_i < 0$ for $i = m+1, \dots, k$ and $\Lambda_i \neq \Lambda_j$ for $i \neq j$. Assume G is of class $C^2(\mathbb{R}^k; \mathbb{R}^k)$ in a neighbourhood of zero and $G(0) = 0$.*

Let $\rho_1(G'(0)) < 1$, then the equilibrium $u \equiv 0$ for (42) with boundary conditions given by (44) is exponentially stable.

In this context, in [3] an analysis of the Lyapunov function for a discrete case was developed. Up to now a discrete counter-part to Theorem 6 has not been proved. A weaker result concerning the L^2-stabilisation was developed in [3].

4.2.2 Stabilisation of a Discrete Problem

The finite volume numerical schemes for boundary L^2-stabilisation of one-dimensional non-linear hyperbolic systems were adopted. In order to facilitate the proofs, we will need additional assumptions discussed below.

The boundary conditions are considered as in (44) with G being a diagonal matrix. To simplify the discussion, let

$$A(u) = \mathrm{diag}(\Lambda_i(u))_{i=1}^m, \;\; \Lambda_i(u) > 0, \;\; \Lambda_i(u) \neq \Lambda_j(u), i \neq j. \tag{46}$$

In the following $\mathbf{u}$ denotes all components of u, i.e., $\mathbf{u} := (u_j)_{j=1}^m$. Let $\delta > 0$ and denote by $M_\delta(0) := \{\mathbf{u} : |u_j| \le \delta, j = 1, \dots, m\}$. Assume δ is sufficiently small such that

$$M_\delta(0) \subset B_\epsilon(0).$$

Further, let Δx denote the cell width of a uniform spatial grid and N the number of cells in the discretisation of the domain $[0, 1]$ such that $\Delta x N = 1$ with cell centres at $x_i = (i + \frac{1}{2})\Delta x$, $i = 0, \dots, N-1$. Further x_{-1} and x_N denote the cell centres of the cells outside the computational domain on the left-hand and right-hand side of the domain, respectively. The interfacial numerical fluxes are computed at cell boundaries $x_{i-1/2} = i\Delta x$ for $i = 0, \dots, N$. The left and right boundary points are at $x_{-1/2}$ and $x_{N-1/2}$, respectively. The temporal grid is chosen such that the CFL condition holds:

$$\lambda \frac{\Delta t}{\Delta x} \leq 1, \; \lambda := \max_{j=1,\ldots,m} \max_{\mathbf{u} \in M_\delta(0)} \Lambda_j(\mathbf{u}) \tag{47}$$

and $t^n = n\Delta t$ for $n = 0, 1, \ldots, K$ where by possibly further reducing Δt, it can be assumed that $K\Delta t = T$. The value of $u_j(x,t)$ at the cell centre x_i is approximated by $u^n_{i,j}$ for $i = 0, \ldots, N-1$ and for each component j at time t^n for $n = 0, 1, \ldots, K$. The left boundary is discretized using $u^n_{-1,j}$. The initial condition is discretized as

$$u^0_{i,j} := \frac{1}{\Delta x} \int_{x_{i-1/2}}^{x_{i+1/2}} u^0_j(x) dx$$

where $u^0_j(x), \; j = 1, \ldots, m$ denotes the components of solution vector $\mathbf{u}$.

Hence, the following discretisation of (42)–(44) for $n = 0, \ldots, K-1$ is introduced:

$$u^{n+1}_{i,j} = u^n_{i,j} - \frac{\Delta t}{\Delta x} \Lambda_j(\mathbf{u}^n_i) \left(u^n_{i,j} - u^n_{i-1,j} \right), \quad i = 0, \ldots, N-1; \tag{48a}$$

$$u^{n+1}_{-1,j} = \kappa_j u^{n+1}_{N-1,j}; \tag{48b}$$

$$u^0_{i,j} = \frac{1}{\Delta x} \int_{x_{i-1/2}}^{x_{i+1/2}} u^0_j(x) dx, \; i = 0, \ldots, N-1; \quad j = 1, \ldots, m; \tag{48c}$$

$$u^0_{-1,j} = \kappa_j u^0_{N-1,j}. \tag{48d}$$

Note that the last condition is the discrete compatibility condition for u^0.

The discrete Lyapunov function at time t^n with positive coefficients μ_j, $j = 1, \ldots, m$ takes the form

$$L^n = \Delta x \sum_{i=0}^{N-1} \sum_{j=1}^{m} \left(u^n_{i,j} \right)^2 \exp(-\mu_j x_i). \tag{49}$$

The numerical result is as follows:

Theorem 7 *Let $T > 0$ and assume (46) holds. For any κ_j, $j = 1 \ldots, m$ such that*

$$0 < \kappa_j < \sqrt{\frac{D^{\min}_j}{D^{\max}_j}}, \tag{50}$$

where $\max_{\mathbf{u}\in M_\delta(0)} \frac{\Delta t}{\Delta x}\Lambda_j(\mathbf{u}) =: D_j^{\max} \leq 1$ *and* $\frac{\Delta t}{\Delta x}\Lambda_j(\mathbf{u}_i^n) \geq \min_{\mathbf{u}\in M_\delta(0)} \frac{\Delta t}{\Delta x}\Lambda_j(\mathbf{u}) =: D_j^{\min} > 0$ *the following holds:*

there exists $\mu_j > 0,\ j = 1,\dots,m$ *and* $\delta > 0$ *such that for all initial data* $u_{i,j}^0$ *with*

$$\|u_{i,j}^0\| \leq \delta,\ \|\frac{u_{i,j}^0 - u_{i-1,j}^0}{\Delta x}\| \leq \delta, \text{ and } \|\frac{u_{0,j}^n - u_{-1,j}^n}{\Delta x}\| \leq \delta \exp\left(t^n \max_{\xi\in M_\delta(0)} \|\nabla_{\mathbf{u}}\Lambda_j(\xi)\|_\infty\right)$$

for all $i = 0,\dots,N-1;\ j = 1,\dots,m$, *the numerical solution* $u_{i,j}^n$ *defined by (48) satisfies*

$$L^n \leq \exp(-\nu t^n)L^0,\ n = 0,1,\dots,K \tag{51}$$

for some $\nu > 0$. *Moreover,* $u_{i,j}^n$ *is exponentially stable in the discrete* L^2*-norm*

$$\Delta x \sum_{i=0}^{N-1}\sum_{j=1}^{m}\left(u_{i,j}^n\right)^2 \leq \tilde{C}\exp(-\nu t^n)\Delta x \sum_{i=0}^{N-1}\sum_{j=1}^{m}\left(u_{i,j}^0\right)^2, n = 0,1,\dots,K. \tag{52}$$

The by-product of this analysis is that, assuming (50), the explicit form of bounds and constants such as μ_j, δ, ν are derived and $\tilde{C} = \frac{C_1}{C_0} = \max_{j=1,\dots,m} \exp(\mu_j x_{N-1})$. The grid size $\Delta x, \Delta t$ is not fixed and can be chosen arbitrarily provided that the CFL condition is satisfied.

Remark 1 One observes that the presented result is weaker than the corresponding continuous result obtained in [22, Theorem 2.3]. Therein, *no* assumption on the boundedness of T is required. The continuous Lyapunov function can be shown to be equivalent to the H^2-norm of u. The exponential decay of this Lyapunov function therefore yields the global existence of u.

A simple linear transport is considered [3]:

$$\frac{\partial}{\partial t}\begin{pmatrix}u_1\\u_2\end{pmatrix} + \frac{\partial}{\partial x}\begin{pmatrix}1 & 0\\0 & -1\end{pmatrix}\begin{pmatrix}u_1\\u_2\end{pmatrix} = 0,\ x \in [0,1], t \in [0,T] \tag{53}$$

and subject to the boundary and initial conditions

$$u_1(t,0) = \kappa\, u_2(t,0),\ u_2(t,1) = \kappa\, u_1(t,1),\ \ u_i(0,x) = u_i^0. \tag{54}$$

In this example, the analytical decay rates ν of the Lyapunov function is of interest. In the following numerical computations, a three point scheme given in Eq. (48a) above is used in each component as in Eq. (48). A time horizon fixed at $T = 12$ is taken and constant initial data $u_1^0 = -\frac{1}{2}$ and $u_2^0 = \frac{1}{2}$ are prescribed. The value of the Lyapunov function is computed by $L_n^{\text{exact}} := L^0\exp(-\nu t^n)$ with ν. These values are compared to those of the numerical Lyapunov function L^n. The

Table 1 The number of cells in the spatial domain [0, 1] is denoted by N

N	L^{inf}	L^2	μ	ν
100	4.32E−03	7.48E−04	5.66E−01	5.69E−01
200	2.18E−03	2.67E−04	5.70E−01	5.72E−01
400	1.09E−03	9.49E−05	5.73E−01	5.73E−01
800	5.48E−04	3.36E−05	5.74E−01	5.74E−01
1,600	2.74E−05	1.19E−05	5.75E−01	5.75E−01

L^{inf} denotes the norm $\|(L_n^{\text{exact}})_n - (L^n)_n\|_\infty$ and L^2 the norm $\|(L_n^{\text{exact}})_n - (L^n)_n\|_2$. The CFL constant is equal to one and $\kappa = \frac{3}{4}$ [3]

L^2- and L^∞-difference between both for different choices of the computational grid, boundary damping κ and values of the CFL constant are considered.

The parameter is fixed at $\kappa = \frac{3}{4}$ and the sharpest possible bound is set for the CFL constant, CFL $= 1$. In Table 1, the results for different grid sizes $\Delta x = \frac{1}{N}$ are presented. In the L^2 and L^{inf} the expected first-order convergence of the numerical discretisation is observed. Further, the values of μ and ν also converge towards the theoretical value in the case $\Delta x \to 0$ which are given by $\mu = \ln \kappa^{-2}$= 5.75E-01 and $\nu = \mu$.

In conclusion, we claim that the analysis of the stabilisation process using a three-point scheme gives very good indicators of the parameters to be used in the stabilisation process.

5 Summary

In this chapter, a review of the evolution equations also referred to as conservation laws have been presented. These are hyperbolic partial differential equations. In general, they model flow or transport processes and their applications are abundant. Furthermore, for the interesting cases, these equations are nonlinear and pose a lot of challenges in mathematical analysis. Examples of their peculiarities include the fact that admissible solutions include discontinuous functions. In such cases the concept of a solution is defined in a weak topology. To identify unique solutions physical insight is employed, especially the idea of entropy. For general approximations of solutions, numerical methods are required. All these ideas have been demonstrated in the chapter with a few examples to demonstrate the ideas.

Towards the end of the chapter extensions of applications of conservation laws to networked flows have been discussed. The idea of solving a Riemann problem at a network vertex has been discussed and some computational ideas and results also presented. In addition boundary stabilisation has been discussed. This is the case in which a particular equilibrium profile is desired. To achieve this profile, boundary conditions are manipulated.

This is not a complete presentation in developments involving conservation laws. A vast collection of literature exists including manuscripts which have been cited in the course of the discussion in the chapter. The reader is encouraged to consult these references to have more detailed and relevant ideas on the mathematical analysis, computational methods as well as applications in networked flows, design, shape optimisation, as well as boundary stabilisation just to mention a few.

Acknowledgements The author would like to thank the anonymous reviewer for very constructive comments which have tremendously improved the quality of this chapter. The author would also like to thank the following organisations which funded some of the research presented in this chapter as well as the organisation of the workshop where the topic was presented: International Centre for Pure and Applied Mathematics (CIMPA), International Mathematics Union (IMU), London Mathematical Society—African Mathematics Millennium Science Initiative (LMS-AMMSI), International Centre for Theoretical Physics (ICTP), African Institute of Mathematical Sciences (AIMS), the National Research Foundation (NRF) of South Africa (NRF), University of KwaZulu-Natal, Witwatersrand and Stellenbosch.

References

1. D. Aregba-Driollet, R. Natalini, Discrete kinetic schemes for multidimensional conservation laws. SIAM J. Numer. Anal. **37**, 1973–2004 (2000)
2. M.K. Banda, Variants of relaxed schemes and two-dimensional gas dynamics. J. comp. Appl. Math. **175**, 41–62 (2005)
3. M.K. Banda, M. Herty, Numerical discretization of stabilization problems with boundary controls for systems of hyperbolic conservation laws. Math. Control Relat. Fields **3**, 121–142 (2013)
4. M.K Banda, M. Seaïd, A class of the relaxation schemes for two-dimensional Euler systems of gas dynamics. Lect. Notes Comput. Sci. **2329**, 930–939 (2002)
5. M.K. Banda, M. Seaïd, Higher-order relaxation schemes for hyperbolic systems of conservation laws. J. Numer. Math. **13**, 171–196 (2005)
6. M.K. Banda, M. Seaïd, Teleaga discrete-velocity relaxation methods for large-eddy simulation. Appl. Math. Comput. **182**, 739–753 (2006)
7. M.K Banda, M. Seaïd, A. Klar, L. Pareschi, Compressible and incompressible limits for hyperbolic systems with relaxation.J. Comput. Appl. Math. **168**, 41–52 (2004)
8. M.K. Banda, M. Herty, A. Klar, Coupling conditions for gas networks governed by the isothermal Euler equations. Netw. Heterog. Media **1**, 295–314 (2006)
9. M.K. Banda, M. Herty, A. Klar, Gas flow in pipeline networks. Netw. Heterog. Media **1**, 41–56 (2006)
10. M.K. Banda, A. Klar, M. Seaïd, A lattice Boltzman relaxation scheme for coupled convection-radiation systems. J. Comput. Phys. **226**, 1408–1431 (2007). doi:10:1016/j.jcp.2007.05.030
11. M.K. Banda, A. Klar, L. Pareschi, M. Seaïd, Lattice-Boltzmann type relaxation systems and high order relaxation schemes for the incompressible Navier–Stokes equations. Math. Comput. **77**, 943–965 (2008). doi:10.1090/S0025-5718-07-02034-0
12. M.K. Banda, M. Herty, J.-M.T. Ngnotchouye, Towards a mathematical analysis for drift-flux multiphase flow models in networks. SIAM J. Sci. Comput. **31**, 4633–4653 (2010)
13. M. Ben-Arzti, J. Falcovitz, Recent developments of the GRP method. JSME (Ser. b) **38**, 497–517 (1995)
14. F. Bouchut, *Nonlinear Stability of Finite Volume Methods for Hyperbolic Conservation Laws and Well-Balanced Schemes for Sources*. Frontiers in Mathematics (Birkhäuser, Basel, 2004)

15. A. Bressan, *Hyperbolic Systems of Conservation Laws: The One-Dimensional Cauchy Problem*. Oxford Lecture Series in Mathematics and its applications, vol 20 (Oxford University Press, Oxford, 2005)
16. T. Chang, L. Hsiao, *The Riemann Problem and Interaction of Waves in Gas Dynamics*. Pitman Monographs and surveys in pure appl. math., vol 41 (Wiley, New York, 1989)
17. N.H. Chen, An explicit equation for friction factor in pipes. Ind. Eng. Chem. Fund. **18**, 296–297 (1979)
18. E. Chiodaroli, C. De Lellis, O. Kreml, Global ill-posedness of the isentropic system of gas dynamics. arXiv:1304.0123v2 (2013)
19. R.M. Colombo, M. Garavello, A well–posed Riemann problem for the p-system at a junction. Netw. Heterog. Media **1**, 495–511 (2006)
20. J.-M. Coron, Local controllability of a 1-D tank containing a fluid modelled by the shallow water equations. ESAIM **8**, 513–554 (2002)
21. J.-M. Coron, *Control and Nonlinearity. Mathematical Surveys and Monographs*, vol 136 (American Mathematical Society, Providence, 2007)
22. J.-M. Coron, G. Bastin, B. d'Andréa-Novel, Dissipative boundary conditions for one-dimensional nonlinear hyperbolic systems. SIAM J. Control. Optim. **47**, 1460–1498 (2008)
23. M.G. Crandall, The semigroup approach to first order quasilinear equations in several space dimensions. Israel J. Math. **12**, 108–132 (1972)
24. M.G. Crandall, L. Tartar, Some relations between non expansive and order preserving mapping. Proc. Am. Math. Soc. **78**, 385–390 (1970)
25. Crane Valve Group, Flow of fluids through valves, fittings and pipes, Crane Technical Paper No. 410 (1998)
26. C.M. Dafermos, *Hyperbolic Conservation Laws in Continum Physics*. Grundlehren der Mathematischen Wissenschaften (Fundamental Principles of Mathematical Sciences), 3rd edn, vol 325 (Springer, New York, 2010)
27. C. De Lellis, L. Szekelyhidi Jr., On admissibility criteria for weak solutions on the Euler equations. Arch. Rational Mech. Anal. **195**, 225–260 (2010)
28. K. Ehrhardt, M. Steinbach, Nonlinear gas optimization in gas networks, in *Modeling, Simulation and Optimization of Complex Processes*, eds. by H.G. Bock, E. Kostina, H.X. Pu, R. Rannacher (Springer, Berlin, 2005)
29. V. Elling, A possible counterexample to well posedness of entropy solutions and to Godunov scheme convergence. Math. Comput. **75**, 1721–1733 (2006)
30. R. Eymard, T. Gallouët, R. Herbin, Finite volume methods, in *Handbook of Numerical Analysis*, vol VII (North-Holland, Amsterdam, 2000) pp.713–1020
31. M. Garavello, B. Piccoli, *Traffic Flow on Networks*. AIMS Series on Applied Mathematics, vol 1 (AIMS, Springfield, 2006)
32. J. Glimm, Solutions in the large for nonlinear hyperbolic systems of equations. Commun. Pure Appl. Math. **18**, 697–715 (1965)
33. E. Godlewski, P.-A. Raviart, *Numerical Approximations of Hyperbolic Systems of Conservation Laws* (Springer, New York, 1996)
34. S. Gottlieb, C.-W. Shu, E. Tadmor, Strong stability-preserving high-order time discretization methods. SIAM Rev. **43**, 89–112 (2001)
35. M. Gugat, M. Herty, A. Klar, G. Leugering, V. Schleper, Well-posedness of networked hyperbolic systems of balance laws. Int. Ser. Numer. Math. **160**, 123–146 (2012)
36. A. Harten, M. Hyman, P. Lax, On finite-difference approximations and entropy conditions for shocks. Commun. Pure Appl. Math. **29**, 297–322 (1982)
37. M. Herty, M. Rascle, Coupling conditions for a class of second order models for traffic flow. SIAM J. Math. Anal. **38**, 592–616 (2006)
38. C. Hirsch, *Numerical Computation of Internal and External Flows* (Wiley, New York, 1988)
39. S. Jin, Z. Xin, The relaxation schemes for systems of conservation laws in arbitrary space dimensions. Commun. Pure Appl. Math. **48**, 235–276
40. A. Klar, Relaxation schemes for a Lattice Boltzmann type discrete velocity model and numerical Navier–Stokes limit. J. Comp. Phys. **148**, 1–17 (1999)

41. A. Klar, L. Pareschi, M. Seaïd, Uniformly accurate schemes for relaxation approximations to fluid dynamic equations. Appl. Math. Lett. **16**, 1123–1127 (2003)
42. D. Kröner, *Numerical Schemes for Conservation Laws*. Wiley-Teubner Series in Advances in Numerical Mathematics (Wiley, Chichester, 1997)
43. S.N. Kruzkov, First order quasilinear equations in several independent variables. Math. USSR Sbornik **10**, 217–243 (1970)
44. A. Kurganov, D. Levy, A third-order semi-discrete central scheme for conservation laws and convection-diffusion equations. SIAM J. Sci. Comput. **22**, 1461–1488 (2000)
45. A. Kurganov, E. Tadmor, New high-resolution central schemes for nonlinear conservation laws and convection-diffusion equations. J. Comput. Phys. **160**, 241–282 (2000)
46. P.D. Lax, *Hyperbolic Systems of Conservation Laws and the Mathematical Theory of Shock Waves* (SIAM, Philadelphia, 1973)
47. P. Lax, B. Wendroff, Systems of conservation laws. Commun. Pure Appl. Math. **13**, 217–237 (1960)
48. P.G. LeFloch, *Hyperbolic Systems of Conservation Laws. The Theory of Classical and Nonclassical Shock Waves*. Lectures in Mathematics. ETH Zurich (Birkhäuser, Basel, 2002). ISBN:3-7643-6687-7
49. F. LeFloch, P.A. Raviart, An asymptotic expansion for the solution of the generalized Riemann problem, part I: general theory. Ann. Inst. H. Poincare. Nonlinear Anal. **5**, 179 (1988)
50. R.J. LeVeque, *Finite Volume Methods for Hyperbolic Problems*. Cambridge Texts in Applied Mathematics (Cambridge University Press, Cambridge, 2002)
51. T. Li, B. Rao, Z. Wang, Contrôlabilité observabilité unilatérales de systèmes hyperboliques quasi-linéaires. C. R. Math. Acad. Sci. Paris **346**, 1067–1072 (2008)
52. T.-P. Liu, The entropy condition and admissibility of shocks. J. Math. Anal. Appl. **53**, 78–88 (1976)
53. T.-P. Liu, The determistic version of the Glimm scheme. Commun. Math. Phys. **57**, 135–148 (1977)
54. A. Martin, M. Möller, S. Moritz, Mixed integer models for the stationary case of gas network optimization. Math. Program. Ser. B **105**, 563–582 (2006)
55. V. Milisic, Stability and convergence of discrete kinetic approximations to an initial-boundary value problem for conservation laws. Proc. Am. Math. Soc. **97**, 595–633 (2004)
56. L. Pareschi, G. Russo, Implicit-explicit Runge–Kutta schemes for stiff systems of differential equations. in *Recent Trends in Numerical Analysis*, ed. by T. Brugano, vol 3 (Nova Science, Commack, 2000) pp. 269–284
57. L. Pareschi, G. Russo, Asymptotically SSP schemes for hyperbolic systems with stiff relaxation, in *Hyperbolic problems: Theory, Numerics, Applications: Proceedings of the Ninth International Conference on Hyperbolic Problems Held in Caltech, Pasadena* (2003) pp. 241–255
58. L. Pareschi, G. Russo, Implicit-explicit Runge–Kutta schemes and applications to hyperbolic systems with relaxation. J. Sci. Comput. **25**, 129–155 (2005)
59. B. Perthame, *Kinetic Formulation of Conservation Laws*. Oxford Lecture Series in Mathematics and its applications, vol. 21 (Oxford University Press, Oxford, 2002)
60. S. Schochet, Glimm's scheme for systems with almost-planar interactions. Commun. Partial Differ. Equ. **16**, 1423–1440 (1991)
61. M. Seaïd, High-resolution relaxation scheme for the two-dimensional riemann problems in gas dynamics. Numer. Methods Partial Differ. Equ. **22**, 397–413 (2006)
62. D. Serre, *Systems of Conservation Laws I* (Cambridge University Press, Cambridge, 1999)
63. J. Smoller, *Shock Waves and Reaction-Diffusion Equations* (Springer, New York, 1983)
64. G. Sod, *Numerical Methods for Fluid Dynamics* (Cambridge University Press, Cambridge, 1985)
65. M. Steinbach, On PDE solution in transient optimization of gas networks. J. Comput. Appl. Math. **203**, 345–361 (2007)
66. P.K. Sweby, High resolution schemes using flux limiters for hyperbolic conservation laws. SIAM J. Numer. Anal. **21**, 995–1011 (1984)

67. E. Tadmor, Approximate solutions of nonlinear conservation laws, in *Advanced Numerical Approximations of Nonlinear Hyperbolic Equations*, ed. by A. Quarteroni, Lecture Notes in Mathematics, vol 1697 (Springer, Berlin, 1998)
68. E.F. Toro, *Riemann Solvers and Numerical Methods for Fluid Dynamics: A Practical Introduction* (Springer, Berlin, 2009)
69. B. van Leer, Towards the ultimate conservative difference schemes. V. A second order sequel to Godunov method. J. Comput. Phys. **32**, 101–136 (1979)
70. F.M. White, *Fluid Mechanics* (McGraw-Hill, New York, 2002)
71. G.B. Whitham, *Linear and Nonlinear Waves* (Wiley-Interscience, New York, 1974)
72. P. Woodward, P. Colella, The numerical simulation of two-dimensional fluid flow with strong shocks. J. Comp. Phys. **54**, 115–173 (1984)
73. J. Zhou, M.A. Adewumi, Simulation of transients in natural gas pipelines using hybrid TVD schemes. Int. J. Numer. Methods Fluids **32**, 407–437 (2000)

LECTURE NOTES IN MATHEMATICS Springer

Edited by J.-M. Morel, B. Teissier; P.K. Maini

Editorial Policy (for Multi-Author Publications: Summer Schools / Intensive Courses)

1. Lecture Notes aim to report new developments in all areas of mathematics and their applications - quickly, informally and at a high level. Mathematical texts analysing new developments in modelling and numerical simulation are welcome. Manuscripts should be reasonably selfcontained and rounded off. Thus they may, and often will, present not only results of the author but also related work by other people. They should provide sufficient motivation, examples and applications. There should also be an introduction making the text comprehensible to a wider audience. This clearly distinguishes Lecture Notes from journal articles or technical reports which normally are very concise. Articles intended for a journal but too long to be accepted by most journals, usually do not have this "lecture notes" character.

2. In general SUMMER SCHOOLS and other similar INTENSIVE COURSES are held to present mathematical topics that are close to the frontiers of recent research to an audience at the beginning or intermediate graduate level, who may want to continue with this area of work, for a thesis or later. This makes demands on the didactic aspects of the presentation. Because the subjects of such schools are advanced, there often exists no textbook, and so ideally, the publication resulting from such a school could be a first approximation to such a textbook. Usually several authors are involved in the writing, so it is not always simple to obtain a unified approach to the presentation.
   For prospective publication in LNM, the resulting manuscript should not be just a collection of course notes, each of which has been developed by an individual author with little or no coordination with the others, and with little or no common concept. The subject matter should dictate the structure of the book, and the authorship of each part or chapter should take secondary importance. Of course the choice of authors is crucial to the quality of the material at the school and in the book, and the intention here is not to belittle their impact, but simply to say that the book should be planned to be written by these authors jointly, and not just assembled as a result of what these authors happen to submit.
   This represents considerable preparatory work (as it is imperative to ensure that the authors know these criteria before they invest work on a manuscript), and also considerable editing work afterwards, to get the book into final shape. Still it is the form that holds the most promise of a successful book that will be used by its intended audience, rather than yet another volume of proceedings for the library shelf.

3. Manuscripts should be submitted either online at www.editorialmanager.com/lnm/ to Springer's mathematics editorial, or to one of the series editors. Volume editors are expected to arrange for the refereeing, to the usual scientific standards, of the individual contributions. If the resulting reports can be forwarded to us (series editors or Springer) this is very helpful. If no reports are forwarded or if other questions remain unclear in respect of homogeneity etc, the series editors may wish to consult external referees for an overall evaluation of the volume. A final decision to publish can be made only on the basis of the complete manuscript; however a preliminary decision can be based on a pre-final or incomplete manuscript. The strict minimum amount of material that will be considered should include a detailed outline describing the planned contents of each chapter.
   Volume editors and authors should be aware that incomplete or insufficiently close to final manuscripts almost always result in longer evaluation times. They should also be aware that parallel submission of their manuscript to another publisher while under consideration for LNM will in general lead to immediate rejection.

4. Manuscripts should in general be submitted in English. Final manuscripts should contain at least 100 pages of mathematical text and should always include
   - a general table of contents;
   - an informative introduction, with adequate motivation and perhaps some historical remarks: it should be accessible to a reader not intimately familiar with the topic treated;
   - a global subject index: as a rule this is genuinely helpful for the reader.

   Lecture Notes volumes are, as a rule, printed digitally from the authors' files. We strongly recommend that all contributions in a volume be written in the same LaTeX version, preferably LaTeX2e. To ensure best results, authors are asked to use the LaTeX2e style files available from Springer's web-server at
   ftp://ftp.springer.de/pub/tex/latex/svmonot1/ (for monographs) and
   ftp://ftp.springer.de/pub/tex/latex/svmultt1/ (for summer schools/tutorials).
   Additional technical instructions, if necessary, are available on request from:
   lnm@springer.com.

5. Careful preparation of the manuscripts will help keep production time short besides ensuring satisfactory appearance of the finished book in print and online. After acceptance of the manuscript authors will be asked to prepare the final LaTeX source files and also the corresponding dvi-, pdf- or zipped ps-file. The LaTeX source files are essential for producing the full-text online version of the book. For the existing online volumes of LNM see:
   http://www.springerlink.com/openurl.asp?genre=journal&issn=0075-8434.
   The actual production of a Lecture Notes volume takes approximately 12 weeks.

6. Volume editors receive a total of 50 free copies of their volume to be shared with the authors, but no royalties. They and the authors are entitled to a discount of 33.3 % on the price of Springer books purchased for their personal use, if ordering directly from Springer.

7. Commitment to publish is made by letter of intent rather than by signing a formal contract. Springer-Verlag secures the copyright for each volume. Authors are free to reuse material contained in their LNM volumes in later publications: a brief written (or e-mail) request for formal permission is sufficient.

Addresses:
Professor J.-M. Morel, CMLA,
École Normale Supérieure de Cachan,
61 Avenue du Président Wilson, 94235 Cachan Cedex, France
E-mail: morel@cmla.ens-cachan.fr

Professor B. Teissier, Institut Mathématique de Jussieu,
UMR 7586 du CNRS, Équipe "Géométrie et Dynamique",
175 rue du Chevaleret,
75013 Paris, France
E-mail: teissier@math.jussieu.fr

For the "Mathematical Biosciences Subseries" of LNM:

Professor P. K. Maini, Center for Mathematical Biology,
Mathematical Institute, 24-29 St Giles,
Oxford OX1 3LP, UK
E-mail: maini@maths.ox.ac.uk

Springer, Mathematics Editorial I,
Tiergartenstr. 17,
69121 Heidelberg, Germany,
Tel.: +49 (6221) 4876-8259
Fax: +49 (6221) 4876-8259
E-mail: lnm@springer.com

MIX
Papier aus verantwortungsvollen Quellen
Paper from responsible sources
FSC® C105338
FSC
www.fsc.org

If you have any concerns about our products,
you can contact us on
ProductSafety@springernature.com

In case Publisher is established outside the EU,
the EU authorized representative is:
Springer Nature Customer Service Center GmbH
Europaplatz 3, 69115 Heidelberg, Germany

Printed by Libri Plureos GmbH
in Hamburg, Germany